ANNUAL
REPORTS IN
MEDICINAL
CHEMISTRY
Volume 37

ANNUAL REPORTS IN MEDICINAL CHEMISTRY
Volume 37

Sponsored by the Division of Medicinal Chemistry of the American Chemical Society

EDITOR-IN-CHIEF:

ANNETTE M. DOHERTY

PFIZER GLOBAL RESEARCH & DEVELOPMENT
FRESNES LABORATORIES
FRANCE

SECTION EDITORS

JANET M. ALLEN • WILLIAM J. GREENLEE • WILLIAM K. HAGMANN
JACOB J. PLATTNER • DAVID W. ROBERTSON • GEORGE L. TRAINOR

EDITORIAL ASSISTANTS

NADEGE PINGRAY • SOPHIE BENSEGNOR

2002

ACADEMIC PRESS
An imprint of Elsevier Science

Amsterdam - Boston - London - New York - Oxford - Paris
San Diego - San Francisco - Singapore - Sydney - Tokyo

Academic Press
An Elsevier Science Imprint
525 B Street, Suite 1900, San Diego, California 92101-4495, USA
http://www.academicpress.com

International Standard Book Number: 0-12-040537-7

PRINTED IN THE NETHERLANDS
02 03 04 05 06 07 KR 9 8 7 6 5 4 3 2 1

CONTENTS

I. CENTRAL NERVOUS SYSTEM AGENTS

Section Editor: David W. Robertson, Pfizer Global Research & Development, Ann Arbor, MI

II. CARDIOVASCULAR AND PULMONARY DISEASES

Section Editor: William J. Greenlee, Schering-Plough Research Institute,
                  Kenilworth, New Jersey

III. CANCER AND INFECTIOUS DISEASES

Section Editor: Jacob J. Plattner, Chiron Corporation, Emeryville, California

IV. IMMUNOLOGY, ENDOCRINOLOGY AND METABOLIC DISEASES

Section Editor: William K. Hagmann, Merck Research Laboratories,
Rahway, New Jersey

V. TOPICS IN BIOLOGY

Section Editor: Janet M. Allen, Inpharmatica, London, United Kingdom

VI. TOPICS IN DRUG DESIGN AND DISCOVERY

Section Editor: George L. Trainor, DuPont Pharmaceuticals Company,
                   Wilmington, Delaware

VII. TRENDS AND PERSPECTIVES

Section Editor: Annette M. Doherty, Pfizer Global Research & Development,
                   Fresnes Laboratories, Fresnes, France

CONTRIBUTORS

PREFACE

Annual Reports in Medicinal Chemistry continues to focus on providing timely and critical reviews of important topics in medicinal chemistry together with an emphasis on emerging topics in the biological sciences which are expected to provide the basis for entirely new future therapies.

Volume 37 retains the familiar format of previous volumes, this year with 27 chapters. Sections I - IV are disease-oriented and generally report on specific medicinal agents with updates from Volume 36 on anticoagulants and antibacterial targets. As in past volumes, annual updates have been limited only to the most active areas of research in favor of specifically focussed and mechanistically oriented chapters, where the objective is to provide the reader with the most important new results in a particular field.

Sections V and VI continue to emphasize important topics in medicinal chemistry, biology, and drug design as well as the critical interfaces among these disciplines. Included in Section V, Topics in Biology, are chapters on cellular pathways, NSAIDS and statins in Alzheimer's disease, metalloproteinases, Fc receptor structure and tumor classification. Chapters in Section VI, Topics in Drug Design and Discovery include Ion channel Modulators and Microwave-Assisted Chemistry.

Volume 37 concludes with To Market, To Market - a chapter on NCE and NBE introductions worldwide in 2001. Last, but not least, there is a chapter on Biosimulation and dynamic modeling of biological systems. In addition to the chapter reviews, a comprehensive set of indices has been included to enable the reader to easily locate topics in Volumes 1-37 of this series.

Volume 37 *of Annual Reports in Medicinal Chemistry* was assembled with the superb editorial assistance of Ms. Nadège Pingray and Ms. Sophie Bensegnor and I would like to thank them for their hard work and enduring support. I have continued to work with innovative and enthusiastic section editors and my sincere thanks goes to them again this year. I hope that you the reader will enjoy and profit from reading this volume.

Annette M. Doherty
Fresnes, France
May 2002

SECTION I. CNS AGENTS

Editor: David W. Robertson, Pfizer Global Research & Development
Ann Arbor, MI 48105

Chapter 1. Promise and Progress of Central G-Protein Coupled Receptor Modulators for Obesity Treatments

Brian L. Largent, Albert J. Robichaud and Keith J. Miller
Bristol-Myers Squibb Company, Pharmaceutical Research Institute
Route 141 and Henry Clay Road, Wilmington, DE 19803

Introduction – The prevalence of obesity worldwide has increased dramatically during the last several decades, and today has reached epidemic proportions in Western society (1, 2). The causes of obesity are varied, but clearly both genetic and environmental factors are responsible. This increased incidence has had an associated rise in co-morbidities linked with obesity. Indeed, the significant medical risks of obesity have led to the more appropriate clinical view of obesity as a medical condition rather than simply a cosmetic issue. An association clearly exists between obesity and Type-2 diabetes, coronary heart disease, hypertension, cholelithiasis, sleep apnea, osteoarthritis and certain forms of cancer (1, 3). The incidence of these co-morbidities increases linearly with the Body Mass Index (BMI) (defined as weight [kg]/height [m]^2) and are greatly increased for individuals with a BMI $\geq$ 30 (ca. 208 lbs. at 5'10" height). Even modest long-term weight reduction (5-8%) may produce significant benefits in lowering the incidence of cardiac problems, stroke, and diabetes in obese individuals. Appropriately, significant effort has been expended in identifying drug targets and efficacious compounds for reduction of body weight. Of the most attractive targets, the G-protein coupled receptor (GPCR) class of receptors has dominated in the search for effective anti-obesity medicines.

Two significant prescription weight-loss medications have been approved for long-term treatment in the US and Europe in recent years: sibutramine and orlistat (4). Earlier appetite suppressants such as "fen-phen" (the combination of fenfluramine and phentermine) and dexfenfluramine were withdrawn due to heart valve complications. Neither of the two approved drugs still on the market is considered to be adequately effective or free from side effects that limit compliance.

Increasing the challenge of developing effective anti-obesity agents are the multiple interacting pathways, both neural and metabolic, which have developed through evolution to regulate critical energy balance (3-6). This system appears to be biased for protection against starvation, i.e., allowing for storage of energy (triglycerides in the form of fat) as contrasted to prevention of obesity and its associated co-morbidities. The neural pathways controlling food intake can be divided into two general, opposing groups – orexigenic and anorexigenic pathways that have a positive and negative influence on food intake, respectively. An array of GPCRs is involved in regulating many, if not all, of these pathways. Many of these receptors are expressed in the hypothalamus, an area of the brain considered to be a primary site of energy balance regulation. This report will summarize advancements in our understanding of some of the more promising GPCRs and detail recent progress toward therapeutic agents that modulate these targets.

1

OREXIGENIC RECEPTORS

Receptors that increase appetite are known as orexigenic receptors which, in response to agonist peptides such as neuropeptide Y (NPY) and melanin concentrating hormone (MCH), stimulate food intake by either increasing hunger or delaying the onset of satiety. GPCRs of this class are attractive drug targets since antagonists are required which are usually less challenging from a drug discovery perspective. This report will review progress for several orexigenic receptors, including NPY receptors, MCH receptors and one of the cannabinoid receptors. Other orexigenic receptor targets that will not be discussed, due to limited reports of medicinal chemistry progress, include the receptors for orexin, ghrelin, and galanin (3, 7, 8).

Neuropeptide Y Receptor Family – A substantial body of data implicates NPY, a 36-amino acid neuropeptide, as a major physiological regulator of energy homeostasis. Indeed, NPY is one of the most potent and efficacious orexigenic agents following intracerebroventricular (i.c.v.) administration (9). NPY delivered chronically or overexpressed in the brain of rodents results in increased body weight and physiological changes consistent with obesity (10-12). Importantly, the NPY system appears to be directly influenced by leptin, a primary hormonal regulator of energy homeostasis (13).

While a role for the NPY system in energy balance is accepted widely, the challenge for drug discovery has been to identify which NPY receptor(s) may be the relevant physiological target for the feeding effects of NPY. The NPY receptor family is complex – consisting of six known subtypes Y_1-Y_5 and y_6 which typically signal through the inhibition of cAMP and elevations of intracellular Ca^{++}. Among this receptor family, the Y_1 and Y_5 receptors have been most closely associated with regulating feeding behavior and have been scrutinized as drug targets, although the Y_2 receptor may serve a limited function as well (9, 14). Our understanding of these receptors has been driven by a collection of studies, including pharmacological, receptor localization, and genetic experiments.

Both the Y_1 and Y_5 receptors demonstrate significant expression in various nuclei of the hypothalamus consistent with a role in feeding behavior. An antiserum to the Y_1 receptor has been used to evaluate localization of Y_1-immunoreactivity in the brain showing a discrete and abundant signal within the hypothalamus of rat (15, 16). These data support other findings by receptor autoradiography and *in situ* hybridization of Y_1 receptor in hypothalamus, as well as a moderate and broad expression pattern throughout the brain (17-19). By contrast, the Y_5 receptor seems to have a more restricted expression pattern within the brain but demonstrates significant hypothalamic expression, at least by *in situ* hybridization (19).

Pharmacological experiments have been equivocal as to which receptor subtype is more important in terms of therapeutic intervention for obesity. Selective antagonists for either receptor have been shown to reduce food intake in various paradigms (9). A drawback to many of these studies, however, is the reliance on less physiological paradigms of enhanced food intake such as significant fasting or administration of exogenous NPY, rather than *ad lib* fed animals. In total, it appears that both Y_1 and Y_5 receptors serve an important role in feeding behavior and have a complex interrelationship, yet their relative contribution to energy balance is unclear.

Genetic models have been no more successful in identifying the appropriate NPY receptor targets for obesity. Both the Y_1 and Y_5 receptor knockout (KO) mice demonstrate counterintuitive phenotypes, *i.e.*, weight gain rather than a lean phenotype (9, 20). Furthermore, in one study, no genetic linkage has been established between polymorphisms for Y_1 or Y_5 and morbidly obese patients (21). Indeed the role of NPY as a regulator of food intake *vs.* a starvation response is still in question. To further complicate matters, several data sets suggest a novel NPY receptor is responsible for some of the feeding effects of NPY. For instance, a putative Y_5 selective agonist was shown to still induce food intake in a Y_5 KO mouse (22).

Various issues remain to be resolved before a selective NPY receptor antagonist is successful as an anti-obesity agent. Questions regarding specificity of response need to be addressed – particularly for NPY Y_1 which has been suggested to regulate various functions that could translate into unacceptable side effects: vasoconstriction, anxiogenesis/depression, and pain perception (23-27). Moreover, the most significant development hurdles for compounds directed to NPY receptors are poor pharmacokinetic properties relating to oral bioavailability and brain penetrance.

While this report is focused on more recently characterized compounds, a number of Y_1 antagonists and agonists have been reported over the last decade. More recently, a systematic investigation of peptide NPY dimer analogs has resulted in identification of highly selective and potent Y_1 receptor antagonists, useful in elucidating Y_1 mediated activity (28). A small molecule ligand for Y_1, **1** (J-115814), has been reported and shows excellent affinity (K_i = 1.4 nM) for Y_1 and > 500 fold selectivity over the closely related Y_2, Y_4 and Y_5 receptor subtypes (22). In addition, this 2-aminopyridine analog (when administered i.c.v. or i.v.) dose-dependently suppressed feeding induced by i.c.v. NPY injection in satiated rats (22).

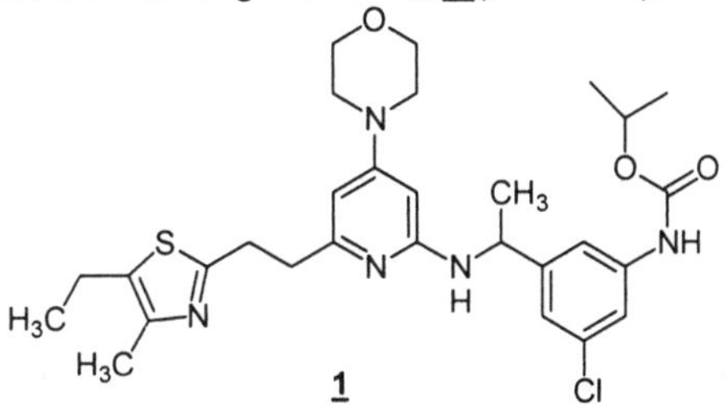

Recent efforts in the area of selective Y_5 receptor antagonists have been encouraged by data supporting Y_5 as a mediator of NPY induced feeding effects. An SAR study of trisubstituted phenylurea derivatives has afforded several ligands, exemplified by **2** and **3**, which showed excellent potency (IC_{50} < 0.1 nM) for the hY_5 receptor and > 10,000 fold selectivity over the hY_1 and Y_2 receptors (29). Another report exemplified the progress made with aminopyrazole derivatives, characterized by **4** (30). This optimized derivative was shown to be modestly potent (IC_{50} = 15 nM) but quite selective over the Y_1 and Y_2 receptors. The study further details the tight SAR with *N*-methyl substitution of either the primary amine or the sulfonamide providing derivatives with > 30-100 fold reduced affinity for the Y_5 receptor.

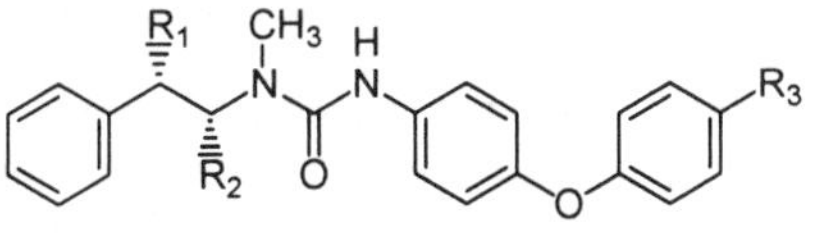

2, R_1 = OH, R_2 = Me, R_3 = F
3, R_1 = H, R_2 = Et, R_3 =H

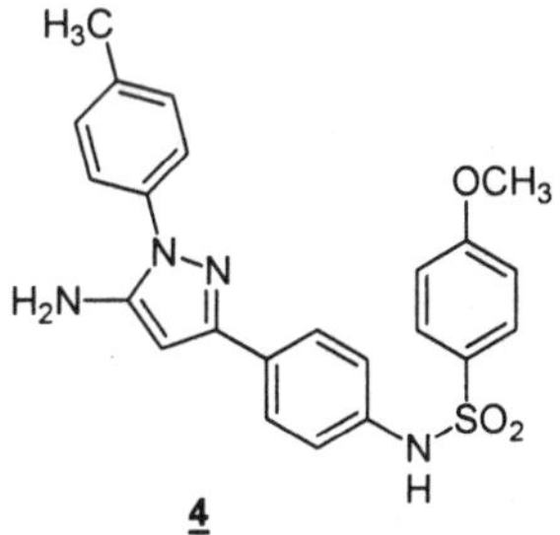

4

A series of four publications, aimed at preparing novel ligands for the Y_5 receptor, demonstrate the variability in structure types identified in chemical library screening approaches to this target (31-34). While optimization of several of these leads, **5** (FR236478), **6** (FR226928), and **7** (FR223118) resulted in potent compounds (IC_{50} values of 0.23-16 nM) with excellent selectivity, pharmacokinetic profiling revealed that they suffered from minimal bioavailability and brain penetrance. Continued screening, however, culminated in the identification of the novel diazabenzazulene derivative **8** (FR252384). This tricyclic ligand demonstrated excellent potency (IC_{50} = 2.3 nM) and was reported to be orally bioavailable and brain penetrant (34).

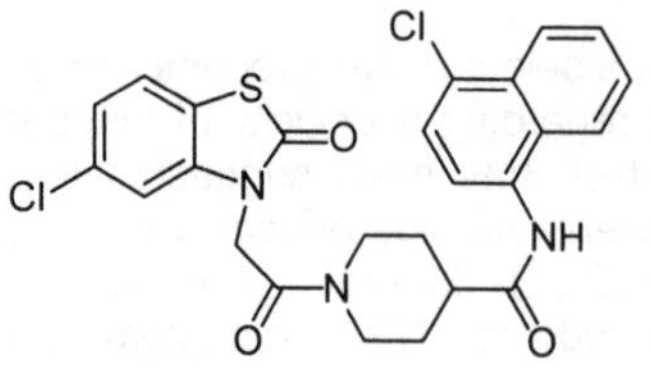

5

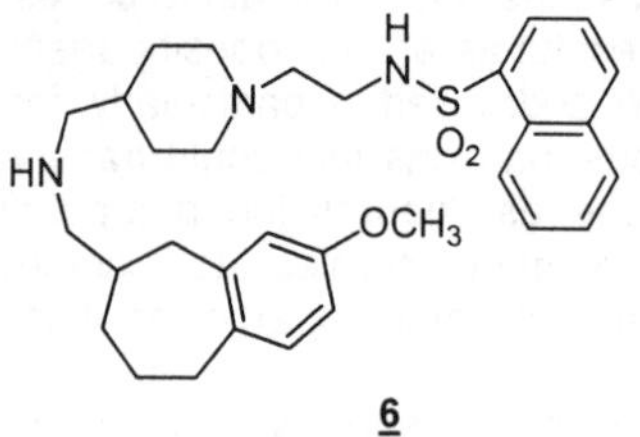

6

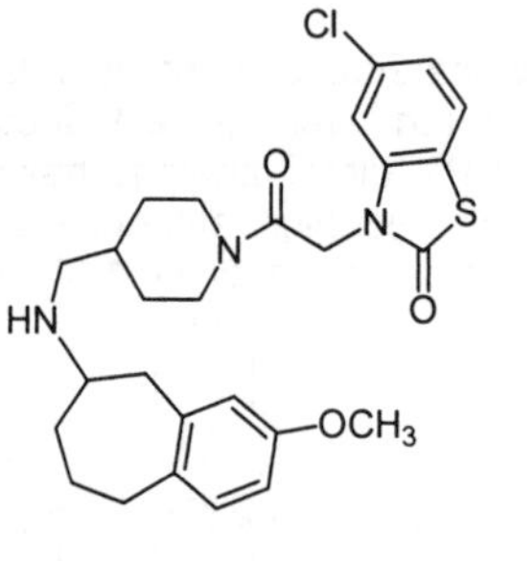

7

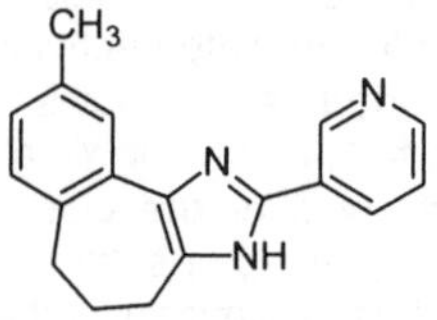

8

Melanin Concentrating Hormone (MCH) Receptor Family - The MCH receptor family has received considerable interest recently with appreciation of the robust biological impact of MCH on feeding behavior. MCH is a 19 amino acid cyclic peptide that is produced in cell bodies of the lateral hypothalamus (7). Administration of MCH i.c.v. increased food intake 1.5 – 3.0 fold (35). Importantly, mice with a KO of the prepro-MCH gene are hypophagic and lean with an apparent selective reduction in adipose tissue (36). By contrast, mice engineered to over-express MCH within the hypothalamus are moderately obese, being 12% heavier than wild-type controls and insulin resistant (37). MCH exerts its actions through two receptor subtypes, MCH_1 and MCH_2 (38, 39). Both subtypes are coupled to elevation in intracellular calcium likely *via* activation of G_q, and there is evidence that MCH_1 may promiscuously couple to $G_{i/o}$ to inhibit adenylate cyclase and elevate arachidonic acid levels (40). Based on their localization within the hypothalamus and other brain regions, both receptors are potential targets for development of antagonists for obesity.

The MCH_1 receptor was first identified as an orphan GPCR termed SLC-1 for which MCH was subsequently demonstrated to be the cognate ligand (41). The receptor is widely distributed throughout the brain, including several hypothalamic nuclei involved in feeding. The MCH_1 receptor may also be found in some peripheral tissues including adipocytes (42). Studies utilizing site-directed mutagenesis have determined that Asp^{123} of the 3rd transmembrane (TM) domain is critical for MCH binding, interacting with Arg^{11} of the MCH peptide, and is a residue also conserved among monoamine receptors (43). The MCH_1 receptor has also been implicated in obesity at both the genetic and pharmacological level. The MCH_1 receptor KO mouse exhibits a lean phenotype akin to that observed in the MCH KO mouse, although the MCH_1 KO animals do not actually weigh less than wild-type animals (44). Interestingly, the KO animals exhibit a significant reduction in white and brown adipose tissue that is the result of an increase in locomotor activity and metabolism, rather than a reduction in food intake.

Progress in identification of selective ligands for the MCH$_1$ receptor has been reported recently. Peptide analogs of MCH have been investigated and shown to be potent radioligands for the MCH$_1$ receptor as well as valuable tools for examination of pharmacology mediated *via* this receptor (45, 46). In the area of small molecule ligands, **9** (T-226292) has shown excellent affinity (5.5 nM) at the hMCH$_1$ receptor with selectivity over the MCH$_2$ receptor (> 1.0 µM) (35). When dosed orally at 30 mg/kg, compound **9** almost completely suppressed food intake induced by i.c.v. injection of MCH. It is not yet known whether this MCH$_1$ selective antagonist decreases *ad lib* feeding or may produce increases in locomotion and metabolism akin to that seen in MCH$_1$ KO animals.

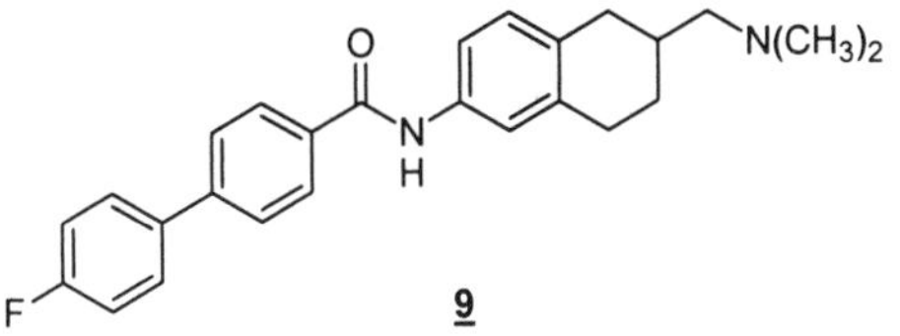

9

The MCH$_2$ receptor was recently cloned and found to be 38% identical to MCH$_1$ with a chromosomal localization (chromosome 6, band 6q16.2-16.3) placing the gene within a region associated with a Prader-Willi like syndrome of obesity (39). MCH$_2$ receptor mRNA is found primarily in the brain with significant expression in the hypothalamus (39). However, there have been no reports of selective MCH$_2$ receptor ligands since the identification of the receptor.

Cannabinoid 1 receptor - The observation that cannabis use increases food intake in users has led to association of the endocannabinoid system with the regulation of feeding behavior (47). The fatty-acid based molecule anandamide has been shown to be an endogenous cannabinoid of the CNS and is under the regulatory influence of leptin within the hypothalamus (48). Subcutaneous administration of anandamide increases food intake by as much as 2.0-fold in rats (49). Two cannabinoid receptors have been identified, CB$_1$ and CB$_2$, both of which are coupled to inhibition of cAMP accumulation (50). The CB$_1$ receptor is primarily localized to the CNS whereas CB$_2$ is found exclusively in the periphery. The CB$_1$ receptor has been implicated in the central regulation of food intake as it is expressed within the hypothalamus and other limbic brain regions associated with reward-based phenomena, including feeding (47). Fasted mice lacking the CB$_1$ receptor exhibit reduced food intake of around 50% compared to wild-type controls over an 18-hour test period. However, the KO animals were not leaner and did not eat less than the wild-type counterparts when fed *ad lib* (48).

A study based on the preparation of derivatives of the CB$_1$ antagonist **10** (SR141716A) was undertaken to investigate the recognition and activation of the cannabinoid receptor (51). Administration of **10** over the course of 14 days reduced food intake and body weight in rats fed *ad lib* (although the decrease in food intake tolerated by day 5) and reversed the hyperphagia produced by anandamide and δ9-tetrahydrocannabinol (52, 53). Compound **10** also reduced food intake acutely in two genetic models of obesity, *ob/ob* and *db/db* mice (48). The CB$_1$ KO mice did not respond to administration of compound **10**, which produced a 50% reduction in food intake in wild-type animals.

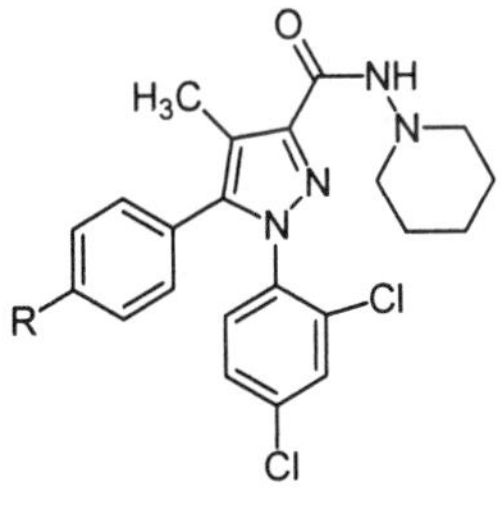

10, R = Cl
11, R = n-pentyl

12, R = OH, n = 3
13, R = Me, n = 6

A systematic examination of the various substituents on the pyrazole core led to the conclusion that the 1- and 5- positions of the heterocycle were responsible for antagonist activity in this series. One C-5 derivative, **11**, showed a 3-fold enhancement in affinity for the CB_1 receptor, relative to **10**. Identification of a novel hydantoin template with moderate affinity as a CB_1 antagonist has also been reported (54). Several derivatives, exemplified by **12** and **13**, had affinities of 70-100 nM for the CB_1 receptor competing with radiolabeled **10**.

ANOREXIGENIC RECEPTORS

Reduction of food intake is mediated by anorexigenic receptors, presumably through increasing satiety. For the most part, these receptors are compelling anti-obesity drug targets since they actively reduce food intake. However, these targets typically require an agonist approach, presenting a more demanding drug discovery challenge. Moreover, issues of tolerance and desensitization are major concerns for agonist therapy. The two anorexigenic systems discussed in this report are the serotonin $5\text{-}HT_{2C}$ receptor and melanocortin-4 receptor. In addition, several other anorexigenic GPCRs have recently garnered attention for drug discovery, but are not addressed in this report: cholecystokinin CCK_1 receptor, corticotrophin releasing factor/urocortin (CRF1/2 receptor) and glucagon-like peptide 1 receptor (3, 4, 6, 7).

$5\text{-}HT_{2C}$ receptor - The neurotransmitter serotonin (5-HT) is a biogenic amine involved in a wide number of central and peripheral processes. 5-HT exerts its action through activation of 14 distinct receptor subtypes (55, 56). Two of these subtypes, the $5\text{-}HT_{1B}$ and $5\text{-}HT_{2C}$ receptors, have been implicated most consistently in regulation of food intake and body weight, and are functionally involved in efficacy of the anti-obesity agent fenfluramine (57-59). $5\text{-}HT_{1B}$ receptors are less attractive as a therapeutic target due to a possible etiological link to primary pulmonary hypertension (60). Increased attention to the $5\text{-}HT_{2C}$ receptor has followed the observation that mice lacking the receptor develop obesity and studies showing the non-selective $5\text{-}HT_{2C}$ agonist mCPP reduced food intake and body weight in clinical studies (61, 62). Obtaining selectivity of $5\text{-}HT_{2C}$ receptor agonists vs. the highly homologous $5\text{-}HT_{2A}$ and $5HT_{2B}$ receptors is crucial as these two subtypes have been implicated in hallucinogenesis and heart valve hypertrophy, respectively (55, 63). The $5\text{-}HT_{2C}$ receptor is linked to elevation of inositol phosphates (via G_q coupling), and cGMP signaling via promiscuous coupling to other G-proteins (64). Additionally, $5\text{-}HT_{2C}$ is believed to cause the release of the obesity-related transcription factor tubby from the cell membrane via the activation of G_q, allowing for its translocation to the nucleus (55). Mice lacking the $5\text{-}HT_{2C}$ receptor develop a late-stage obesity syndrome characterized by an increase in weight of 20% compared to wild-type controls and accompanied by insulin and leptin resistance (65). In rodents, stimulation of the $5\text{-}HT_{2C}$ receptor appears to delay meal onset and reduce meal duration, rather than altering metabolic function (66).

As with any agonist ligand the development of tolerance is a concern, but several studies have shown that tolerance to the anorectic action of $5\text{-}HT_{2C}$ agonists did not occur. Rats chronically treated with the $5\text{-}HT_{2C}$ agonists mCPP, **14** (Ro 60-0175) and fenfluramine have reduced levels of feeding and body weight to controls (by 10-15%) that is sustained for a period of 14 days (67). Further, the selective $5\text{-}HT_{2C}$ antagonist **15** (SB-242084) blocked effects of fenfluramine on food intake (59). There have been a number of nonselective 5-HT$_{2C/2B/2A}$ ligands identified in the past decade, but only recently have truly selective (> 100 fold) 5-HT$_{2C}$ agonists been realized (56, 68). The first of these, **16** (IL639), has been reported to be orally bioavailable and efficacious in a rat chronic feeding assay (12 weeks) with no indication of tolerance. This biaryl indoline derivative has K$_i$ values of 5.2 nM, 1510 nM, and 1440 nM at the human 5-HT$_{2C}$, 5-HT$_{2B}$, and 5-HT$_{2A}$ receptors, respectively (69). For additional information on 5-HT$_{2C}$ receptor pharmacology, see Chapter 3 of this Volume.

14

15

16

<u>Melanocortin-4 receptor (MC$_4$)</u> - The melanocortin system has been a focus of obesity researchers since the discovery of its involvement in the obese phenotype of the Agouti mouse. The melanocortin-4 receptor (MC$_4$) is one of five melanocortin receptor subtypes. The endogenous ligand for the MC$_4$ receptor is α-melanocyte stimulating hormone (α-MSH), a 13 amino acid peptide derived from pro-opiomelanocortin (POMC). Binding of α-MSH leads to activation of adenylate cyclase *via* coupling to G$_s$ (70). This receptor is unique because an endogenous antagonist, Agouti Related Peptide (AGRP), also exists (*vide infra*)(71). Individuals with mutations to the MC$_4$ receptor are obese, reaching BMI levels greater than 50, with mutations occurring at a rate of 4% in the morbidly obese (72, 73). Individuals lacking the ability to produce α-MSH due to a mutation within the gene for POMC are also morbidly obese (74). Peptide-like agonists and antagonists of α-MSH have been shown to reduce and enhance feeding, respectively (75). Thus, MT-II is a non-selective small peptide agonist of the MC$_4$ receptor that reduces food intake in normal animals and is inactive in mice lacking the MC$_4$ receptor. SHU9119 is a similar small peptide that exhibits antagonist activity at the MC$_4$ receptor and stimulates food intake when given intracisternally (76). Relevant to possible additional indications or side effects, it has been noted that MC$_4$ receptor agonism can cause penile erection (70). While MC$_4$ is the primary receptor associated with the regulation of food intake, MC$_3$ and MC$_5$ may also play a role in the anorectic activity of α-MSH (7, 70).

Achievement of selectivity for an MC$_4$ agonist is paramount due to involvement of the other MC receptor subtypes in peripheral functions such as pigmentation and immune response (77). There has been recent progress in identification of selective MC$_4$ ligands. Macrocyclic peptide MC$_4$ agonists with > 100 fold selectivity over the closely related MC$_3$ and MC$_5$ receptors have been reported (78, 79). Recently disclosed is identification of the small molecule agonist, **17**, which is reported to possess a K$_i$ of 1.1 nM at hMC$_4$ and > 200 fold selectivity over hMC$_1$, hMC$_3$ and hMC$_5$ receptors (80).

17

<u>Agouti-related peptide (AGRP)/Melanocortin-4 Receptor</u> - AGRP is an endogenous antagonist (*vide supra*) to the MC$_4$ receptor (71). The peptide is co-localized with NPY in neurons of the arcuate nucleus within the hypothalamus. The C-terminal portion of the peptide (87-132) is believed to be necessary for optimal binding and contains a five-fingered spider-toxin motif with an eight amino acid portion mimicking α-MSH. While AGRP is competitive with α-MSH for binding at the MC$_4$ receptor, some regions of receptor interaction for the two peptides are apparently different (81). The binding of

AGRP may also be dependent upon interactions with attractin (*mahogany* – an obesity suppressing mutation) and syndecan-3 (71, 82).

Due to the differential binding regions for AGRP and α-MSH, it may be possible to block AGRP inhibition of α-MSH activity as an approach to reducing food intake. This strategy requires the synthesis of an AGRP antagonist to disinhibit the natural level of agonism provided by α-MSH. Recently, 1-aryl-2-piperazinylethyl piperazine and indole analogs, exemplified by **18** (IC_{50} = 49 nM at AGRP/MC$_4$), have been reported to be antagonists of AGRP (83).

18

Conclusion – A wealth of attractive obesity therapy targets are currently available for drug discovery efforts. Some of these targets are well-studied and validated with preclinical proof of concept, *e.g.*, NPY, CB$_1$ and 5-HT$_{2C}$ receptors. Others are at an early stage but hold significant promise such as MC$_4$ and MCH receptors. Obesity is a serious medical issue with a highly deleterious impact on the individual and society and effective pharmacotherapy is lacking. Recent discoveries involving central GPCR drug targets and progress in small molecule development hold great promise for significant advances in the important area of anti-obesity therapeutics.

References

1.   P.G. Kopelman, Nature, 404, 635 (2000).
2.   A.M. Prentice, Br. Med. Bull., 60, 51 (2001).
3.   V.E.F. Crowley, G.S.H. Yeo and S. O'Rahilly, Nat. Drug Discov., 1, 276 (2002).
4.   J. Proietto, B.C. Fam, D.A. Ainslie and A.W. Thorburn, Expert Opin. Invest. Drugs, 9, 1317 (2000).
5.   B.M. Spiegelman and J.S. Flier, Cell, 104, 531 (2001).
6.   M. Chiesi, C. Huppertz and K.G. Hofbauer, Trends Pharmacol. Sci., 22, 247 (2001).
7.   W.S. Dhillo and S.R. Bloom, Curr. Opin. Pharmacol., 1, 651 (2001).
8.   A. Inui, Neuroscience, 2, 1 (2001).
9.   H.A. Wieland, B.S. Hamilton, B. Krist and H.N. Doods, Exp. Opin. Invest. Drugs, 9, 1327 (2000).
10.   N. Zarjevski, I. Cusin, R. Vettor, F. Rohner-Jeanrenaud and B. Jeanrenaud, Endocrinology, 133, 1753 (1993).
11.   P.D. Raposinho, D.D. Pierroz, P. Broqua, R.B. White, T. Pedrazzini and M.L. Aubert, Mol. Cell Endocrinol., 185, 195 (2001).
12.   T. Kaga, A. Inui, M. Okita, A. Asakawa, N. Ueno, M. Kasuga, M. Fujimiya, N. Nishimura, R. Dobashi, Y. Morimoto, I.M. Liu and J.T. Cheng, Diabetes, 50, 1206 (2001).
13.   Y. Zhang, R. Proenca, M. Maffel, M. Barone, L. Leopold and J.M. Friedman, Nature, 372, 425 (1994).
14.   P. Naveilhan, H. Hassani, J.M. Canals, A.J. Ekstrand, A. Larefalk, V. Chajlant, E. Arenas, K. Gedda, L. Svensson, P. Thoren and P. Ernfors, Nat. Med., 5, 1188 (1999).
15.   K. Fuxe, B. Tinner, L. Caberlotto, B. Bunnemann and L.F. Agnati, Neurosci. Lett., 225, 49 (1997).
16.   K. Migita, A.D. Loewy, T.V. Ramabhadran, J.E. Krause and S.M. Waters, Brain Res., 889, 23 (2001).
17.   P.J. Larsen, S.P. Sheikh and J.D. Mikkelsen, J. Recept. Res., 15, 457 (1995).
18.   R.M.C. Parker and H. Herzog, Eur. J. Neurosci., 11, 1431 (1999).
19.   M.M. Durkin, M.W. Walker, K.E. Smith, E.L. Gustafson, C. Gerald and T.A. Branchek, Exp. Neurology, 165, 90 (2000).
20.   A.A. Butler and R.D. Cone, Trends Genet., 17, 550 (2001).
21.   C. Roche, P. Boutin, C. Dina, G. Gyapay, A. Basdevant, J. Hager, B. Guy-Grand, K. Clement and P. Froguel, Diabetologia, 40, 671 (1997).
22.   A. Kanatani, M. Hata, S. Mashiko, A. Ishihara, O. Okamoto, Y. Haga, T. Ohe, T. Kanno, N. Murai, Y. Ishii, T. Kukuroda, T. Fukami and M. Ihara, Mol. Pharmacol., 59, 501 (2001).
23.   J. You, L. Edvinsson and J. R. M. Bryan, J. Cereb. Blood Flow Metab., 21, 77 (2001).

24. J.P. Redrobe, Y. Dumont, A. Fournier and R. Quirion, Neuropsychopharmacology, 26, 615 (2002).
25. A. Kask, E. Vasar, L.T. Heidmets, L. Allikmets and J.E. Wikberg, Eur. J. Pharmacol., 414, 215 (2001).
26. P.R. Brumovsky, T.-J. Shi, H. Matsuda, J. Kopp and M.J. Villar, Exp. Neurol., 174, 1 (2002).
27. J.Z. Wang, T. Lundeberg and L.C. Yu, Brain Res., 893, 264 (2001).
28. A. Balasubramaniam, V.C. Dhawan, D.E. Mullins, W.T. Chance, S. Sheriff, M. Guzzi, M. Prabhakaran and E.M. Parker, J. Med. Chem., 44, 1479 (2001).
29. C. Fotsch, J.D. Sonnenberg, N. Chen, C. Hale, W. Karbon and M.H. Norman, J. Med. Chem., 44, 2344 (2001).
30. C.P. Kordik, C. Luo, B.C. Zanoni, S.L. Dax, J.J. McNally, T.W. Lovenberg, S.J. Wilson and A.B. Reitz, Bioorg. Med. Chem. Lett., 11, 2283 (2001).
31. Y. Satoh and C. Hatori, Bioorg. Med. Chem. Lett., 12, 1009 (2002).
32. H. Itani, H. Ito, Y. Sakata, Y. Hatakeyama, H. Oohashi and Y. Satoh, Bioorg. Med. Chem. Lett., 12, 757 (2001).
33. H. Itani, H. Ito, Y. Sakata, Y. Hatakeyama, H. Oohashi and Y. Satoh, Bioorg. Med. Chem. Lett., 12, 799 (2002).
34. S. Tabuchi, H. Itani, Y. Sakata, H. Oohashi and Y. Satoh, Bioorg. Med. Chem. Lett., 12, 1171 (2002).
35. S. Takekawa, A. Asami, Y. Ishihara, J. Terauchi, K. Kato, Y. Shimomura, M. Mori, H. Murakoshi, K. Kato, N. Suzuki, O. Nishimura and M. Fujino, Eur. J. Pharmacol., 438, 129 (2002).
36. M. Shimada, N.A. Tritos, B.B. Lowell, J.S. Flier and E. Maratos-Flier, Nature, 396, 670 (1998).
37. D.S. Ludwig, N.A. Tritos, J.W. Mastaitis, R. Kulkarni, E. Kokkotou, J. Elmquist, B. Lowell, J.S. Flier and E. Maratos-Flier, J. Clin. Invest., 107, 379 (2001).
38. J. Chambers, R.S. Ames, D. Bergsma, A. Muir, L.R. Fitzgerald, G. Hervieu, G.M. Dytko, J.J. Foley, J. Martin, W.-S. Liu, J. Park, C. Ellis, S. Ganguly, S. Konchar, J. Cluderay, R. Leslie, S. Wilson and H.M. Sarau, Nature, 400, 261 (1999).
39. A.W. Sailer, H. Sano, Z. Zeng, T.P. McDonald, J. Pan, S.-S. Pong, S.D. Feighner, C.P. Tan, T. Fukami, H. Iwaasa, D.L. Hreniuk, N.R. Morin, S.J. Sadowski, M. Ito, M. Ito, A. Bansal, B. Ky, D.J. Figueroa, Q. Jiang, C.P. Austin, D.J. MacNeil, A. Ishihara, M. Ihara, A. Kanatani, L.H.T.V.d. Ploeg, A.D. Howard and Q. Liu, Proc. Natl. Acad. Sci., 98, 7564 (2001).
40. Y. Saito, H.-P. Nothacker and O. Civelli, Trends Exp. Med., 11, 299 (2000).
41. L.F. Kolakowski, B.P. Jung, T. Nguyen, M.P. Johnson, K.R. Lynch, R. Cheng, H.H.Q. Heng, S.R. George and B.F. O'Dowd, FEBS, 398, 253 (1996).
42. R.L. Bradley, E.G. Kokkotou, E. Maratos-Flier and B. Cheatham, Diabetes, 49, 1073 (2000).
43. D. MacDonald, N. Murgolo, R. Zhang, J.P. Durkin, X. Yao, C.D. Strader and M.P. Graziano, Mol. Pharmacol., 58, 217 (2000).
44. D.J. Marsh, D.T. Weingarth, D.E. Novi, H.Y. Chen, M.E. Trumbauer, A.S. Chen, X.-M. Guan, M.M. Jiang, Y. Feng, R.E. Camacho, Z. Shen, E.G. Frazier, H. Yu, J.M. Metzger, S.J. Kuca, L.P. Shearman, S. Gopal-Truter, D.J. MacNeil, A.M. Strack, D.E. MacIntyre, L.H.T.V.d. Ploeg and S. Qian, Proc. Natl. Acad. Sci., 99, 3240 (2002).
45. V. Audinot, C. Lahaye, T. Suply, P. Beauverger, M. Rodriguez, J.-P. Galizzi, J.-L. Fauchere and J.A. Boutin, Br. J. Pharmacol., 133, 371 (2001).
46. V. Audinot, P. Beauverger, C. Lahaye, T. Suply, M. Rodriguez, C. Ouvry, V. Lamamy, J. Imbert, H. Rique, J.-L. Nahon, J.-P. Galizzi, E. Canet, N. Levens, J.-L. Fauchere and J.A. Boutin, J. Biol. Chem., 276, 13554 (2001).
47. R. Mechoulam and E. Fride, Nature, 410, 763 (2001).
48. V. DiMarzo, S.K. Goparaju, L. Wang, J. Liu, S. Batkai, Z. Jarai, F. Fezza, G.I. Miura, R.D. Palmiter, T. Sugiura and G. Kunos, Nature, 410, 822 (2001).
49. C.M. Williams and T.C. Kirkham, Psychopharmacology, 143, 315 (1999).
50. E.S. Onaivi, A. Chakrabarti and G. Chaudhuri, Prog. Neurobiol., 48, 275 (1996).
51. J.L. Wiley, R.G. Jefferson, M.C. Grier, A. Mahadevan, R.K. Razdan and B.R. Martin, J. Pharmacol. Exp. Ther., 296, 1013 (2001).
52. G. Colombo, R. Agabio, G. Diaz, C. Lobina, R. Reali and G.L. Gessa, Life Sci., 63, PL113 (1998).
53. C.M. Williams and T.C. Kirkham, Pharmacol., Biochem. Behav., 71, 341 (2002).
54. F. Ooms, J. Wouters, O. Oscari, T. Happaerts, G. Bouchard, P.-A. Carrupt, B. Testa and D.M. Lambert, J. Med. Chem., 45, 1748 (2002).

55. J.R. Raymond, Y.V. Mukhin, A. Gelasco, J. Turner, G. Collinsworth, T.W. Gettys, J.S. Grewal and M.N. Garnovskaya, Pharmacol. Ther., 92, 179 (2001).

56. A.J. Robichaud and B.L. Largent, Annu. Rep. Med. Chem., 35, 11 (2000).

57. J.J. Lucas, A. Yamamoto, K. Scearce-Levie, F. Saudou and R. Hen, J. Neurosci., 18, 5537 (1998).

58. S.P. Vickers, P.G. Clifton, C.T. Dourish and L.H. Tecott, Psychopharmacology, 143, 309 (1999).

59. S.P. Vickers, C.T. Dourish and G.A. Kennett, Neuropharmacology, 41, 200 (2001).

60. M.R. MacLean, P. Herve, S. Eddahibi and S. Adnot, Br. J. Pharmacol., 131, 161 (2000).

61. P.A. Sargent, A.L. Sharpley, C. Williams, E.M. Goodall and P.J. Cowen, Psychopharmacology, 133, 309 (1997).

62. A.E.S. Walsh, K.A. Smith, A.D. Oldman, C. Williams, E.M. Goodall and P.J. Cowen, Psychopharmacology, 116, 120 (1994).

63. L.W. Fitzgerald, T.C. Burn, B.S. Brown, J.P. Patterson, M.H. Corjay, P.A. Valentine, J.-H. Sun, J.r. Link, I. Abbaszade, J.M. Hollis, B.L. Largent, P.R. Hartig, G.F. Hollis, P.C. Meunier, A.J. Robichaud and D.W. Robertson, Mol. Pharmacol., 57, 75 (2000).

64. G.L. Alberts, J.F. Pregenzer, W.B. Im, P.G. Zaworski and G.S. Gill, Eur. J. Pharmacol., 383, 311 (1999).

65. K. Nonogaki, A.M. Strack, M.F. Dallman and L.H. Tecott, Nature Med., 4, 1152 (1998).

66. P.G. Clifton, M.D. Lee and C.T. Dourish, Psychopharmacology, 152, 256 (2000).

67. S.P. Vickers, K.R. Benwell, R.H. Porter, M.J. Bickerdike, G.A. Kennett and C.T. Dourish, Br. J. Pharmacol., 130, 1305 (2000).

68. J.J. Chambers, D.M. Kurrasch-Orbaugh, M.A. Parker and D.E. Nichols. J. Med. Chem., 44, 1003 (2001).

69. A. Robichaud, W. Chen, C. McClung, D. Clark, W. Deng, E. Brondyke-Calvello, I.S. Mitchell, R.P. Taekyu, L.W. Fitzgerald, D.S. Conklin, J.F. McElroy, K.W. Rohrbach, K.J. Miller, B.L. Largent and D. Robertson, 221st ACS National Meeting, San Diego, CA, April 1-5, 2001.

70. J.E.S. Wikberg, Exp. Opin. Ther. Patents, 11, 61 (2001).

71. D.M. Dinulescu and R.D. Cone, J. Biol. Chem., 275, 6695 (2000).

72. C. Vaisse, K. Clement, E. Durand, S. Hercberg, B. Guy-Grand and P. Froguel, J. Clin. Invest., 106, 253 (2000).

73. S. Farooqi, G.S.H. Yeo, J.M. Keogh, S. Aminian, S.A. Jebb, G. Butler, T. Cheetham and S. O'Rahilly, J. Clin. Invest., 106, 271 (2000).

74. H. Krude, H. Biebermann, W. Luck, R. Horn, G. Brabant and A. Gruters, Nat. Genetics, 19, 155 (1998).

75. S.C. Benoit, M.W. Schwartz, J.L. Lachey, M.M. Hagan, P.A. Rushing, K.A. Blake, K.A. Yagaloff, G. Kurylko, L. Franco, W. Danhoo and R.J. Seeley, J. Neurosci., 20, 3442 (2000).

76. T. Adage, A.J.W. Scheurink, S.F. de Boer, K. de Vries, J. Pieter-Konsman, F. Kuipers, R.A.H. Adan, D.G. Baskin, M.W. Schwartz and G. van Dijk, J. Neurosci., 21, 3639 (2001).

77. A. Catania, L. Airaghi, G. Colombo and J.M. Lipton, Trends Exp. Med., 11, 304 (2000).

78. M.A. Bednarek, T. MacNeil, R. Tang, R.N. Kalyani, L.V.d. Ploeg and D.H. Weinberg, Biochem. Biophys. Res. Commun., 286, 641 (2001).

79. M.A. Bednarek, T. MacNeil, R.N. Kalyani, R. Tang, L.V.d. Ploeg and D. Weinberg, J. Med. Chem., 44, 3665 (2001).

80. R.P. Nargund, I. Sebhat, Z. Ye, K. Barakat, D. Weinberg, T. MacNeil, R. Kalyani, W. Martin, D. Cashen, H. Chen, J. Drisko, R. Mosley, T. Fong, R. Stearns, R. Miller, R. Tamvakopoulos, L. Colwell, A. Strack, X. Shen, C. Tan, S.-S. Pong, A. Howard, A. Sailer, G. Hickey, E. MacIntyre, L. Van der Ploeg and A. Patchett, 222nd American Chemical Society, Chicago, IL, August 26-30, 2001.

81. C. Haskell-Luevano, R.D. Cone, E.K. Monck and Y.-P. Wan, Biochemistry, 40, 6164 (2001).

82. O. Reizes, J. Lincecum, Z. Wang, O. Goldberger, L. Huang, M. Kaksonen, R. Ahima, M.T. Hinkes, G.S. Barsh, H. Rauvala and M. Bernfield, Cell, 106, 105 (2001).

83. P. Arasasingham, C. Fotsch, T. Jenkins, T. Lee, J.F. Kincaid, Y. Bo, X. Ouyang, M.H. Norman, M.J. Kelly, K.L. Stark and B. Karbon, 222nd American Chemical Society, Chicago, IL, August 26-30, 2001.

Chapter 2. Attention Deficit Hyperactivity Disorder: Pathophysiology and Design of New Treatments

Shelly A. Glase, David W. Robertson and Lawrence D. Wise
Pfizer Global Research and Development
Ann Arbor, MI 48105

Introduction - Attention deficit hyperactivity disorder (ADHD) is a common psychiatric disorder in children, and estimates of its incidence generally range from 3-5%; however, much higher incidences are sometimes cited (1,2). The actual incidence is a controversial subject, and recent studies suggest some of the differences in reported rates are based upon stringency of criteria used in diagnosing the disorder (2). Despite evidence this disorder has a relatively stable incidence and prevalence across ethnic groups and national boundaries, it is recognized and treated primarily within the United States (3). ADHD is a heterogeneous behavioral disorder with multiple possible etiologies including neuroanatomic and neurochemical abnormalities, genetic polymorphisms, CNS insults that occur pre- or postnatally or in childhood, and environmental factors.

In this report we will consider some of the current understanding of the biology and treatments for ADHD. We will also detail some of the mechanistic avenues that have been reported for discovery of potential new medicines to treat this disorder.

BIOLOGY OF ADHD

Our understanding of this disorder has evolved since it's somewhat nebulous definition in the 1930's as "minimal brain damage", to the point physicians can now readily define and diagnose the disorder using standardized neuropsychiatric assessments (Table 1).

Table 1. Evolution in Diagnosis of ADHD

Year	Description	Reference
1930	Minimal Brain Damage	4
1960	Minimal Brain Dysfunction	5
1968	Hyperkinetic Reaction of Childhood	6
1980	Attention Deficit Disorder	7
1987	ADHD	8
1994	AD/HD	9

The standard psychiatric diagnostic criteria for ADHD include frequent manifestation of: 1) inattention to details; 2) difficulty sustaining attention; 3) not listening; 4) failing to finish tasks; 5) difficulty with organizing; 6) avoiding tasks that require sustained attention; 7) losing things; 8) being easily distracted; and 9) forgetfulness (9). Core symptom areas of the disorder include inattention and impulsivity/hyperactivity, and the symptoms vary in pervasiveness, frequency of occurrence, and the degree to which they impair functioning of the affected individual. As is the case with most psychiatric diseases, multiple subtypes of ADHD are recognized including a form of the disorder that is dominated by inattention, a rarer form that is dominated by hyperactivity and impulsivity, and the most common combined type where diagnostic criteria are met for both inattention and hyperactivity/impulsivity. The presence of ADHD subtypes may complicate genetic analyses and pharmacological treatment of the disorder (vide infra).

Although ADHD is often associated with school-age children, it is increasingly being recognized in preschool children (3-6 years old) and there is documented impaired function and similar patterns of co-morbid psychopathology in these younger children (10,11). In addition, ADHD commonly persists into adulthood, although adults often develop coping skills to overcome some of the difficulties associated with the disorder. The life-long prevalence of ADHD, and treatment of the disorder in adults has been reviewed (12). At present, psychostimulants remain first-choice treatment strategies for adults with ADHD *(vide infra)*.

The neuroanatomic and neurochemical bases of ADHD continue to be extensively studied, and the attentional networks centered in the prefrontal cortex including subregions important in executive control, alerting functions, and selective attention have received considerable attention. Abnormalities of executive function are observed in ADHD patients and these may arise from a dysregulation of behavioral inhibition systems. In terms of neurochemical abnormalities, evidence suggests dysfunction of dopaminergic and noradrenergic systems in patients with ADHD. Studies in spontaneously hypertensive rats (SHR), an animal model of ADHD-related behaviors, suggest these animals have an imbalance between the dopaminergic and noradrenergic systems relative to normal Wistar-Kyoto (WKY) control rats (13).

<u>Genetics</u> - A growing body of literature suggests ADHD has a substantial genetic component, but molecular genetic studies to elucidate genes etiologically involved in the disease have not produced incontrovertible findings, as is expected due to the polygenic architecture of ADHD and the variety of subtypes of the disorder (14). Because of the long-recognized role of dopamine dysfunction in the disease, considerable attention has been focused upon receptors, enzymes and transporters involved in dopaminergic pathways, and we will delineate results from a few recent studies. No coding region sequence polymorphisms were identified in the dopamine receptor D2 gene in ADHD patients (15). A transmission disequilibrium test of the Ser9/Gly dopamine D3 receptor (DRD3) gene polymorphism in adult ADHD suggests the DRD3 gene does not play a major role in etiology of the disorder (16). The most commonly reported (17) molecular genetic abnormalities involve the dopamine transporter gene (DAT 1) on chromosome 5 and the dopamine receptor D4 gene (DRD4) on chromosome 11. While a number of laboratories have demonstrated an association between the DRD4 seven-repeat allele and ADHD, a number of laboratories have failed to confirm this finding. In a subgroup analysis (18) of children with ADHD and concurrent conduct disorder symptoms, there was evidence of association between DRD4 and ADHD ($p = 0.05$ for the seven-repeat allele). Further studies using well-defined populations of ADHD patients may clarify the role of DRD4 in the disorder.

The dopamine β-hydroxylase (DBH) gene, which encodes the enzyme that catalyzes the conversion of dopamine to norepinephrine, has recently been studied in ADHD. Haplotype relative risk analysis of the DBH TaqI restriction site polymorphism showed a preferential transmission of the TaqI A2 allele ($p = 0.03$), partially replicating previous findings demonstrating an association between the DBH TaqI A2 allele and ADHD (19). A recent study suggested a linkage between ADHD and a VNTR polymorphism in the MAOA gene, but not the MAOB gene (20).

While serotonin (5-HT) does not appear to play a major role in the pathophysiology of ADHD, there appears to be a significant association of polymorphisms in the 5-HT transporter (5HTT) promoter region, with a higher frequency of the long variant allele in patients with high scores in the Wender Utah Rating scale. However, there was no association between the 5HTT polymorphism and personality traits using other clinical rating scales (21).

At the level of expressed protein, the dopamine transporter has been reported to be elevated in adults with ADHD compared to normal controls (22). However, a recent positron emission tomography study has not confirmed these results (23).

Comorbidities - In addition to symptoms of ADHD, this disorder is often complicated by presence of a variety of additional psychiatric co-morbidities including major depression, anxiety disorders, and conduct disorders (10, 24). Therefore, the significant use of stimulants such as methylphenidate to treat ADHD presents a therapeutic dilemma: while stimulants unequivocally improve the core symptoms of ADHD, they may sometimes exacerbate co-morbid symptoms of anxiety and depression. Individuals with ADHD also experience a higher incidence of sleep disorders (25).

In both children and adults, ADHD may enhance vulnerability to substance abuse, including nicotine, alcohol, and illicit drugs. It has been suggested that early treatment of ADHD in children may reduce substance abuse (26).

Current Pharmacologic Treatments - First-line therapies for ADHD are stimulants such as methylphenidate (**1**), and amphetamine preparations; pemoline, an additional stimulant, is not widely used due to a rare liver toxicity. Stimulants have been extensively studied and clearly ameliorate the core symptoms of ADHD, including inattention, impulsivity and hyperactivity (27). Psychostimulants are equally effective in treating younger children and adolescents with ADHD (28).

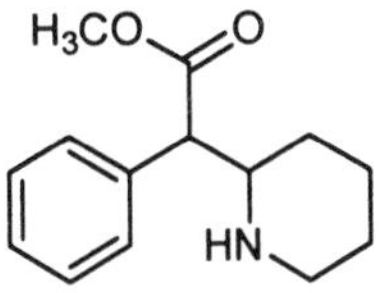

1

The most widely prescribed stimulant, methylphenidate, has been in use for over 40 years. Although methylphenidate was approved in the U.S. during the 1960s, data on pharmacokinetic properties of the drug were not reported until the 1980s. Pharmacokinetic and pharmacodynamic drug interactions of the psychostimulants methylphenidate, amphetamine and pemoline have been reviewed (29). Understanding of methylphenidate's pharmacokinetic properties led to development and launch of several modified release forms of this medicine that permit once-daily administration, obviating the often-embarrassing administration of a dose of stimulant during the school day. A number of different forms of methylphenidate are currently available, including methylphenidate immediate-release, methylphenidate extended-release, and methylphenidate immediate-release/extended-release (30). There has been some controversy regarding comparative efficacy of extended-release preparations of methylphenidate versus immediate-release preparations. However, recent results suggest once-daily modified-release forms are as effective as immediate-release preparations of methylphenidate. Placebo-controlled, double-blind, randomized clinical trials suggest once-daily administration of modified-release methylphenidate to children with ADHD is effective and safe in controlling ADHD symptoms throughout the school day (31). In another recently published study, OROS methylphenidate dosed once-daily and immediate-release methylphenidate dosed thrice-daily did not differ significantly from each other in treatment of ADHD symptoms, and both were superior to placebo (32). Moreover, enantiomerically pure forms of methylphenidate have been developed and launched; new delivery systems and forms of methylphenidate have been reviewed (33).

Although stimulant medicines are now well established pharmacotherapies for ADHD, they do have numerous limitations: they are scheduled drugs which creates issues of abuse liability, diversion and medico-legal concerns; approximately 30% of patients do not respond to or cannot tolerate these medicines, and there are concerns about effects of these medicines on appetite, mood, and growth.

NEW THERAPEUTIC APPROACHES

Mechanistic approaches that have received greatest attention in recent years include development of selective norepinephrine (NE) reuptake inhibitors, histamine H₃ receptor antagonists, and various nicotinic receptor agonists. In the following sections of this report, we delineate advances in discovery of agents in these mechanistic areas that may provide new medicines for treatment of ADHD.

Monoamine Reuptake Inhibitors – Tricyclic anti-depressants that inhibit the re-uptake of norepinephrine, such as desipramine (**2**) and nortriptyline (**3**), have efficacy in treating ADHD (34,35). The same side effect issues that impair utility of these medicines in treating depression limit their use in treating ADHD, including anticholinergic and antihistaminergic side effects. Moreover, effects of these drugs on cardiac conduction is of concern, particularly regarding use of these medicines in fairly healthy children. Because of some documented cases of sudden death in children treated with tricyclic antidepressants, their use in treating ADHD has fallen into disfavor.

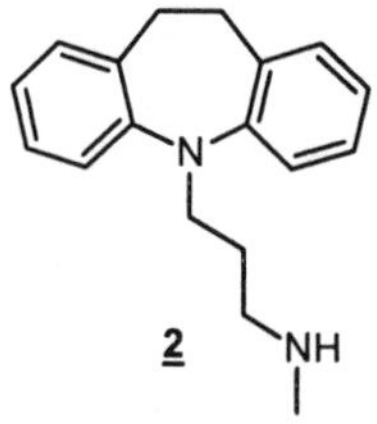

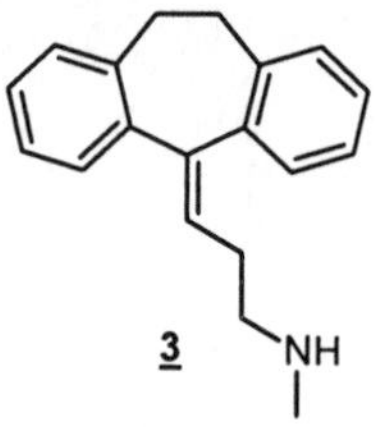

Atomoxetine (**4**), currently under development for treating ADHD in adults and children, was one of the earliest described selective norepinephrine reuptake inhibitors that does not interact with the postsynaptic receptor systems that produce side-effect liabilities of tricyclic anti-depressants (36,37). Interestingly, it is a close structural analog of fluoxetine (**5**), a prototypic selective serotonin reuptake inhibitor (SSRI). Like fluoxetine, atomoxetine has weak affinities for adrenergic, histaminergic and cholinergic receptors (37).

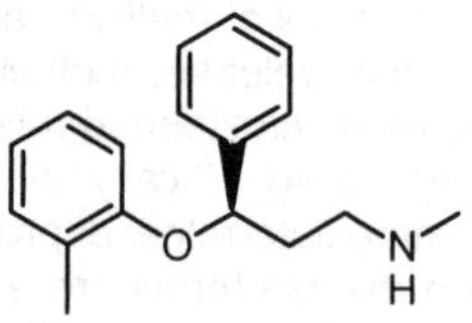

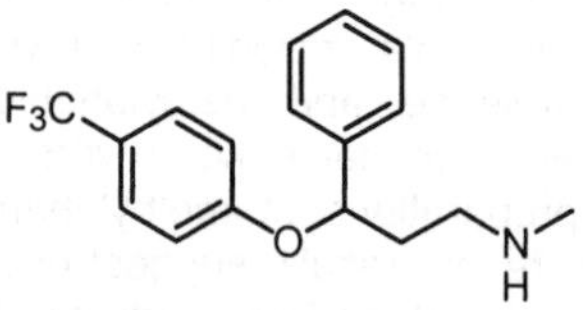

In rat synaptosomal preparations, atomoxetine was shown to be selective for NE reuptake transporters when compared to dopamine (DA) and serotonin (5-HT) transporters; K_i values were 1.9, 1600 and 750 nM, respectively (37). However as shown in Table 2, in radioligand binding assays with human transporters transfected into HEK293 cells, the NE selectivity was significantly reduced (38). Although inhibition of 5-HT reuptake in ADHD patients may alleviate some co-morbid psychiatric conditions, enhancement of serotonergic tone does not improve core ADHD symptoms.

Table 2. Transporter affinities of monoamine reuptake inhibitors (K_D, nM)

Compound	NE	DA	5-HT
Atomoxetine	2.0	1080	8.9
Fluoxetine	240	3600	0.81
Methyphenidate	234	24	44,000
Desipramine	0.83	3190	17.6

CYP2D6 is the major enzyme responsible for formation of atomoxetine's major metabolite 4-hydroxyatomoxetine.   Several additional P450 enzymes were also found to contribute, to a smaller extent, to formation of this oxidative metabolite (39). One potential complication of atomoxetine is a bimodal pattern of metabolism, with the presence of slow and rapid metabolizers. Pharmacokinetic evaluation in healthy adult males demonstrated plasma half-lives of approximately 4 hours in rapid metabolizers and 19 hours in poor metabolizers (40). The full impact of this bimodal metabolic pattern remains an unresolved issue.

In a Phase II trial, atomoxetine was shown to be well-tolerated and effective in alleviating symptoms in 11 of 21 (52%) adults with well-characterized ADHD. This response rate was equivalent to that observed for methylphenidate and desipramine (41). A study in children and adolescents (aged 8 to 18) with ADHD indicated that atomoxetine was superior to placebo in reducing core symptoms.   Currently available agents (psychostimulants) were not included in this trial, so a comparison of the degree of symptom reduction could not be made. As with the adult trial, treatment with atomoxetine was safe and well-tolerated (42).

Reboxetine, a selective NE reuptake inhibitor, is in use in several European countries for depression. Reboxetine is a racemic compound and exists as a mixture of (-)-*R,R* and (+)-*S,S* enantiomers (**6**); as shown in Table 3, the *S,S* enantiomer is more potent and selective than the *R,R* antipode (43). Two patent applications claim reboxetine as a treatment for ADHD (44,45). Moreover, the *S,S* enantiomer in particular is claimed for a variety of conditions that would benefit from a selective NE reuptake inhibitor, including ADHD (43).

Table 3. Transporter affinities of reboxetine enantiomers (K_i, nM)

reboxetine	NE	5-HT
racemic	1.6	129
R,R	7	104
S,S	0.23	2937

6   **7**

Manifaxine (**7**), a NE and DA reuptake inhibitor, is in development for ADHD and obesity. This compound is 4-fold selective for NE over DA with no 5-HT reuptake inhibition.   Phase II clinical trials are underway for these indications (46). (*R*)-Didesmethylsibutramine (**8**), a metabolite of the obesity drug sibutramine, has potent activity at all three monoamine reuptake sites; K_i values were 3, 26 and 31 nM for NE, DA and 5-HT uptake inhibition, respectively (47). This compound has been studied in clinical trials for depression and ADHD. O-Desmethylvenlafaxine (**9**), the major metabolite of the antidepressant venlafaxine, exhibited similar potency to the parent in inhibiting the uptake of NE and 5-HT, but not DA (48). This compound and other derivatives of venlafaxine are claimed in WO 0059851 as potential treatments for depression and affective disorders such as ADHD (49).

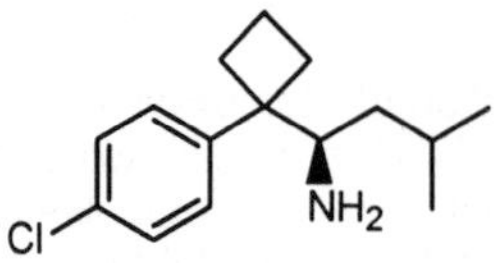

8 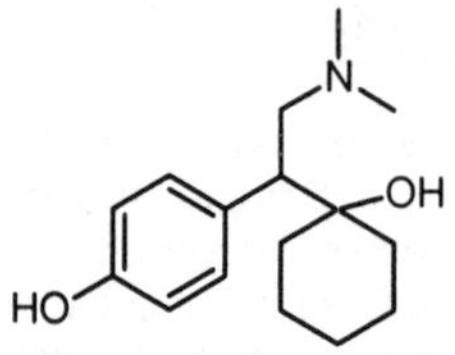   **9** 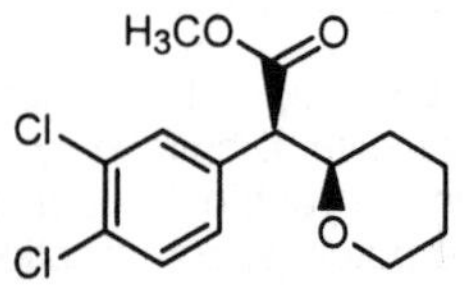   **10**

Like methylphenidate, tetrahydropyranyl esters such as **10** have high affinity for dopamine reuptake transporters (50). The *R,R* enantiomer in particular (IC_{50} values = 17 and >10,000 nM for DA and 5-HT uptake inhibition, respectively) is comparable to methylphenidate (IC_{50} values = 17.2 and >100,000 nM for DA and 5-HT reuptake inhibition, respectively).

Histamine H_3 Antagonists - The histamine H_3 receptor was discovered in 1983 and the receptor gene was cloned in 2000 (51,52). Histamine H_3 antagonists have been proposed for a number of CNS indications including cognition enhancement and treatment of ADHD. Two recent review articles discuss the potential of histamine H_3 receptor ligands in treatment of CNS disorders (53,54).

The imidazole GT-2331 (**11**) was the first histamine H_3 receptor antagonist to be characterized (55). A repeated acquisition aversion of an avoidance task using spontaneously hypertensive rat (SHR) pups was reported for evaluating the cognition/attention-enhancing potential of H_3 receptor antagonists. In this model, compound **11** significantly enhanced performance of the SHR pups at 1 mg/kg sc; similar effects were seen with methylphenidate (56). This compound was reported to be well tolerated in Phase I clinical studies, and adverse events were generally mild and CNS related (57).

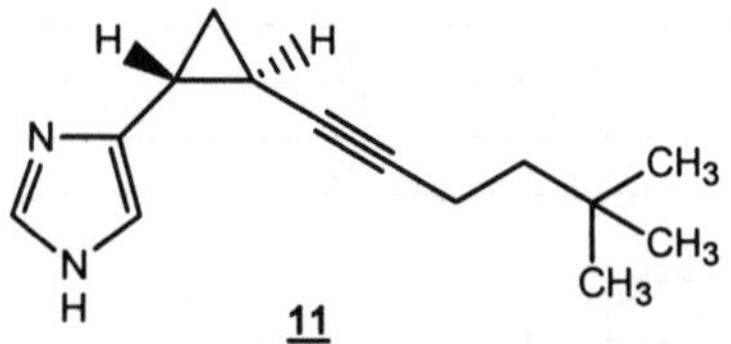

11

While earlier H_3 compounds were generally imidazole analogues, with some complications from inhibition of cytochrome P450 enzymes, more recent efforts are focused on non-imidazole structures. Several series of histamine H_3 antagonists with low nM receptor affinity have also been identified in recent patents. Representative examples are the biphenyl compounds **12** and **13**, with K_i values of 2.84 and 3.8 nM, respectively (58, 59).

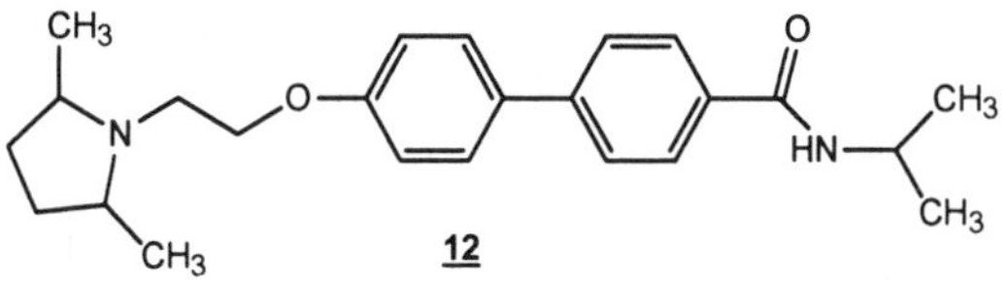

12

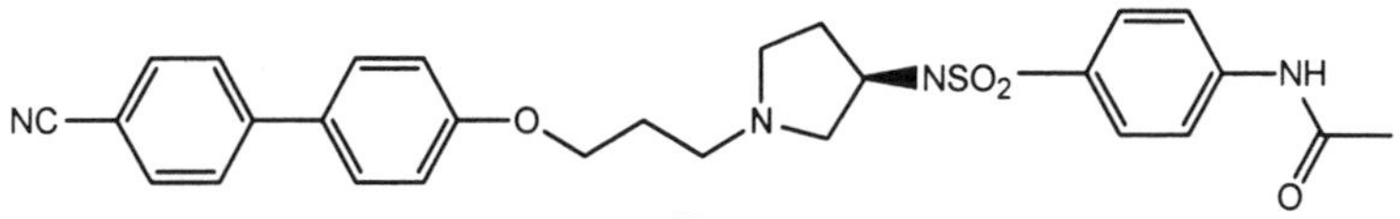

13

Structurally related aryloxyalkylamines, **14-17**, have also been described as histamine H_3 antagonists with K_i values of 0.2, 0.3, 1 and 7 nM, respectively (60-63).

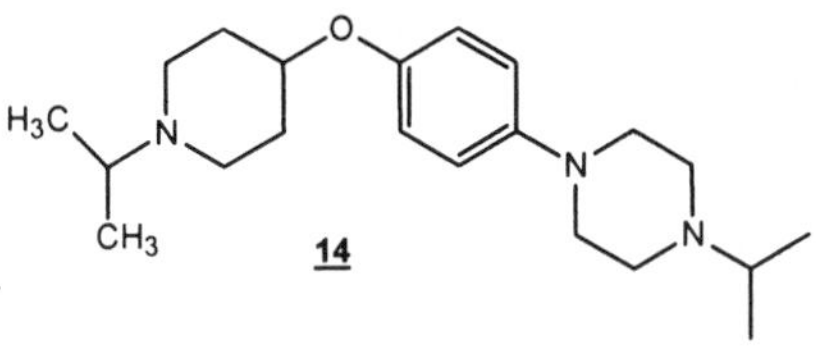

14

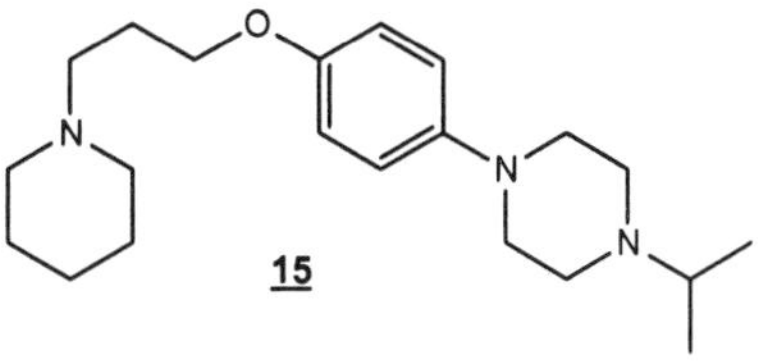

15

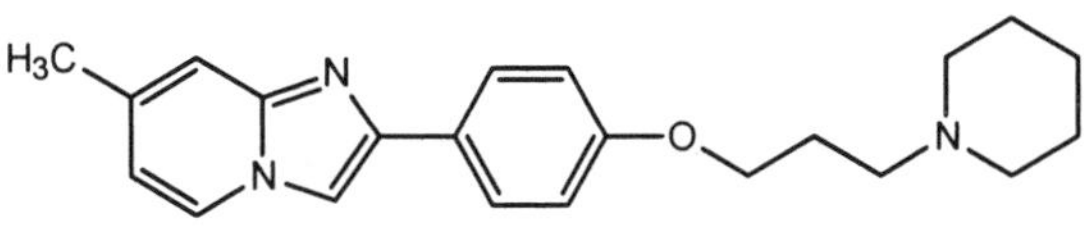

16

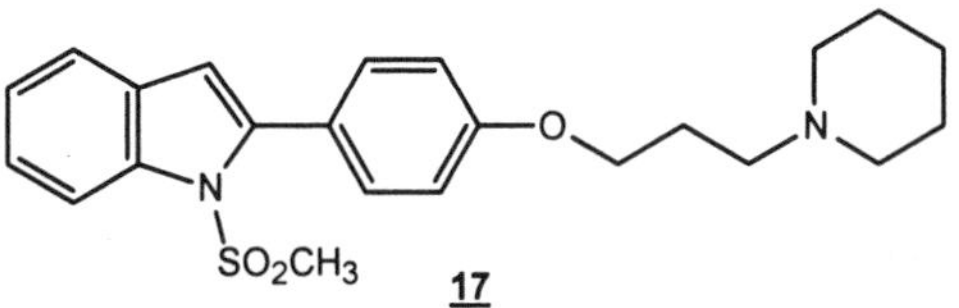

17

A series of pyrrolylalkylphenyl derivatives was also described (64). A key compound, **18**, was reported to bind to the histamine H_3 receptor with a K_i value of 0.41 nM.

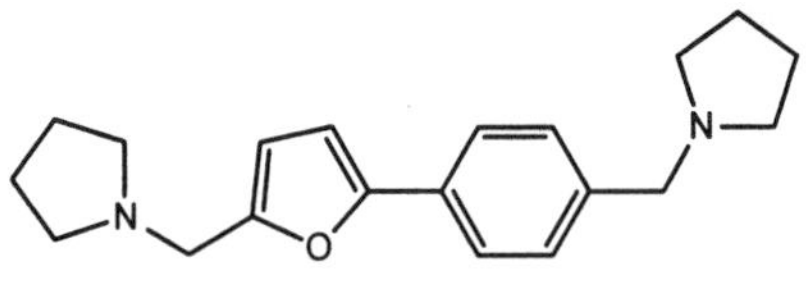

18

Nicotinic Receptor Agonists – Adolescents and adults with ADHD smoke more than the general population. Some have suggested this represents a form of self-medication with nicotine. Nicotinic acetylcholine receptor agonists potentiate release of dopamine as well as norepinephrine, serotonin, acetylcholine, glutamate and γ-aminobutyric acid, neurotransmitters associated with learning and memory. Nicotine skin patch treatments improve attentiveness in adults with ADHD (65,66). ABT-089 (**19**), a nicotinic acetylcholine receptor agonist with good oral bioavailability, enhances cognition in rodent and primate models (67,68). When compared to (-)-nicotine, ABT-089 displayed similar potency and efficacy in facilitating acetylcholine release, but was 25% less potent and only 70% as efficacious in stimulating DA release (68). A series of quinuclidine substituted heteroarylamides with agonist activity at the α7 nicotinic acetylcholine receptor are described (69). Compound 20, with a K_i value of 2 nM, is a representative example from this patent.

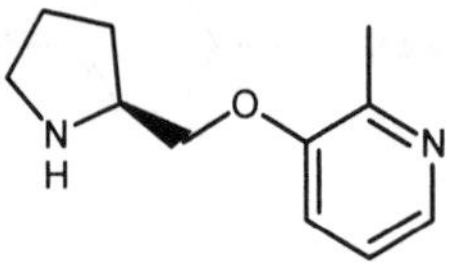

19 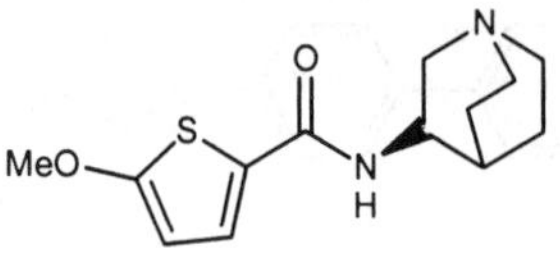   20

Conclusion – ADHD is a psychiatric disorder that has become well-characterized over the past several decades. However, even in the U.S., the primary country where the disorder is diagnosed and treated routinely, ADHD has stimulated considerable controversy. As has been the case with other psychiatric disorders such as depression and anxiety, once the stigmas associated with ADHD subside, and as we learn more about the pathophysiology and subtypes of the disorder, new medicines may be developed and launched to supplement long-used stimulants such as methylphenidate. Because of the substantial public health issues created by ADHD, new therapies should have significant societal impact.

References

1. J.C. Anderson, S. Williams, R. McGee, and P.A. Silva, Arch. Gen. Psychiatry, 44, 69 (1987).
2. W.J. Barberesi, S.K. Katusic, R.C. Colligan, V.S. Pankratz, A.L. Weaver, K.J. Weber, D.A. Mrazek, and S.J. Jacobsen, Arch. Pediatr. Adolesc. Med., 156, 217 (2002).
3. L.S. Goldman, M. Genel, R.J. Bezman, and P.J. Slanetz, J. Am. Med. Assoc., 279, 1100 (1998).
4. T.J. Spencer, Arch. Neurol., 59, 314 (2002).
5. J. Biederman, R. Russell, J. Soriano, J. Wozniak, and S.V. Faraone, J. Affect. Disord., 51 101 (1998).
6. Diagnostic and Statistical Manual of Mental Disorders (2nd ed. [DSM-II]); American Psychiatric Association, 1968.
7. Diagnostic and Statistical Manual of Mental Disorders (3rd ed. [DSM-III]); American Psychiatric Association, 1980.
8. Diagnostic and Statistical Manual of Mental Disorders (revised 3rd ed. [DSM-IIIR]); American Psychiatric Association, 1987.
9. Diagnostic and Statistical Manual of Mental Disorders (4th ed. [DSM-IV]); American Psychiatric Association, 1994.
10. T.E. Wilens, J. Biederman, S. Brown, S. Tanguay, M.C. Monuteaux, C. Blake and T.J. Spencer, J. Amer. Acad. Child Adol. Psych., 41, 262 (2002).
11. D.F. Connor, J. Dev. Behav. Pediatr., 23 (1 Suppl), S1-9 (2002).
12. J.P. Horrigan, Exp. Opin. Pharmacother., 2, 573 (2001).
13. V.A. Russell, Behav. Brain Res., 130, 191 (2002).
14. A. E. Doyle and S.V. Faraone, Current Psychiatry Reports, 4, 146 (2002).
15. R.D. Todd and E.A. Lobos, Am. J. Med. Genet., 114, 34 (2002).
16. P. Muglia, U. Jain, and J.L. Kennedy, Behav. Brain Res., 130, 91 (2002).
17. M. Gill, G. Daly, S. Heron, Z. Hawi, and M. Fitzgerald, Mol. Psychiatry, 2, 311 (1997).
18. J. Holmes, A. Payton, J. Barrett, R. Harrington, P. McGuffin, M. Owen, W. Ollier, J. Worthignton, M. Gill, A. Kirley, Z. Hawi, M. Fitzgerald, P. Asherson, S. Curran, J. Mill, A. Gould, E. Taylor, L. Kent, M. Craddock, and A. Thapar, Am. J. Med. Genet., 114, 150 (2002).
19. T. Roman, M. Schmitz, G.V. Polanczyk, M. Eizirik, L.A. Rohde, and M.H. Hutz, Am. J. Med. Genet., 114, 154 (2002).

20. S. Jiang, R. Xin, S. Lin, Y. Qian, G. Tang, D. Wang, and X. Wu, Am. J. Med. Genet., 105, 783 (2001).
21. W. Retz, J. Thome, D. Blocher, M. Baader, and M. Rosler, Neurosci. Lett., 319, 133 (2002).
22. D.D. Dougherty, A.A. Bonab, T.J. Spencer, S.L. Rauch, B.K. Madras, and A.J. Fischman, Lancet, 354, 2132 (1999).
23. C.H. van Dyck, D.M. Quinlan, L.M. Cretella, J.K. Staley, R.T. Malison, R.M. Baldwin, J.P. Seibyl, and R.B. Innis, Am. J. Psychiatry, 159, 309 (2002).
24. K.D. Gadow and E.E. Nolan, J. Child Psychol. Psychiatry, 43, 191 (2002).
25. J.J.S. Kooij, H.A.M. Middelkoop, K. van Gils, and J.K. Buitelaar, J. Clin. Psychiatry, 62, 952 (2001).
26. J. J. Wilson and F. R. Levin, Curr. Psychiatry Rep., 3, 497 (2001).
27. A. J. Zametkin and M. Ernst. N. Engl. J. Med., 340, 40 (1999).
28. R. L. Findling, E.J. Short, and M.J. Manos, J. Am. Acad. Child Adolesc. Psychiatry, 40, 1441 (2001).
29. J.S. Markowitz and K.S. Patrick, Clin. Pharmacokinet., 40, 753 (2001).
30. A. Marchetti, R. Magar, H. Lau, E.L. Murphy, P.S. Jensen, C.K. Connors, R. Findling, E. Wineburg, I. Carotenuto, T.R. Einarson, and M. Iskedjian, Clin. Ther., 23, 1904 (2001).
31. L.L. Greenhill, R.L. Findling, and J. M. Swanson, Pediatrics, 109, E39 (2002).
32. M.L. Wolraich, L.L. Greenhill, W. Pelham, J. Swanson, T. Wilens, D. Palumbo, M. Atkins, K. McBurnett, O. Bukstein, and G. August, Pediatrics, 108, 883 (2001).
33. S.R. Pliszka, Expert Opin. Investig. Drugs, 10, 1797 (2001).
34. J. Biederman, R.J. Baldessarini, V. Wright, D. Knee, and J. S. Harmatz, J. Am. Acad. Child Adolesc. Psychiatry, 28, 777 (1989).
35. T.E. Wilens, J. Biederman, E. Mick, and T.J. Spencer, J. Nerv. Ment. Dis., 183, 48 (1995).
36. J.H. Heiligenstein and G.D. Tollefson, Eur. Patent EP 721777-A2 (1996).
37. D.T. Wong, P.G. Threlkeld, K.L. Best, and F.P. Bymaster, J. Pharmacol. Exp. Ther., 222, 61 (1982).
38. M. Tatsumi, K. Groshan, R.D. Blakely, and E. Richelson, Eur. J. Pharmacol., 340, 249 (1997).
39. B.J. Ring, J.S. Gillespie, J.A. Eckstein, and S.A. Wrighton, Drug Metab. Dispos. 30, 319 (2002).
40. N.A. Farid, R.F. Bergstrom, E.A. Ziege, C.J. Parli, and L. Lemberger, J. Clin. Pharmacol. 25 (4), 296 (1985).
41. T. Spencer, J. Biederman, T. Wilens, J. Prince, M. Hatch, J. Jones, M. Harding, S.V. Faraone, and L. Seidman, Am. J. Psychiatry, 155, 693 (1998).
42. D. Michelson, D. Faries, J. Wernicke, D. Kelsey, K. Kendrick, R. Sallee, and T. Spencer, Pediatrics, 108, E83 (2001).
43. E.H.F. Wong, S. Ahmed, R.C. Marshall, R. McArthur, D.P. Taylor, L. Birgerson, and P. Cetera, PCT Patent Appl. WO 0101973 (2001).
44. J.H. Heiligenstein, PCT Patent Appl. WO 9915177 (1999).
45. F. Hassan, J.M. McCall, D.P. Taylor, P.F. Von Voigtlander, and E.H.F. Wong, PCT Patent Appl. WO 9952531 (1999).
46. J.L. Kelley, D.L. Musso, G.E. Boswell, F.E. Soroko, and B.R. Cooper, J. Med. Chem., 39, 347 (1996).
47. C.M. Mendel, T.B. Seaton, and S.P. Weinstein, U.S. Patent 6 372 798 (2002).
48. E.A. Muth, J.A. Moyer, J.T. Haskins, T.H. Andree, and G.E.M. Husbands, Drug Dev. Res., 23, 191 (1991).
49. T.P. Jerussi and C.H. Senanayake, PCT Patent Appl. WO 0059851 (2000).
50. P.C. Metzler, P. Blundell, P. Wang, and B.K. Madras, PCT Patent Appl. WO 0232842 (2002).
51. J.M. Arrang, M. Garbarg, and J.C. Schwartz, Nature, 302, 832 (1983).
52. Z. Yuan, D. Michalovich, H. Wu, K.B. Tan, G.M. Dytko, I.J. Mannan, R. Boyce, J. Alston, L.A. Tierney, X. Li, N.C. Herrity, L. Vawter, H.M. Sarau, R.S. Ames, C.M. Davenport, J.P. Hieble, S. Wilson, D.J. Bergsma, and L.R. Fitzgerald, Mol. Pharmacol., 59, 434 (2001).
53. M.J. Tozer and S.B. Kalindjian, Exp. Opin. Ther. Patents, 10, 1045 (2000).
54. P.L. Chazot and V. Hann, Curr. Opin. Invest. Drugs, 2, 1428 (2001).
55. C.E. Tedford, J.G. Phillips, R. Gregory, G.P. Pawlowski, L. Fadnis, M.A. Khan, S.M. Ali, M.K. Handley, and S.L. Yates, J. Pharmacol. Exp. Ther., 289, 1160 (1999).
56. G.B. Fox, J.B. Pan, T.A. Esbensade, Y.L. Bennani, L.A. Black, R. Faghih, A.A. Hancock, and M.W. Decker, Behav. Brain Res., 131, 151 (2002).
57. M.T. Halpern, Curr. Opin. CNS Invest. Drugs, 1, 524 (1999).
58. Y.L. Bennani and R. Faghih, U.S. Patent 6 316 475 (2001).

59. Y.L. Bennani, R. Faghih, W.J. Dwight, A. Vasudevan, and S.E. Scott, PCT Patent Appl. WO 0206223 (2002).
60. R. Apodaca, N.I. Carruthers, C.A. Dvorak, C.R. Shah, and W. Xiao, PCT Patent Appl. WO 0212190 (2002).
61. R. Apodaca, N.I. Carruthers, C.A. Dvorak, D. Rudolph, C.R. Shah, and W. Xiao, PCT Patent Appl. WO 0212214 (2002).
62. J.G. Breitenbucher, N.I. Carruthers, X. Li, L.C. McAtee, C. R. Shah, and R.L. Wolin. PCT Patent Appl. WO 0174815 (2001).
63. J.G. Breitenbucher and W. Chai, PCT Patent Appl. WO 0174773 (2001).
64. M. Bogenstaetter, W. Chai, and A.K. Kwok, PCT Patent Appl. WO 0212224 (2000).
65. E.D. Levin, C.K. Conners, D. Silva, W. Canu, and J. March, Exp. Clin. Psychopharmacol., 9, 83 (2001).
66. E.D. Levin, Nicotinic Receptors in the Nervous System, 251 (2002).
67. M.W. Decker, A.W. Bannon, P. Curzon, K.L. Gunther, J. D. Brioni, M.W. Holladay, N. Lin, Y. Li, J.F. Daanen, J.J. Buccafusco, M.A. Prendergast, W.J. Jackson, and S.P. Arneric, J. Pharmacol. Exp. Ther., 283, 247 (1997).
68. N. Lin, D.E. Gunn, K.B. Ryther, D.S. Garvey, D.L. Donnelly-Roberts, M.W. Decker, J.D. Brioni, M.J. Buckley, A.D. Rodrigues, K.G. March, D.J. Anderson, J.J. Buccafusco, M.A. Prendergast, J.P. Sullivan, M. Williams, S.P. Arneric, and M.W. Holladay, J. Med. Chem., 40, 385 (1997).
69. J.K. Myers, B.N. Rogers, V.E. Groppi, D.W. Piotrowski, A.L. Bodnar, E.J. Jacobsen, and J.W. Corbett, PCT Patent Appl. WO 0217358 (2002).

Chapter 3: 5-HT$_{2C}$ Receptor Modulators: Progress in Development of New CNS Medicines

Lawrence W. Fitzgerald and Michael D. Ennis
Structural, Analytical, and Medicinal Chemistry and CNS Discovery Research
Pharmacia Corporation, Kalamazoo, MI 49007

Introduction - The decades-old discovery of serotonin (5-HT) as a vasoconstrictor and prominent neurotransmitter (1-3), has since sparked an intense research effort that has elucidated the diverse roles of 5-HT in mediating sensory, motor, and behavioral processes. Although 5-HT-containing neurons comprise only a rather circumscribed region of mammalian hindbrain, the diversity of 5-HT actions is attributed to the vast network of 5-HT projections throughout the brain coupled with a surprisingly large number of 5-HT receptor subtypes (4,5). Whereas many 5-HT receptors had been initially categorized on the basis of classic pharmacological methods, the advent of molecular cloning technology has elaborated, and at times re-classified these receptor subtypes. The 5-HT receptor family is comprised of seven subfamilies and fourteen distinct receptor subtypes grouped according to pharmacology, sequence homology, and signal transduction criteria (6-8).

The important role of 5-HT in neuropsychiatric maladies has been validated by translation of 5-HT pharmacology to therapeutic benefit in treatment of an array of diseases such as anxiety disorders, depression, schizophrenia, migraine, chemotherapy-induced emesis, and appetite control (9,10). In fact, among the best-selling CNS medicines worldwide are drugs (e.g., selective serotonin reuptake inhibitor (SSRIs), and atypical antipsychotics) that broadly modulate central 5-HT functions (9).

Substantial progress has been made in identifying selective modulators for various 5-HT receptor subtypes (10,11). The search for selective ligands for the 5-HT$_{2C}$ receptor has been particularly difficult because of the close sequence homology of the 5-HT$_{2C}$ receptor to closely related receptors, the 5-HT$_{2A}$ and 5-HT$_{2B}$ receptors (12). While the largely nonselective pharmacological probes of the past may have, at times, obscured our understanding of the 5-HT$_{2C}$ receptor, the availability of a 5-HT$_{2C}$ "knockout" mouse and new selective ligands has recently provided improved clarity. In this report, we will describe important recent advances toward discovery of selective 5-HT$_{2C}$ modulators, and their use as both pharmacological probes of 5-HT$_{2C}$ function and therapeutic agents.

Receptor structure - The 5-HT$_{2C}$ receptor contains three introns and seven predicted transmembrane spanning domains, although an unprecedented eighth domain has also been postulated for this receptor (13). One splice variant has been identified in multiple species, although it is of dubious functional relevance as it is truncated and devoid of a 5-HT-binding site (14). The 5-HT$_{2C}$ receptor is the only known G protein-coupled receptor (GPCR) described to date that undergoes RNA editing (15). This is a process in which adenosine-to-inosine RNA editing events at five nucleotide positions alters the amino acid coding potential with the putative second intracellular loop of the protein. By virtue of the importance of this region in G protein coupling, RNA editing serves as diversity-generating process to alter pharmacological and regulatory processes of the receptor. Editing of the 5-HT$_{2C}$ transcripts has been shown to augment agonist affinities and functional potencies, constitutive receptor activity, and the specificity of G-protein activation (16-18).

Receptor Distribution - Unlike the 5-HT$_{2A}$ and 5-HT$_{2B}$ receptor subtypes, the 5-HT$_{2C}$ receptor is distributed almost exclusively in the brain. It was first identified by radioligand binding methods because of its high density in the choroid plexus, a tissue lining the cerebral ventricles involved in the production of cerebral spinal fluid (19). 5-HT$_{2C}$ receptor protein and mRNA is also localized to a number of cortical and sub-cortical brain structures including the cortex, hippocampus, nucleus accumbens/caudate nucleus, substantia nigra, cerebellum, amygdala, and discrete nuclei of the thalamus and hypothalamus (20-22). This localization reflects the various neural functions with which it is linked, such as cognition, emotional processes, movement, and appetite control.

Signal Transduction - The 5-HT$_{2C}$ receptor has long been recognized to activate phospholipase C (PLC) that catalyzes the formation of phosphatidylinositol-4,5-bisphosphate (PIP$_2$) to inositol 1,4,5-triphosphate (IP$_3$) and diacylglycerol (23,24). IP$_3$ then serves a second messenger in the cytosol to stimulate release of Ca^{2+} from the endoplasmic reticulum, whereas diacylglycerol remains membrane-bound to activate phospolipase C (25,26). Both Ca^{2+} and phospholipase C regulate a variety of other intracellular proteins that augment cellular functioning. An elegant peptide blocking strategy recently confirmed the long-suspected role of G$_q$ G-proteins in mediating this signaling cascade upon 5-HT$_{2C}$ receptor activation by 5-HT in native choroid plexus cells (27).

The 5-HT$_{2C}$ receptor can also activate phospholipase A$_2$ and subsequent formation of arachidonic acid metabolites (28). A comparative pharmacological assessment of agonists regarding their propensity to activate PLA$_2$ versus PLC pathways revealed another intriguing property of 5-HT$_{2C}$ receptors. Some agonists preferentially activated one pathway over the other providing evidence for "agonist-directed trafficking" of agonist receptor stimuli (29).

The 5-HT$_{2C}$ receptor may also couple to production of cGMP, inhibition and stimulation of nitric oxide, and activation of cAMP production pathway (8, 30,31). Independent and convergent regulation of these diverse signaling pathways leads to acute augmentation of diverse enzymes, ion channels, and transporter processes as well as longer-term adaptations in gene expression (8). Aside from these traditional "second messenger" signaling processes is the direct association of 5-HT$_{2C}$ receptors with other proteins, such as multivalent scaffolding proteins. The c-terminus of the 5-HT$_{2C}$ receptor was shown to selectively interact with the PDZ domain of scaffolding protein MUPP1 (32). This discovery may lead to the understanding of entirely new modes of intracellular signaling apart from G-protein mediated processes. For example, deletion of the PDZ domain of the 5-HT$_{2C}$ receptor blocks the phosphorylation and re-sensitization of the receptor (33).

POTENTIAL CLINICAL APPLICATIONS

Obesity - The role of the 5-HT$_{2C}$ receptor in regulation of appetite is based on several keys lines of evidence. First, the 5-HT$_{2C}$ receptor is enriched in nuclei of the hypothalamus that regulate satiety. 5-HT$_{2C}$ agonists inhibit food intake even after chronic administration in rodents (34,35). Whereas selective antagonists block these effects, paradoxically, they alone do not increase food intake and weight gain (12,36,37). 5-HT$_{2C}$ "knockout" mice developed delayed-onset obesity and hyper-insulinemia, and are insensitive to the hypophagic effects of fenfluramine (38,39). Finally, nonselective 5-HT$_{2C}$ agonists reduce food intake and weight in humans (40).

Fenfluramine, at least partly, owes its anorectic actions (via 5-HT release and blockade of 5-HT re-uptake) to stimulation of 5-HT$_{2C}$ receptors, whereas its major metabolite norfenfluramine is a potent 5-HT$_{2C}$ agonist (41,42). Fenfluramine was removed from the U.S. market in 1997 because of its association with an increased incidence of valvular heart disease (VHD). It had been also associated with an increased risk for developing primary pulmonary hypertension (PPH). Because of the paucity of the 5-HT$_{2C}$ receptors in peripheral tissues, the 5-HT$_{2C}$ receptor is unlikely involved with these illnesses. Fenfluramine's ability to serve as a 5-HT transporter substrate is hypothesized to contribute to PPH whereas the mitogenic 5-HT$_{2B}$ agonist properties of norfenfluramine have been implicated in VHD (43-45). Selective 5-HT$_{2C}$ agonists, devoid of these cross-reactive interactions, hold the prospect of being safe and therapeutically effective agents for the long-term treatment of obesity. For additional information on role of GPCR modulators in treatment of obesity, see Chapter 1 of this volume.

Epilepsy - Epilepsy refers generically to a multiplicity of many distinct disorders characterized by recurrent seizures resulting from massive neuronal discharge. Support for the potential utility of 5-HT$_{2C}$ agonists as anti-epileptic medicines is afforded by the discovery that 5-HT$_{2C}$ "knockout" mice exhibit sporatic spontaneous seizures characterized by lower focal seizure thresholds, increased focal seizure excitability, and facilitated propagation within the forebrain (46,47). Wild-type mice recapitulated the mutant phenotype when pre-treated with a nonselective 5-HT$_{2C}$ antagonist prior to electro-shock testing. Pharmacological studies in rodents also suggest that stimulation of 5-HT$_{2C}$ receptors with a nonselective 5-HT$_{2C}$ agonist affords seizure protection, which can be attenuated by pre-treatment with a 5-HT$_{2C/2B}$ antagonist (48).

Schizophrenia - Atypical antipsychotic drugs are more effective than typical antipsychotic drugs in treating negative symptoms of schizophrenia, while producing fewer extra-pyramidal side effects. The potential contribution of the 5-HT$_{2C}$ receptor to the "atypical" profile is suggested because these drugs exhibit moderate-to-high affinities (49) and inverse agonist properties for the receptor (50). The 5-HT$_{2C}$ receptor also regulates cortical and sub-cortical dopamine (DA) function (51). 5-HT$_{2C}$ agonists inhibit mesolimbic DA but not nigrostriatal function, whereas antagonists have the converse effect. To the extent that reduced mesocorticolimbic DA function has been hypothesized in schizophrenia (52), disinhibition of this pathway by 5-HT$_{2C}$ antagonists/inverse agonists may underlie the unique effects of atypical agents on ameliorating negative symptoms of this disease (50).

Anxiety Disorders - 5-HT$_{2C}$ antagonists have shown anxiolytic activity in several animal models (53-55). Interestingly, 5-HT$_{2C}$ agonists have also shown activity in models of panic and obsessive-compulsive disorder, two types of anxiety disorders in humans (56,57). Clinical support for the 5-HT$_{2C}$ mechanism is provided by a recent report that deramciclane (vide infra) was efficacious and well-tolerated in a controlled Phase 2 trial in GAD patients (58). 5-HT$_{2C}$ agonists/antagonists may also be useful for the treatment of depression as suggested by some preclinical animal models, and the known disinhibitory role of 5-HT$_{2C}$ antagonists on mesolimbic DA function (51,57).

5-HT$_{2C}$ RECEPTOR AGONISTS

The close sequence homology among members of the 5-HT$_2$ receptor subfamily has made identification of agonists selective for the 5-HT$_{2C}$ receptor subtype a formidable challenge. Further complicating this area is the existence of multiple RNA edited isoforms, different sources of receptor (e.g, rat vs human), and different methods for determining ligand affinities (e.g., agonist vs. antagonist radioligands). These factors combine to make comparisons of 5-HT$_{2C}$ affinities and selectivities reported across different laboratories a difficult endeavor. For example, a series of isotryptamines has been reported to contain full agonists at the 5-HT$_{2C}$ receptor with excellent selectivity over 5-HT$_{2A}$ receptor (59). The difluoroindole 1 reportedly possessed potent 5-HT$_{2C}$ affinity (Ki =1 nM) and 100-fold selectivity versus the 5-HT$_{2A}$. receptor. Re-examination of the binding profile of 1 from another laboratory, however, indicated that the 5-HT$_{2A}$/5-HT$_{2C}$ receptor selectivity was at best only 10-fold (60). This discrepancy is likely the result of the initial data being derived using antagonist radioligands as opposed to the more appropriate to use of agonist radioligands that label the agonist high-affinity state of the receptor (60). Whether inspired by **1** or not, much of the recent activity in the search for selective 5-HT$_{2C}$ agonists that is described below has focused on tryptamine and isotryptamine derivatives.

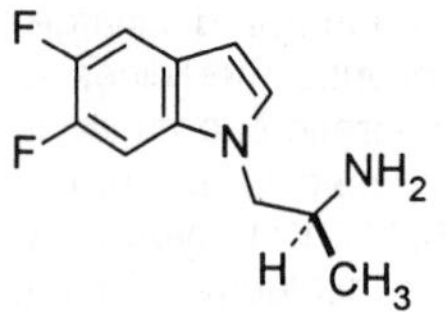

1 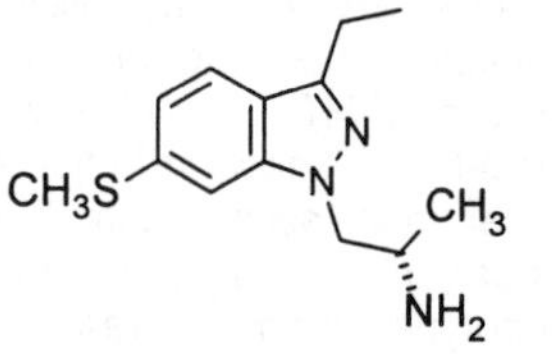   **2** 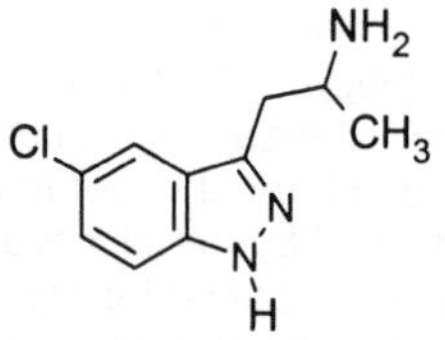   **3**

Indazoles such as **2** and **3** have recently been claimed as 5-HT$_{2C}$ agonists useful as anti-obesity agents. Compound **2** is reported to have a 5-HT$_{2C}$ Ki of 37 nM (61) and an efficacy (relative to 5-HT) of 81% (EC$_{50}$ = 8 nM). This compound is non-selective, however, displaying affinity for the 5-HT$_{2B}$ receptor (47 nM) and is a partial agonist at 5-HT$_{2A}$ (48% efficacy, EC$_{50}$ = 285 nM). Compound **3** is a partial agonist at both 5-HT$_{2C}$ and 5-HT$_{2A}$ receptors with efficacies of 51% (EC$_{50}$ = 55 nM) and 46% (EC$_{50}$ = 346 nM), respectively (62).

A number of tricyclic tryptamine and isotryptamines have also been reported recently. The isotryptamine compounds **4** and **5** are claimed as direct acting 5-HT$_2$ receptor ligands for use in therapy, particularly as anti-obesity agents (63). Compound **4** is a potent 5-HT$_{2C}$ agonist (Ki = 19 nM, 72% efficacy) but displays poor selectivity relative to either 5-HT$_{2A}$ (Ki = 48 nM) or 5-HT$_{2B}$ (Ki = 31 nM). Improved 5-HT$_{2C}$/5-HT$_{2A}$ selectivity was achieved in this series by halogen substitution on the aromatic ring (a recurring strategy across templates), although this resulted in an interesting switch of absolute stereochemistry for the more potent isomer from (R) to (S). The fluoro analog **5** is reported to possess over 8-fold selectivity for 5-HT$_{2C}$ over 5-HT$_{2A}$ receptors (Ki values = 65 nM vs 550 nM, respectively). However, **5** still exhibits significant potency at 5-HT$_{2B}$ receptors (Ki = 161 nM). The closely related tryptamine **6** displayed levels of affinity, selectivity, and efficacy similar to **5** when tested as the racemate (Ki values, 5-HT$_{2C}$ = 81 nM, 5-HT$_{2A}$ = 448 nM; 5-HT$_{2B}$ 122 nM; 69% efficacy)(64). Azatryptamine compounds such as **7** have also been reported recently, but no biological data were disclosed (65).

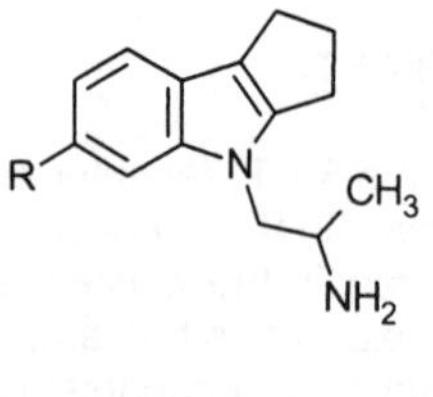

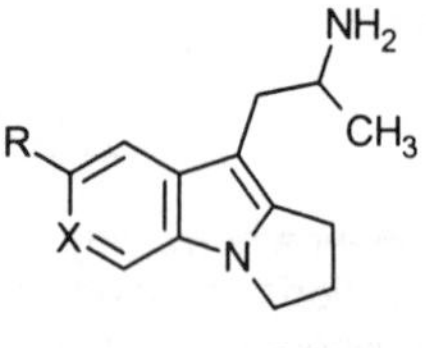

4 R = H, *R*-isomer   **6** R = Br, X = CH
5 R = F, *S*-isomer   **7** R = H, X = N

The tricyclic tryptamine **8** is reported to have remarkable selectivity for the 5-HT$_{2C}$ receptor relative to the 5-HT$_{2A}$ receptor (66). When tested as the racemate, **8** showed potent 5-HT$_{2C}$ receptor affinity (Ki = 18 nM) and 130-fold selectivity over 5-HT$_{2A}$ receptors (Ki = 2,340 nM). Illustrating the dramatic effect that minor structural changes can have on binding profiles is the methylated analog **9**, which exhibited a 10-fold loss of 5-HT$_{2C}$ affinity (Ki = 194 nM) and significant reduction in 5-HT$_{2A}$ selectivity (Ki = 954 nM, 5-fold selectivity) relative to **8**. Cyclization of the pendant aminoethyl side-chain of **8** onto the C-2 position gives rise to the tetracycle **10**. These compounds have recently been reported as 5-HT ligands useful for treating obesity and CNS disorders (67). The closely related azepinoindole **11** (PNU-22394) has been described as a 5-HT$_{2C}$ agonist (Ki = 18.8 nM, 83% efficacy) that has recently been reported to decrease feeding in rats and produce a weight loss in humans (68). Although dose-related clinical side effects observed included headache, anxiety, nausea, and vomiting, these effects were dramatically reduced following four days of dosing. No hallucinations were observed despite the fact that **11** is also a potent, high efficacy 5-HT$_{2A}$ agonist (5-HT$_{2A}$ Ki = 19 nM, 64% efficacy). Compound **11** also has excellent affinity at 5-HT$_{2B}$ receptors (Ki = 28.5 nM).

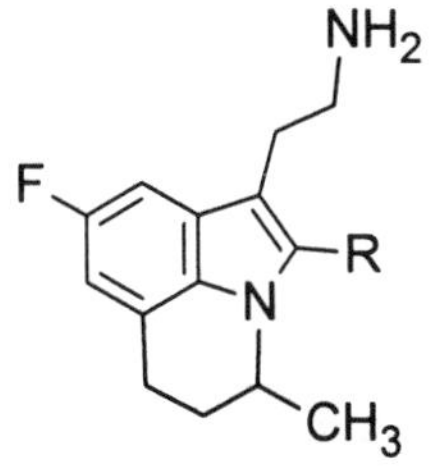

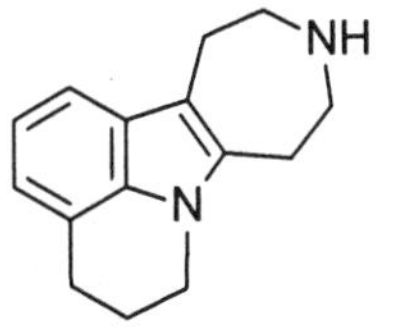

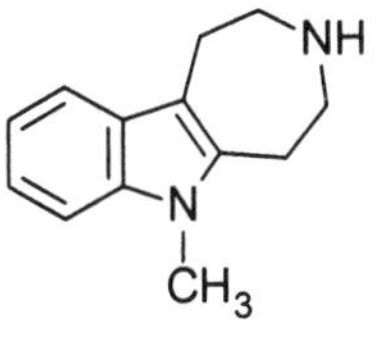

8 R = H
9 R = Me   **10**   **11**

Both indoles and indolines have provided potent 5-HT$_{2C}$ ligands, with the indolines often providing improved selectivity. Moreover, a remarkable effect on selectivity has been noted (69) for the 7-position substituent on the indoline **12**. Whereas the 7-chloro compound **12** displays a modest 5-fold selectivity for 5-HT$_{2C}$ (5-HT$_{2C}$ Ki = 55 nM, 62% efficacy; 5-HT$_{2A}$ Ki = 252 nM), the 7-bromo analog **13** possesses a 14-fold selectivity with little loss in either affinity or efficacy at 5-HT$_{2C}$ receptors (5-HT$_{2C}$ Ki = 77 nM, 59% efficacy; 5-HT$_{2A}$ Ki = 1092 nM). The pyridyl analog **14** is a very potent agonist with high intrinsic activity (Ki = 9 nM, 93% efficacy) but poor selectivity (5-HT$_{2A}$ Ki = 45 nM, 5-HT$_{2B}$ Ki =12 nM) (70). The napthyl analog **15** also displays poor selectivity (5-HT$_{2C}$ = 107 nM, 5-HT$_{2B}$ Ki = 39 nM). Relative to **14** or **15**, the benzodioxane compound **16** retains modest potency and high intrinsic activity at the 5-HT$_{2C}$ receptor (Ki = 70 nM, 90% efficacy) with an attenuated 5-HT$_{2B}$ receptor affinity (Ki = 218 nM) (71).

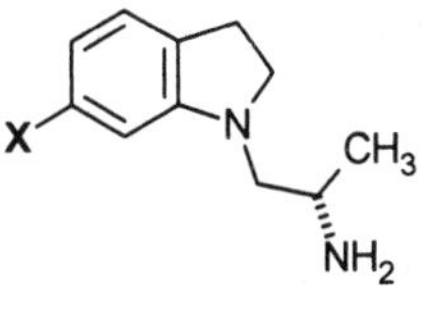

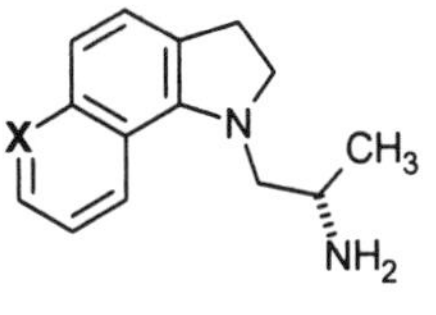

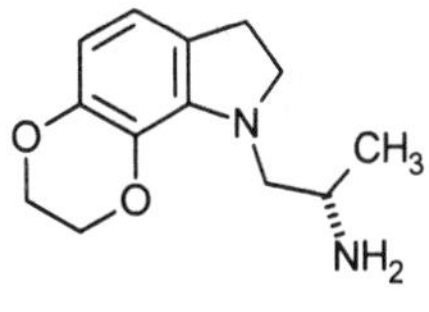

12 X = Cl
13 X = Br   **14** X = N
                      15 X = CH$_2$ **16**

Tricyclic indolines such as **17** and **18** are reported to be potent 5-HT$_{2C}$ agonists (72,73). It is worth noting that these cyclized isotryptamine indoline derivatives contain an imbedded aryl-piperazine substructure. The 7-chloroindoline **17** is reported to have efficacy at 5-HT$_{2C}$ receptors of 87% (EC$_{50}$ = 3 nM) and at 5-HT$_{2A}$ of 59% (EC$_{50}$ = 100 nM), whereas the 7-chloro-4-methyl analog **18** shows enhanced potency at 5-HT$_{2C}$ receptors (97% efficacy, EC$_{50}$ = 0.4 nM) but is also active at 5-HT$_{2A}$ receptors (60% efficacy, Ki = 19 nM). Structurally related to these compounds are the ring-expanded azepinoindoles and azepinoindolines **19**, recently claimed as 5-HT ligands for treatment of CNS diseases (74).

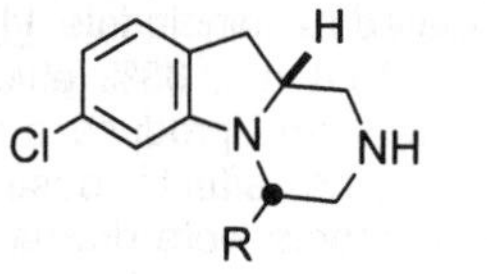

17 R = H   **19**
18 R = CH$_3$

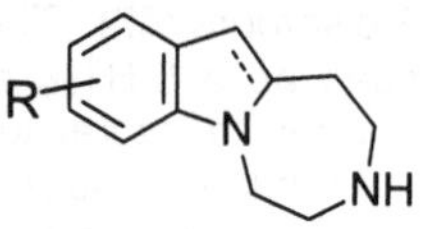

Although highly selective 5-HT$_{2C}$ receptor antagonists have been described (*vide infra*), achieving highly selective and efficacious 5-HT$_{2C}$ agonists has been very difficult. The indole-based templates described above contain several potent and highly efficacious 5-HT$_{2C}$ agonists, but selectivities against 5-HT$_{2A}$ and 5-HT$_{2B}$ receptors are generally quite modest (5-10-fold). The SAR revealed by recent structures has trended toward polycyclic indoline-based templates with aromatic substitution. That selectivity is even possible within this highly homologous family of receptors has recently been demonstrated by disclosure of the structure and biological activity of a family of γ-carbolines that are claimed to afford both selective 5-HT$_{2C}$ agonists and 5-HT$_{2A}$ antagonists (75-77). In what may be the most significant recent advance in the search for selective 5-HT$_{2C}$ agonists, the binding profiles, efficacy, and *in vivo* data for the tetracyclic indoline compounds **20** and **21** (compounds IK264 and IL639, respectively) have recently been reported (77). Compound **20** displays excellent affinity for the 5-HT$_{2C}$ receptor (Ki =10.5 nM), nearly 40-fold selectivity over 5-HT$_{2A}$ (Ki = 406), and nearly 20-fold selectivity over 5-HT$_{2B}$ (Ki = 209 nM). Even more impressive is the binding profile for **21**, which possesses excellent 5-HT$_{2C}$ affinity (Ki = 5.2 nM) and selectivities over both 5-HT$_{2A}$ and 5-HT$_{2B}$ receptors of well over 250-fold (5-HT$_{2A}$ Ki = 1,440 nM, 5-HT$_{2B}$ Ki = 1,510 nM). Interestingly, both compounds are partial agonists (Intrinsic activities = 46% for **20** and 57% for **21**) with functional EC$_{50}$ values significantly rightward-shifted from the radioligand inhibition curves (EC$_{50}$ = 250 nM and 464 nM, respectively). Both **20** and **21** were orally bioavailable and efficacious in a chronic model of feeding in male rats in studies up to 12 weeks with no sign of tolerance to the anorectic effect (78). The discovery of these highly selective 5-HT$_{2C}$ partial agonists represents a significant step forward in the area of 5-HT research, and should help further our understanding of the pharmacological roles of the 5-HT$_{2C}$ receptor.

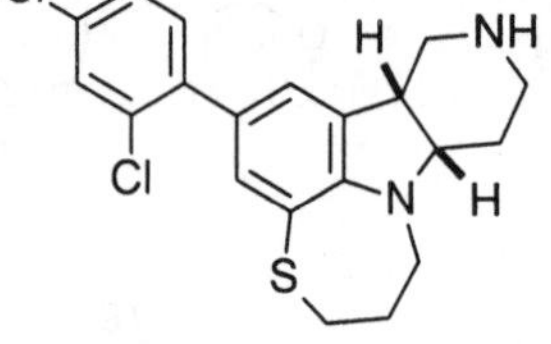

20 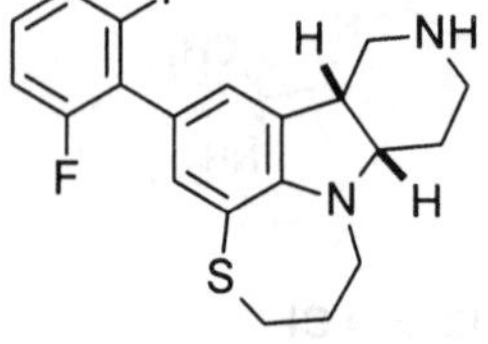   **21**

Although all the compounds shown above can be considered to be structurally inspired by 5-HT (*e.g.*, tryptamine or isotryptamine related), there have been claims of selective and non-selective 5-HT$_{2C}$ agonists derived from aryl piperazines.   Indeed, one of the original non-selective 5-HT$_{2C}$ agonists is mCPP (**22**), a compound that was seminal in establishing some of the early pharmacology of the 5-HT$_2$ receptor subtype. Selectivity has been reported for the aryl piperazines **23** and **24**, both of which are very potent at 5-HT$_{2C}$ receptors (Ki = 2.4 and 0.8 nM, respectively) and are claimed to be agonists with EC$_{50}$ values of 8.5 and 7.8 nM, respectively (79). Although both compounds have good affinity at 5-HT$_{2A}$ receptors (Ki values = 38 and 28 nM), their potent 5-HT$_{2C}$ receptor affinities provides receptor selectivity of 16- and 36-fold, respectively.   The tetrahydro-1H-pyrazino[1,2-a]quinoxaline **25** has also been described as a potent full agonist at the 5-HT$_{2C}$ receptor (Ki = 3 nM, 100% efficacy, EC$_{50}$ = 8 nM) but no selectivity data were provided (80). When administered to fasted rats, **25** produced a dose-dependent decrease in food intake with ED$_{50}$ values of 2 mg/kg i.p. and 10 mg/kg p.o. (81).

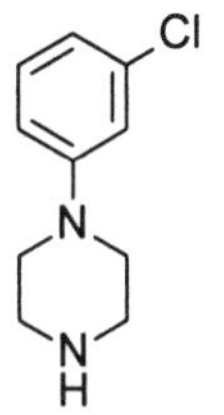

	23 R = methyl	
22	**24** R = ethyl	**25**

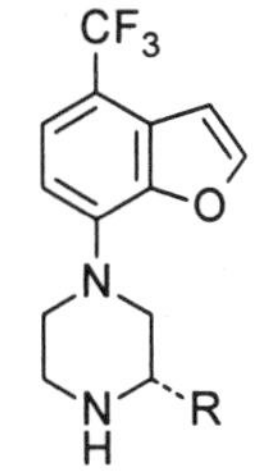

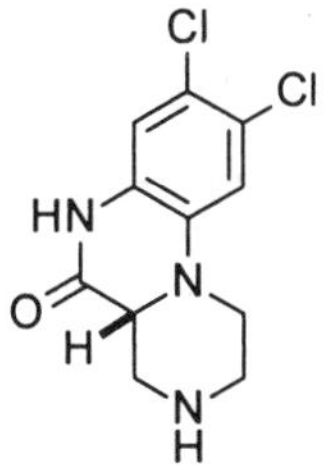

5-HT$_{2C}$ RECEPTOR ANTAGONISTS

Relative to the robust activity in the search for novel 5-HT$_{2C}$ agonists, few disclosures of 5-HT$_{2C}$ antagonists have appeared recently. The clinical development of SB-243213 (**26**) targeting anxiety and depression has been confirmed (82), and additional pre-clinical pharmacology has been published (83). SB-243213 is a potent (Ki ~ 1 nM) and selective (>100-fold relative to numerous enzymes, ion channels and receptors, including 5-HT$_{2A}$

26

and 5-HT$_{2B}$ receptors) inverse agonist that blocks mCPP-induced hypolocomotion (ID$_{50}$ = 1.1 mg/kg p.o) with a long duration of action (>8h).   Compound **26** is orally active in both the social interaction and Geller-Seifter conflict tests at 0.3 and 0.5 mg/kg, respectively, when administered 1h pre-test. Chronic administration of **26** (4.5 or 9 mg/kg, p.o., b.i.d., 14 days) did not result in tolerance in the social interaction test, nor did it result in withdrawal anxiogenesis.   In contrast to observations from the 5-HT$_{2C}$ "knockout" mouse, **26** did not increase body weight or affect seizure threshold.   At 10 mg/kg p.o, **26** increased

27

deep slow wave sleep (SWS) quantity and decreased paradoxical sleep (PS) quantity by 27% and 35% respectively in the rat (84). This effect on rat sleep profile is similar to that seen with the SSRI paroxetine, although the mechanisms underlying these pharmacological changes may be different for these compounds (84).

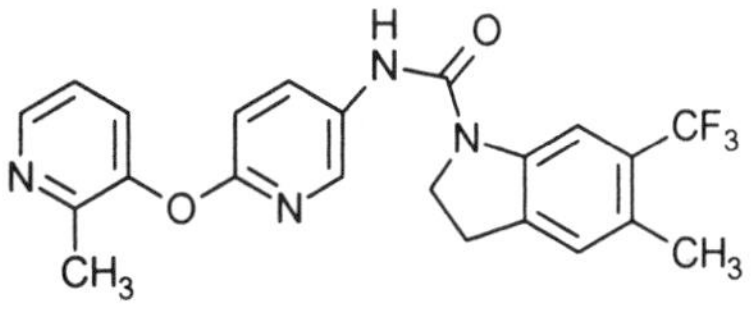

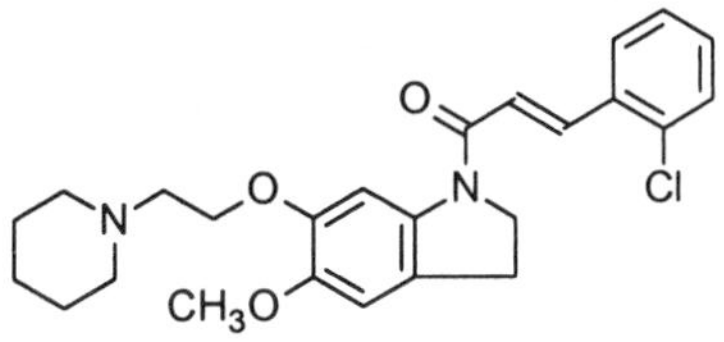

The evolution of selective 5-HT$_{2C}$ antagonists related to **26** continues with the disclosure of compounds such as **27**. These compounds have recently been claimed as 5-HT$_{2C}$ receptor antagonists having potent affinity (pKi = 7.5-9.8) and utility for the treatment of CNS disorders, particularly anxiety and depression (85).

New information on nonselective 5-HT$_{2C}$ receptor deramciclane (**28**) has appeared reporting its effectiveness in treating generalized anxiety disorder, as well as method of use claims for treating cognitive and obsessive-compulsive disorders (58,86-88). In a randomized placebo-controlled double-blind study with 212 patients, a significant decrease in HAM-A scores at week 8 was reported for groups receiving 30 and 60 mg/day (58,86). In addition, **28** was reportedly well tolerated and safe with no evidence of withdrawal symptoms following abrupt discontinuation of treatment.

28

Conclusions – Pharmacology mediated *via* 5-HT$_{2C}$ receptors has enjoyed increased attention as significant achievements in receptor biology and medicinal chemistry has pointed to opportunities for therapeutic intervention in several disease states for which there is significant unmet medical need (*e.g.*, anxiety, depression, and obesity). The recent clinical advancement of selective 5-HT$_{2C}$ ligands may further elucidate the role of 5-HT$_{2C}$ receptors in treatment of obesity and CNS disorders

References

1. I.H. Page, Physiol. Rev., 34, 563 (1954).
2. D.W. Wooley and E. Shaw, Proc. Natl. Acad. Sci., USA, 40, 228 (1954).
3. A. Dahlström and K. Fuxe, Acta. Psychiatr. Scand., 62, 1 (1964).
4. L. Uphouse, Neurosci. & Behav. Rev., 21, 679 (1997).
5. N.M. Barnes and T. Sharp, Neuropharmacology, 38, 1083 (1999).
6. P.R. Hartig, "Molecular Biology and Transduction Characteristics of 5-HT Receptors", Springer, 1997, Vol. 129 (7).
7. D. Hoyer, D.E. Clarke, J.R. Fozard, P.R. Hartig, G.R. Martin, E.J. Mylecharane, P.R. Saxena, and P.P. Humphrey, Pharmacol. Rev., 46, 157 (1994).
8. J.R. Raymond, Y.V. Mukin, A. Gelasco, J. Turner, G. Collinsworth, T.W. Gettys, J.S. Grewal, and M.N. Garnovskaya , Pharmacol. & Ther., 92,179 (2001).
9. B.J. Jones and T.P. Blackburn, Pharmacol, Biochem. Behav., 71, 555, 2002.
10. A.J. Robichaud and B.L. Largent, Annu. Rep. Med.Chem., 35, 11, (2000).
11. L.M. Gaster and F.D. King, Annu. Rep. Med. Chem., 33, 21, (1998).
12. M. Isaac, Drugs of the Future, 26, 383, (2001).
13. L. Yu, H. Nguyen, H. Le, L.J. Bloem, C.A. Kozak, B.J. Hoffman, T.P. Snutch, H.A. Lester, N. Davidson, and H. Lubbert , Mol. Brain Res., 11, 143, (1991).
14. H. Canton, R.B. Emeson, E.L. Barker, J.R. Backstrom, J.T. Lu, M.S. Chang, E. Sanders-Bush, Mol. Pharmacol., 50, 799, (1996).
15. C.M. Burns, H. Chu, S.M. Rueter, L.K. Hutchinson, H. Canton, E. Sanders-Bush, and R.B. Emeson, Nature, 387, 303, (1997).
16. C.M. Niswender, S.C. Copeland, K. Herick-Davis, R.B., Emeson, and E. Sanders-Bush J. Biol. Chem., 274, 9472, (1999).
17. L.W. Fitzgerald, G. iyer, D.S. Conklin, C.M. Krause, A. Marshall, J.P. Patterson, D.P. Tran, G.J. Jonak, and P.R. Hartig, Neuropsychopharmacology, 21, 82S, (1999).
18. R.D. Price, D.M. Weiner, M.S. Chang, and E. Sanders-Bush, J. Biol. Chem., 276, 44663, (2001).
19. A. Pazos, R. Cortes, and J.M. Palacios, Brain Res. 346, 231 (1985).
20. B.J. Hoffmann and E. Mezey, FEBS Lett., 247, 453, 1987.
21. J.F. Lopez-Gimenez, G. Mengod, J.M. Palacios, and M.T. Vilaro, Synapse, 42, 12, (2001).
22. D.A. Clemett, T. Punhani, M.S. Duxon, T.P. Blackburn, and K.C. Kone, Neuropharmacololology, 39, 123, (2000).
23. P.J. Conn and E. Sander-Bush, J. Pharmacol. Exp. Ther., 242, 552, (1987).
24. L.W. Fitzgerald, D.S. Conklin, C.M. Krause, A.P. Marshall, J.P. Patterson, D.P.

Tran, G. Iyer, W.A. Kostich, B.L. Largent, and P.R. Hartig, J. Neurochem., <u>72</u>, 2127,(1999).

25. R.H.P. Porter, K.R. Benwell, H. Lamb, C.S. Malcolm, N.H. Allen, D.F. Revell, D.R. Adams, and M.J. Sheardown, Brit. J. Pharamacol., <u>128</u>, 13, (1999).

26. J.C. Jerman, S.J. Brough, T. Gager, M.Wood, M.C. Coldwell, D. Smart, and D.N. Middlemiss, Eur. J. Pharmacol., <u>414</u>, 23, (2001).

27. M.S. Chang, L. Zhang, J.P. Tam, and E. Sanders-Bush, J. Biol. Chem., <u>275</u>, 7021, (2000).

28. C.C. Felder, R.Y. Kanterman, A.L. Ma, and J. Axelrod. Proc. Natl. Acad. Sci., <u>87</u>, 2187, (1990).

29. K.A. Berg, S. Maayani, J. Goldfarb, C. Scaramelli, P.Leff, and W.P. Clarke, Mol. Pharmacol., <u>54</u>, 94 (1998).

30. M. Marcoli, G. Maura, M. Tortarolo, and M. Raiteri, J. Neurochem., 69, 427. (1997).

31. V.L. Lucaites, D.L. Nelson, D.B Wainscott, and M. Baez, Life Sci., 59, 1081, (1996).

32. C. Becamel, A. Figge, S. Poliak, A. Dumuis, E. Peles, J. Bockaert, H. Lubbert, and C. Ullmer, J. Biol., Chem., <u>275</u>, 12974 (2001).

33. J.R. Backstrom, M.S. Chang, H. Chu, C.M. Niswender, E. Sander-Bush, J. Biol. Chem., <u>275</u>, 23620, (2000).

34. S.P. Vickers, M.J. Bickerdike, and C.T. Dourish, Neurosci. News, <u>2</u>, 22, (1999).

35. P.G. Clifton, M.D., Lee, and C.T. Dourish, Psychopharmacology, <u>152</u>, 256, (2000).

36. K.C.F. Fone, R. H. Austin, I.A. Topham, G.A. Kennett, and T. Punhani, Brit. J. Pharmacol., <u>123</u>, 1707, (1998).

37. M.J. Bickerdike, S.P. Vickers, and C.T. Dourish, Diabetes Obes. Metab., <u>1</u>, 207, (1999).

38. L.H. Tecott, L.M. Sun, S.F. Akana, A.M. Strack, D.H. Lowenstein, M.F. Dallman, and D. Julius, Nature, <u>374</u>, 542, (1995).

39. S.P. Vickers, G.A. Kennett, C.T. Dourish, and L.H. Tecott, Psychopharmacology, 143, <u>309</u>, (1999).

40. P.A. Sargent, A.L. Sharpley, C. Williams, E.M. Goodall, and P.J. Cowen, Psychopharmacology, <u>133</u>, 309, (1997).

41. G. Curzon, E.L. Gibson, and A.O. Oluyomi, Trends Pharmacol. Sci., <u>18</u>, 21, (1998).

42. S.P. Vickers, C.T. Dourish, and G.A. Kennett, Neuropharmacology, <u>41</u>, 200, (2001).

43. L.W. Fitzgerald, T.C. Burn, B.S. Brown, J.P. Patterson, M.H. Corjay, P.A. Valentine, J.H. Sun, J.R. Link, I. Abbaszade, J.M. Hollis, B.L. Largent, P.R. Hartig, G.F. Hollis, P.C. Meunier, A.J. Robichaud, and D.W. Robertson, Mol. Pharmacol., <u>57</u>, 75, (2000).

44. R.B. Rothman, M.H. Baumann, J.E. Savage, L. Rauser, A. McBride, S. Hufisein, and B.L. Roth, Circulation, <u>102</u>, 2836, (2000).

45. R.B. Rothman and M.H. Baumann, Pharmacol. Biochem. Behav., <u>71</u>, 825, (2002).

46. L.K. Heisler, H.M. Chu, and L.H. Tecott, Ann. NY Acad., Sci., <u>861</u>, 74, (1998).

47. C.D. Applegate and L.H. Tecott, Exp. Neurol., <u>154</u>, 522. (1998).

48. N. Upton, T. Stean, D. Middlemiss, T. Blackburn, and G. Kennett, Eur. J. Pharmacol., <u>359</u>, 33, (1998).

49. B.L. Roth, D. Roland, D. Ciaranello, and H.Y. Meltzer, J. Pharmacol. Exp. Ther., <u>260</u>, 1361, (1992).

50. K. Herrick-Davis, E. Grinde, and M. Teitler, J. Pharmacol. Exp. Ther., <u>295</u>, 226, (2000).

51. V. Di Matteo, M. Cacchio, C. Di Giulio, and E. Esposito, Pharmacol. Biochem. Behav., <u>71</u>, 727 (2002).

52. A. Deutch, B. Moghaddam, R.B. Innis, J.H. Krystal, G.K. Aghajanian, B.S. Bunney, and D.S. Charney, Schizophren. Res., <u>4</u>, 121, (1991).

53. J.R. Martin, T.M. Ballard, and G.A. Higgins, Pharmacol. Biochem. Behav. 71, 615 (2002).

54. M.D. Wood, C. Reavill, B. Trail, A. Wilson, T. Stean, G. Kennet, S. Lighthowler, T.P. Blackburn, D.T. Thomas, and T.L. Gager, Neuropharmacology, <u>41</u>, 186, (2001).

55. I. Gacsályi, E. Schmidt, I., Gyertyán, E. Vaser, A. Lang, A. Fekete, J. Hietala, E. Syvälahti, P. Tuomainen, and P.T. Männistö, Drug Dev. Res., <u>40</u>, 333, (1997).

56. F. Jenck, J.-L. Moreau, H. Berendsen, M. Boes, C.L.E. Broekkamp, J.R. Martin, J. Wichmann, and A.M.L. Van Delft, Eur. Neuropsychopharmacol. <u>8</u>, 161 (1998).

57. J.R. Martin, M. Bös, F. Jenck, J.-L. Moreau, V. Mutel, A.J. Sleight, J. Wichmann, J.S. Andrews, H.H. G. Berendsen, C.L.E. Broekkamp, G.S.F. Ruigt, C. Köhler, A.M.L. Van Delft, J. Pharmacol. Exp. Ther., <u>286</u>, 913, (1998).

58. H. Naukkararinen, Eur. Neuropsychopharmacology, 11(Suppl. 3), S301, (2001).

59. M. Bos, F. Jenck, J.R. Martin, J.-L. Moreau, A.J. Sleight, J. Wichmann and U. Widmer, J. Med. Chem., <u>40</u>, 2762 (1997).

60. J. Chang-Fong, J. Addo, M. Dukat, C. Smith, N.A. Mitchell, K Herrick-Davis, M. Teitler and R. A. Glennon, Bioorg. Med. Chem. Lett., <u>12</u>, 155 (2002).

61. D.R. Adams, J.M. Bentley, J.R.A. Roffey, R.J. Hamlyn, A.R. George, WO Patent 0012481 (2000).
62. D.R. Adams, J.R.A. Roffey, C.E. Dawson, WO Patent 0012482 (2000).
63. J.M. Bentley, F.R.A. Roffey, J.E.P Davidson, H.L. Mansell, R.J. Hamlyn, I.A. Cliffe, D.R. Adams and N.J. Monck, WO Patent 0112603-A1 (2001).
64. D.R. Adams, J.M. Bentley, J.R.A. Roffey, R.J. Hamlyn, S. Gaur, M.A.J. Duncton, J.E.P. Davidson, M.J. Bickerdike, I.A. Cliffe and H.L. Mansell, WO Patent 0012510 (2000).
65. J.M. Bentley, P. Hebeisen and S. Taylor, WO Patent 0166548 A1 (2001).
66. M. Issac, A. Slassi, A. O'Brien, L. Edwards, N. MacLean, D. Bueschkens, D.K. Lee, K. McCallum, I. De Lannoy, L. Demchyshyn and R. Kamboj, Bioorg. Med. Chem. Lett., 10, 919, (2000).
67. J.B. Hester, B.N. Rogers, E.J. Jacobsen, M.D. Ennis, B.A. Acker and S.L. Vandervelde, WO Patent 0064899 A1 (2000).
68. R.B. McCall, S.R. Franklin, D.K. Hyslop, C.S. Knauer, C.L. Chio, C.L. Haber and L.W. Fitzgerald, Soc. Neurosci. Abst., 2001.
69. D.R. Adams, J.M. Bentley, J.R.A. Roffey, R.J. Hamlyn, S. Gaur, M.A.J. Duncton, D. Bebbington, N.J. Monck, C.E. Dawson, R.M. Pratt and A.R. George, WO Patent 0012475 (2000).
70. D.R. Adams, J.M. Bentley, J.R.A. Roffey, C.D. Bodkin, H.L. Mansell, A.R. George and I.A. Cliffe, WO Patent 0012502 (2000).
71. J.R.A. Roffey, J.E.P. Davidson, H.L. Mansell, R.J. Hamlyn and D.R. Adams, WO Patent 0112602 A1 (2001).
72. D.R. Adams, J.M. Bentley, J. Davidson, M.A.J. Duncton and R.H.P. Porter, WO Patent 0044753 (2000).
73. J.M. Bentley, P. Hebeisen, M. Muller, H. Richter, S. Roever, P. Mattei and S. Taylor, WO Patent 0210169 A1 (2002).
74. M.D. Ennis, R.L. Hoffman, N.B. Ghazal and R.M. Olson, WO Patent 0172752 A2 (2001).
75. A.J. Robichaud, T. Lee, W. Deng, I.S. Mitchell, W. Chen, C.D. McClung, E.J.B. Calvello and D.M. Zawrotny, WO Patent 0077001 A1 (2000).
76. A.J. Robichaud, T. Lee, W. Deng, I.S. Mitchell, M.G. Yang, S. Haydar, W. Chen, C.D. McClung, E.J.B. Calvello and D.M. Zawrotny, WO Patent 0077002 A1 (2000).
77. A.J. Robichaud, T. Lee, W. Deng, I.S. Mitchell, S. Haydar, W. Chen, C.D. McClung, E.J.B. Calvello and D.M. Zawrotny, WO Patent 0077010 A2 (2000).
78. A.J. Robichaud, W. Chen, C. McClung, M. Stranz, D. Clark, W. Deng, E.B. Calvello, I.S. Mitchell, T. Lee, P. Rajagopalan, D.W. Robertson, B.F. Molino, S.N. Haydar, M.P. Conlon, B.C. Duffy, J. Liao, N.G. Pawlush, L.W. Fitzgerald, D.S. Conklin, C.M. Krause, J.F. McElroy, K.W. Rohrbach, K.J. Miller and B.L. Largent, Abstracts of Papers, MEDI-105, 221st ACS National Meeting, San Diego, CA, 2001.
79. K. Briner, T.P. Burkholder, M.L. Heiman and D.L.G. Nelson, WO Patent 0109123 A1 (2001).
80. A.L. Sabb, G.S. Welmaker and J.A. Nelson, WO Patent 0035922 (2000).
81. G.S. Welmaker, J.A. Nelson, J.E. Sabalski, A.L. Sabb, J.R. Potoski, D. Graziano, M. Kagan, J. Coupet, J. Dunlop, H. Mazandarani, S. Rosenzweig-Lipson, S. Sukoff and Y. Zhang, Bioorg. Med. Chem. Lett., 10, 1991 (2000).
82. S.M. Bromidge Abstracts of Papers, MEDI-304, 222nd ACS National Meeting, Chicago, IL, United States, August 26-30, 2001.
83. M.D. Wood, C. Reavill, B. Trail, A. Wilson, T. Stean, G.A. Kennett, S. Lightowler, T.P. Blackburn, D. Thomas, T.L. Gager, G. Riley, V. Holland, S.M. Bromidge, I.T. Forbes and D.N. Middlemiss, Neuropharmacology, 41, 186 (2001).
84. M.I. Smith, D.C. Piper, M.S. Duxon and N. Upton, Pharmacol. Biochem. Behav., 71, 599 (2002).
85. S.M. Bromidge, P.J. Lovell, S.F. Moss and H.T. Serafinowska, WO Patent 0214273 A1 (2002).
86. H. Kanerva and O. Maki-Ikola, WO Patent 0168067 A2 (2001).
87. O. Maki-Ikola, U.S. Patent 6,335,371 (2002).
88. O. Maki-Ikola, M. allin, J. Rouru, L. Lehtonen and M. Jaskari, U.S. Patent 6,335,372 (2002).

Chapter 4. Emerging Themes in Alzheimer's Disease Research: Paradigm Shift in Drug Discovery

Todd E. Morgan[1] and Grant A. Krafft[2]

School of Gerontology, University of Southern California, Los Angeles, CA 90089[1]

Acumen Pharmaceuticals, Glenview, IL 60025[2]

Introduction - This year marks the 95th anniversary of the first published case of Alzheimer's disease (AD), a disease currently afflicting more than 12 million people worldwide. Over the past 15 years, the pace of AD research has accelerated dramatically, and the discovery focus has moved from cholinergic enhancement medicines to drugs that interfere with amyloid. Recent results have emerged to challenge the core dogma of the prevailing "amyloid cascade" hypothesis which invokes amyloid plaque deposition as the cause of AD. The present report examines this central hypothesis, highlighting recent results that implicate non-fibrillar Aβ oligomers rather than fibrils as the molecular pathogens in AD, and evaluating those drug discovery approaches that are poised to capitalize on these new findings.

HISTORICAL BACKGROUND

In 1907, Bavarian psychiatrist Alois Alzheimer described pathology and symptoms of a 51 year-old woman, Auguste D., who had suffered progressive cognitive decline, learning and memory deficits, and paranoid and delusional behavior (1). Alzheimer used a silver stain on cortical tissue samples to reveal prevalent and highly unusual lesions that he called "neurofibrillary tangles" and "senile plaques", and these lesions remain the basis for definitive post-mortem diagnosis of Alzheimer's disease (AD). Alzheimer's initial report did not discuss any definitive cause for AD, but he did observe (his italics): ***"Plaques are not the cause of senile dementia, but only an accompanying feature of senile involution of the central nervous system"*** (2).

Alzheimer's early exoneration of amyloid plaques apparently escaped most AD researchers, including Glenner and Wong, who in 1984 identified the 39-43 residue amyloid β peptide as the major plaque component (3). Shortly thereafter, synthetic Aβ peptides were shown to assemble into fibrils with staining properties identical to senile plaques (4), and in 1990, synthetic Aβ was shown to be neurotoxic (5). With the Aβ peptide sequence in hand, several groups identified the gene encoding the Aβ precursor protein (APP) in 1987, setting the stage for discovery of APP mutations responsible for certain familial AD cases (6-9). Based on these findings, the "amyloid cascade" hypothesis was articulated in 1992, implicating accumulated amyloid plaques and fibrils as the cause of AD (10).

A close corollary to the amyloid cascade hypothesis is inflammation mediated by glial activation (11). Activated glial cells are often closely associated with amyloid plaque deposition, while inflammatory mediators such as IL-1 can increase Aβ production (12-15). This can lead to more plaques, leading to more activated glia, setting up an accelerating cascade. A number of studies suggest that glial activation exacerbates fibril toxicity, and one study even demonstrated that Aβ fibril toxicity *requires* glial activation, arguing for an essential link between amyloid plaques, inflammation and AD (16,17).

The 20th century closed with amyloid deposition and inflammation dominating the AD mechanistic landscape, but as the 21st century began, a major challenge to these two widely-held beliefs has emerged. Soluble neurotoxic oligomers of Aβ 1-42 known as Aβ-derived diffusible ligands (ADDLs), first described in 1998, recently have been identified in impaired transgenic AD mice and in AD brain, and shown to interfere directly with neuronal long-term potentiation (LTP) and to exert region-specific neurotoxicity in CA1 (18-24). These results establish ADDLs as the likely molecular pathogens in AD and provide a unifying explanation for many confounding observations previously not reconciled by the amyloid cascade hypothesis.

THE ROLE OF Aβ AND THE AMYLOID CASCADE HYPOTHESIS

The definitive molecular etiology of AD has been elusive. Strong evidence suggests the main constituent of the amyloid plaque, Aβ, plays a prominent role in disease progression. This evidence comes from multiple lines of investigation. For example, Aβ levels are modulated by all four identified genetic components associated with AD: mutations in APP, presenilin-1 (PS-1) & PS-2 and the inheritance of the epsilon 4 allelle of apolipoprotein E (apoE4) (9). Early onset familial AD (FAD) mutations in PS and some FAD APP mutations result specifically in increases of Aβ 1–42 production in cell culture and in transgenic mouse models (25-27). Although apoE4 does not increase Aβ production, it is less efficient in clearing Aβ 1-42 from the brain, due to its lower Aβ 1-42 binding affinity (28). APP, the precursor to the Aβ peptide, is localized on the 21st chromosome and Downs syndrome (trisomy 21) patients develop AD-like symptoms and pathologies at an early age (29). Evidence also comes from *in vitro* and *in vivo* experiments demonstrating that Aβ is neurotoxic (5, 30). Thus, the Aβ 1-42 peptide plays a critical role in the pathogenesis of AD.

A connection between Aβ and neurofibrillary tangles has not been fully established, however, several studies have demonstrated that Aβ specifically up-regulates paired helical filaments (PHF) specific tau phosphorylation (31,32). Recent data on cyclin-dependent kinase 5 (Cdk5) supports such a connection (33,34). When Cdk5 associates with its regulatory subunit, p35, Cdk5/p35 kinase activity is initiated, a requirement for neurite growth. p25, a truncated fragment of p35, causes Cdk5 to be constitutively activated and mislocalized *in vivo*. Cdk5/p25 kinase phosphorylates tau and causes morphological degeneration (including the appearance of PHF) and profound apoptotic cell death of primary neurons. p25 is produced and accumulates in brains of patients with AD (33). Moreover, Aβ 1-42 induces conversion of p35 to p25 in primary cortical neurons (34).

Aβ is derived by proteolytic cleavage of β-APP, an integral membrane protein of unknown function. It is highly expressed in the brain and APP knockout mice show age-related cognitive deficits and neuropathology (35). Recent data suggest that APP may normally serve as a membrane cargo receptor for kinesin-1. Kinesin-1 is responsible for ATP-dependent movement of vesicular cargoes within neurons (36,37). Whether this, or other yet to be identified functions, require proteolytic processing of APP is not known. However, Aβ is detected in cognitively normal individuals, suggesting APP processing occurs in the absence of disease (38-40).

For generation of Aβ, proteolytic processing of APP is essential. APP is cleaved by sequential actions of three unique proteases, α-, β-, and γ-secretases (41). Each secretase cleaves at a unique site to generate several different APP derivatives. The Aβ peptide is produced by sequential activity of β- then γ-secretase which produce peptides ranging from 40-43 amino acids. β-secretase (also referred to as β-site APP-cleaving enzyme, BACE) has been identified and BACE knockout mice produce reduced Aβ levels (42). On the other hand, the exact identity of γ-secretase remains unclear. The γ-secretase cleavage sites lie within the transmembrane region of APP. This unusual transmembrane proteolytic activity has recently been shown to require the presenilins (PS). However, PS are also involved in cleavage of the developmental protein, Notch (43-46). Thus, the phenotype of the PS-1 knockout is late embryonic lethal, making this transgenic ineffective for examining the *in vivo* role of PS-1 in Aβ generation (47,48). Recently, postnatal, neuron-specific PS-1 deficient mice have been shown to possess reduced $A\beta$ levels which further validates the essential role of PS-1 in γ–secretase activity (49,50). β- and γ-secretases are potential therapeutic targets because they are essential for $A\beta$ generation. However, since APP is not the only substrate for these enzymatic activities, side effects of inhibiting either activity may prevent their ultimate therapeutic use. Moreover, any potential therapy that modifies APP processing may affect the normal functioning of APP and its metabolic derivates (*vide infra*).

The cellular source and site of generation of $A\beta$ are not known, although proteolytic processing of APP can occur during and after APP intracellular trafficking so its metabolic derivatives (including $A\beta$) can be released into vesicle lumens and the extracellular space. In AD brains extracellular $A\beta$ appears as compact, fibrillar deposits and amorphous, non-fibrillar, diffuse deposits (9). Compact amyloid plaques, also known as neuritic amyloid plaques, are localized to the limbic and association cortices and are associated with dystrophic neurites. The limbic and association cortices are regions severely damaged in AD. On the other hand, diffuse amyloid plaques are prevalent throughout the brain, and are not associated with neuritic elements. In addition, diffuse amyloid plaques appear before neuritic amyloid plaques in brains from Downs syndrome patients and APP transgenic mice (29,51). These observations may support the view that diffuse amyloid plaques are precursors of the neuritic amyloid plaque.

Based on these observations, and because fibrillar $A\beta$ preparations were neurotoxic to cultured neurons, it was postulated the fibrillization of $A\beta$ into these amyloid plaques caused AD-related dementia (the amyloid cascade hypothesis). However, careful AD neuropathology studies have not been able to correlate the density of neuritic amyloid plaques and neurodegeneration or clinical dementia (52,53). Recent studies demonstrate that the severity of neurodegeneration correlates with small, stable oligomers of $A\beta$ (54,55). In addition, recent studies in APP transgenic mice suggest that amyloid plaque deposition is not required for synaptic loss or learning/memory deficits (20,56). Thus, the classical amyloid cascade hypothesis needs to be re-evaluated.

OLIGOMERIC $A\beta$ CAUSES NEUROTOXICITY

The amyloid cascade hypothesis was first challenged in a 1995 report that demonstrated the neurotoxicity of a non-fibrillar $A\beta$ preparation (57). Soon, numerous laboratories reported identification and neurotoxicity of small, non-fibrillar $A\beta$ assemblies (19,58-61). One such preparation consisting of ADDLs killed mature neurons in organotypic hippocampal slice cultures at nanomolar concentrations (19). Fyn, a protein tyrosine kinase in the *src*-family, was required for this toxicity implicating a specific signaling pathway underlying ADDL toxicity. ADDL toxicity may involve generation of reactive oxygen species, superoxide or peroxynitrite through a specific signaling pathway, because ADDLs reversibly inactivate the free radical-sensitive enzyme, aconitase, prior to initiating neurotoxicity (62). In addition, ADDLs inhibit LTP, a classical model of synaptic plasticity and memory (19,22).

More recent studies have demonstrated that cell-derived $A\beta$ oligomers have biological activity *in vivo*. Intracerebroventricular microinjection of conditioned medium from APP overexpressing cells completely blocked LTP in the CA1 hippocampal region. $A\beta$ oligomers similar to ADDLs were the culprit because conditioned medium devoid of monomeric $A\beta$ was active. Monomeric $A\beta$ was removed with an insulin-degrading enzyme. Plus, conditioned medium devoid of oligomeric $A\beta$ was inactive. Here, oligomerization was prevented with a selective γ-secretase inhibitor, which selectively blocked $A\beta$ dimer and trimer formation at doses that allow appreciable monomer production (23).

Several transgenic mouse models of AD support the concept that soluble Aβ oligomers may be the primary source of synaptic dysfunction. A comparison of multiple lines of transgenic mice that overexpress wild-type or mutated human APP reveals a dissociation between Aβ plaque deposition and synaptic dysfunction and loss (20,63). Aβ plaque deposition only occurred in the mouse lines possessing the mutated APP transgene, while loss of synaptophysin immunoreactivity occurred in both wild type (no plaques) and mutant lines (plaques). Synaptophysin is a presynaptic axonal terminal marker. Importantly, there is good correlation between cognitive decline and loss of synaptophysin-immunoreactive presynaptic terminals in AD-vulnerable regions of the brain (64-66).

These observations were extended by another research group using a different AD transgenic mouse model (56). This AD transgenic mouse exhibits early histopathological features of AD and forms Aβ deposits but not plaques. Here, spatial learning abilities were tested using the Morris water maze (MWM) model which is sensitive to hippocampal damage. The MWM model tests the ability of the animal to spatially learn and remember the location of an escape platform using environmental cues. Transgenic AD mice showed significant deficits that increased in severity with age as compared to wild-type mice.

There is convincing evidence that oligomeric Aβ species exist *in vivo*. Water soluble Aβ oligomers have been detected in human brain and are elevated 12-fold in AD brain (67). Using antibodies that preferentially bind oligomeric species of Aβ, Lambert and colleagues demonstrated the presence of ADDLs in temporal and frontal cortices from AD brain, but not in control brains or AD cerebellum, a region not susceptible to AD-related damage (61). Oligomeric Aβ species were detected in CSF samples (59). Chinese hamster ovary cells stably transfected with APP cDNA express SDS-stable Aβ species including monomer, dimer, and trimer, which were identified by sequencing and affinity for specific Aβ antibodies. These Aβ oligomeric species may assemble shortly after Aβ synthesis in intracellular vesicles as they are detected in cell lysates and isolated microsomes and they co-migrate with proteins associated with subcellular vesicles (59).

These observations strongly implicate oligomeric Aβ assemblies (e.g., ADDLs) as the neurotoxins underlying the pathophysiology of AD. If true, therapies aimed at inhibiting their assembly or activity, or at elevating their clearance from the brain would be particularly effective. Specific cyclodextrin derivatives, but not cyclodextrin itself, were shown to be effective at inhibiting ADDL formation (61). Moreover, complete immuno-neutralization of ADDLs was demonstrated using antibodies raised specifically against ADDL immunogens. The possibility of effective immune intervention targeting Aβ and ADDLs is stimulating another remarkable shift in AD research directions, as discussed in the following sections.

IMMUNIZATION AND AD

AD research took a dramatic leap forward when it was shown that immunization of APP transgenic mice with aggregated Aβ resulted in attenuated AD-like pathology (68). This particular vaccination regimen was effective in preventing Aβ-plaque formation in young animals and reducing the progression of plaque deposition in older mice. Numerous labs have confirmed and extended this observation using different mouse models and immunization paradigms (69-72). Aβ inoculations caused activation of microglia (73) and antibody-mediated phagocytosis of Aβ by microglia led to its degradation (69). Removal of existing Aβ deposits by microglia after plaque decoration by anti-Aβ antibodies has also been demonstrated using a novel *in vivo* multi-photon imaging technology (74), suggesting that microglial activation may enable Aβ plaque phagocytosis.

Aβ immunization was also shown to have beneficial functional consequences (70-72,75). For example, passive immunization with the mouse monoclonal Aβ antibody m266 resulted in a rapid reversal of memory deficits in APP transgenic mice. Within 24 hours of administration, improved performance in an object recognition memory test and a holeboard learning task was observed. This reversal of memory deficits was not associated with reduced Aβ deposition. An Aβ-antibody complex was detectable in plasma and CSF, however, the beneficial memory effects occurred well before any meaningful reduction of total Aβ (72).

Recall that nanomolar concentrations of oligomeric Aβ assemblies (e.g., ADDLs) are capable of rapid inhibition of information storage mechanisms (LTP, *vide supra*, 19), and that complete *in vitro* protection from ADDL neurotoxicity was mediated by anti-ADDL antibodies (61). It is possible that the beneficial effects of Aβ immunization in various rodent memory models may result from removal or direct immuno-neutralization of these oligomeric Aβ species. This supports the argument for development of immunotherapies against oligomeric Aβ species (61).

THERAPEUTIC APPROACHES: INTERSECTING THE Aβ OLIGOMER AXIS

Indictment of Aβ oligomeric assemblies (e.g., ADDLs) as the molecular pathogens in AD establishes a new conceptual framework for devising effective therapeutic and preventative approaches. In the broadest sense, intervention breaks down into three categories: 1) block Aβ 1-42 production to minimize ADDL formation; 2) enhance clearance of Aβ 1-42 and/or ADDLs; 3) design blockers of ADDL assembly or activity. The following sections provide an overview of current and potential approaches that intersect these Aβ oligomer possibilities.

Blocking Aβ Production - Many current drug discovery approaches focus on blocking Aβ production, some of which may ultimately be effective in AD therapy. Lowering the Aβ 1-42 monomer concentration significantly will reduce the rate of ADDL formation, establishing a different equilibrium that disfavors oligomeric assemblies. Lowering ADDL levels will result in direct reduction of aberrant neuronal signaling (21, 23). Additional advantages may accrue for inhibitors that specifically block Aβ 1-42, because Aβ 1-42 forms oligomers more favorably than Aβ 1-40 (76). Aβ 1-40 does not assemble into stable oligomers, yet at high concentrations, oligomeric Aβ 1-40 forms and they can be chemically crosslinked to generate stable structures (77).

Secretase Inhibitors - The most direct approach to blocking Aβ production involves inhibition of the secretase enzymes. Good progress has been made towards discovery of BACE inhibitors (78) and γ-secretase inhibitors (79), and a number of compounds are reported to be in pre-clinical or Phase I human clinical trials. Blocking BACE should lead to reduction of both Aβ 1-42 and Aβ 1-40, and the major challenge for medicinal chemists is generating sufficient potency in a molecule that can penetrate the CNS to generate inhibitory concentrations of drug. Studies of transgenic mice engineered with a disrupted BACE gene appear to develop without any detectable defects (42), suggesting chronic inhibition of BACE activity may be an acceptable therapeutic approach. For BACE inhibitors, there is the question relating to potential consequences of blocking all Aβ production, because it is possible that Aβ 1-40 has important normal functions related to adhesion and neuronal plasticity (80,81).

Blockage of γ-secretase on a chronic basis may also result in unacceptable side effects, because this enzyme appears to be involved in processing of other important proteins such as Notch (43-46). Perhaps PS-1 and the protein complex with which it is associated, is not the only enzyme responsible for the C-terminal Aβ cleavage. This possibility may enable design of effective inhibitors that lack limiting side effects (82,83). A recent report described γ-secretase inhibitors that lowered Aβ production without cleaving Notch, although questions have been raised regarding the actual target of these inhibitors (84,85). Another recent report demonstrated that the NSAIDs, ibuprofen, indomethacin, and sulindac sulphide, specifically lowered Aβ 1-42 and increased Aβ 1-38, apparently by modulating γ-secretase, and without any effect on APP. This result has generated intense interest, because this profile of activity is highly attractive for an Aβ 1-42 lowering agent (86).

Modulating APP - Several approaches to modulating APP could result in effective lowering of Aβ, such as blocking or reducing the synthesis of APP (87). This might be accomplished at the transcriptional level or at some post-transcriptional step. The drug phenserine, originally of interest as a cholinesterase inhibitor, has been shown to reduce the synthesis of APP, and consequently levels of Aβ. It has been reported that phenserine interacts with APP mRNA at an IL-1 response element, and this element may provide the mechanism by which IL-1 increases Aβ (88,89). There is continuing medicinal chemistry activity to generate phenserine analogues with optimized properties (90). A recent report has described a ribozyme capable of reducing APP mRNA, but no data were reported on reduction of actual Aβ levels (91). There is evidence suggesting that various neurotransmitters, neuro-immunophilins and cycloxygenase inhibitors also may modulate APP synthesis (92). The cycloxygenase inhibitor effect may occur by blocking prostaglandin E2, which has been shown to elevate APP synthesis (93).

Another potentially viable approach involves upregulating the α-secretase enzyme activity that cleaves APP within the Aβ sequence, potentially lowering overall Aβ production (94). Three muscarinic receptor subtypes, M1, M3, and M5, are coupled to upregulated shedding of sAPP, suggesting that muscarinic agonists may elevate α-secretase activity (95,96). Several metalloprotease enzymes have been shown to be capable of carrying out this cleavage, and recent work suggests that two or more enzymes in the ADAM family have this activity (97). It appears that activation of PKC epsilon results in upregulation of α-secretase activity, although which particular enzyme is involved was not established (98). The utility of this approach has not been established in any of the transgenic AD models to this point, and verification in a clinical setting may first emerge from data on one of the muscarinic compounds currently being studied (99).

Enhancing ADDL or Aβ 1-42 Clearance - One of the most direct intervention approaches would involve activating specific clearance of ADDLs or the Aβ 1-42 that forms them. Clearance of Aβ was one of the motivations for the initial Aβ immunization approaches discussed above, although the vaccination with fibrils clearly also targeted removal or reduction of plaques. It is likely that targeting plaques via anti-fibril antibodies played a role in generating the severe CNS inflammation suffered by human subjects receiving AN-1792, which resulted in recent termination of that clinical study (100). Anti-Aβ antibodies were expected to bind and clear Aβ in the peripheral circulation, accelerating Aβ efflux from the CNS and ultimately reducing brain Aβ levels. Reduction of Aβ was a consequence of some Aβ immunization protocols in transgenic AD mouse models (68-71), but several immunization experiments led to cognitive protection with no reduction in brain Aβ levels (e.g., 72). As suggested above, this raises the clear likelihood that Aβ oligomers (e.g., ADDLs) are the targets for some of the generated antibodies. The recent description of rapid deficit reversal by injection of the m266 antibody indicated that "a soluble brain Aβ species" was its target, and it is likely only binding to soluble Aβ oligomers (e.g., ADDLs) would result in such rapid cognitive improvement. This is supported by the demonstrated neuroprotective ability of ADDL-directed antibodies (61).

Blocking ADDL Assembly or Activity - Direct interference with the assembly or activity of ADDLs represents another highly attractive strategy. The recent results that certain cyclodextrin compounds interfere with ADDL assembly and toxicity suggest that small molecules with ADDL-blocking ability and the ability to penetrate the CNS may be promising candidates for human clinical trials (61). It is possible that certain compounds designed to be blockers of fibril assembly also will have the ability to block ADDLs. However, compounds that convert plaque deposits back to Aβ monomer, without interfering with oligomer assembly, could exacerbate brain damage (101). The ability to interfere directly with aberrant neuronal processes activated by ADDLs is also potentially attractive. Because one of the earliest ADDL activities is disruption of neuronal LTP (19,22), compounds that compensate for this blockage, such as nicotinic agonists may be effective therapeutics (102). Other signaling pathways that are neuron selective also are clearly involved in ADDL neurotoxicity, such as the Fyn kinase pathway, and compounds that are able to interfere with processes downstream of Fyn activity also may hold promise (18,19,22).

Conclusion - The paradigm shift from blocking Aβ plaques and fibrils to blocking small soluble Aβ toxins such as ADDLs represents a reversal in dogma and an opportunity for generating highly effective AD therapeutics. It has taken more than 90 years to accumulate compelling scientific evidence to support Alzheimer's original supposition that plaques are not the cause of AD. Clear evidence implicating ADDLs as the molecular pathogens in AD sets the stage for therapeutics that can block AD progression, prevent AD onset, and most importantly, potentially reverse existing AD deficits.

References

1. A. Alzheimer, Allg. Z. Psychiatr. Psych.-Gerichtl. Med. 64, 146 (1907).
2. A. Alzheimer, Z. Gesamte Neurol. Psychiatr., 4, 356 (1911) Engl. translation: H. Förstl and R. Levy, Hist. Psychiatry, 2, 71 (1991).
3. G.G. Glenner and C.W. Wong, Biochem. Biophys. Res. Commun., 120, 885 (1984).
4. E.M. Castano, J. Ghiso, F. Prelli, P.D. Gorevic, A. Migheli and B. Frangione, Biochem. Biophys. Res. Commun., 141, 782 (1986).
5. B.A. Yankner, L.K. Duffy and D.A. Kirschner, Science, 250, 279 (1990).
6. A. Goate, M.C. Chartier-Harlin, M. Mullan, J. Brown, F.Crawford, L. Fidani, L. Guiffra, A. Haynes, N. Irving, L. James, R. Mant, P. Newton, K. Rooke, P. Roques, C. Talbot, M. Pericak-Vance, A. Roses, R. Williamson, M. Rossor, M. Owen and J. Hardy, Nature, 349, 704 (1991).
7. M.C. Chartier-Harlan, F. Crawford, HI Houlden, A. Warren, D. Hughes, L. Fidani, A. Goate, M. Rossor, P. Roques, J. Hardy and M. Mullan, Nature, 353, 844 (1991).
8. M. Mullan, F. Crawford, K, Axelman, H. Houlden, L. Lilius, B. Winblad and L. Lannfelt, Nature-Genetics, 1, 345 (1992).
9. D.J. Selkoe, Physiol. Rev. 81, 741 (2001).
10. J.A. Hardy and G.A. Higgins, Science, 256, 184 (1992).
11. H. Akiyama, S. Barger, S. Barnum, B. Bradt, J. Bauer, G.M. Cole, N.R. Cooper, P. Eikelenboom, M. Emmerling, B.L. Fiebich, C.E. Finch, S. Frautschy, W.S.T. Griffin, H. Hampel, M. Hull, G. Landreth, L-F. Lue, R. Mrak, I.R. Mackenzie, P.L.McGeer, M.K. O'Banion, J. Pachter, G. Pasinetti, C. Plata-Salaman, J. Rogers, R. Rydel, Y. Shen, W. Streit, R. Strohmeyer, I. Tooyoma, F.L. Van Muiswindeil, R. Veerhuis, D. Walker, S. Webster, B. Wegrzyniak, G. Wenk and T. Wyss-Coray, Neurobiol. Aging, 21, 383 (2000).
12. P.L. McGeer, S. Itagaki, H, Tago and E.G. McGeer, Neurosci. Lett., 79, 195 (1987)

13. S. Itagaki, P.L. McGeer, H. Akiyama, S. Zhu and D Selkoe, J. Neuroimmunol., 24, 173 (1989)
14. S. Haga, K. Akai and T. Ishii, Acta Neuropathol (Berl)., 77, 569 (1989)
15. J.G. Sheng, K. Ito, R.D. Skinner, R.E. Mrak, C.R. Rovnaghi, L.J. and W.S.T. Griffin, Neurobiol. Aging, 17, 761 (1996).
16. L. Meda, M.A. Cassatella, G.I. Szendrei, L. Otvos, P. Garon, M. Willalba, D. Ferrari and F. Rossi, Nature, 374, 647 (1995)
17. D. Guilian, L.J. Haverkamp, J. Yu, W. Karshin, D. Tom, J. Li, A. Kazanskaia, J. Kirkpatrick and A.E. Roher, J. Biol. Chem., 273, 29719 (1998).
18. G.A. Krafft, W.I. Klein, B.A. Chromy, M.P. Lambert, C.E. Finch, T.E. Morgan, P. Wals, I. Rozovsky and A. Barlow, US Patent # 6,218,506 (2001, published in 1998).
19. M.P. Lambert, A.K. Barlow, B.A. Chromy, C. Edwards, R. Freed, M. Liosatsos, T.E. Morgan, I. Rozovsky, B. Trommer, K.L. Viola, P. Wals, C. Zhang, C.E. Finch, G.A. Krafft and W.L. Klein, Proc. Natl. Acad. Sci. USA, 95, 6448 (1998).
20. L. Mucke, E. Masliah, G-Q. Yu, M. Mallory, E.M. rockenstein, G. Tatsuno, K. Hu, D. Kholodenko, K. Johnson-Wood and I. McConlogue, Jr. Neurosci., 20, 4050 (2000).
21. Y. Gong, K.L. Viola, L. Chang, M.P. Lambert, C.E. Finch, G.A. Krafft and W.L. Klein, Soc. Neurosci. Meeting Abstract, (2001). Manuscript submitted.
22. H.W. Wang, J.F. Pastemak, H. Kuo, H. Ristic, M.P. Lambert, B. Chromy, K.L. Viiola, W.L. Klein, W.B. Stine, G.A. Krafft and B.L. Trommer, Brain Res., 924, 133 (2002).
23. D.M. Walsh, I. Klyubin, J.V. Fadeeva, W.K. Cullen, R. Anwyl, M.S. Wolfe, M.J. Rowan and D.J. Selkoe, Nature, 416, 535 (2002).
24. W.L. Klein, G.A. Krafft and C.E. Finch, Trends in Neuroscience, 24, 219 (2001).
25. M. Citron, T. Oltersdork, C. Haass, L. McConlogue, A. Hung, P. Seubert, C. Vigo-Pelfrey, I. Lieberburg and D.J. Selkoe, Nature, 360, 673 (1992).
26. K. Duff, C. Eckman, C. Zehr, X. Yu, C-M. Prada, J. Perez-Tur, M. Hutton, L. Bue, Y. Harigaya, D. Yager, D. Morgan M.N. Gordon, L. Holcomb, L. Refolo, B. Zenk, J. Hardy, and S. Younkin, Nature, 383, 710 (1996).
27. M. Citron, D. Westaway, W. Xia, G. Carlson, T.A. Diehl, D. Kholodenka, R. Motter, R. Sherrington, B. Perry, H. Yao, R. Strome, I. Lieberburg, J. Rommens, S. Kim, D. Schenk, P. Fraser, P. St. George-Hyslop and D.J. Selkoe, Nature-Med., 3, 67 (1997).
28. M. Gearing, H. Mori and S.S. Mirra, Ann. Neurol., 39, 395 (1996).
29. C.A. Lemere, J.K. Blustzjan, H. Yamaguchi, T. Wisniewski, T.C. Saido and D.J. Selkoe, Neurobiol. Dis., 3, 16 (1996).
30. C.J. Pike, D. Burdick, A.J. Walencewicz, C.G. Glabe and C.W. Cotman, J. Neurosci., 13, 1676 (1993).
31. M.P. Lambert, G. Stevens, S. Sabo, K. Barber, G. Wang, W. Wade, G. Krafft, S. Snyder, T.F. Holzman and W.L. Klein, J. Neurosci. Res., 39, 377 (1994).
32. J. Busciglio, A. Lorenzo, J. Yeh and B.A. Yankner, Neuron, 14, 879 (1995).
33. G.N. Patrick, L. Zukerberg, M. Nikolic, S. Monte, P. Dikkes and L-H. Tsai, Nature, 402, 615 (1999).
34. M-S. Lee, Y. Kwon, M. Li, J. Peng, R. Friedlander and L-H. Tsai, Nature, 405, 360 (2000).
35. G.R. Dawson, G.R. Seabrook, H. Zheng, D.W. Smith, S. Graham, G. O'Dowd, B.J. Bowery, S. Boyce, M.E. Trumbauer, H.Y. Chen, L.H. Van der Ploeg and D.J. Sirinathsinghji, Neuroscience, 90, 1 (1999).
36. A. Kamal, G.B. Stokin, Z. Yang, C-H. Xia and S.B. Goldstein, Neuron, 28, 449 (2000).
37. A. Kamal, A. Almenar-Queralt, J.F. LeBlanc, E.A. Roberts and S.B. Goldstein, Nature, 414, 643 (2001).
38. C. Haas, M. Schlossmacher, A.Y. Hung, C. Vigo-Pelfrey, A. Mellon, B. Ostaszewski, I. Lieberburg, E.H. Koo, D. Schenk, D. Teplow and D.J. Selkoe, Nature, 359, 322 (1992).
39. P. Seubert, C. Vigo-Pelfrey, F. Esch, M. Lee, H. Dovey, D. Davis, S. Sinha, M.G. Schlossmachder, J. Whaley, C. Swindlehurst, R. McCormack, R. Wolfert, D.J. Selkoe, I. Lieberburg and D. Schenk, Nature, 359, 325 (1992).
40. M. Shoji, T.E. Golde, J. Ghiso, T.T. Cheung, S. Estus, L.M. Shaffer, Z. Cai, D.M. McKay, R. Tintner, B. Frangione and S.G. Younkin, Science, 258, 126 (1992).
41. W.P. Esler and M.S. Wolfe, Science, 293, 1449 (2001).
42. Y. Luo, B. Bolon, S. Kahn, B.D. Bennett, S. Babu-Khan, P. Denis, W. Fan, H. Kha, J. Zhang, Y. Gong, L. Martin, J.C. Louis, Q. Yan, W.G. Richards, M. Citron and R. Vassar, Nat. Neurosci., 4, 231 (2001).

43. Y. Ye, N. Lukinova and M.E. Fortini, Nature, 398, 525 (1999).
44. G. Struhl and I. Greenwald, Nature, 398, 525 (1999).
45. B. De Strooper, W. Annaert, P. Cupers, P. Saftig, K. Craessaerts, J.S. Mumm, E.H. Schroeter, V. Schrijvers, M.S. Wolfe, W.J. Ray, A. Goate and R. Kopan, Nature, 398, 518 (1999).
46. W. Song, P. Nadeau, M. Yuan, X. Yang, J. Shen and B.A. Yankner, Proc. Natl. Acad. Sci. USA, 96, 6959 (1999).
47. J. Shen, R.T. Bronson, D.F. Chen, W. Xia, D.J. Selkoe and S. Tonegawa, Cell, 89, 629 (1997).
48. P.C. Wong, H. Zheng, H. Chen, M.W. Becher, D.J. Sirinathsinghji, M.E. Trumbauer, H.Y. Chen, D.L. Price, L.H. Van der Ploeg and S.S. Sisodia, Nature 387, 288 (1997).
49. H. Yu, C.A. Saura, S.Y. Choi, L.D. Sun, X. Yang, M. Handler, T. Kawarabayashi, L. Younkin, B. Fedeles, M.A. Wilson, S. Younkin, E.R. Kandel, A. Kirkwood and J. Shen, Neuron, 31, 713 (2001).
50. I. Dewachter, D. Reverse, N. Caluwaerts, L. Ris, C. Kuiperi, C. Van den Haute, K. Spittaels, L. Umans, L. Serneels, E. Thiry, D. Moechars, M. Mercken, E. Godaux and F. Van Leuven, J. Neurosci., 22, 3445 (2002).
51. D. Games, D. Adams, R. Alessandrini, R. Barbour, P. Berthelette, C. Blackwell, T. Carr, J. Clemens T, Donaldson, F, Gillespie, T. Guido, S. Hagoplan, K. Johnson-Wood, K. Khan, M. Lee, P. Leibowitz, I. Lieberburg, S. Little, E. Masliah, L. McConlogue, M. Montoya-Zavala, L. Mucke, L. Paganini, E. Penniman, M. Power, D. Schenk, P. Seubert, B. Snyder, F. soriano, H. Tan, J. Vitale, W. Wadsorth, B. Wolozin and J. Zhao, Nature, 373, 523 (1995).
52. D.F. Swaab, P.J. Lucassen, A.P. van de Nes, R. Ravid and A. Salehi in "Connections, Cognition, and Alzheimer's Disease," B.T. Hyman, C. Duyckaerts and Y. Christen, Eds., Springer-Verlag, Berlin/New York, 1997, p83.
53. J. Hardy and K. Gwinn-Hardy, Science, 282, 1075 (1998).
54. L-F. Lue, Y-M. Kuo, A.E. Roher, L. Brachova, Y. Shen, L. Sue, T. Beach, J.H. Kurth, R.E. Rydel and J. Rogers, Amer. J. Path., 155, 853 (1999).
55. C.A. McLean, R.A. Cherny, F.W. Fraser, S.J. Fuller, M.J. Smith, K. Beyreuter, A.I. Bush and C.L. Masters, Ann. Neurol., 46, 860 (1999).
56. M. Koistinaho, M. Ort, J.M. Cimadevilla, R. Vondrous, B. Cordell, J. Koistinaho, J. Bures and L.S. Higgins, Proc. Natl. Acad. Sci. USA, 98, 14675 (2001).
57. T. Oda, P. Wals, H. Osterburg, S.A. Johnson, G. Pasinetti, T.E. Morgan, I. Rozovsky, W.B. Stine, S.W. Snyder, T.F. Holzman, G. Krafft and C.E. Finch, Exp. Neurol., 136, 22 (1995).
58. A.E. Roher, M.O. Chaney, Y.M. Kuo, S.D. Webster, W.B. Stine, L.J. Haverkamp, A.S. Woods, R.J. Cotter, J.M. Tuohy, G.A. Krafft, B.S. Bonnel and M.R. Emmerling, J. Biol. Chem., 271, 20631 (1996).
59. D.M. Walsh, B. Tseng, R. Rydel, M. Podlisny and D.J. Selkoe, Biochem., 39, 10831 (2000).
60. D.M. Hartley, D.M. Walsh, C.P. Ye, T. Diehl, S. Vasquez, P.M. Vassilev, D.B. Teplow and D.J. Selkoe, J. Neurosci., 19, 8876 (1999).
61. M.P. Lambert, K.L. Viola, B.A. Chromy, L. Chang, T.E. Morgan, J. Yu, D.L. Venton, G.A. Krafft, C.E. Finch and W.L. Klein, J. Neurochem., 79, 595 (2001).
62. V.D. Longo, K.L. Viola, W.L. Klein and C.E. Finch, J. Neurochem., 75, 1977 (2000).
63. A.Y. Hsia, E. Masliah, L. McConlogue, G-Q. Yu, G. Tatsun, K. Hu, D. Kholodenko, R.C. Malenka, R.A. Nicoll and L. Mucke, Proc. Natl. Acad. Sci. USA, 96, 3228 (1999).
64. E. Masliah, M. Mallory, L. Hansen, R. DeTeresa, M. Alford and R. Terry R, Neurosci. Lett., 174,67 (1994).
65. D.W. Dickson, H.A.Crystal, C. Bevona, W. Honer, I. Vincent, P. and Davies, Neurobiol. Aging, 16, 285 (1995).
66. C.I. Sze, J.C. Troncoso, C. Kawas, P. Mouton, D.L. Price and L.J. Martin, J. Neuropathol. Exp. Neurol., 56, 933 (1997).
67. Y-M. Kuo, M.R. Emmerling, C. Vigo-Pelfres, T.C. Kasunic, J.B. Kirkpatrick, G.H. Murdoch, M.J.Ball and A.E. Roher, JBC, 271, 4077 (1996).
68. D. Schenk, R. Barbour, W. Dunn, G. Gordon, H. Grajeda, T. Guido, K. Hu, J. Huang, K. Johnson-Wood, K. Khan, D. Kholodenko, M. Lee, Z. Liao, I. Lieberburg, R. Motter, L. Mutter, F. Soriano, G. Shopp, N. Vasquez, C. Vandevert, S. Walder, M. Wogulis, T. Yednock, D. Games and P. Seubert, Nature, 400, 173 (1999).
69. F. Bard, C. Cannon, R. Barbour, R-L. Burke, D. Games, H. Grajeda, T. Guido, K. Hu, J. Huang, K. Johson-Wood, K. Khan, D. Kholodenko, M. Lee, I. Lieberburg, R. Motter, M. Nguye, F. Soriano, N. Vasquez, K. Weiss, B. Welch, P. Seubert, D. SchenK and T. Yednock, Nature, 6, 916 (2000).

70. C. Janus, J. Pearson, J. McLaurin, P.M. Mathews, Y. Jiang, S.D. Schmidt, M.A. Chishti, P. Horne, D. Heslin, J. French, H.T.J. Mount, R.A. Nixon, M. Mercken, C. Bereron, P.E. Fraser, P. St. George-Hyslop and D. Westaway, Nature, 408, 979 (2000).

71. D. Morgan, D.M. Diamond, P.E. Gottschall, K.E. Ugen, C. Dickey, J. Hardy, K. Duff, P. Jantzen, G. DiCarlo, D.M. Wilcock, K.Connor, J. Hatcher, C. Hope, M. Gordon and G.W. Arendash, Nature, 408, 982 (2000).

72. J-C. Dodart, K.R. Bales, K.S. Gannon, S.J. Greene, R.B. DeMattos, C. Mathis, C.A. DeLong, S. Wu, X. Wu, D.M. Holtzman and S.M. Paul, Nature-Neurosci., 5, 452 (2002).

73. D.M. Wilcock, M.N. Gordon, K.E. Ugen, P.E. Gottschall, G. Dicarlo, C. Dickey, K. Boyett, P. Jantzen, K. Connor, J. Melachrino, J. Hardy and D. Morgan, DNA Cell Bio., 20, 731 (2001).

74. B.J. Bacskai, S.T. Kajdasz, R.H. Christie, C. Carter, D. Games, P. Seubert, D. Schenk and B.T. Hyman, Nature-Medicine, 7, 369 (2001).

75. G.W. Arendash, M.N. Gordon, D.M. Diamond, L.A. Audtin, J.M. Hatcher, P. Jantzen, G. Dicarl, D. Wilcock and D. Morgan D, DNA Cell Bio., 20, 737 (2001).

76. W. B. Stine, L. Kulans, B. Chromy, C. Park, M. Lambert, G. Priebe, C. E. Finch, W. L. Klein and G. A. Krafft, Society for Neuroscience Abstract, (1998).

77. G. Bitan, A. Lomakin and D.B. Teplow, J. Biol. Chem., 276, 35176 (2001).

78. S. Roggo, Curr. Top. Med. Chem., 2, 359 (2002).

79. M.S. Wolfe, Curr. Top. Med. Chem., 2, 371 (2002).

80. J. Wu, R. Anwyl and M.J. Rowan, Eur. J. Pharmacol., 284, R (1995).

81. M.A. Kowalska and K. Badellino K, Biochem. Biophys. Res. Commun., 205, 1829 (1994).

82. C. Yu, S.H. Kim, T. Ikeuchi, H. Xu, L. Gasparini, R. Wang and S.S. Sisodia SS, J. Biol. Chem., 276, 43756 (2001).

83. S.S. Sisodia SS and P.H. St George-Hyslop. Nat. Rev. Neurosci. 3, 281 (2002).

84. A. Petit, F. Bihel, C. Alves da Costa, O. Pourquie, F. Checler and J.L. Kraus, Nat. Cell Biol., 3, 507 (2001).

85. W.P. Esler, C. Das, W.A. CampbelL, W.T. Kimberly, A.Y. Kornilova, T.S. Diehl, W. Ye, B.L. Ostaszewski, W. Xia , D.J. Selkoe and M.S. Wolfe, Nat. Cell Biol., 4, E110 (2002).

86. S. Weggen, J.L. Eriksen, P. Das, S.A. Sagi, R. Wang, C.U. Pietrzik, K.A. Findlay, T.E. Smith, M.P. Murphy, T. Bulter, D.E. Kang, N. Marquez-Sterling N, T.E. Golde and E.H. Koo, Nature, 414, 212 (2001).

87. R.M. Nitsch, R.J. Wurtman and J.H. Growdon, Ann. N. Y. Acad. Sci., 777, 175 (1996).

88. K.T. Shaw, T. Utsuki, J. Rogers, Q.S. Yu, K. Sambamurti, A. Brossi, Y.W. Ge, D.K. Lahiri and N.H. Greig, Proc. Natl. Acad. Sci. USA., 98, 7605 (2001).

89. J.T. Rogers, L.M. Leiter, J. McPhee, C.M. Cahill, S.S. Zhan, H. Potter and L.N. Nilsson, J. Biol. Chem., 274, 6421 (1999).

90. Q. Yu, H.W. Holloway, J.L. Flippen-Anderson, B. Hoffman, A. Brossi and N.H. Greig, J. Med. Chem. 44, 4062 (2001).

91. N. Dolzhanskaya, J. Conti, G. Merz and R.B. Denman, Mol. Cell. Biol. Res. Com. 4, 239 (2000).

92. R.K. Lee and R.J. Wurtman, Ann. N. Y. Acad. Sci. 920, 261 (2000).

93. R.K. Lee, S. Knapp and R.J. Wurtman, J. Neurosci., 19, 940 (1999).

94. J. Mills and P.B. Reiner, J. Neurochem. 72, 443 (1999).

95. R.M. Nitsch, B.E. Slack, R.J. Wurtman and J.H. Growdon, Science, 258, 304 (1992).

96. J.D. Buxbaum, M. Oishi, H.I. Chen, R. Pinkas-Kramarski, E.A. Jaffe, S.E. Gandy and P. Greengard, Proc. Natl. Acad. Sci. USA., 89, 10075 (1992).

97. B.E. Slack, L.K. Ma and C.C. Seah, Biochem. J. 357, 787 (2001).

98. G. Zhu, D. Wang, Y.H. Lin, T. McMahon, E.H. Koo and R.O. Messing, Biochem. Biophys. Res. Commun., 285, 997 (2001).

99. A.D. Korczyn, Expert. Opin. Investig. Drugs, 9, 2259 (2000).

100. K. Birmingham and S. Frantz, Nat. Med., 8, 199 (2002).

101. B. Permanne, C. Adessi, G.P. Saborio, S. Fraga, M.J. Frossard, J.V. Dorpe, I. Dewachter, W.A. Banks, F.V. Leuven and C. Soto, FASEB J., Apr 10, epub. (2002).

102. K.T. Dineley, K.A. Bell, D. Bui and J.D. Sweatt, J. Biol. Chem., publ. as 10.1074/jbc.M200066200 (2002).

SECTION II. CARDIOVASCULAR AND PULMONARY DISEASES

Editor: William J. Greenlee, Schering Plough Research Institute, Kenilworth,NJ

Chapter 5: Recent Advances in Pulmonary Hypertension Therapy

Russell A. Bialecki
AstraZeneca Pharmaceuticals
1800 concord Pike, Wilmington, DE 19850-5437

<u>Introduction</u> – Pulmonary hypertension is functionally defined as inappropriately high pulmonary arterial pressure for a given level of blood flow through the lungs. Normally, the pulmonary circulation is a low-pressure, high-flow system. Pulmonary artery systolic pressure at rest in healthy individuals is 18 to 25 mm Hg, with a mean pulmonary pressure of 12 to 16 mm Hg. This low pressure is due to the large cross-sectional area of the pulmonary circulation. Normal pulmonary circulation accommodates increased blood flow, as during exercise, without increases in pulmonary arterial pressure by recruitment of under-perfused blood vessels and distention of patent vessels. In addition, smooth muscle tone in the media of pulmonary arterioles is lower and the amount of smooth muscle in pulmonary resistance vessels is less than that of systemic arteries. Increases in pulmonary blood flow, left atrial or pulmonary venous pressure, blood viscosity, and decreases in the cross-sectional area of the pulmonary arterial lumen can all contribute to increased pulmonary arterial pressure. The clinical presentation of pulmonary hypertension is often associated with several of these changes.

Pulmonary hypertension is a major complication of several pulmonary disorders including emphysema and chronic bronchitis and can have many potential causes (1,2). A rigorous diagnostic classification of the various forms of pulmonary hypertension based on common clinical features has been developed to standardize diagnosis and treatment (3). However, the disease is conveniently classified here as (a) primary pulmonary hypertension (PPH; i.e., unexplained) or (b) secondary to various lung, cardiovascular and other diseases (e.g., chronic bronchitis, hypoxia, emphysema, cardiac and extrathoracic conditions).

Irrespective of cause, pulmonary hypertension is defined clinically by mean pulmonary arterial pressures >25 mm Hg at rest or >30 mm Hg during exercise and an increase in pulmonary vascular resistance (ratio of change in blood pressure over flow). *Cor pulmonale*, a characteristic of advanced disease, is defined as enlargement of the right ventricle resulting from disorders of the respiratory system. Pulmonary hypertension invariably precedes *cor pulmonale* and leads to right heart failure. *Cor pulmonale* is the third most common cardiac disorder (after coronary and hypertensive heart disease) in persons more than 50 years of age (4).

When outlining recent advances in the development of new therapies for pulmonary hypertension, it is useful to understand what is known about its pathogenesis. This review provides cursory background information about the disease with a major focus on recent therapeutic advances. For a greater understanding of the disease, the reader is referred to several in-depth reviews on the pathophysiology and clinical presentation of pulmonary hypertension (5-7).

Primary pulmonary hypertension is rare (estimated incidence 1-2 per million) and is associated with pronounced increases (>60 mm Hg) in mean pulmonary arterial pressure. In childhood, the disease affects both genders equally. After puberty, it is more common in women than in men (1.7: 1); it is most prevalent in persons 20 to 40 years of age; and the disease has no racial bias. PPH may occur sporadically or in familial clusters but the clinical and pathologic features of the disease are similar. Six percent of patients with PPH have a positive family history (8). Inheritance is autosomal dominant with variable penetrance, and occurs in a 2:1 female-to-male ratio. Genetic anticipation, or a tendency for decreasing age of onset in subsequent generations, has been described (9). A gene associated with PPH has been localized to chromosome 2q31-32 (10).

Until recently, median life expectancy for patients with PPH was two to three years after diagnosis. Predictors of survival include performance in the six-minute walk test (11), pulmonary artery pressure, right atrial pressure, cardiac index and mixed venous oxygen saturation (12). For reasons unknown, the presence of Raynaud's disease is associated with a worse prognosis. Survival is longest in patients who have cardiac output >2.5 L/min and respond favorably to acute vasodilator challenge (13). In contrast, cardiac output <2.5 L/min, left untreated, predicts a survival of only several months. The disease is progressive and has no cure.

Secondary pulmonary hypertension, with modest increases in mean pulmonary arterial pressures (<40 mm Hg), is more prevalent but under diagnosed. Reliable estimates of incidence of this condition are difficult to obtain because numerous underlying pathologies contribute to the disease. In both primary and secondary pulmonary hypertension, a decrease in lung diffusion capacity correlates with increased risk of mortality. The major stimulus for increased pulmonary vascular resistance in patients with secondary pulmonary hypertension stemming from emphysema and chronic bronchitis has traditionally been viewed as alveolar hypoxia, with a direct relationship between pulmonary hypertension and the extent of underlying disease (14,15). In normal persons, pulmonary vasoconstriction is an important control mechanism for ventilation/perfusion matching throughout the lung. However, long-term hypoxic exposure induces vascular remodeling and other changes (e.g., decreased cardiac output) associated with a poor prognosis.

Although treatment strategies for the different types of pulmonary hypertension vary, several common pathophysiologic hallmarks exist. Chronic increases in pulmonary blood pressure are associated with vasoconstriction and remodeling of the vascular bed. Anatomical changes include: (a) muscularization of pulmonary arterioles; (b) abnormally high deposition of collagen and elastin within the vascular media and adventitia; (c) medial smooth muscle cell hypertrophy and hyperplasia; and (d) differentiation of precursor cells into a mature myofibroblast phenotype. Intimal, adventitial, and medial thickening narrows the vascular lumen. Arterial thrombosis, intimal fibrosis, and foci of inflammatory cells are often times evident. In addition, plexiform lesions or characteristic formations of abnormally dilated, thin-walled arteriovenous connections, are frequently found in PPH.

Recent studies have demonstrated decreased expression of the voltage-dependent potassium channel Kv1.5 subunit (16, 17). This observation is important because this ion channel has been implicated in modulating hypoxic pulmonary vasoconstriction. Interestingly, high concentrations of fenfluramine, an agent that increases the risk of pulmonary hypertension, decreases expression of the Kv1.5 subunit when administered to normal smooth muscle cells in culture (18).

TREATMENT

The principal aim of therapy is to attenuate the pathogenesis and initiate regression of the disease. Historically, vasodilator substances have been the most studied. Presently available therapeutic interventions for this disease include the use of oxygen if the patient is hypoxemic ($PaO_2 \leq 55$ mm Hg or oxygen saturation $\leq 90\%$). Diuretics are also considered if there is evidence of right heart failure with peripheral edema and hepatic congestion. More than one diuretic is required in refractory cases and patients may be treated with intravenous administration of these agents. However, the use of diuretics must be monitored because excessive diuresis decreases cardiac output and reduces the effectiveness of vasodilators.

Digoxin, a cardiac ionotrope, produces a modest increase in cardiac output in patients with pulmonary hypertension and right ventricular failure (19). This treatment can cause a significant reduction in circulating norepinephrine without detectable effects on baroreceptor responsiveness.

Several clinical studies have shown that anticoagulant therapy is associated with improved survival (20,21). Patients with PPH are prone to thromboembolism because of sluggish pulmonary blood flow, dilated right heart chambers, venous insufficiency and relative physical inactivity. On the basis of a lung scan showing nonuniformity of pulmonary blood flow, warfarin is given to PPH patients as concurrent therapy. Survival rates of patients receiving the anticoagulant were 47% after 3 years compared to 31% in patients that did not receive warfarin. Although warfarin has never been examined in a prospective, randomized, long-term clinical trail, the use of anticoagulant is recommended based on the potential for thrombosis.

Calcium channel blockers are an effective vasodilator therapy in some patients with PPH. Rich *et al* demonstrated that patients with PPH who showed a significant decrease in pulmonary arterial pressure and pulmonary vascular resistance with acute challenge of calcium channel blockers had improved survival when treated chronically with high doses of these agents compared to controls (22). However, only 20% of patients with PPH demonstrate marked pulmonary vasodilatation in response to an acute challenge with vasodilators. Thus, only a minority of patients receives benefit from this therapy.

Circulating concentrations of prostacyclin (PGI_2), are decreased in patients with PPH due to a reduction in prostacyclin synthase (23, 24). Release of PGI_2 has been shown to be important in (a) maintaining the pulmonary vasculature in a relaxed state; (b) preventing vascular smooth muscle cell hypertrophy and hyperplasia; and (c) decreasing platelet aggregation. Thus, replacement therapy logically follows. In fact, the introduction of intravenous infusion of PGI_2, or epoprostanol marked a major advance in the treatment of PPH. Continuous intravenous infusion of epoprostanol improves exercise tolerance, cardiac output, and pulmonary hemodynamic indices in patients with both primary and secondary forms of pulmonary hypertension (11, 25). Epoprostanol therapy also increases survival in patients with PPH. In some patients, pulmonary arterial pressure decreases occur despite the absence of an acute vasodilator response, suggesting a beneficial remodeling effect on the pulmonary vasculature with chronic therapy. However, the chronic intravenous infusion of epoprostanol is associated with inherent difficulties including dosage problems, infection of the central venous catheter and possible interruption in therapy. The average cost of treatment is $60,000USD per year. Taken together, these shortcomings have stimulated efforts to produce stable analogs of PGI_2 that can be delivered via more convenient routes at less cost.

Surgical procedures including atrial septostomy (opening a hole in the septum), thromboendarterectomy (removal of clot and arterial intima), as well as heart-lung, double lung, or single lung transplantation presently provide a successful form of therapy for PPH (26,27). However organ donations are scarce and transplantation is associated with additional possible complications including operative mortality, transplant rejection, and infections. The three-year survival rate for patients with PPH who undergo lung transplantation is approximately 68% (27). A unique surgical procedure known as graded balloon dilation atrial septosomy has been reported to immediately reduce right ventricular end-diastolic pressure, increase cardiac index, and increase survival (28). However, more studies are necessary before this experimental surgical procedure can be widely adopted.

A marked increase in our understanding of the pathogenesis of pulmonary hypertension has occurred recently and has begun to translate to effective treatments for many patients that have this condition. Table 1 summarizes the reported status of new therapies for the treatment of pulmonary hypertension from preclinical development through commercial launch. Some compounds represent novel therapeutic approaches whereas others epitomize improvements based on existing treatments.

Table 1. Classification and Development Stage of Potential Therapies for Pulmonary Hypertension.

Classification	Drug	Compound Number	Status
Prostacyclin analogs	Treprostinil sodium	1	Preregistration
	Iloprost	2	Launched
	Beroprost	3	Registered
Nitric oxide donors	GEA3175	4	Preclinical
	NCX900	NA[1]	Preclinical
Endothelin receptor antagonists:	Bosentan	5	Registered
	PABSA	6	Preclinical
	Sitaxsentan sodium	7	Phase III
	TA0201	8	Preclinical
	Darusentan	9	Preclinical
	SB247083	10	Preclinical
	PD156707	11	Preclinical
	PD180988	12	Preclinical
	PD164800	13	Preclinical
Thromboxane A_2 receptor antagonist /synthase inhibitor	Terbogrel	14	Phase II
Phosphodiesterase V Inhibitors	Sildenafil	15	Launched for ED[2]
	E4010	16	Preclinical
	T1032	17	Preclinical

[1]NA, not available; [2]ED; erectile dysfunction

Prostacyclin Analogs - Trials evaluating the therapeutic utility of subcutaneous, inhaled, and oral delivery of prostacyclin analogs are underway or have been recently completed. Treprostinil sodium (UT-15) **1** is a long acting analog of PGI_2 analog. The measured activity half-life ($t_{1/2}$) of treprostinil sodium and PGI_2 is 53-87 minutes and 2 minutes, respectively. Both compounds are potent vasodilators, inhibit platelet aggregation and attenuate smooth muscle cell proliferation. Treprostinil sodium is amenable to subcutaneous dosing using a pager-sized MiniMed microinfusion devise. In a recent 12-week trial, 470 patients receiving treprostinil sodium showed decreased total pulmonary resistance, improved exercise capacity in the six-minute walk test, and improved symptom profile compared with placebo controls (29). Improvements in the six-minute walk test were dose-dependent with more impressive effects noted in patients with increased severity of the disease. Most patients experience pain and/or erythemia at the infusion site. In

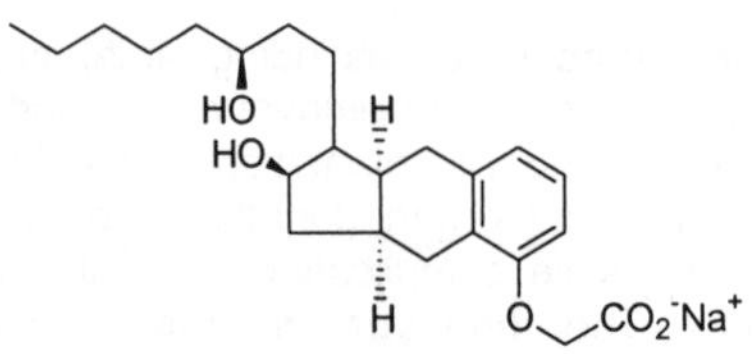

1

some cases, pain was severe and limited dose escalation. No successful resolution to this problem has been found. An advisory panel for the FDA has recommended approval of treprostinil sodium with a final FDA decision pending.

Inhalational and oral PGI_2 analogs are in end-stage development. Iloprost **2**, a stable PGI_2 analog administered by inhalation, has an activity $t_{1/2}$ of 30 minutes and has shown encouraging results in clinical trials. Twenty-four patients with PPH who had New York Heart Association (NYHA) class III or IV limitations (IV being worst) demonstrated significant improvements in exercise capacity and pulmonary hemodynamics over a one-year treatment period (30). A randomized, placebo controlled, double blind study is underway. However, the anticipated treatment paradigm is cumbersome (inhalation approximately nine times daily) and may limit the practicality of this approach.

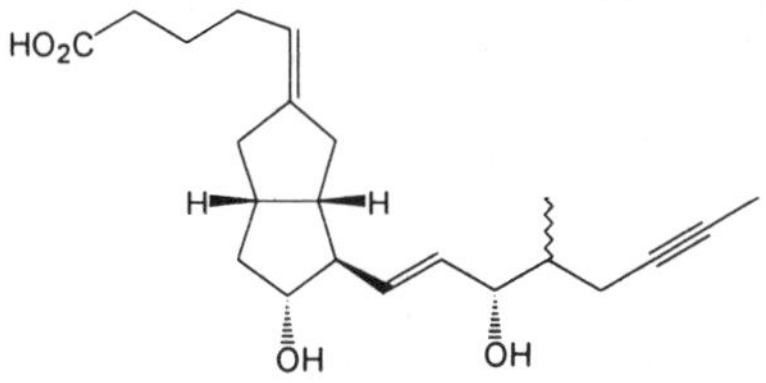

2 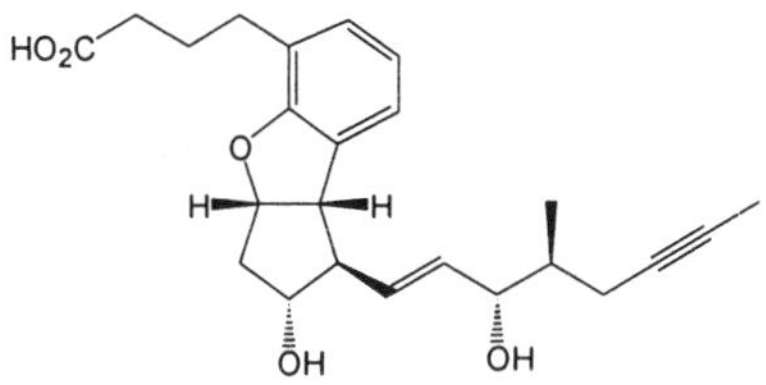   **3**

Beraprost **3**, an orally active PGI_2 analog, has also shown promise as a treatment for PPH. It represents an equimolar mixture of two diastereoisomers, APS314 and APS315, each of which has two optical isomers. The compound is rapidly absorbed after oral administration. Administration of beraprost at 40 µg/dose, po, t.i.d. for 4 weeks produced pharmacokinetic values of 122 and 294 pg/ml and half-lives of 38 and 32 minutes for the two diastereoisomers APS314d and APS315d, respectively. Significant improvements were observed in cardiopulmonary hemodynamics and NYHA functional class in 34 PPH patients in an open labeled uncontrolled dosing study. Beraprost also improved the survival of PPH patients (31). A large, multicenter Phase III clinical trial is ongoing to determine the effect of beraprost on PPH patients with NYHA class II/III pulmonary arterial hypertension.

Nitric Oxide Donors – Normal pulmonary endothelial cells utilize nitric oxide synthase and L-arginine to synthesize nitric oxide (NO). This unstable free radical causes relaxation of underlying vascular smooth muscle by stimulating soluble guanylate cyclase and guanosine 3', 5'-cyclic monophosphate (cGMP) formation. Nitric oxide is a primary determinant of pulmonary vascular tone (32), inhibits pulmonary arterial smooth muscle cell hypertrophy and hyperplasia, and prevents platelet aggregation (33). Decreased nitric oxide synthase has been noted in patients with PPH (34). Delivered via inhalation in the gaseous form, NO causes pulmonary vasodilatation without systemic effects (35). The binding of gaseous NO to hemoglobin and subsequent quenching via formation of methemoglobin confers selectivity.

Although chronic ambulatory inhalation of NO is feasible (36), technical difficulties associated with constant inhalational delivery prevail. To circumvent these limitations, preclinical development of orally active NO donor compounds is in progress. GEA 3175 **4** is a mesoionic oxatriazole imine. Little is known about the mechanism of NO release from this compound. However, by analogy to the liberation of NO from a well-studied class of NO donors, the sydnonimines, the process is postulated to involve a pH-independent oxidative reaction (37). Isolated tissue pharmacology studies indicate that GEA 3175 relaxes

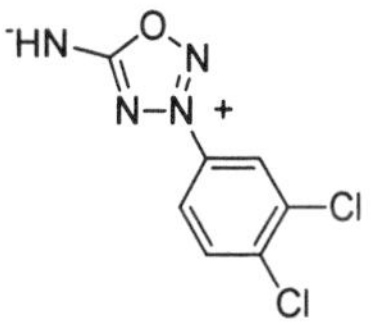

4

precontracted rat aorta, porcine coronary artery, and rabbit jugular vein (EC_{50} values = 5, 3 and 2 nmol/L, respectively). Intravenous administration of GEA 3175 (300 µg/kg) to conscious cannulated rats decreases blood pressure comparable to sodium nitroprusside (30 µg/kg). In comparison, chronic intravenous administration of GEA 3175 (33-66 µg/kg/hr for 7 days) has no significant effect on blood pressure. Confounding reports indicate that GEA 3175 inhibits thrombin-induced (38) but not adenosine 5'-diphosphate-induced (39) platelet aggregation. Preclinical development of NCX900, an NO-NSAID with potential for treating pulmonary hypertension, is also ongoing.

A lack of NO donor selectivity for the pulmonary circuit may preclude commercial development as a potential therapy for pulmonary hypertension. High systemic plasma levels of NO can cause hypotension, negative ionotropic effects, and increased bleeding times due to inhibition of platelet aggregation and adhesion. The same chemical attributes that allow for NO liberation *in vivo* such as thermal and chemical instability make it difficult to modulate organ-specific release of NO. In addition, potentially toxic metabolites (e.g., peroxynitrite, nitrogen dioxide, methemaglobin) can be formed.

Endothelin Receptor Antagonists – Endothelins (ET) are potent vasoconstrictor and mitogenic peptides released from vascular endothelial cells. Although three ET isoforms have been identified, only ET-1 has been implicated in the pathogenesis of pulmonary hypertension. Two major receptor subtypes have been described; ET_A is selective for ET-1 and ET_B possesses equal affinity for all three ET isoforms (40-43). Increased expression and release of ET-1 occurs in nearly all forms of pulmonary hypertension including those associated with alveolar hypoxia (44-47). Plasma levels of ET-1 correlate with disease severity (48). Increased expression of ET-1 mRNA may account for elevated plasma concentrations in human pulmonary hypertension (44,49) and rodent models of the disease (50,51). Decreased clearance of ET-1 via the pulmonary circulation also contributes to increases in plasma levels of the peptide in patients with PPH (52).

Studies of mixed ET_A/ET_B receptor and ET_A receptor-selective antagonists in animal models support the hypothesis that ET receptor blockade will attenuate the pathogenesis of pulmonary hypertension (53-57). Bosentan **5,** a sulfonamide, is a competitive mixed ET_A/ET_B receptor antagonist. The compound has favorable pharmacokinetic parameters in man (oral bioavailability 49% and plasma $t_{1/2}$ 3-5 hr). The first placebo-controlled study with bosentan in patients with PPH or pulmonary hypertension secondary to scleroderma showed improved exercise capacity, cardiopulmonary hemodynamics and NYHA functional class at 12 weeks (58). The six-minute walk distance increased by 70 meters in the treated group compared to controls. World Health Organization functional class improved from III to II in nine patients in the treated group compared with one in the placebo group. A larger multicenter, randomized, double blind, placebo controlled study was recently completed in which 213 patients were evaluated for 16 weeks. Preliminary results indicate that bosentan significantly improved scores in the six-minute walk test (59). The major adverse effect of bosentan is increased liver transaminase activity. Thus, liver function must be monitored regularly. A minor incidence of anemia has also been demonstrated. Deleterious drug interactions occur with bosentan and cyclosporin A or glibenclamide. The recommended dosing regimen is 62.5 mg, b.i.d. for 4 weeks followed by maintenance with 125 mg, b.i.d.

A structurally distinct mixed ET_A/ET_B receptor antagonist, PABSA **6**, has shown promise in preclinical studies. PABSA competitively inhibits binding of ^{125}I-ET1 to ET_A and ET_B receptors with nanomolar affinity. For comparison, the potency of PABSA for the ET_A receptor is 87-fold greater than bosentan (60). Orally administered PABSA (10-100 mg/kg) effectively reduces blood pressure in deoxycorticosterone acetate-salt sensitive rats, spontaneously hypertensive rats, and stroke prone spontaneously hypertensive rats for $\geq$ 24 hr (60). Efficacy of PABSA in patients with PPH remains to be determined.

Selective ET_A receptor antagonists have been developed to treat pulmonary hypertension. Sitaxsentan **7** and TA0201, **8** represent orally active thiophene sulfonamide and benzenesulfonamide derivatives that competitively antagonize ^{125}ET-1 binding to the ET_A receptor (IC_{50}= 1400 and 6 pM, respectively). These compounds show high bioavailability (~90%) in various species. Although the plasma $t_{1/2}$ value for sitaxsentan is greater (~6-7 hr) than that of TA0201 (~0.9 hr), the pharmacodynamic activity of both compounds is long-lasting (61,62). Separate efficacy studies of sitaxsentan and TA0201 demonstrated prevention and/or reversal of pulmonary hypertension in rodent models (63,64). Early clinical trials with sitaxsentan have shown acutely improved cardiopulmonary hemodynamics in patients with pulmonary hypertension secondary to congestive heart failure (65). The trial was stopped early because statistical significance was achieved with a smaller number of patients than anticipated. Two additional clinical trials employing patients with PPH demonstrated sustained effects and sitaxsentan was well tolerated at several doses (66, 67).

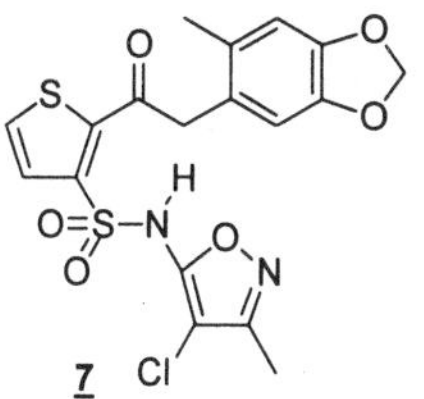

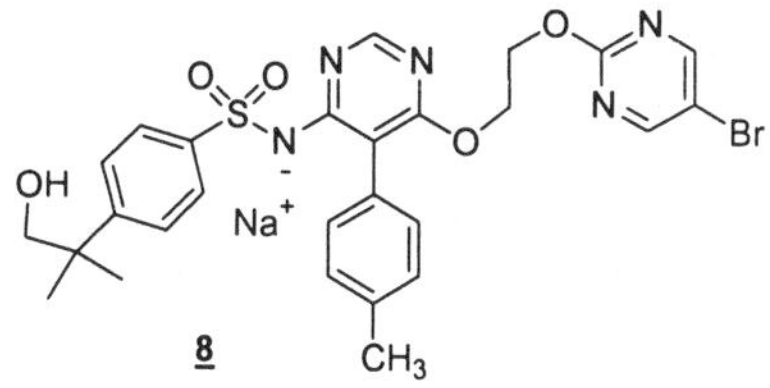

Additional orally active ET_A receptor selective antagonists include the propionic acid derivatives darusentan **9** and SB247083 **10**. Although structurally divergent from the sulfonamides, these compounds demonstrated nanomolar binding affinity for the cloned human ET_A receptor and good bioavailability (90% and 60%, respectively). Darusentan and SB247083, studied separately, showed efficacy in animal models of pulmonary hypertension (68,69).

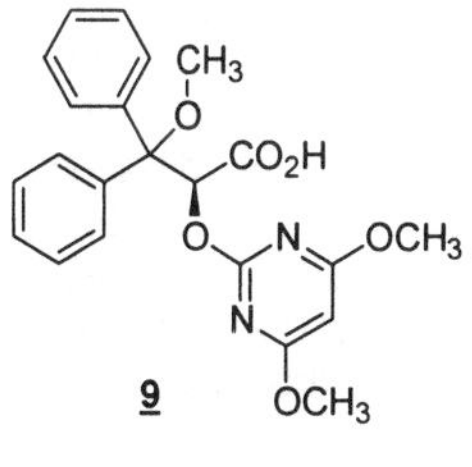

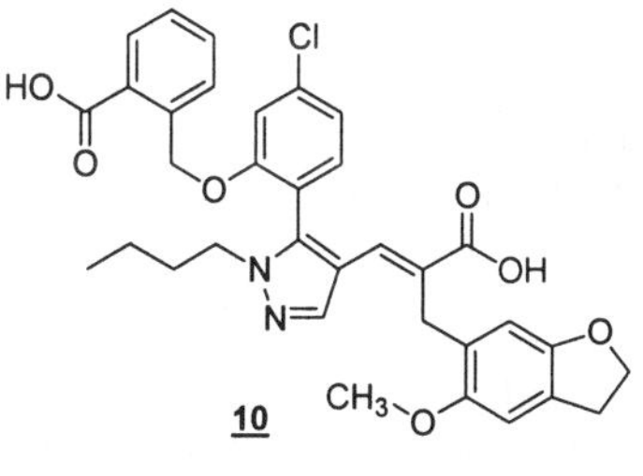

PD156707 **11** represents one of the first orally active, high affinity (IC_{50} = 0.3 nM) butenolide ET_A receptor selective antagonists (70). This compound shows good oral efficacy in animal models of pulmonary hypertension (71). However, food decreases absolute bioavailability from 67% to 20%, reduces Cmax 68% and increases $t_{1/2}$ 3.5 times. The benzothiazines PD180988 **12** and PD164800 **13** represent novel structures that also show high affinity for the ET_A receptor with the former compound causing dose-dependent inhibition of hypoxia-induced pulmonary vasoconstriction in lamb (72). It is uncertain whether these compounds show improved pharmacokinetic properties compared with PD156707.

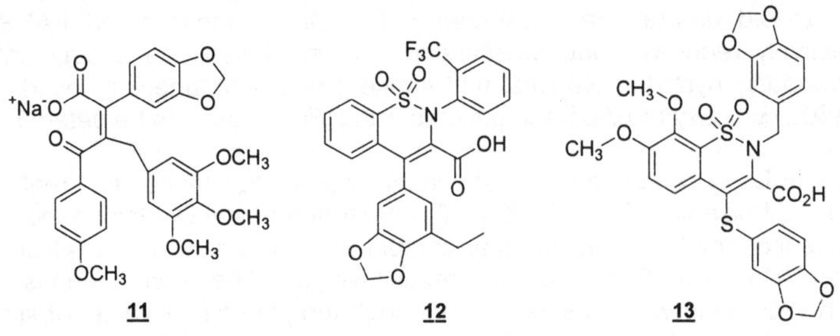

11   **12**   **13**

Thromboxane A$_2$ Antagonist/Synthase Inhibitor - Thromboxane A$_2$ (TxA$_2$), a metabolite of arachidonic acid, stimulates vasoconstriction and platelet aggregation. Increased production of TxA$_2$ has been noted in patients with PPH (73). Consequently, blockade of either the TxA$_2$ /endoperoxide receptor or TxA$_2$ synthesis may decrease vasoconstriction and thromboembolism associated with disease. Despite the effectiveness of TxA$_2$ antagonists or synthase inhibitors in diverse animal models of thrombosis and pulmonary hypertension (74), tests of clinical efficacy have proven disappointing. The lack of synthase inhibitor efficacy in man was thought to be due partly to incomplete enzyme inhibition and/or to the accumulation of the endoperoxide, PGH$_2$, a thromboxane receptor agonist (75). More recently, the combination of a TxA$_2$/endoperoxide receptor antagonist with a TxA$_2$ synthase inhibitor has shown improvement over selective compounds (76).

Terbogrel **14**, a guanidine-substituted hexenoic acid derivative, is a potent blocker of the TxA$_2$/endoperoxide receptor (IC$_{50}$ = 11 nM) on washed human platelets and inhibits TxA$_2$ synthase (IC$_{50}$ = 11 nM) activity (77). Oral Terbogrel is readily absorbed with a bioavailability of 30% and elimination t$_{1/2}$ of 7.5-10 hr. Results from efficacy studies of terbogrel in patients with PPH have yet to be disclosed.

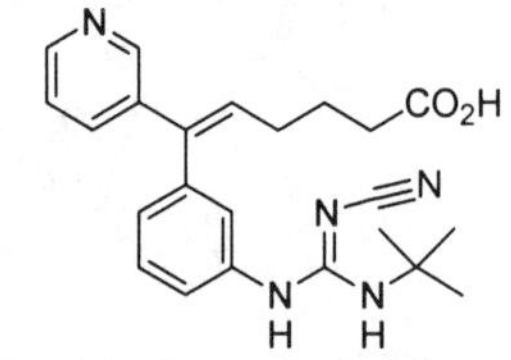

14

Phosphodiesterase 5 Inhibitors – Mammalian phosphodiesterases (PDE's) represent a superfamily of at least 11 enzymes (78). Although PDE's differ in amino acid sequence, substrate specificities, modes of regulation and tissue distribution, they share the common feature of inactivating cGMP and/or adenosine 3', 5'-monophosphate (cAMP). The PDE5 variant inactivates cGMP specifically and is inhibited by experimental and clinically available agents. In addition to its high concentration in the copora cavernosa, PDE5 is abundant in platelets as well as vascular, tracheal and visceral smooth muscle (79). In animal studies, PDE5 inhibitors cause selective pulmonary vasodilatation and potentiate the duration and magnitude of NO-mediated pulmonary vasodilatation (80-82). Early clinical studies showed efficacy of the nonselective PDE5 inhibitor, dipyridamole, to lower pulmonary artery pressure in patients with pulmonary hypertension secondary to COPD (83). Results from independent studies show that patients that did not respond to inhaled NO responded to the combination of NO plus dipyridamole (84). In addition, the combination of NO plus dipyridamole caused greater pulmonary vasodilatation than either agent alone (85).

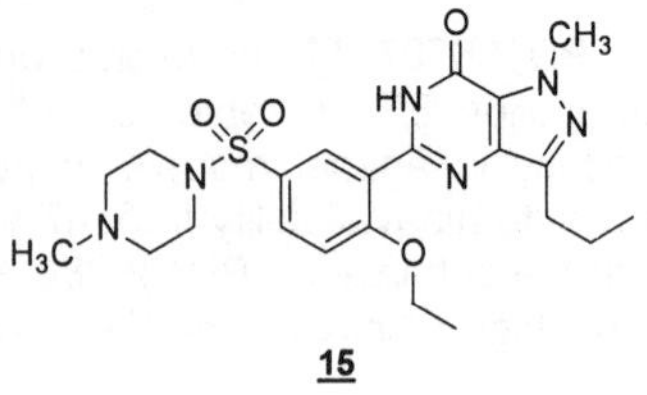

Sildenafil **15**, a selective PDE5 inhibitor containing a pyrazolopyrimidine nucleus, is marketed for erectile dysfunction. Efficacy of this drug has been shown in

15

drug has been shown in animal models of acute pulmonary hypertension (86,87). Clinical case study reports indicate that sildenafil (2 mg/kg, q.i.d.), administered at doses well below the 50 mg amount used to treat erectile dysfunction, substituted for intravenous prostacyclin in a child with pulmonary hypertension (88). Low dose sildenafil also blunted rebound pulmonary hypertension after withdrawal of inhaled NO from infants and an adolescent (89,90). High dose sildenafil (100 mg, 5 times per day) modestly improved cardiopulmonary hemodynamics in a 21 year-old male with PPH (91). Sildenafil also enhanced NO-induced pulmonary vasodilatation in a patient with severe interstitial fibrosis (92). Although it is tempting to initiate empirical therapy with sildenafil in patients with pulmonary hypertension, caution must be exercised because only limited and anecdotal clinical reports describing the use of sildenafil have appeared. Appropriately designed clinical trials are required to determine the potential of sildenafil as therapy for pulmonary hypertension. Sildenafil is contraindicated for patients with sepsis and in patients receiving nitrovasodilators (e.g., nitroglycerin or sodium nitroprusside) because of the potential for severe hypotension and pulmonary shunt.

Two structurally diverse PDE5 inhibitors are being developed for the treatment of pulmonary hypertension. E4010 **16**, a substituted phthalazine, is orally active and potently inhibits (Ki = 92 pM) PDE5 in human platelets. In conscious hypoxic pulmonary hypertensive rats, a single oral administration of E4010 (1.0 mg/kg) decreased mean pulmonary arterial pressure without significantly changing other cardiopulmonary parameters (93). T1032 **17**, an isoquinolone, also potently inhibits (Ki = 1 nM) PDE5 (94). The efficacy of E4010 and T1032 in human disease remains to be determined.

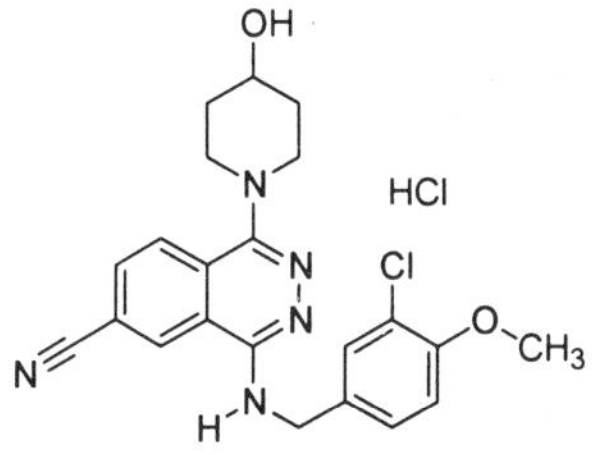

16 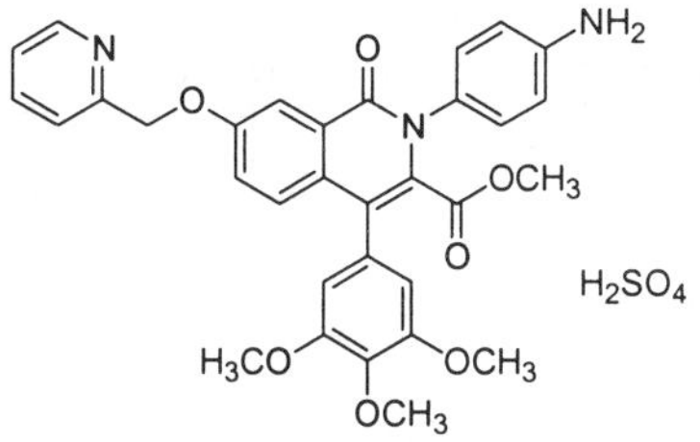   **17**

Conclusion - Recent advances in our understanding of the cellular and molecular mechanisms underpinning the development of pulmonary hypertension is leading to more effective therapies with better drug-like properties. The identification of newer molecular targets may allow us to reverse what was previously considered to be an 'irreversible' disease.

References

1. I.P. Williams, M.J. Boyd, A.M. Humberstone, A.G. Wilson and F.J. Millard, Br. J. Dis. Chest, 78, 211 (1984).
2. E. Weitzenblum, A. Sautegeau, M. Ehrhart, M. Mammosser, C. Hirth and E. Roegel, Am. Rev. Respir. Dis., 130, 993 (1984).
3. S. Rich, World Symposium on Primary Pulmonary Hypertension. Evian, France, 1998. *www.who.int/ncd/cvd/pph.html*
4. H.I. Palevsky and A.P. Fishman, J. Am. Med. Assoc., 263, 2347 (1990).
5. J.C. Wanstall and T. Jeffery, Drugs, 56, 989 (1998).
6. T.D. Nauser and S.W. Stites, Am. Fam. Physician, 63, 1789 (2001).
7. R.A. Bialecki. Curr. Opin. Cardiovasc. Pulmon. Renal Invest. Drugs, 1, 458 (1999)

8. S. Rich, D.R. Dantzker, S.M. Ayers, E.H. Bergofsky, B.H. Brundage, K.M. Detre, A.P. Fishman, R.M. Goldring, B.M. Groves, S.K. Koerner, P.C. Levy, L.M. Reid, C.E. Vreim and G.W. Williams, Ann. Intern. Med., 107, 216 (1987).
9. J.E. Loyd, M.G. Butler, T.M. Foroud, P.M. Conneally, J.A. Phillips 3rd and J.H. Newman, Am. J. Respir. Crit. Care Med., 152, 93 (1995).
10. W.C. Nichols, D.L. Koller, B. Slovis, T. Foroud, V.H. Terry, N.D. Arnold, D.R. Siemieniak, L. Wheeler, J.A. Phillips 3rd, J.H. Newman, P.M. Conneally, D. Ginsburg and J.E. Loyd, Nat. Genet., 15, 277 (1997).
11. R.J. Barst, L.J. Rubin, W.A. Long, M.D. McGoon, S.Rich, D.B. Badesch, B.M. Groves, V.F. Tapson, R.C. Bourge, B.H. Brundage, S.K. Koerner, D. Langleben, C.A. Keller, S. Murali, B.F. Uretsky, L.M. Clayton, M.M. Jöbsis, S.D. Blackburn, D. Shortino and J.W. Crow, N. Engl. J. Med., 334, 296 (1996).
12. G.E. D'Alonzo, R.J. Barst, S.M. Ayers, E.H. Bergofsky, B.H. Brundage, K.M. Detre, A.P. Fishman, R.M. Goldring, B.M. Groves, J.T. Kernis, P.S Levy, G.G. Pietra, L.M. Reid, J.T. Reeves, S. Rich, C.E. Vreim, G.W. Williams and M. Wu, Ann. Intern. Med., 115, 343 (1991).
13. O. Sitbon, M. Humbert, J.L. Jagot, O. Taravella, M. Fartoukh, F. Parent, P. Herve, and G. Simonneau, Eur. Respir. J. 12, 265 (1998).
14. L.M. Reid, Am. Rev. Respir. Dis., 119, 531 (1979)
15. L.M. Reid, Chest, 89, 279 (1986).
16. J.X. Yuan, A.M. Aldinger, M. Juhaszova, J. Wang, J.V. Conte Jr., S.P. Gaine, J.B. Orens and L.J. Rubin, Circulation, 98, 1400 (1998).
17. E.D. Michelakis and E.K. Weir, Clin. Chest. Med., 22, 419 (2001)
18. J. Wang, M. Juhaszova, J.V.J. Conte, S.P. Gaine, L.J. Rubin and J.X. Yuan, Lancet, 352, 290 (1998).
19. S. Rich S, M. Seidlitz, E. Dodin, D. Osimani, D. Judd, D. Genthner, V. McLaughlin and G. Francis, Chest, 114, 787 (1998).
20. V. Fuster, P. Steel, W.D. Edwards, B.J. Gersh, M.D. McGoon and R.L. Frye, Circulation, 70, 580 (1984).
21. S. Rich, E. Kaufmann and P.S. Levy, N. Engl. J. Med., 327, 76 (1992).
22. E.K. Weir, L.J. Rubin, S.M. Ayers, E.H. Bergofsky, B.H. Brundage, K.M. Detre, C.G. Elliott, A.P. Fishman, R.M Goldring, B.M. Groves, J.T. Kernis, S.K. Koerner, P.S. Levy, G.G. Pietra, L.M. Reid, S. Rich, C.E. Vreim, G.W. Williams and M. Wu, Am. Rev. Respir. Dis., 140, 1623 (1989).
23. B.W. Christman, C.D. McPherson, J.H. Newman, G.A. King, G.R. Bernard, B.M. Groves and J.E. Loyd, N. Engl. J. Med., 327, 70 (1992)
24. R.M. Tuder, C.D. Cool, M.W. Geraci, J. Wang, S.H. Abman, L. Wright, D. Badesch and N.F. Voelkel, Am. J. Respir. Crit. Care Med., 159, 1925 (1999).
25. D.B. Badesch, V.F. Tapson, M.D. McGoon, B.H. Brundage, L.J. Rubin, F.M.Wigley, S. Rich, R.J. Barst, P.S. Barrett,K.M. Kral, M.S. Jöbsis, J.E. Loyd, S. Murali, A. Frost, R. Girgis, R.C. Bourge, D.D. Ralph, C.G. Elliott, N.S. Hill, D. Langleben, R.J. Schilz, V.V. McLaughlin, I.M. Robbins, B.M. Groves, S. Shapiro, T.A. Medsger, S.P. Gaine, E. Horn, J.C. Decker and K. Knobil, Ann. Intern. Med., 132, 425 (2000).
26. T.W. Higenbottam, D. Spiegelhalter, J.P. Scott, V. Fuster, A.T. Dinh-Xuan, N. Caine and J. Wallwork, Br. Heart J., 70, 366 (1993).
27. M.K. Pasque, E.P. Trulock, J.D. Cooper, A.N. Triantafillou, C.B. Huddleston, M. Rosenbloom, S. Sundaresan, J.L. Cox and G.A. Patterson, Circulation, 92, 2252 (1995).
28. J. Sandoval, J. Gaspar, T. Pulido, E. Bautista, M.L. Martinez-Guerra, M. Zeballos, A. Palomar and A. Gomez, J. Am. Coll. Cardiol., 32, 297 (1998).
29. G. Simonneau, R.J. Barst, N. Galie, R. Naeije, S. Rich, R.C. Bourge, A. Keogh, R. Oudiz, A. Frost, S.D. Blackburn, J.W. Crow and L.J. Rubin, Am. J. Respir. Crit. Care Med., 165, 800 (2000).
30. M.M.Hoeper, M. Schwarze, S. Ehlerding, A. Adler-Schuermeyer, E. Spiekerkoetter, J. Niedermeyer, M. Hamm and H. Fabel, N. Engl. J. Med., 342, 1866 (2000).
31. N. Nagaya, M. Uematsu, Y. Okano, T. Satoh, S. Kyotani, F. Sakamaki, N. Nakanishi, K. Miyatake and T. Kunieda, J. Am. Coll. Cardiol., 34, 1188 (1999).
32. M.G. Perrson, L.E. Gustafsson, N.P. Wiklund, S. Moncada, and P. Hedqvist, Acta Physiol. Scand., 140, 449 (1990).
33. S. Singh and T.W. Evens, Euro. Respir. J., 10, 699 (1997).
34. A. Giaid and D. Saleh, N. Engl. J. Med., 333, 214 (1995).
35. J. Pepke-Zaba, T.W. Higenbottam, A.T. Dinh-Xuan, D. Stone and J. Wallwork, Lancet, 338, 1173 (1991).

36. Y. Katayama, T.W. Higenbottam, G. Cremona, S. Akamine, E.A. Demoncheaux, A.P. Smith and T.E Siddons, Circulation, 98, 2429 (1998).
37. K. Schönafinger, Farmaco, 54, 316 (1999)
38. M. Grenegard, M.C. Gustafsson, R.G. Andersson and T. Bengtsson, Br. J. Pharmacol., 118, 2140 (1996).
39. P.A. Whiss and R. Larsson, Hemostasis, 28, 260 (1998).
40. E.R. Martin, B. Brenner and B.J. Ballerman, J. Biol. Chem., 265, 14044 (1990).
41. Y. Masuda, H. Miyazaki, M. Kondoh, H. Watanabe, M. Yanagisawa, T. Masaki and K. Murakami, FEBS Lett., 257, 208 (1989).
42. R.A. Bialecki, C.S. Fisher, W.W. Murdoch, H.G. Barthlow and D.L. Bertelsen, Am. J. Physiol. 272, 211 (1997)
43. D.W.P. Hay, M.A. Luttmann, W.C. Hubbard and B.J. Undem, Br. J. Pharmacol., 110, 1175 (1993).
44. A. Giaid, M. Yanagisawa, D. Langleben, R.P. Michel, R. Levy, H. Shennib, S. Kimura, T. Masaki, W.P. Duguid and D.J. Stewart, N. Engl. J. Med., 328, 1732 (1993).
45. D. Stewart, R.D. Levy, P. Cernacek and D. Langleben, Ann. Intern. Med., 114, 464 (1991).
46. H. Chang, G.-J. Wu, S.-M. Wang and C.-R. Hung, Ann. Thorac. Surg., 55, 450 (1993).
47. R.J. Cody, G.J. Haas, P.F. Binkley, Q. Capers and R. Kelley, Circulation, 85, 504 (1992).
48. N. Galie, F. Grigioni, L. Bacchi-Reggiani, G.P. Ussia, R. Parlangeli, P. Catanzariti, S. Boschi, A. Branzi and B. Magnani, Eur. J. Clin. Invest., 26, 273 (1996).
49. A. Giaid, R.P. Michel, D.J. Stewart, M. Sheppard, B. Corrin and Q. Hamid, Lancet, 341, 1550 (1993).
50. T.J. Stelzner, R.F. O'Brien, M. Yanagisawa, T. Sakurai, K. Sato, S. Webb, M. Zamora, I.F. McMurtry and J.H. Fisher. Am. J. Physiol., 262, L614 (1992).
51. B.M. Wilkes, C.M. Macica and P.F. Mento, Am.J. Physiol., 267, E242 (1994).
52. J. Dupuis, P. Cernacek, J.-C. Tardif, D.J. Stewart, G. Gosselin, I. Dyrda, R. Bonan and J. Crepeau, Am. Heart J., 135, 614 (1998).
53. S.T. Bonvallet, M.R. Zamora, K. Hasunuma, K. Sato, N. Hanasato, D. Anderson and T.J. Stelzner, Am.J. Physiol., 266, H1327 (1994).
54. R.A. Bialecki, C.S. Fisher, B.M. Abbott, H.G. Barthlow, R.G. Caccese, R.B. Stow, J. Rumsey and W. Rumsey, Pulm. Pharmacol. Ther., 12, 303 (1999).
55. S.-J. Chen, Y.-F. Chen, T.J. Opgenorth, J.L. Wessale, Q.C. Meng, J. Durand, V.S. DiCarlo and S. Oparil, J. Cardiovasc. Pharmacol., 29, 713 (1997).
56. S. Eddahibi, B. Raffestin, M. Clozel, M. Levame and S. Adnot, Am. J. Physiol., 268, H828 (1995).
57. S.-J. Chen, Y.-F. Chen, J. Durand, V.S. DiCarlo and S. Oparil, J. Appl. Physiol., 79, 2122 (1995).
58. R.N. Channick, G. Simonneau, O. Sitbon, I.M. Robbins, A. Frost, V.F. Tapson, D.B. Badesch, S. Roux, M. Rainisio, F. Bodin and L.J. Rubin, Lancet, 358, 1113 (2001).
59. V.V. McLaughlin, Expert Opin. Pharmacother., 3, 159 (2002).
60. T. Iwasaki, S. Mihara, T. Shimamura, M. Kawakami, Y. Hayasaki-Kajiwara, N. Naya, M. Fujimoto and M. Nakajima, Cardiovasc. Pharmacol. 34, 139 (1999).
61. J.R. Wu-Wong, Current Opin. Invest. Drugs, 2, 531 (2001).
62. H. Morimoto, N. Ohashi, H. Shimadzu, E. Kushiyama, H. Kawanishi, T. Hosaka, Y. Kawase, K. Yasuda, K. Kikkawa, R. Yamanuchi-Kohno and K. Yamada, J. Med. Chem., 44, 3369 (2001).
63. S.J. Chen, T. Brock, F. Stavros, I. Okun, C Wu, F. Chan, S. Mong, R.A.F. Dixon, S. Oparil and Y.-F. Chen, FASEB J., 10, A104 (1996).
64. M. Ueno, T. Miyauchi, S. Sakai, K. Goto and I. Yamaguchi, J. Cardiovasc. Pharmacol., 36, S305 (2000).
65. M.M. Givertz, W.S. Colucci, T.H. LeJemtel, S.S. Gottlieb, J.M. Hare, M. T. Slawsky, C. V. Leier, E. Loh, J.M. Nicklas and B.E. Lewis, Circulation, 101, 2922 (2000).
66. D.A. Calhoun, K. Renfroe and A.B Alper, Circulation, 102, 18, Abstr. 2029 (2000).
67. R.J. Barst, S. Rich, E.M. Horn, V.V. McLaughlin, D. Kerstein, A.C. Wilditz, J.D. McFarlin and R.A.F. Dixon, Circulation, 102, 18 Abstr. 2076 (2000).
68. S. Prié, T.K. Leung, P. Cernacek, J.W. Ryan and J. Dupuis, J. Pharmacol. Exp. Ther. 282, 1312 (1997).
69. D.C. Underwood, S. Bochnowicz, R.R. Osborn, M.A. Luttmann, C.S. Louden, T.K. Hart, J.D. Elliott and D.W. Hay, Pulm. Pharmacol. Ther., 12, 13 (1999).

70. W.C. Patt, J.J. Edmunds, J.T. Repine, K.A. Berryman, B.R. Reisdorph, C. Lee, M.S. Plummer, A. Shahripour, S.J. Haleen, J.A. Keiser, M.A. Flynn, K.M. Welch, E.E. Reynolds, R. Rubin, B. Tobias, H. Hallak and A.M. Doherty, J. Med. Chem., 40,1063 (1997).
71. S. Haleen, R. Schroeder, D. Walker, E. Quenby-Brown, K. Welch, H. Hallak, A. Uprichard and J. Keiser, J. Cardiovasc. Pharmacol.,31 Suppl 1, S331 (1998).
72. Y. Coe, S.J. Haleen, K.M. Welch and F. Coceani, J. Cardiovasc. Pharmacol., 5 Suppl 1, S331 (2000).
73. B.W. Christman, C.D. McPherson, J.H. Newman, G.A. King, G.R. Bernard, B.M.Groves and J.E. Loyd, N. Engl. J. Med., 327, 70 (1992).
74. T. Nagata, Y. Uehara, K. Hara, K. Igarashi, H. Hazama, T. Hisada, K. Kimura, A. Goto and M. Omata, Respirology, 2, 283 (1997).
75. G.A. Fitzgerald, I.A.G. Reilly and A.K. Pederson, Circulation, 72, 1194 (1985).
76. I.G. Fiddler and P. Lumley, Circulation, 81, 169 (1990).
77. R. Soyka, B.D. Guth, H.M. Weisenberger, P. Luger and T.H. Muller, J. Med. Chem., 42, 1235 (1999).
78. D.M. Essayan, J. Allergy Clin. Immunol., 108, 671 (2001).
79. M.D. Cheitlin, A.M. Hutter Jr., R.G. Brindiss, P. Ganz, S. Kaul, R.O. Russell Jr. and R.M. Zusman, Circulation, 99, 168 (1999).
80. D.A. Braner, J.R. Fineman, R. Chang and S.J. Soifer, Am. J. Physiol., 264, H252 (1993).
81. T.J. McMahon, L.J. Ignarro and P.J. Kadowitz, J. Appl. Physiol. 74, 1704 (1993).
82. K.G. Thusu, F.C Morin 3rd, J.A. Russell and R.H. Steinhorn, Am.J. Respir. Crit. Care Med. 152, 1605 (1995).
83. G. Nenci, M. Berrettini, T. Todisco, V. Costantini and P. Parise, Respiration, 53, 13 (1988).
84. D.A Fullerton, J. Jaggers, F. Piedalue, F.L. Grover and R.C. McIntyre Jr., J. Thorac. Cardiovasc. Surg., 113, 363 (1997).
85. J.W. Ziegler, D.D. Ivy, J.W. Wiggens, J.P. Kinsella, W.R. Clarke and S.H. Abman, Am.J. Respir. Crit. Care Med., 158, 1388 (1998).
86. J. Weimann, R. Ullrich, J. Hromi, Y. Fujino, M.W. Clark, K.D. Bloch and W.M. Zapol, Anesthesiology, 92, 1702 (2000).
87. F. Ichinose, J. Erana-Garcia, J. Hromi, Y. Raveh, R. Jones, L. Krim, M.W. Clark, J.D. Winkler, K.D. Bloch and W.M. Zapol, Crit. Care Med., 29, 1000 (2001).
88. D. Abrams, I. Schulze-Neick and A.G. Magee, Heart, 84, E4 (2000).
89. A.M. Atz and D.L. Wessel, Anesthesiology, 91, 307 (1999).
90. G. Mychaskiw, V. Sachdev and B.J. Heath, J. Clin. Anesth., 13, 218 (2001).
91. S. Prasad, J. Wilkinson and M.A. Gatzoulis, N. Engl. J. Med., 343, 1342 (2000).
92. L.M. Bigatello, D. Hess, K.C. Dennehy, B.D. Medoff and W.E. Hurford, Anesthesiology, 92, 1827 (2000).
93. N. Hanasato, M. Oka, M. Muramatsu, M. Nishino, H. Adachi and Y. Fukuchi, Am. J. Physiol., 277, L225 (1999).
94. T. Ukita, Y. Nakamura, A. Kubo, Y. Yamamoto, Y. Moritani, K. Saruta, T. Higashijima, J. Kotera, M. Takagi, K. Kikkawa and K. Omori, J. Med. Chem., 44, 2204 (2001).

Chapter 6. Phosphodiesterase 5 Inhibitors

Andrew W. Stamford
Schering-Plough Research Institute
2015 Galloping Hill Rd, Kenilworth, NJ, 07033

Introduction - The cyclic nucleotides cAMP and cGMP are principal components of cell signal transduction pathways. Intracellular degradation of these key second messengers is carried out by phosphodiesterases (PDEs). The PDE superfamily currently comprises 11 PDE subfamilies (1-3). In general, PDEs are dimeric proteins with each monomer consisting of a N-terminus involved in allosteric regulation by cyclic nucleotides or other signaling molecules, a catalytic domain that is highly conserved across families, and a regulatory C-terminus. Cyclic nucleotide specificity and catalytic activity of PDEs varies across families. PDE5, PDE6, and PDE9 hydrolyze cGMP specifically; PDE4, PDE7, and PDE8 are cAMP specific, and PDE1, PDE2, PDE3, PDE10, and PDE11 show dual specificity. PDE subfamilies are differentially distributed among tissues and are discretely localized at the subcellular level. Moreover, PDE activity can be regulated by cyclic nucleotides and by other cell signaling molecules. Specific localization of PDEs and regulation of PDE activity is likely to promote tight control over cyclic nucleotide concentrations, and to facilitate rapid changes in cyclic nucleotide levels in response to cell stimuli. Given that precise regulation of cyclic nucleotide levels is critical for proper cell function, PDE isozyme selectivity is an important consideration in the development of PDE inhibitors as therapeutic agents.

PDE5 is widely distributed in smooth muscle, lung and platelets. Inhibitors of PDE5 were originally investigated as antihypertensive and antianginal agents, and their antiplatelet activity is well known (4,5). The advent of the selective PDE5 inhibitor sildenafil (Viagra^{TM}) for the treatment of erectile dysfunction (ED), has stimulated tremendous interest in the development of new PDE5 inhibitors, and has provided opportunities to explore new therapeutic applications for PDE5 inhibitors (6). This review will discuss PDE5 inhibitors that are under clinical development for the treatment of ED, as well as advances in the medicinal chemistry of selective PDE5 inhibitors that have been reported over the past year. Reviews have been published describing developments in the medicinal chemistry of PDE5 prior to this time frame and this review will attempt to complement the most recent review of the area (4,7,8). Potential therapeutic indications for PDE5 inhibitors other than for ED will also be reviewed.

PDE5 Molecular Biology and Biochemistry - PDE5 is encoded as a single gene that is expressed as two alternative splice variants, PDE5A1 and PDE5A2 (9-11). Human PDE5A1 is highly expressed in pancreas, and moderately expressed in lung and small intestine. The human PDE5A2 variant is highly expressed in heart, kidney, lung, pancreas, placenta, colon, ovary, prostate, small intestine and testis, and moderately expressed in brain, skeletal muscle, spleen and thymus (11). Recently three PDE5 isoforms were identified in human corpus cavernosum, however the functional significance of the distinct isoforms is not known (12). Recent developments in the biochemistry of PDE5 have been reviewed (13). PDE5 exists as a homodimer. Each subunit contains a catalytic domain highly conserved across all PDEs and two allosteric cGMP binding sites. The catalytic domain comprises a cGMP binding site and a Zn^{2+} binding motif that is likely to be involved in catalysis (13,14).

In smooth muscle cells, elevation of cGMP levels results from activation of guanylate cyclase by NO or atrial natriuretic peptide. cGMP activates protein kinase G (PKG) which in turn phosphorylates other protein targets, ultimately eliciting smooth muscle relaxation (13). Recent studies have shed light on regulation of PDE5 activity in intact smooth muscle cells. When cGMP levels are elevated, cGMP binds to the allosteric sites of PDE5 permitting phosphorylation of ser92 by PKG. This stimulates PDE5 activity, as well cGMP binding to PDE5 allosteric sites, and may provide a negative feedback mechanism for cGMP and PKG when cGMP levels become excessive (15,16). In penile smooth muscle cells, binding of cGMP to PDE5 allosteric sites might also be a physiologically relevant mechanism for decreasing intracellular cGMP levels, and may provide a dynamic reservoir for cGMP (15,17). Phosphorylation of PDE5 does not affect inhibition by competitive PDE5 inhibitors (15). Dephosphorylation of PDE5 by myosin light chain phosphatase has been demonstrated in smooth muscle cells. Therefore phosphorylation / activation of PDE5 by PKG and subsequent dephosphorylation / deactivation of PDE5 by phosphatases may represent key regulatory elements of smooth muscle contraction / relaxation cycles (16).

PDE5 as a Therapeutic Target for Erectile Dysfunction - ED has been defined as the inability to achieve or maintain an erection of the penis sufficient to permit satisfactory sexual intercourse (18). The physiology of penile erection and the risk factors associated with ED have been reviewed (19). The neuronal NO/cGMP pathway is the principal signaling pathway mediating penile smooth muscle relaxation and hence erection. PDE5 is the predominant cGMP PDE in the penis. Hydrolysis of cGMP to GMP by PDE5 terminates cGMP signaling and promotes restoration of penile flaccidity. Inhibition of PDE5 in the corpus cavernosum smooth muscle cells results in the elevation of cGMP, thereby potentiating the actions of the NO/cGMP pathway.

Sildenafil **1**, a potent, selective PDE5 inhibitor, is an effective treatment for ED of

organic (eg diabetes, cardiovascular disease, spinal cord injury) and mixed organic – psychogenic origin, and its pharmacological and clinical profile has been reviewed (20,21). Side effects of sildenafil include visual disturbance which has been attributed to inhibition of PDE6, the photoreceptor PDE. Since PDE1 is expressed in the vasculature, inhibition of PDE1 by sildenafil may contribute to the cardiovascular side effects of flushing and potentiation of the hypotensive effect of NO donor medications. Consequently, a major focus of the development of second-generation PDE5 inhibitors is improvement of the PDE isozyme selectivity profile with respect to PDE6 and PDE1.

PDE5 Inhibitors Under Clinical Development for ED - Tadalafil (IC-351, **2**) is

currently in advanced Phase 3 clinical trials for the treatment of ED (22). Tadalafil is a potent inhibitor of human corpus cavernosum PDE5 (IC_{50} = 2 nM) with a selectivity ratio of 780 for PDE5 over PDE6 and of >10,000 for PDE5 over PDEs 1-4 and PDEs 7-10, and is also an inhibitor of human corpus cavernosum PDE11 (IC_{50} = 37 nM) (22-24). In randomized, double-blind placebo-controlled Phase 2 clinical trials of men suffering mild to moderate ED, tadalafil at an oral dose of 10 mg significantly improved the ability to achieve successful sexual intercourse. In a Phase 3 clinical trial of diabetic men with ED, tadalafil administered at a dose of 20 mg resulted in significantly improved erections relative to placebo control (22). Significant differences in the clinical profile of tadalafil compared to sildenafil are the lack of reports of visual disturbance, consistent with improved selectivity of tadalafil for PDE5 over PDE6, and its considerably longer half-life (17.5 hr).

Vardenafil (BAY 38-9456, **3**), a subnanomolar-potent PDE5 inhibitor (IC_{50} = 0.7 nM) that bears structural resemblance to sildenafil (25), is also is in advanced Phase 3 clinical trials for ED. When tested in the same assays, vardenafil was a significantly more potent PDE5 inhibitor than sildenafil, with a higher selectivity ratio for PDE5 over PDEs 1 and 6 (26). Vardenafil was also more potent than sildenafil in

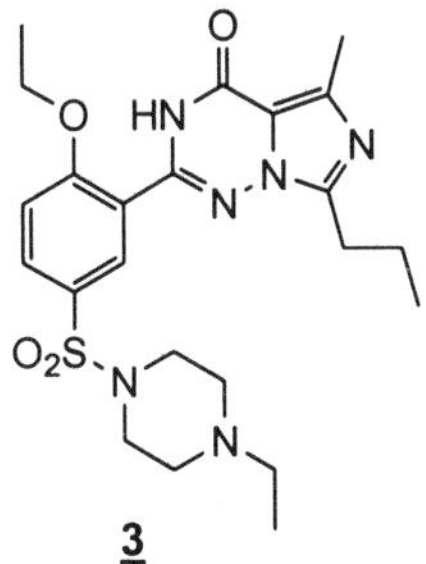

PDE	IC_{50} nM **3**	IC_{50} nM **sildenafil**
1C	121	350
2A	>10,000	>10,000
3B	2,680	>10,000
4B	1,910	2,900
5A	0.81	12
6	11	49
7B	4,600	>10,000
8	>10,000	>10,000
9A	3,370	>10,000
10	1,000	3,800

potentiation of sodium nitroprusside (SNP) - induced relaxation of isolated pre-contracted human trabecular smooth muscle strips. In a conscious rabbit model, vardenafil (1 mg/kg po) produced transient erections, which were potentiated by coadministration of a NO donor, and was more potent than sildenafil (26,27). In man, vardenafil has a rapid onset of action (T_{max} = 0.7 hr) and a half-life of 3 – 5 hr consistent with on demand dosing (27). In a placebo-controlled, randomized, double-blind phase 2 clinical trial of men with ED of mixed etiology, vardenafil (5, 10, and 20 mg) taken at least 1 hr before sexual intercourse produced a significant improvement in erectile function at all dose levels (28). The most frequently reported adverse effects were headache and flushing, occurring in 12% and 16% respectively of subjects taking the 10 mg dose. In a phase 3 randomized, double-blind multi-center study of diabetic men with mild-to-severe ED, vardenafil at doses of 10 and 20 mg also improved erectile function relative to placebo control (29).

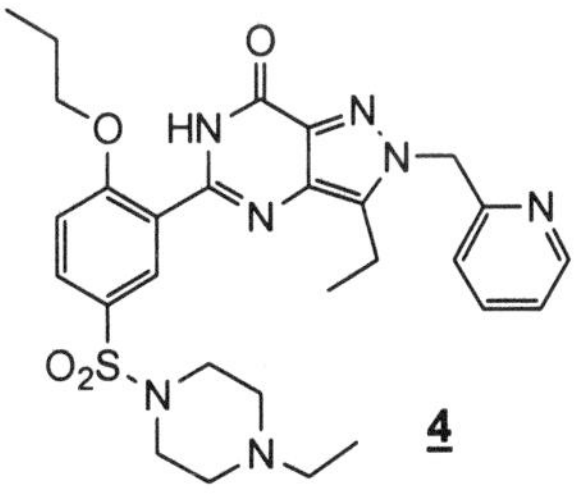

UK-343,664 (**4**) like sildenafil is a pyrazolopyrimidinone PDE5 inhibitor, and was advanced into clinical trials (30). Compound **4** inhibited PDE5 with an IC_{50} = 1.1 nM, and selectivity over other PDEs was reported to be > 100-fold (24). Thus **4** represents a significant improvement in selectivity for PDE5 with respect to PDE6 compared to sildenafil. Single dose pharmacokinetics of **4** in man has been reported (30). Plasma levels and exposure of **4** increased supra-proportionally with oral dose, and T_{max} was unusually dose-dependent. Compound **4** displayed high affinity for human P-glycoprotein (PGP) and the non-proportional pharmacokinetics of **4** was attributed in part to PGP activity *in vivo*.

Recent Medicinal Chemistry Developments - A recent patent application describes pyrazolopyrimidinone derivatives related to sildenafil in which the phenyl substituent was replaced by a 3-pyridyl substituent. A representative compound is **5** which inhibited human PDE5 with an IC_{50} of 1.7 nM and was 224-fold selective for PDE5 over bovine PDE6 (31). Replacement of the methoxyethyl residue of **5** by other polar or basic substituents afforded PDE5 potency and selectivity over PDE6 similar to that of **5**. An additional disclosure describes analogues of **5** wherein replacement of the piperazinesulfonyl residue by a variety of groups was tolerated, exemplified by **6** (PDE5 IC_{50} = 0.9 nM), however PDE isozyme selectivity data was not reported (32). Related patent applications describe analogs in which variations of the pyrazolopyrimidinone nucleus are disclosed. For example, purinone **7** inhibited PDE5 by 80% at a concentration of 10 nM (33), imidazotriazinone **8** exhibited a PDE5 IC_{50} < 100 nM (34), and pyrrolopyrimidinone **9** inhibited PDE5 with IC_{50} = 0.59 nM (35). Additional disclosures claim phenyl-substituted imidazotriazinones, triazolotriazinones and isoxazolopyrimidinones represented by **10**, **11** and **12**, respectively, however no specific PDE5 inhibition data was reported (36-38).

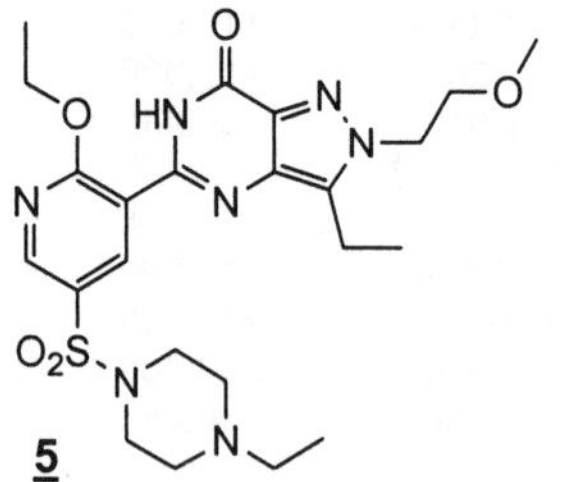

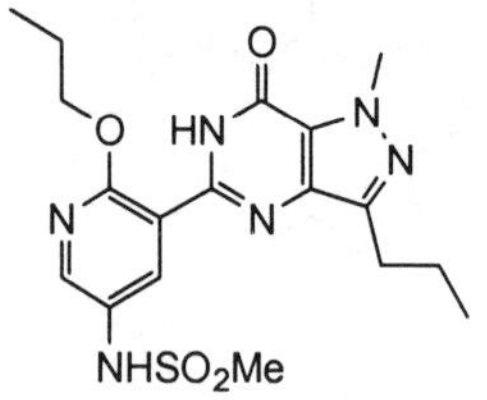

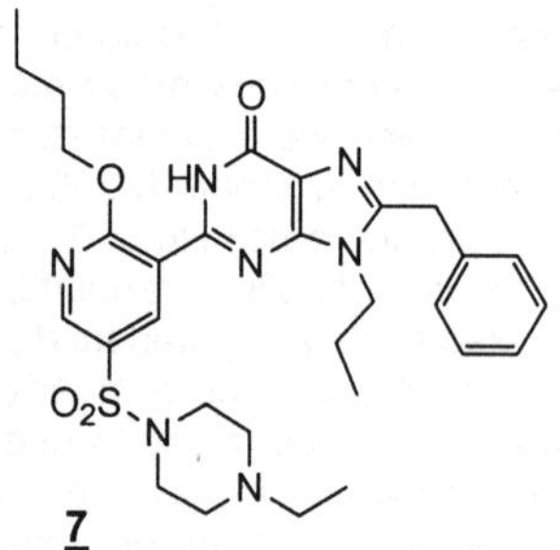

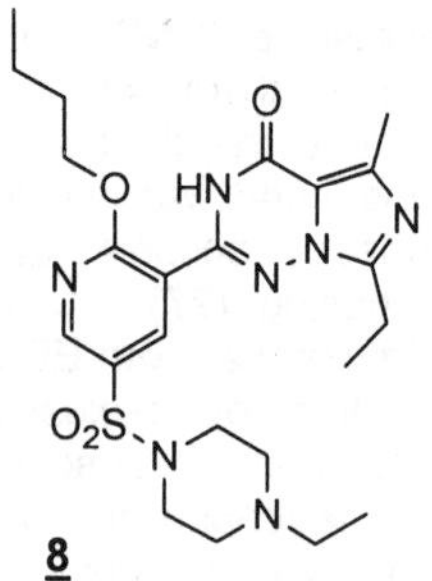

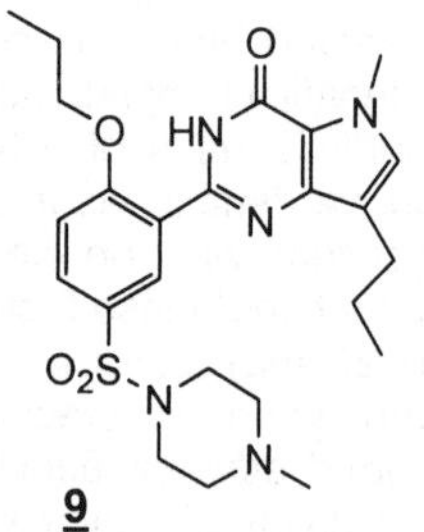

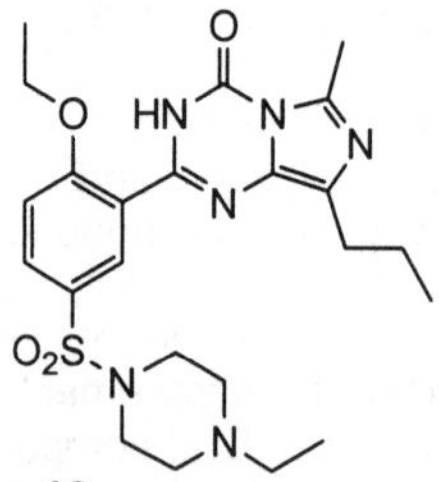

Modifications to the ethoxy and sulfonamido substituents of sildenafil have been explored in an effort to enhance PDE5 potency and selectivity (39-41). A series of sildenafil analogues which incorporated the ethoxy oxygen atom of sildenafil into a benz-fused heterocyclic ring, exemplified by **13**, resulted in decreased PDE5 potency and diminished selectivity for PDE5 over PDE6 in comparison to sildenafil (39). Replacement of the sulfonamido group by acylamino groups (eg **14**), or appendage of a carboxylic acid onto the sulfonamido residue (eg **15**) afforded PDE5 inhibitors that were more potent than sildenafil, but with similar or reduced isozyme selectivity (41,42). DA-8159 **16** is a sildenafil analogue that is claimed to exhibit

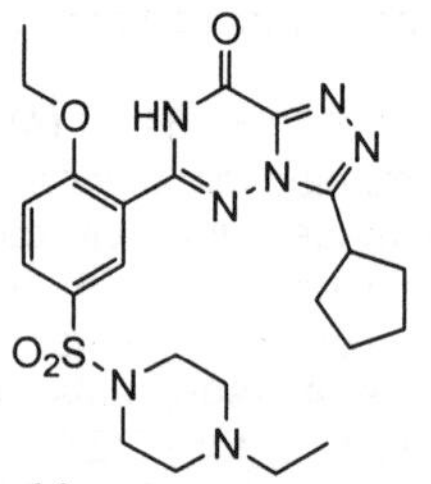

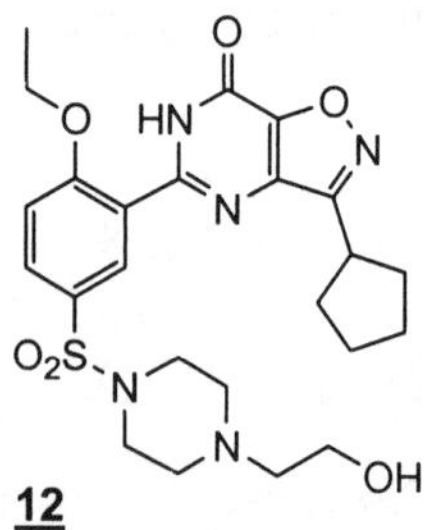

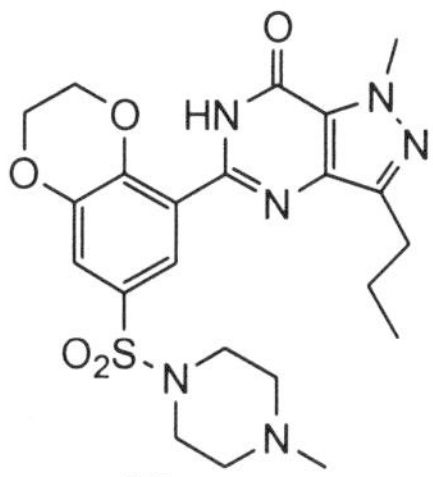

13 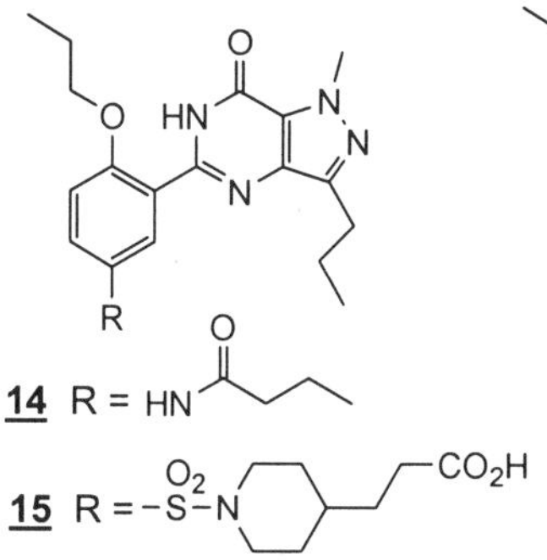   **14** R =

15 R =

superior isozyme selectivity and a longer half-life than sildenafil (42,43), however no specific data was reported. In a rat model of penile erection, orally administered **16** increased the frequency of erections with potency equivalent to sildenafil. Compound **16** was also active in an anaesthetized dog model of erectile activity, and the potency and duration of **16** in this model was stated to be comparable to that of sildenafil (42).

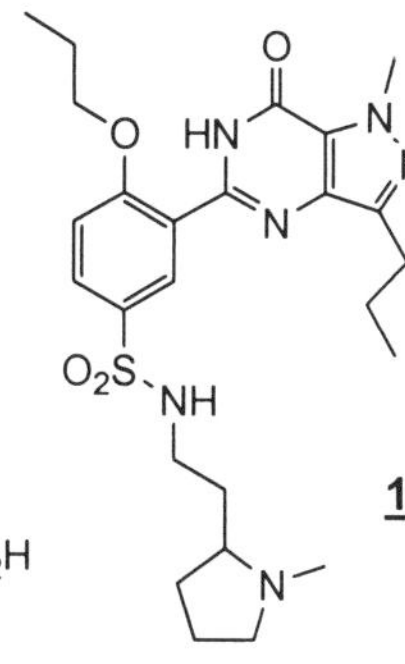

PDE5 inhibitors **17** and **18** based on a pyrazolopyridopyrimidone template have recently been described (44). Conceptually, **17** and **18** were derived from PDE5 inhibitors such as **19** that showed enhanced selectivity for PDE5 over other PDE isozymes, most notably PDE1 and PDE6 (45). Compound **17** inhibited human platelet PDE5 with an IC_{50} = 0.31 nM, did not significantly inhibit PDEs 1-4, and exhibited 160-fold selectivity for PDE5 with respect to PDE6 (44).

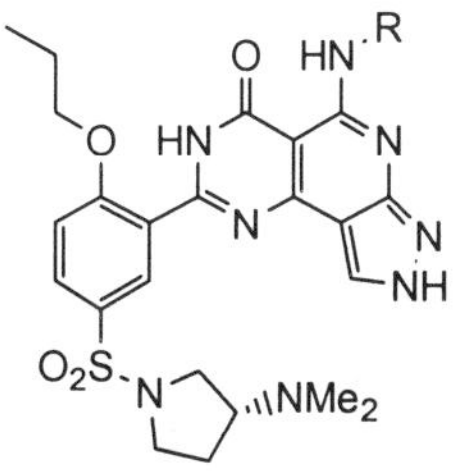

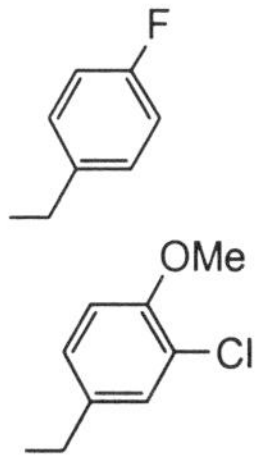

Manipulation of a screening lead related to the known PDE5 inhibitor **20** resulted in the identification of **21** (PDE5 IC_{50} = 38 nM) suitable for further optimization (46). Parallel synthesis of amide libraries derived from **21** revealed that amides containing H-bonding functionality exhibited superior selectivity, exemplified by **22**. Under the same assay conditions **22** (PDE5 IC_{50} = 0.8 nM) was a 2-fold more potent inhibitor of PDE5 than sildenafil. Compound **22** displayed a significantly improved PDE isozyme selectivity profile for PDE5 with respect to PDE1 and PDE6, exhibiting 6,400-fold selectivity and 35-fold selectivity respectively, in comparison to selectivity of 160-fold and 7-fold respectively for sildenafil. *In vitro*, **22** potentiated relaxation of rabbit corpus cavernosum tissue with potency equivalent to sildenafil. In rats and dogs, **22** showed good oral bioavailability (F = 33% and 66%, respectively) with half-life (1.2 hr and 2.3 hr respectively) consistent with an on-demand dosing regime in both species (46).

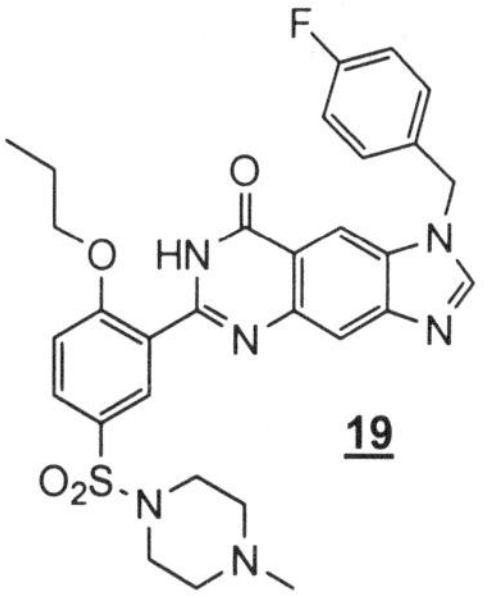

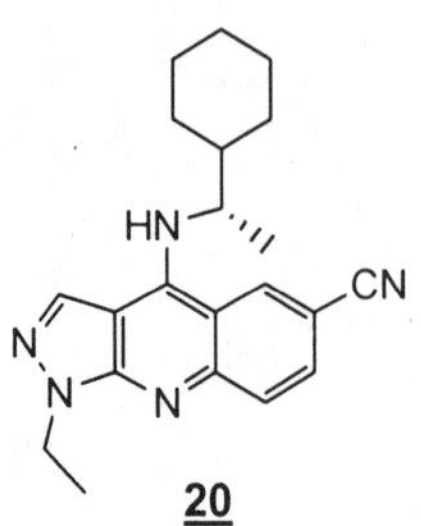

20 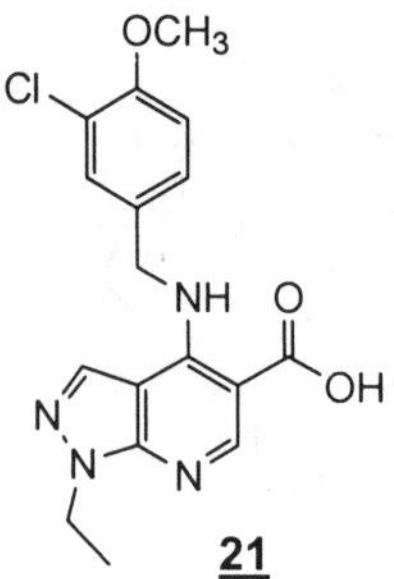 **21** 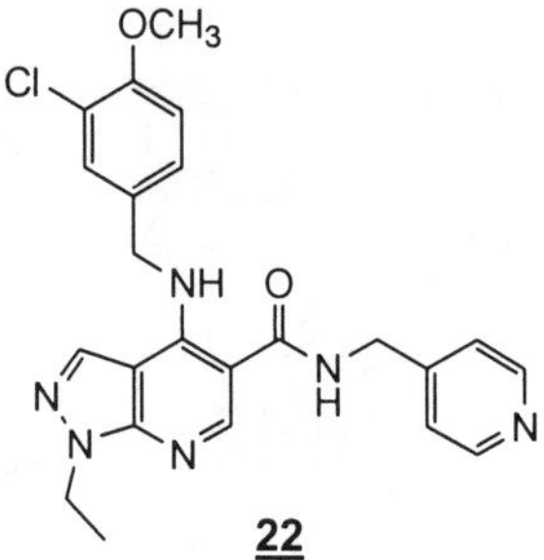 **22**

Selective PDE5 inhibitors based on a 4-aryl-1-isoquinolone template **23** have recently been described (47). Consideration of a naphthalene carboxamide pharmacophore present in an earlier series of PDE5 inhibitors in relation to the cGMP structure were the design elements that inspired the isoquinolone scaffold. Incorporation of a 4-anilino substituent at the 2-position was optimal for PDE5 potency, affording **24**. Derivatization of the 3-carboxy group as a methyl ester was shown to be optimal, while removal of the 6-methoxy substituent, **25**, was tolerated. Replacement of the 7-methoxy substituent as a 2-pyridylmethoxy group to improve water solubility resulted in T-1032 (**26**) with enhanced PDE5 potency (IC_{50} = 1 nM). With respect to modification of the 4-aryl substituent, replacement of the 4-methoxy group with bromo **27** (PDE5 IC_{50} = 0.22 nM) or methyl **28** (PDE5 IC_{50} = 0.39 nM) conferred enhanced potency. Despite the reduced potency of **26** as a PDE5 inhibitor compared to **27** and **28**, compound **26** (EC_{30} = 7.9 nM) was an order of magnitude more potent than **27** and **28** in an *in vitro* rabbit corpus cavernosum tissue relaxation assay, and was equipotent with sildenafil. The more potent vasorelaxatory effect of **26** compared to **27** and **28** was ascribed to its lower logD, affording enhanced solubility and improved cell permeability (47).

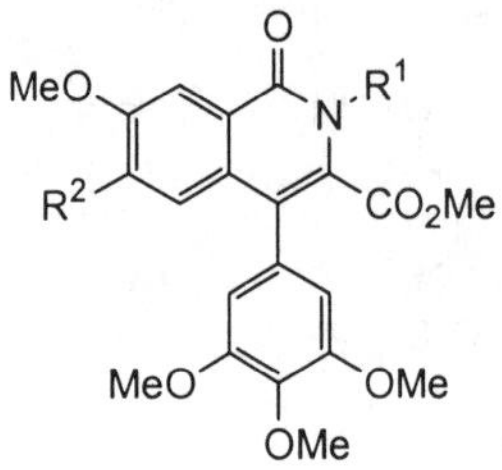

23 R^1 = H
    R^2 = OMe

24 R^1 = 4-anilino
    R^2 = OMe

25 R^1 = 4-anilino
    R^2 = H

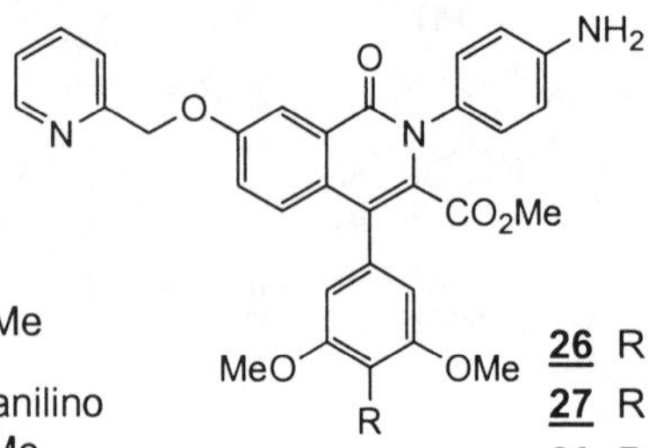

26 R = OMe
27 R = Br
28 R = Me

Compound **26** inhibited canine PDE5 with an IC_{50} = 1 nM, and was a less potent inhibitor of canine PDE1 (IC_{50} = 3.0 μM) and an equipotent inhibitor of canine PDE6 (IC_{50} = 28 nM) compared to sildenafil under the same assay conditions (48). *In vitro*, **26** potentiated the SNP-induced relaxation of isolated canine and rabbit corpus cavernosum, and of isolated rat aorta, in a concentration-dependent manner (49,50). Intravenously or intraduodenally administered **26** potentiated erections elicited by pelvic nerve stimulation in dogs with potency equivalent to sildenafil (49).

FR226807 (**29**) is a structurally novel PDE5 inhibitor identified by screening (51). Inhibition by **29** in comparison with sildenafil against a panel of human PDEs was reported. Compound **29** inhibited PDE5 (IC_{50} = 1.1 nM) and PDE6 (IC_{50} = 20 nM) with potency similar to that of sildenafil, but was a more selective inhibitor of PDE5 with respect to PDE1 than sildenafil. In dogs, intravenously administered **29** potentiated the increase in cavernosal pressure elicited by pelvic nerve stimulation more potently than sildenafil, and was reported to induce a less pronounced decrease in arterial blood pressure compared with sildenafil. A patent application disclosed benzimidazolones and imidazopyridones represented by **30** (IC_{50} < 10 nM) as inhibitors of human platelet PDE5 (52).   Use of imidazopyridopyrazinones exemplified by **31** (PDE5 IC_{50} = 2 nM) has been claimed for treatment of ED (53).

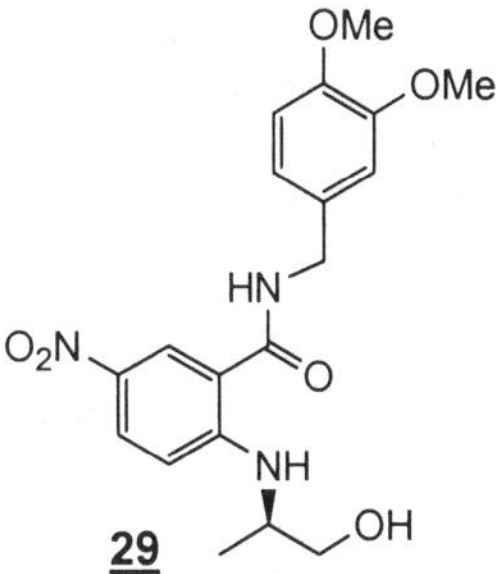

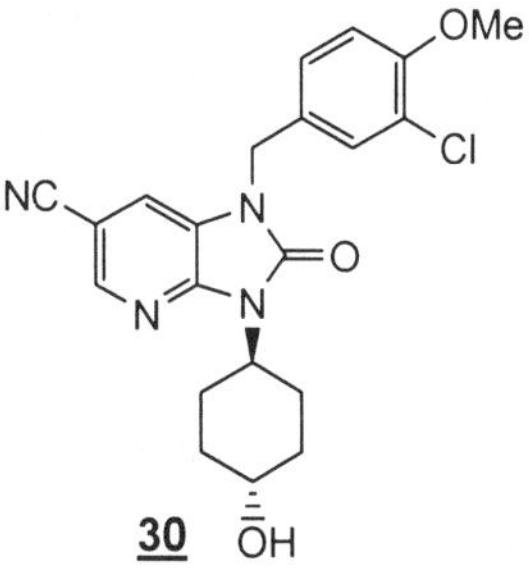

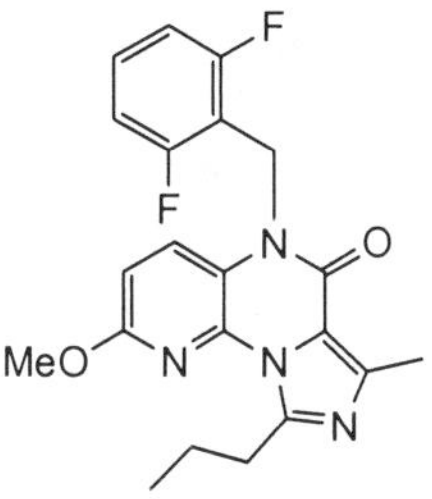

The imidazoquinazolinethione KF31327 (**32**) is an extremely potent non-competitive inhibitor of PDE5 (IC_{50} = 74 pM) isolated from canine trachea (54). The inhibition profile of **32** in comparison to sildenafil was determined against canine trachea PDEs 1-5. The PDE5 inhibitory potency of **32** was significantly greater than that of sildenafil, a competitive inhibitor, but **32** was less selective than sildenafil for PDE5 with respect to PDE3. Despite the greater PDE5 potency of **32**, both sildenafil and **32** at a concentration of 10 nM produced similar inhibition of nitroglycerin-induced aggregation of rabbit platelets.   In contrast to sildenafil, **32** (1 μM) also inhibited collagen-induced platelet aggregation by a cGMP-independent pathway, which was attributed to elevation of cAMP by inhibition of PDE3 (54).

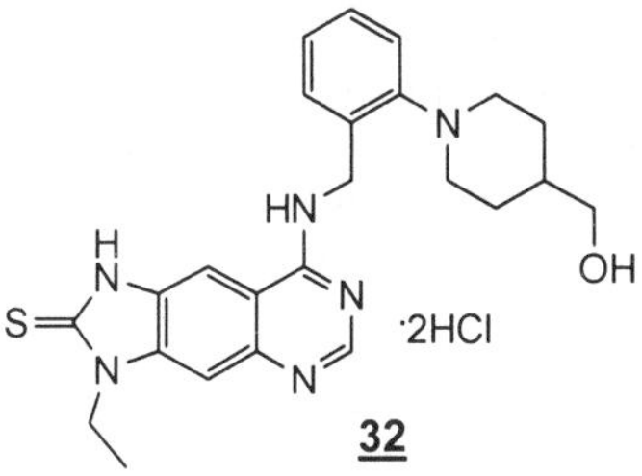

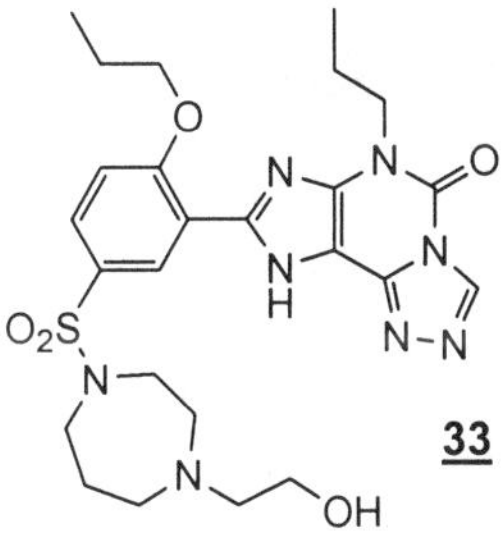

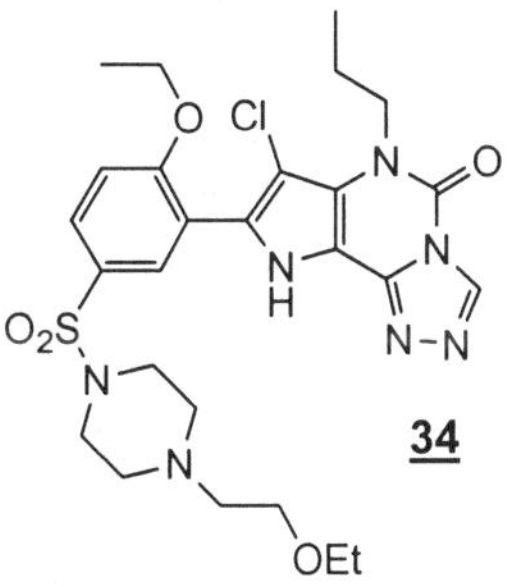

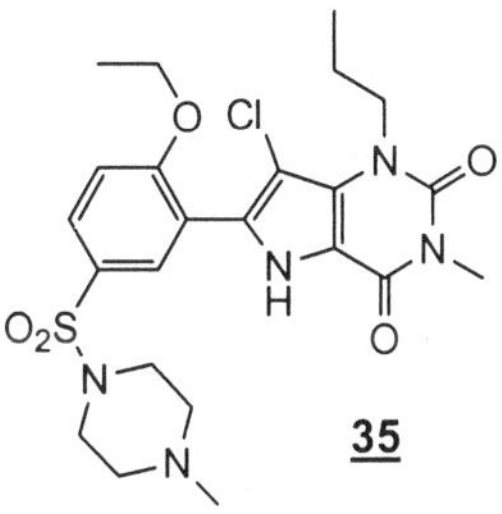

A series of phenyl-substituted triazolopurinones exemplified by **33** was recently reported to be potent, selective PDE5 inhibitors (55).   Compound **33** potently inhibited human platelet PDE5 (IC_{50} = 0.34 nM). Substitution of the pyrimidone N by lower alkyl groups was tolerated, and sulfonamide modifications produced inhibitors that varied in potency by 15-fold. Variation of the heterocyclic core resulted in a related series of pyrrolotriazolopyrimidone PDE5 inhibitors exemplified by **34**, which was reported to inhibit PDE5 with an IC_{50} = 42 pM (56). Halogenated analogues in the latter series exhibited subnanomolar PDE5 IC$_{50}$'s for a variety of sulfonamido residues. The structural class represented by **34** is related to a pyrimidinedione series of PDE5 inhibitors exemplified by **35** which inhibited PDE5 with an IC$_{50}$ of 3.4 nM (57).

A patent application recently disclosed a pyrroloquinolone class of PDE5 inhibitors represented by **36**, which was reported to inhibit human PDE5 with IC$_{50}$ = 0.19 nM (58). The compound was also stated to be active in an *in vivo* model of erectile function, however specific data was not reported. The 4-methoxyphenyl group is not essential for potency, and replacement of the methoxy group by water-solubilizing groups is also tolerated. Modification of the pyridone ring by O- or N-methylation resulted in loss of potency (58).

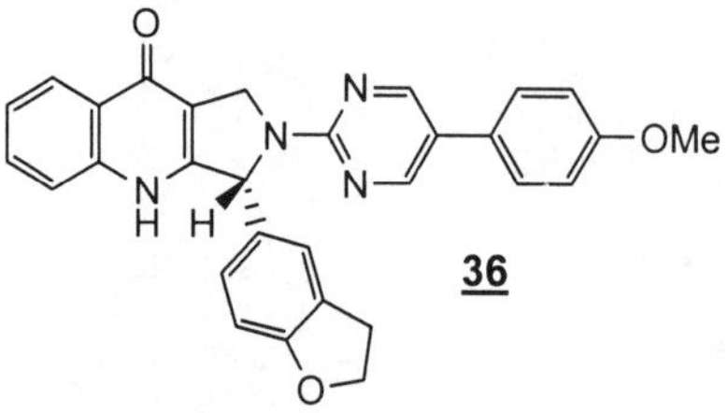

36

Appending a 3-chloro-4-methoxyphenylmethylamino substituent onto a heterocyclic scaffold has been a recurrent strategy for the identification of PDE5 inhibitors. The thienopyrimidine **37** potently inhibited human PDE5 (IC$_{50}$ = 0.55 nM) and did not significantly inhibit PDEs 1-4 (59). Replacement of the amido group by a variety of other substituents was tolerated with respect to PDE5 potency, and the tetrahydropyridine could be rendered non-basic without loss of potency. Compound **37** induced relaxation of isolated pre-contracted rat aorta with an EC$_{50}$ of 2.7 nM compared to an EC$_{50}$ of 6.1 nM for sildenafil (59). An independent group has also exploited a thienopyrimidine scaffold for the development of PDE5 inhibitors. In a recent patent application the hemodynamic effects of a benzothienopyrimidine PDE5 inhibitor EMD 221829 (**38**) in comparison with sildenafil were reported in dogs and rats (60). Intraduodenally administered **38** (1 mg/kg), was reported to potentiate erectile function in anaesthetized dogs with potency similar to that of sildenafil. Interestingly **38**, unlike sildenafil, had no effect on blood pressure or heart rate of anaesthetized dogs when co-administered intravenously (1 mg/kg) with the NO donor isosorbitol dinitrate (50 µg/kg/hr). Similar results were obtained in

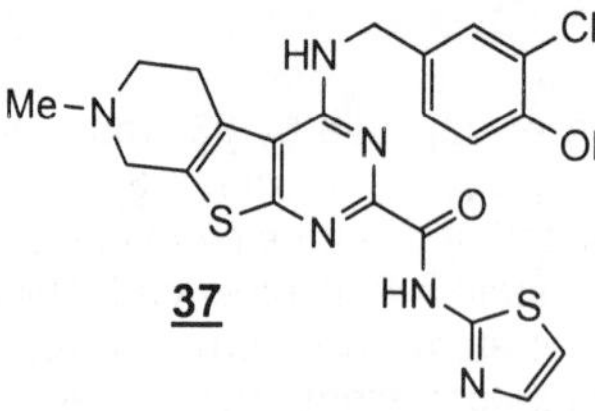

37 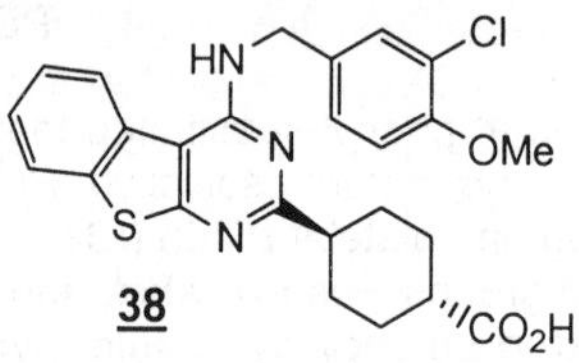   **38**

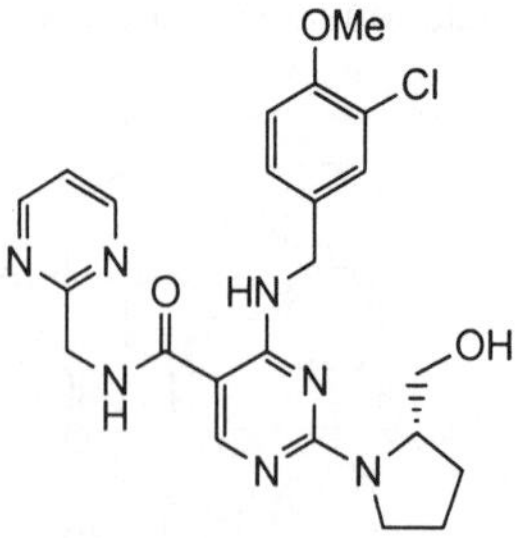

39 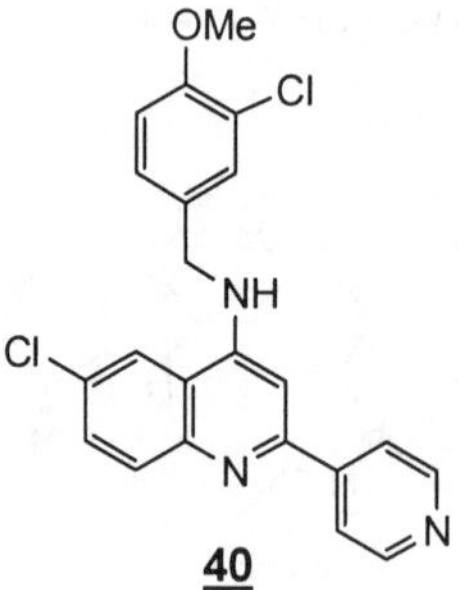   **40** 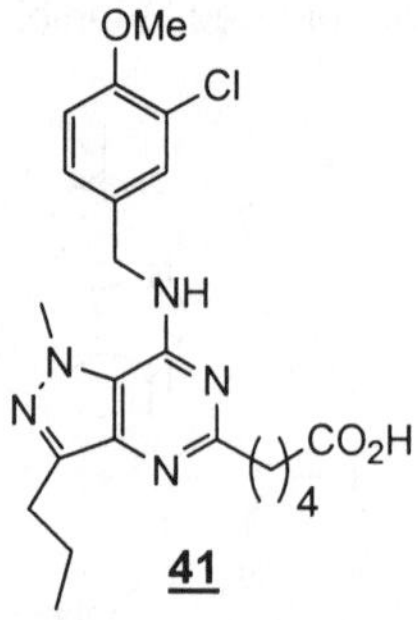   **41**

normotensive and spontaneously hypertensive rats (60). Oral bioavailability of the ethanolamine salt of **38** in dogs was 41-50% (61). The PDE5 potency and PDE isozyme selectivity of **38** was not reported. Two patent applications disclosed trisubstituted six-membered aza heterocycles, predominantly pyrimidine derivatives exemplified by **39**, as PDE5 inhibitors (62,63). Again, the 3-chloro-4-methoxyphenylmethylamino substituent features prominently in the disclosed structures, however no PDE5 potency or selectivity data was reported. Other recent patent applications disclose aminoquinolines and pyrazolopyrimidines, represented by **40** and **41** respectively, as PDE5 inhibitors (64,65). Compound **40** was reported to inhibit PDE5 (IC_{50} = 0.24 nM) selectively over PDEs 1-4 (64). No specific PDE5 inhibition data was reported for **41** (65).

Xanthine derivatives substituted at C8 by an isoquinolin-4-ylmethyl residue have recently been reported as potent, selective PDE5 inhibitors (66). Compounds **42** and **43** inhibited human platelet PDE5 with IC_{50}'s of 2 nM and 2.8 nM respectively, demonstrating that considerable structural variation of the N3 substituent is compatible with PDE5 inhibition. Compounds in this disclosure were stated to be selective for PDE5 over PDE1 and PDE6, however no isozyme selectivity data was reported.

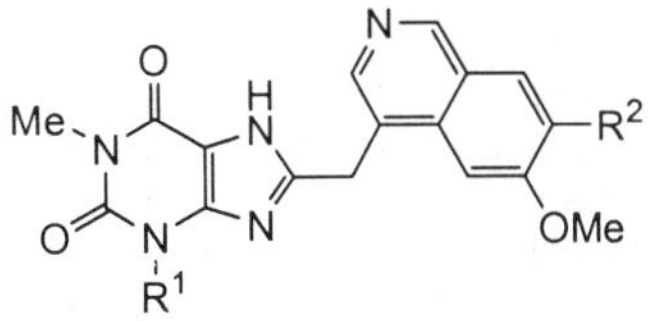

42 R^1 = $(CH_3)_2CHCH_2$, R^2 = OMe

43 R^1 = -CH_2-⟨⟩-$NHSO_2NMe_2$, R^2 = H

OTHER POTENTIAL THERAPEUTIC INDICATIONS

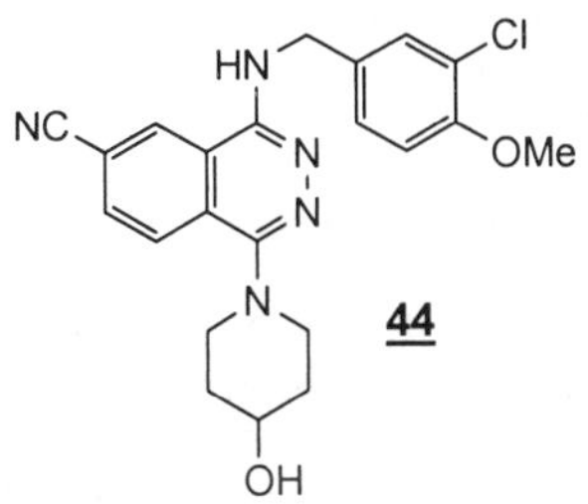

Primary Pulmonary Hypertension – PDE5 inhibitors are known to be selective pulmonary artery vasodilators in animal models of pulmonary hypertension (67). E-4010 (**44**) is a recently discovered potent, selective and orally active PDE5 inhibitor that has been evaluated in such models (68). Compound **44** (0.1% in diet) was reported to improve mortality of rats with pulmonary hypertension induced by monocrotaline. This outcome was also associated with reduced ventricular hypertrophy and increased lung cGMP levels, but not cAMP levels (69). In chronically hypoxic rats, **44** (0.01%, 0.1% in diet) also attenuated an increase in pulmonary arterial pressure (PAP) without affecting systemic arterial blood pressure, and evoked reductions in ventricular hypertrophy and pulmonary arterial remodeling (70). Sildenafil is under clinical evaluation for the treatment of pulmonary hypertension. In a randomized, double-blind study in healthy adult men, a 56% increase in PAP induced by hypoxia in the placebo treated group was almost completely absent in the sildenafil (100 mg) treated group (71). In a small study of subjects with primary pulmonary hypertension, sildenafil was dosed cumulatively up to 100 mg (72). Sildenafil treatment reduced mean PAP by 6.4 mm Hg and markedly reduced pulmonary vascular resistance, with maximal effects after the first 25 mg dose. The combination of inhaled iloprost, a prostanoid vasodilator, and oral sildenafil produced a lowering of mean PAP greater than by either agent alone, with no effect on heart rate or systemic arterial pressure (72). Sildenafil also attenuated the rise in right ventricular systolic pressure induced by chronic hypoxia in wild-type mice, and to a lesser extent by the eNOS/NO pathway, suggesting that sildenafil does not operate exclusively by the eNOS/NO pathway. Right ventricular hypertrophy and pulmonary vascular remodeling were inhibited by sildenafil in hypoxic wild-type mice, but not in eNOS deficient mice (71). Overall the data suggest that PDE5 inhibitors alone or in combination with other agents may prove beneficial in the management of primary pulmonary hypertension.

Diabetic Gastropathy – NO is an important messenger in the regulation of intestinal function, and neuronal NO synthase gene deletion in mice results in delayed gastric emptying and loss of NO-mediated non-adrenergic non-cholinergic neurotransmission (73). This phenotype is mimicked in genetically and streptozotocin-induced diabetic mice. Sildenafil restored gastric emptying in diabetic mice, which was attributed to potentiation of depleted NO levels in the pylorus and it was suggested that PDE5 inhibitors might be effective in the treatment of diabetic gastropathy.

Insulin Resistance and Hyperlipidemia – Recent studies have explored a potential role of the NO/cGMP signaling pathway in glucose utilization by skeletal muscle. The nitrate donor SNP and the PDE5 inhibitor zaprinast, albeit at relatively high concentrations, increased cGMP levels and promoted glucose utilization in muscle tissue from insulin-sensitive (lean Zucker) rats (74). The response to both zaprinast and SNP was impaired in muscle isolated from insulin-resistant (obese Zucker *fa/fa*) rats and it was proposed that impairment in the cGMP/NO signaling pathway contributes to insulin resistance (74). It has also been demonstrated that SNP increased both glucose transport and expression of the glucose transporter GLUT-4 at the cell surface in isolated rat muscle (75), however the role of cGMP in glucose metabolism by skeletal muscle is controversial (76). The effect of sildenafil on serum triglyceride levels was evaluated in a 28-day clinical trial in non-diabetic human subjects with risk factors for insulin resistance syndrome (77). Sildenafil at doses of 10, 25 and 50 mg reduced serum triglycerides by 40%, 31%, and 12% from baseline values (mg/dl) of 255, 213, and 191 respectively, compared to an 11% drop from a baseline of 185 mg/dl for placebo. In a separate study of diabetic subjects, chronic treatment with sildenafil was stated to significantly improve glucose control, however no specific data was presented (77). In *ob/ob* mice, sildenafil treatment for 5 days caused a reduction in triglyceride and glucose levels.

Conclusion - The large number of reports of PDE5 inhibitors in the recent literature reflects the high level of interest in PDE5 as a therapeutic target. While the majority of early PDE5 inhibitors owe some structural similarity to cGMP, structural diversity of more recently developed PDE5 inhibitors has burgeoned. Following the clinical success of sildenafil, the newer PDE5 inhibitors vardenafil and tadalafil are in advanced clinical trials while a number of PDE5 inhibitors are in earlier stages of clinical development for ED. Potential therapeutic applications of PDE5 inhibitors other than for ED continue to be explored, and PDE5 inhibitors show promise for the treatment of pulmonary hypertension. Intriguing data that suggests that PDE5 inhibitors might be therapeutically beneficial in the treatment of insulin resistance and type 2 diabetes requires further study to more fully elucidate a role of the cGMP signaling pathway in glucose metabolism, and to confirm the anti-diabetic effects of PDE5 inhibitors in the clinic.

References

1. J.A. Beavo, Physiol.Rev., 75, 725 (1995).
2. S.H. Soderling and J.A. Beavo, Curr.Opin.Cell Bio., 12, 174 (2000).
3. K. Yuasa, J. Kotera, K. Fujishige, H. Michibata, T. Sasaki and K. Omori, J.Biol.Chem., 275, 31469 (2000).
4. M. Czarniecki, H.-S. Ahn and E.J. Sybertz, Ann.Rep.Med.Chem., 31, 61 (1996).
5. J. Geiger, Exp.Opin.Invest.Drugs, 10, 865 (2001).
6. M. Boolell, M.J. Allen, S.A. Ballard, S. Gepi-Attee, G.J. Muirhead, A.M. Naylor, I.H. Osterloh and C. Gingell, Int.J.Impot.Res., 8, 47 (1996).

7.   G.N. Maw, Ann.Rep.Med.Chem., 34, 71 (1999).
8.   D.P. Rotella, Drugs Fut., 26, 153 (2001).
9.   L.M. McAllister-Lucas, W.K. Sonnenburg, A. Kadlecek, D. Seger, H. LeTrong, J.L. Colbran, M.K. Thomas, K.A. Walsh, S.H. Francis, J.D. Corbin and J.A. Beavo, J.Biol.Chem., 268, 22863 (1993).
10.   K. Loughney, T.R. Hill, V.A. Florio, L. Uher, G.J. Rosman, S.L. Wolda, B.A. Jones, M.L. Howard, L.M. McAllister-Lucas, W.K. Sonnenburg, S.H. Francis, J.D. Corbin, J.A. Beavo and K. Ferguson, Gene, 216, 139 (1998).
11.   J. Kotera, K. Fujishige, Y. Imai, E. Kawai, H. Michibata, H. Akatsuka, N. Yanaka and K. Omori, Eur.J.Biochem., 262, 866 (1999).
12.   C.S. Lin, A. Lau, R. Tu and T.F. Lue, Biochem.Biophys.Res.Commun., 268, 628 (2000).
13.   J.D. Corbin and S.H. Francis, J.Biol.Chem., 274, 13729 (1999).
14.   S.H. Francis, I.V. Turko, K.A. Grimes and J.D. Corbin, Biochemistry, 39, 9591 (2000).
15.   J.D. Corbin, I.V. Turko, A. Beasley and S.H. Francis, Eur.J.Biochem., 267, 2760 (2000).
16.   S.D. Rybalkin, I.G. Rybalkina, R. Feil, F. Hofmann and J.A. Beavo, J.Biol.Chem., 277, 3310 (2002).
17.   V.K. Gopal, S.H. Francis and J.D. Corbin, Eur.J.Biochem., 268, 3304 (2001).
18.   NIH Consensus Conference. Impotence. NIH Consensus Development Panel on Impotence., JAMA 270, 83 (1993).
19.   T.F. Lue, N.Engl.J.Med., 342, 1802 (2000).
20.   M.B. Noss, G.J. Christ and A. Melman, Drugs Today, 35, 211 (1999).
21.   E.G. Boyce and E.M. Umland, Clin.Ther., 23, 2 (2001).
22.   H. Porst, Int.J.Impot.Res., 14, Suppl. 1, S57 (2002).
23.   L.A. Sorbera, L. Martin, P.A. Leeson and J. Castaner, Drugs Fut., 26, 15 (2001).
24.   R.W. Baxendale, C.P. Wayman, L. Turner and S.C. Phillips, J.Urol., 165, Suppl., 223, (2001).
25.   H. Haning, U. Niewohner, T. Schenke, M. Es-Sayed, G. Schmidt, T. Lampe and E. Bischoff, Biorg.Med.Chem.Lett., 12, 865 (2002).
26.   I. Saenz de Tejada, J. Angulo, P. Cuevas, A. Fernandez, I. Moncada, A. Allona, E. Lledo, H.G. Korschen, U. Niewohner, H. Haning, E. Pages and E. Bischoff, Int.J.Impot.Res., 13, 282 (2001).
27.   E. Bischoff, U. Niewoehner, H. Haning, M. Es-Sayed, T. Schenke and K.H. Schlemmer, J.Urol., 165, 1316 (2001).
28.   H. Porst, R. Rosen, H. Padma-Nathan, I. Goldstein, F. Giuliano, E. Ulbrich and T. Bandel, Int.J.Impot.Res., 13, 192, (2001).
29.   J. Pryor, Int.J.Impot.Res., 14, Suppl. 1, S65 (2002).
30.   S. Abel, K.C. Beaumont, C.L. Crespi, M.D. Eve, L. Fox, R. Hyland, B.C. Jones, G.J. Muirhead, D.A. Smith, R.F. Venn and D.K. Walker, Xenobiotica, 31, 665 (2001).
31.   M.E. Bunnage, K.M. Devries, L.J. Harris, P.C. Levett, J.P. Mathias, J.T. Negri, S.D.A. Street and A.S. Wood, WO Patent 0127113 (2001).
32.   C.M.N. Allerton, C.G. Barber, G.N. Maw and D.J. Rawson, WO Patent 0127112 (2001).
33.   G.N. Maw and D.J. Rawson, Eur. Patent EP1092718 (2001).
34.   G.N. Maw, Eur. Patent EP1092719 (2001).
35.   D.-K. Kim, J.Y. Lee, D.H. Ryu, N.K. Lee, S.H. Lee, N.-H. Kim, J.-S. Kim, J.H. Ryu, J.-Y. Choi, G.-J. Im, W.-S. Choi, T.K. Kim and H. Cha, WO Patent 0160825 (2001).
36.   U. Niewohner, H. Haning, T. Lampe, M. Es-Sayed, G. Schmidt, E. Bischoff, K. Dembowsky, E. Perzborn and K.-H. Schlemmer, WO Patent 0147928 (2001).
37.   U. Niewohner, H. Haning, T. Lampe, M. Es-Sayed, G. Schmidt, E. Bischoff, K. Dembowsky, E. Perzborn and K.-H. Schlemmer, WO Patent 0147929 (2001).
38.   U. Niewohner, H. Haning, T. Lampe, M. Es-Sayed, G. Schmidt, E. Bischoff, K. Dembowsky, E. Perzborn and K.-H. Schlemmer, WO Patent 0147934 (2001).
39.   D.-K. Kim, N. Lee, J.Y. Lee, D.H. Ryu, J.-S. Kim, S.-H. Lee, J.-Y. Choi, K. Chang, Y.-W. Kim, G.-J. Im, W.-S. Choi, T.-K. Kim, J.-H. Ryu, N.H. Kim and K. Lee, Biorg.Med.Chem., 9, 1609 (2001).
40.   D.-K. Kim, D.H. Ryu, N. Lee, J.Y. Lee, J.-S. Kim, S. Lee, J.-Y. Choi, J.-H. Ryu, N.H. Kim, G.-J. Im, W.-S. Choi and T.-K. Kim, Biorg.Med.Chem., 9, 1895 (2001).
41.   D.-K. Kim, J.Y. Lee, N. Lee, D.H. Ryu, J.-S. Kim, S. Lee, J.-Y. Choi, J.-H. Ryu, N.H. Kim, G.-J. Im, W.-S. Choi and T.-K. Kim, Biorg.Med.Chem., 9, 3013 (2001).
42.   T.Y. Oh, K.K. Kang, B.O. Ahn, M. Yoo and W.B. Kim, Arch.Pharm.Res., 23, 471 (2000).
43.   H.J. Shim, E.J. Lee, J.H. Kim, S.H. Kim, J.W. Kwon, W.B. Kim, S.-W. Cha and M.G. Lee, Biopharm.Drug Dispos., 22, 109 (2001).
44.   Y. Bi, P. Stoy, L. Adam, B. He, J. Krupinski, D. Normandin, R. Pongrac, L. Seliger, A. Watson and J.E. Macor, Biorg.Med.Chem.Lett., 11, 2461 (2001).

45. D.P. Rotella, Z. Sun, Y. Zhu, J. Krupinski, R. Pongrac, L. Seliger, D. Normandin and J.E. Macor, J.Med.Chem., 43, 5037 (2000).
46. G. Yu, H.J. Mason, X. Wu, J. Wang, S. Chong, G. Dorough, A. Henwood, R. Pongrac, L. Seliger, B. He, D. Normandin, L. Adam, J. Krupinski and J.E. Macor, J.Med.Chem., 44, 1025 (2001).
47. T. Ukita, Y. Nakamura, A. Kubo, Y. Yamamoto, Y. Moritani, K. Saruta, T. Higashijima, J. Kotera, M. Takagi, K. Kikkawa and K. Omori, J.Med.Chem., 44, 2204 (2001).
48. J. Kotera, K. Fujishige, H. Michibata, K. Yuasa, A. Kubo, Y. Nakamura and K. Omori, Biochem. Pharmacol., 60, 1333 (2000).
49. T. Noto, H. Inoue, T. Ikeo and K. Kikkawa, J.Pharmacol.Exp.Ther., 294, 870 (2000).
50. M. Takagi, H. Mochida, T. Noto, K. Yano, H. Inoue, T. Ikeo and K. Kikkawa, Eur.J.Pharmacol., 411, 161 (2001).
51. N. Hosogai, K. Hamada, M. Tomita, A. Nagashima, T. Takahashi, T. Sekizawa, T. Mizutani, Y. Urano, A. Kuroda, K. Sawada, T. Ozaki, J. Seki and T. Goto, Eur.J.Pharmacol., 428, 295 (2001).
52. K. Sawada, T. Inoue, Y. Sawada and T. Mizutani, WO Patent 0105770 (2001).
53. D. Marx, N. Hofgen, U. Egerland, S. Szelenyi and T. Kronbach, WO Patent 0168097 (2001).
54. R. Hirose, H. Okumura, A. Yoshimatsu, J. Irie, Y. Onada, Y. Nomoto, H. Takai, T. Ohno and M. Ichimura, Eur.J.Pharmacol., 431, 17 (2001).
55. J. Gracia Ferrer, J. Feixas Gras, J.M. Prieto Soto, A. Vega Noverola and B. Vidal Juan, WO Patent 0107441 (2001).
56. B. Vidal Juan, C. Esteve Trias, J. Gracia Ferrer and J.M. Prieto Soto, WO Patent 0212246 (2002).
57. B. Vidal Juan, J. Gracia Ferrer, J.M. Prieto Soto and A. Vega Noverola, WO Patent 0194350 (2001).
58. Z. Sui, M.J. Macielag, J. Guan, W. Jiang and J.C. Lanter, WO Patent 0187882 (2001).
59. H. Yamada, N. Umeda, S. Uchida, Y. Shiinoki, H. Horikoshi and N. Mochizuki, Eur. Patent EP1167367 A1 (2002).
60. M. Brandle, T. Ehring and C. Wilm, WO Patent 0164192 (2001).
61. S. Schreder, C. Wildner and K. Schamp, WO Patent 0151052 (2001).
62. K. Yamada, K. Matsuki, K. Omori and K. Kikkawa, WO Patent 0183460 (2001).
63. K. Yamada, K. Matsuki, K. Omori and K. Kikkawa, WO Patent 0119802 (2001).
64. N. Umeda, K. Ito, S. Uchida and Y. Shiinko, WO Patent 0112608 (2001).
65. R. Jonas, H.-M. Eggenweiler, P. Schelling, M. Christadler and N. Beier, WO Patent 0118004 (2001).
66. G. Bhalay, S.P. Collingwood, R.A. Fairhurst, S.F. Gomez, R. Naef and D.A. Sandham, WO Patent 0177110 (2001).
67. A.H. Cohen, K. Hanson, K. Morris, B. Fouty, I.F. McMurtry, W. Clarke and D.M. Rodman, J.Clin.Invest., 97, 172 (1996).
68. N. Watanabe, H. Adachi, Y. Takase, H. Ozaki, M. Matsukura, K. Miyazaki, K. Ishibashi, H. Ishihara, K. Kodama, M. Nishino, M. Kakiki and Y. Kabasawa, J.Med.Chem., 43, 2523 (2000).
69. K. Kodama and H. Adachi, J.Pharmacol.Exp.Ther., 290, 748 (1999).
70. N. Hanasato, M. Oka, M. Muramatsu, M. Nishino, H. Adachi and Y. Fukuchi, Am.J.Physiol., 277, L225 (1999).
71. L. Zhao, N.A. Mason, N.W. Morrell, B. Kojonazarov, A. Sadykov, A. Maripov, M.M. Mirrakhimov, A. Aldashev and M.R. Wilkins, Circulation, 104, 424 (2001).
72. H. Wilkens, A. Guth, J. Konig, N. Forestier, B. Cremers, B. Hennen, M. Bohm and G.W. Sybrecht, Circulation, 104, 1218 (2001).
73. C.C. Watkins, A. Sawa, S. Jaffrey, S. Blackshaw, R.K. Barrow, S.H. Snyder and C.D. Ferris, J.Clin.Invest., 106, 373 (2000).
74. M.E. Young and B. Leighton, Biochem.J., 329, 73 (1998).
75. G.J. Etgen, D.A. Fryburg and E.M. Gibbs, Diabetes, 46, 1915 (1997).
76. J.S. Stamler and G. Meissner, Physiol.Rev., 81, 209 (2001).
77. D.A. Fryburg, E.M. Gibbs and N.P. Koppiker, WO Patent 0213798 (2002).

Chapter 7. Antagonists of VLA-4

George W. Holland, Ronald J. Biediger, and Peter Vanderslice
Texas Biotechnology Corporation
7000 Fannin, Houston, TX 77030

Introduction - Integrins are a family of heterodimeric cell surface receptors consisting of an α and a β subunit (1). Different α and β subunits can associate to yield a family of no less than 23 members each with a distinct pattern of ligand selectivity. The integrin $\alpha4\beta1$ is also known as VLA-4 (very late antigen–4). VLA-4 is expressed on the surface of most leukocyte or related cell types including lymphocytes, eosinophils, monocytes, basophils, and mast cells (2,3). More recently, the integrin has been detected on neutrophils in certain inflammatory settings (4,5). The ligands that bind to VLA-4 include the inducible cell surface receptor vascular cell adhesion molecule-1 (VCAM-1) (6) and the extracellular matrix molecule fibronectin *via* its alternatively spliced connecting segment-1 (CS-1) domain (7). The interaction of VLA-4 with VCAM-1 and/or fibronectin is thought to be key to leukocyte adhesion, migration and activation. These processes are key to the normal immune response as well as the progression of various inflammatory and autoimmune diseases. Accordingly, VLA-4 (and related integrins) is of intense current interest as a drug discovery target by a number of research groups.

In this chapter we summarize the highlights of work in the area of VLA-4 biology and small molecule antagonists that have been published since last reviewed in this Annual (8,9). A more comprehensive review on VLA-4 antagonists has recently published (10).

Recent Biology - Over the last ten years, monoclonal antibodies to the $\alpha4$ subunit as well as inhibitory peptides based on VLA-4 ligands have been used in animal models to implicate the integrin in several inflammatory diseases including asthma (11), contact hypersensitivity, multiple sclerosis, diabetes, inflammatory bowel disease, arthritis and allograft rejection (11-19). Early studies identified the VLA-4 ligand VCAM-1 as a homing receptor for monocytes in an animal model of atherosclerosis (20). More recently, gene disruption techniques were used to generate mice that express less than 10% of the VCAM-1 found in normal wild-type mice (21). When crossed with mice deficient for the LDL receptor, the resulting offspring showed a significant reduction in atherosclerotic lesion size as compared to those mice that were deficient in the LDL receptor alone. Interestingly, mice deficient for both the ICAM-1 and the LDL receptor genes did not show a reduction in lesion area relative to the controls (21).

A growing body of evidence suggests that VLA-4 and its ligands may be involved in the vaso-occlusion events associated with sickle cell disease (22-24). A recent study using human sickle cells in a rat model of retinal vaso-occlusion demonstrated that TNF-α induced sickle cell retention could be greatly reduced by a VLA-4 peptide antagonist (25). Such antagonists may prove useful in limiting the duration of the painful crisis episodes common to sickle cell patients. In addition, to the extent that TNF-α contributes to the angiogenic component of diseases such as sickle cell disease or rheumatoid arthritis, it is interesting to consider that a VLA-4 antagonist may provide additional benefit in retarding the neovascularization process (26).

A series of studies in which mice were generated containing "knock-outs" of the genes for the $\alpha4$ integrin subunit or VCAM-1 demonstrated a clear role for VLA-4 and VCAM-1 in embryonic development (27-29). This suggests the potential for reproductive toxicity with the use of VLA-4 antagonists. Both types of knockout mice display an embryonic lethal phenotype characterized by a failure in chorioallontoic fusion resulting in defects in placental development. Additional abnormalities were observed in the developing heart. A recent study investigated the potential reproductive consequences of VLA-4 antagonist treatment (30). Early stage rat embryos were cultured following microinjection of a α4 antibody or a series of synthetic small molecule antagonists **1 – 3**. The embryos treated with the antibody showed abnormal chorioallontoic fusion as might be expected based on the gene knockout studies. Two of the three small molecule antagonists tested yielded similar results. One of the compounds however had no effect on chorioallontoic fusion and the embryos appeared normal, similar to those treated with the inactive control compound **4**. Although it is unclear why the antagonists would yield different results, it is interesting that the one antagonist that did not show any toxic effects binds to only the high affinity state of the integrin whereas the other two compounds could bind to both the low and high affinity forms (30). Regardless of the reason, this data suggests that not all VLA-4 antagonists will necessarily have reproductive toxicology issues. Clearer insights into this issue can be expected as more VLA-4 antagonists complete preclinical testing and enter clinical trials.

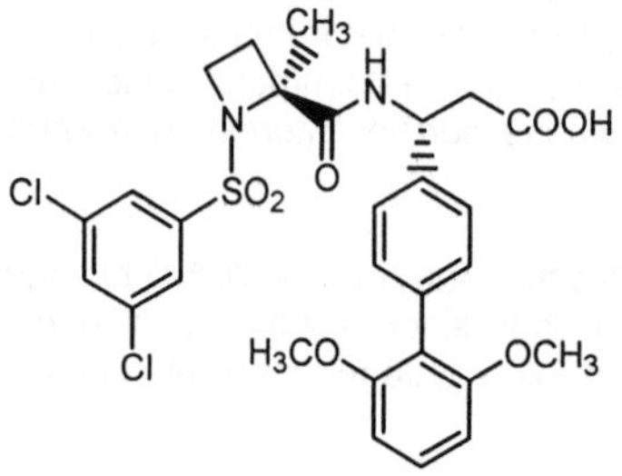

1

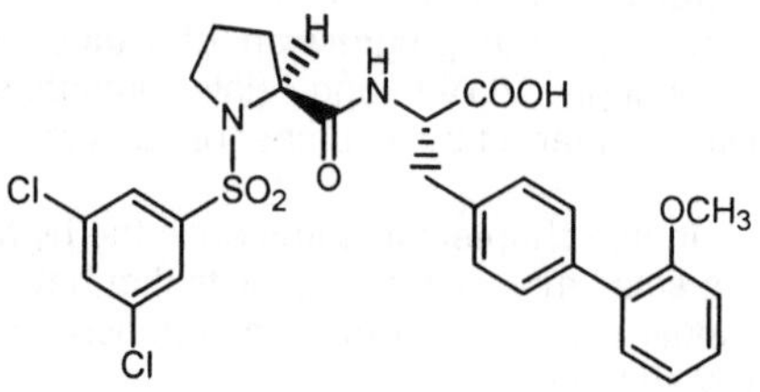

2

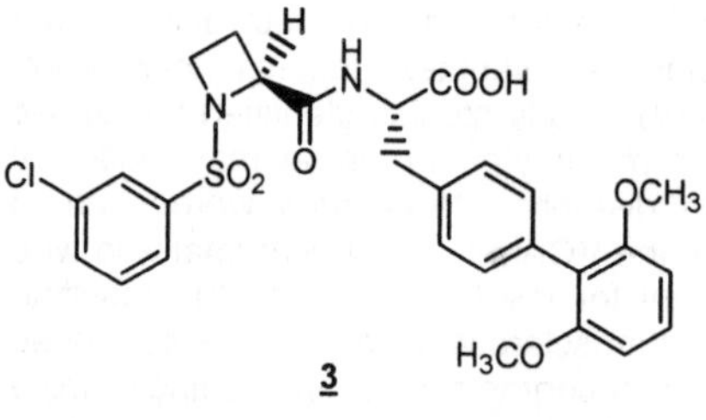

3

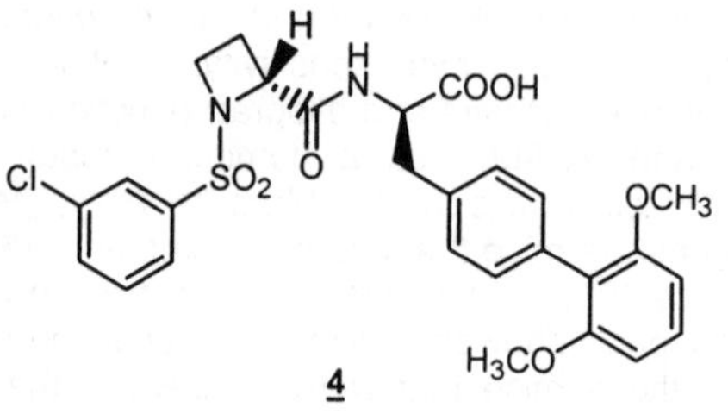

4

<u>Small Molecule Antagonists</u> - To date, most of the progress on the discovery of small molecule antagonists of VLA-4 has been *via* combinatorial and classical medicinal chemistry techniques (8-10). Some structural information for *de novo* design was generated from NMR studies on cyclic peptides antagonists, such as **5** and **6**, which were early scaffolds presumed to mimic the active RGD and LDV conformations respectively (31-33). No crystal structural information on αvβ1 has yet appeared. Recently however, the crystal structure of the extracellular segment of a related integrin, αvβ3, in complex with a cyclic pentapeptide ligand **7**, cyclo(RGDf-mV), has been reported (34). The cyclic peptide binds at the major interface of the αv and β3 subunits and each residue of the RGD sequence is revealed to be extensively involved in interactions with the binding surface. An *in silico* approach, homology modeling, is becoming an increasingly useful technique for drug design and recently a β1 homology model based on the I domain of the integrin CD11B/CD18 with bound Mg^{+2} was published (35). This model was used to rationalize the activity of both the cyclic peptide **6** and a more advanced small

molecule antagonist **8**. In addition, this model offers some rationale for the 1000-fold selectivity of **8** for VLA-4 vs α4β7.

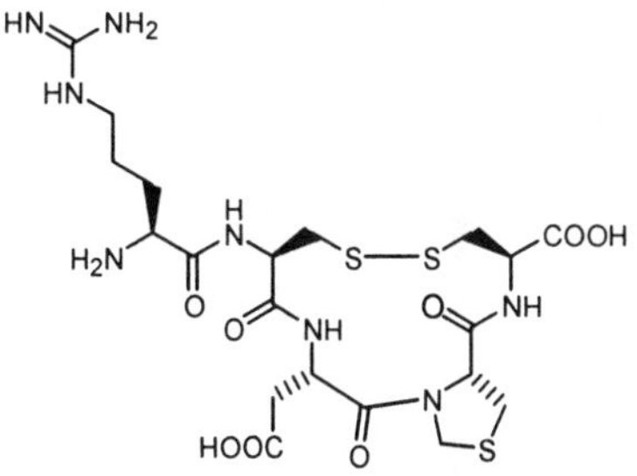

Linear LDV analogs have been extensively explored by a number of research groups as this approach was particularly facilitated by combinatorial techniques. It was discovered that N-capping of the LDV sequence with a phenylacetyl moiety gave analogs with increased potency. Additionally, great latitude was found to be available for the choice of carboxylic acid bearing C-caps (10). Subsequent work lead to the incorporation of a phenylureidophenylacetyl moiety as the N-cap to yield antagonists of low nanomolar activity in VLA-4/VCAM and CS-1 binding assays. At least three compounds from this series have now entered clinical trials. One compound IVL-745 (**9**) is reported to be under development as an inhaled antiasthmatic (36-40). The dihydroindole compound **10** is reported to be in phase II clinical trials for multiple sclerosis and asthma (stereochemistry not disclosed). Early on, Bio 1211 (**11**) was taken to phase II studies as an inhaled antiasthmatic but has since been dropped from development. Subsequently, much of the recent effort on this class of antagonists has been focused on strategies to improve the bioavailability and pharmacokinetic profiles as exemplified by compound **12**. This compound is reported to have 59% bioavailability in rat when administered at 10 mg/kg, p.o. (41).

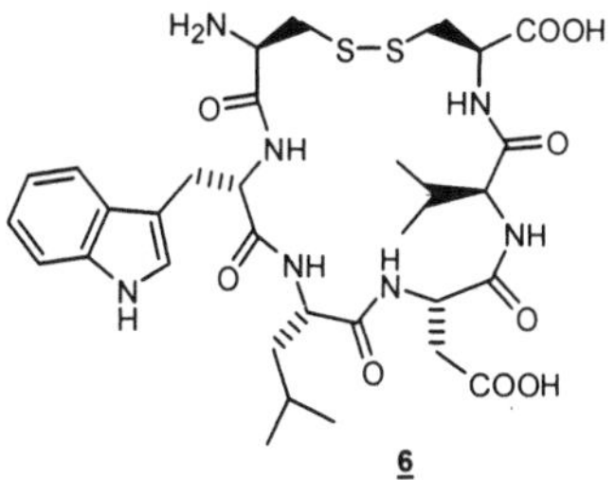

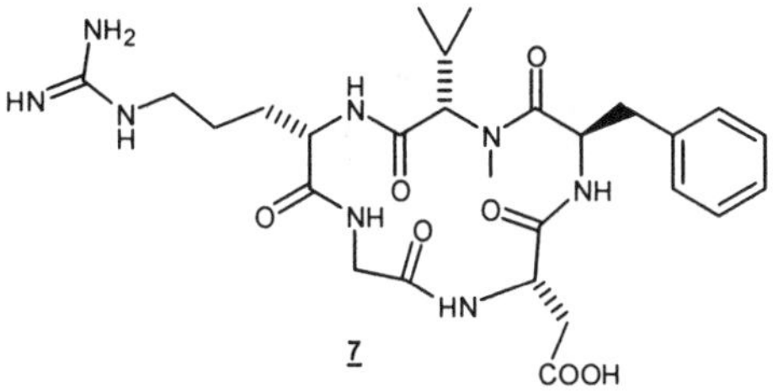

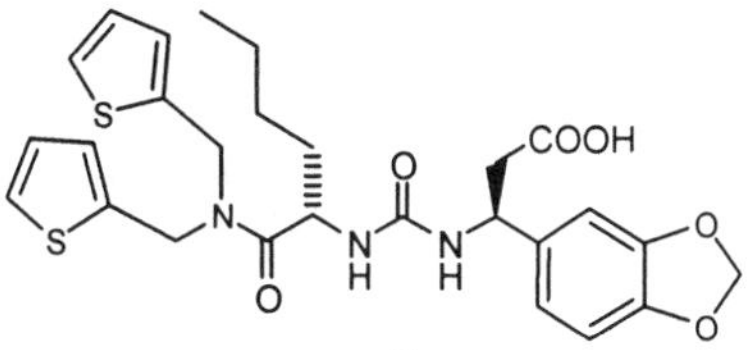

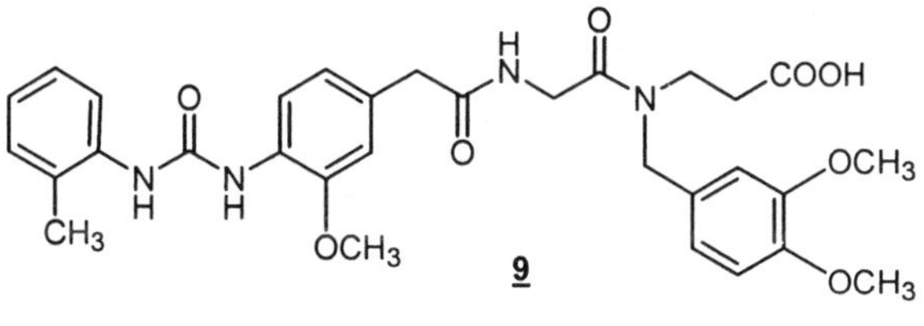

The recent patent literature contains additional examples of phenylureido analogs that contain modifications that would be expected give enhanced potency and pharmacological properties (42, 43). Since no data is presented, the effect of these modifications is not immediately known. Some specific examples include the N-morpholinoethyl analogs **13-15**, the dihydroindole **16**, and the tetrahydroquinolines **17** and **18** (stereochemistry not defined).

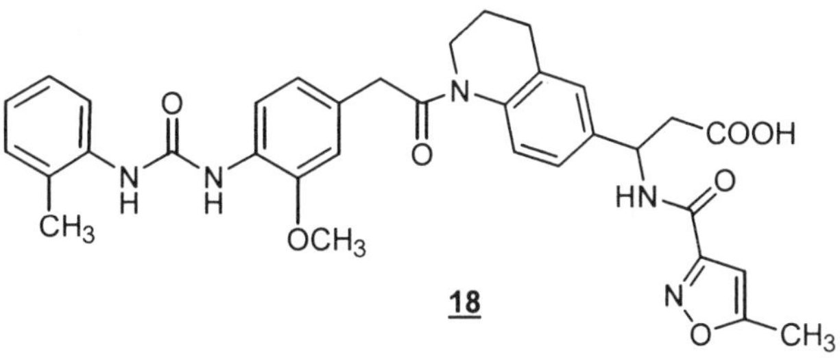

18

Finally, as exemplified by **19** and **20**, a series of pyridone and imidazole based analogs have appeared in which the diphenyl urea moiety is either mimicked or masked by a 2-(phenylamino)benzoxazole (44). These compounds are reported to be active in both the ovalbumin delayed hypersensitivity and collagen induced arthritis mouse models but specific activities are not delineated.

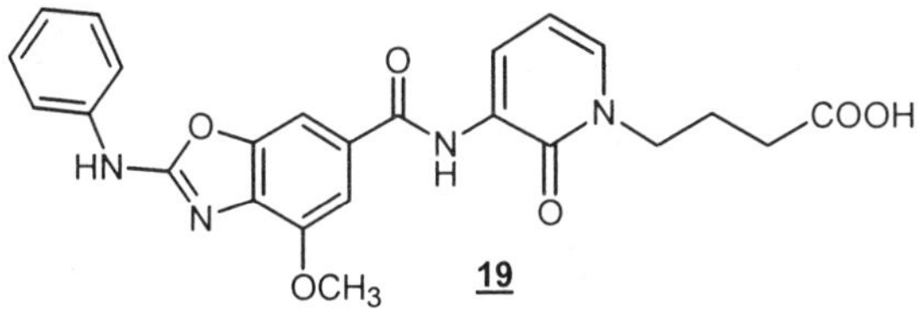

19

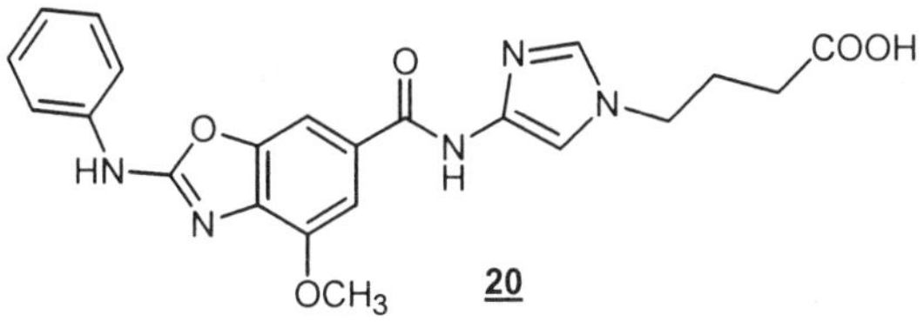

20

High throughput screening of compound libraries has lead to the discovery of N- and C-capped L-phenylalanines with VLA-4 antagonist activity. These early leads were rapidly explored by high throughput synthetic techniques and several groups have now selected compounds from this class of antagonists for clinical development (10). In general, a wide range of substitution is permitted as an N-cap at the para-position of the phenyl ring. The 2,6-dichlorophenyl moiety has been utilized as a key structural feature by several groups to obtain low nanomolar active compounds. Compound **21** appears to be a hybrid of several different structural types and is reported to have an IC_{50} of 0.01 nM in a VCAM expressing CHO / HL-60 cell binding assay (45). R-411 (**22**) was the most active (IC_{50} = 12 nM, Ramos cell) of a series of related imide and lactam derivatives prepared to probe its bioactive conformation (46). This study serves to illustrate the complexities of interpreting SAR data derived with flexible binding motifs. 2,6-Dichlorobenzyl ethers of tyrosine, e.g. **23** (IC_{50} = 35 nM), were found to yield compounds which retained activity (47).

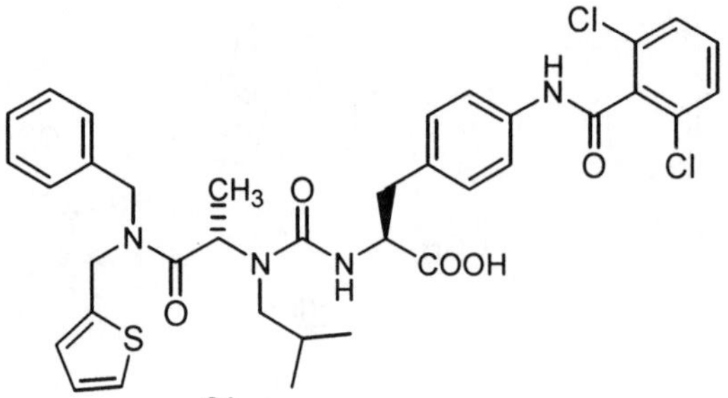

21

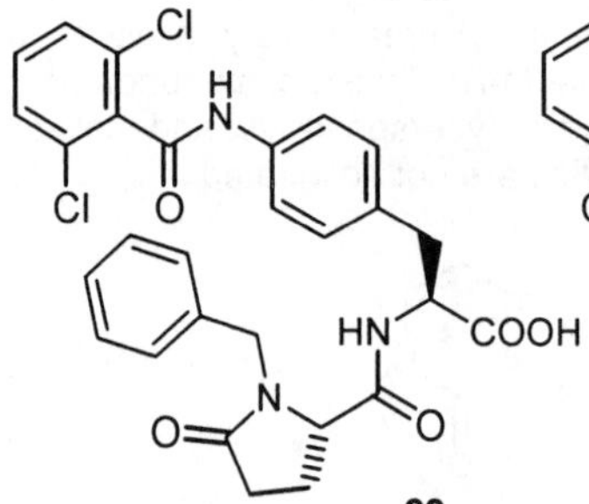

22

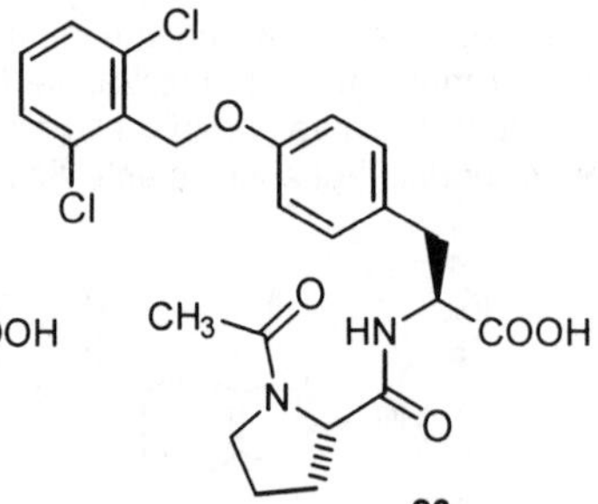

23

Several groups have explored sulfonamide derivatives such as **24** and **25**. Compound, **24**, was evaluated in a pharmacokinetic study in four species, rat, beagle dog, rhesus monkey and sheep. The compound was bioavailable across all species (23% in rhesus monkey -59% in rat) but was rapidly cleared. In rat, **24** is excreted mainly unchanged in the bile with the acyl glucuronide being the only major metabolite identified (48). Another group working with sulfonamide **25** observed that incorporation of a piperidine into the molecule could be used to modulate protein binding (49,50).

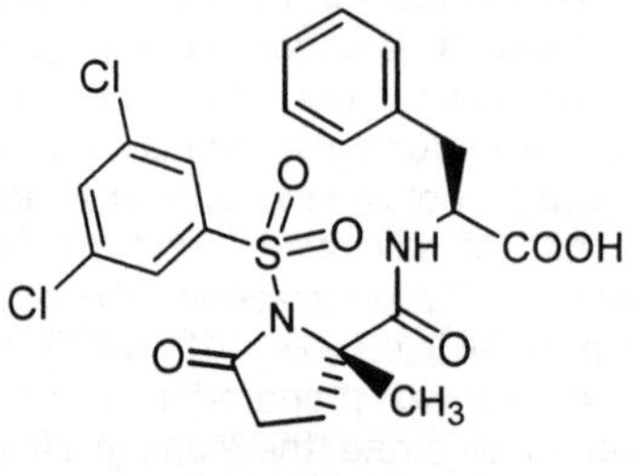

24

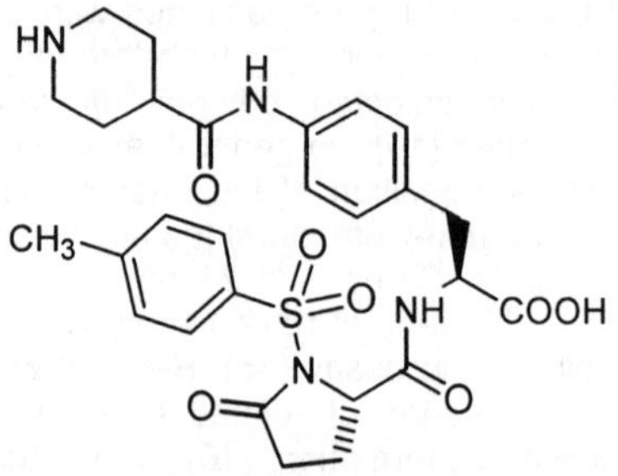

25

More recently it has been found that the phenylalanine can be elaborated to biphenyl derivatives. The 2,6-dimethoxy derivative **26** was the one of the earliest examples reported (51,52). Some of the more recent examples from the patent literature include **27**, **28** and **29** but no biological data has yet been reported (53-55).

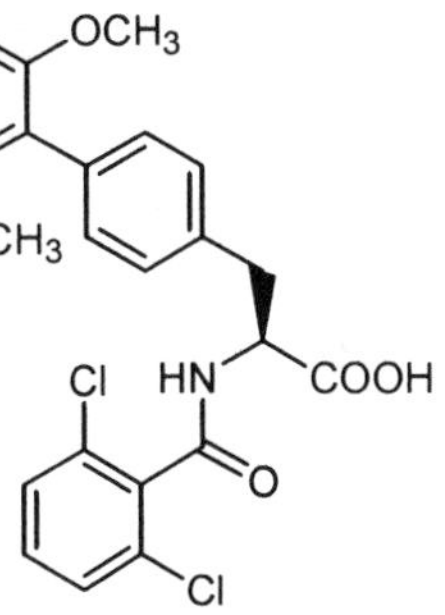

26

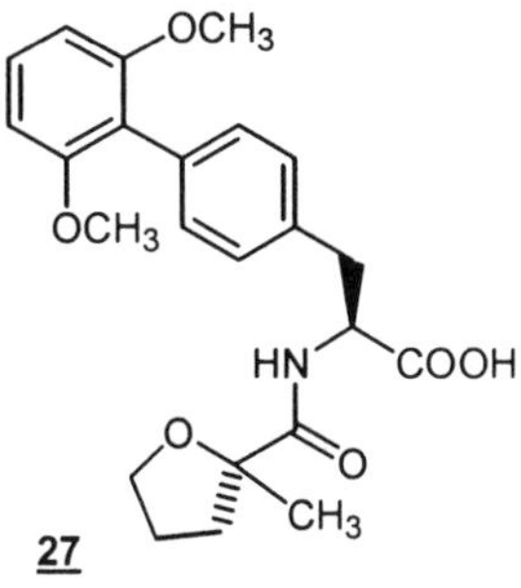

27

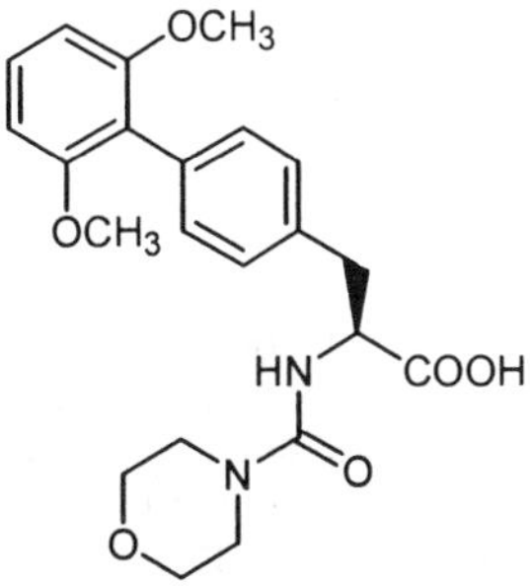

28

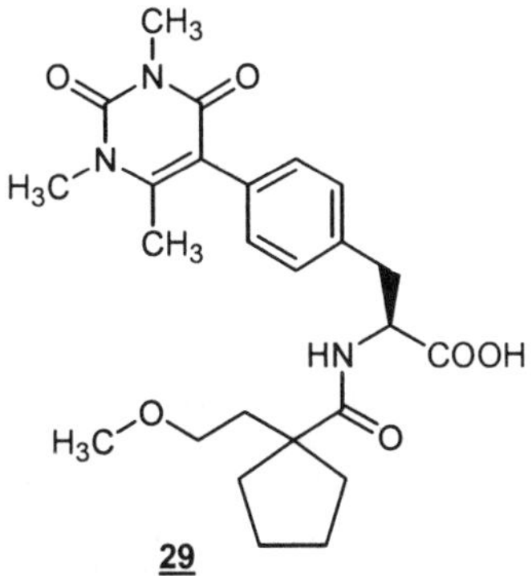

29

The recent patent literature reveals a new series of antagonists, e.g. **30**, in which the phenylalanine moiety is replaced by an unnatural amino acid containing a 2-oxopyrimidine core (56). Additionally, the N-cap moiety is constructed so as to form a urea linkage, preferably utilizing a 4-substituted piperidine as exemplified by **31** - **33**. Analog **30** is reported to have an IC$_{50}$ of 0.97 nM in an VLA-4-IgG /CS-1 binding assay.

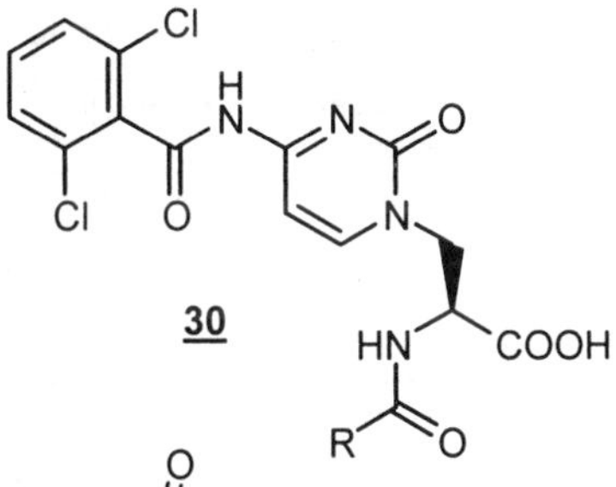

30

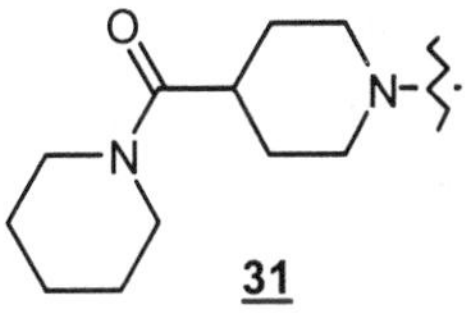

31

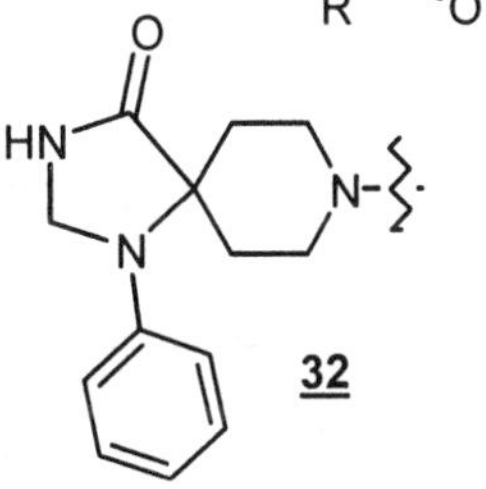

32

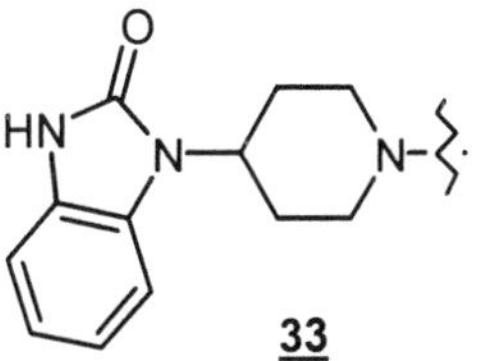

33

Since a carboxylic acid is essential for activity and is a key structural feature of all classes of VLA-4 antagonists, it is to be expected that metabolism will occur *via* acylglucuronidation and reports to this effect are beginning to appear (48).

Unfortunately, standard carboxylic mimics that may be resistant to glucuronidation have not proven to be active compounds (33). However, the patent literature discloses compound **34** which contains an acylsulfonamide as a carboxylic mimic and this may be the first example of a non-carboxylic acid containing analog with activity (57).

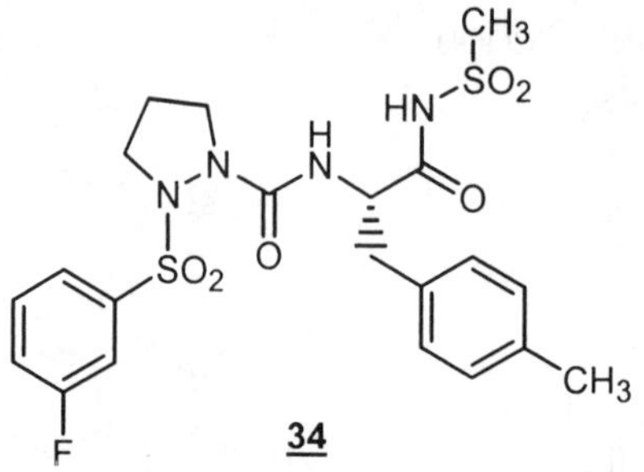

34

Summary – The role of adhesion molecules, and especially the interactions of VLA-4/VCAM-1, in the pathogenesis and progression of inflammatory and autoimmune disease continues to be elucidated (58). SAR and *in silico* studies are beginning to bring clarity to selective inhibition between α4β1 and α4β7 and our understanding of the relationship between agonist binding and the activation state of the target cells has increased. Because a carboxylic acid moiety is a requirement for activity in all class of antagonists, finding molecules with an acceptable bioavailability and pharmacokinetic profile has been especially challenging. Rapid clearance, high protein binding and acylglucuronide formation remain as significant obstacles. Even so, much progress has been made in the search for clinically useful antagonists of VLA-4 since last reviewed in this Annual. Although we do not yet know all of the biological and structural details, at least seven compounds have been taken into clinical evaluation (Table 1) and a number of others are as yet undisclosed.

Table 1: Small Molecule VLA-4 Antagonists in Clinical Development

Compound	Code	Clinical Phase	Delivery	Clinical Indications
9	IVL-745	II	inhaled	asthma
10		II	inhaled	inflammation, asthma, multiple sclerosis
11	Bio-1211	discontinued	oral / inhaled	Inflammatory bowel disease, multiple sclerosis
22	R-411	II	oral	asthma, others
NSA	Bio-1515		oral	inflammatory bowel disease, asthma
NSA	1031	II	inhaled	asthma
NSA	Bio-1272			asthma

NSA : No Structure Available

References

1. R.O. Hynes, Cell, 69, 11 (1992).
2. M.E. Hemler, C. Huang, Y. Takada, L. Scharz, J.L. Strominger and M.L. Clabby, J. Biol. Chem., 262, 11478 (1987).
3. B.S. Bochner, F.W. Luscinskas, M.A. Gimbrone, W. Newman, S.A. Sterbinsky, C.P. Derse-Anthony, D.Klunk and R.P. Schleimer, J. Exp. Med., 173, 1553 (1991).
4. P. Kubes, X.F. Niu, C.W. Smith, M.E. Kehrli, P.H. Reinhardt. R.C. Woodman, FASEB J., 9, 1103 (1995).
5. T.B. Issekutz, M. Miyasaka, A.C. Issekutz, J. Exp. Med., 183, 2175 (1996).
6. M.J. Elices, L. Osborn, Y. Takada, C. Crouse, S. Luhowskyj, M.E. Hemler and R.R. Lobb, Cell, 60, 577 (1990).
7. E.A. Wayner, A. Garcia-Pardo, M.J. Humpheries, J. A. McDonald and W.G. Carter, J. Cell Biol., 109, 1321(1989).
8. J.J. Pinwinski, N-Y Sih and M.M. Billah, Ann. Reprts Med. Chem., 34, 61 (1999).
9. S.P. Adams and R.R. Lobb, Ann. Reports Med. Chem., 34, 179 (1999).
10. J.W. Tilley and A. Sidduri, Drugs of the Future, 26 (10), 985 (2001).
11. W.M. Abraham, M.W. Seilczak, A. Ahmed, A. Cortes, I.T. Lauredo, J. Kim, B. Pepinsky, C.D. Benjamin, D.R. Leone, R.R. Lobb, R.R and P.F. Weller, J. Clin. Invest., 93, 776 (1994).
12. P.L. Chisholm, C.A. Williams and R.R. Lobb, Eur. J. Immunol., 23, 682 (1993).
13. T.A. Yednock, C. Cannon, L.C. Fritz, F. Sanchez-Madrid, L. Steinman and N. Karin, Nature, 356, 63 (1992).
14. J.L. Baron, E. Reich, I. Visintin and C.A. Janeway, J. Clin. Invest., 93, 1700 (1994).
15. L.C. Burkly, A. Jakubowski and M. Hattori, Diabetes, 43, 529 (1994).
16. D.K. Podolsky, R.R. Lobb, N. King, C.D. Benjamin, B. Pepinsky, P. Sehgal and M. deBeaumont, J. Clin. Invest., 92, 372 (1993).
17. S.M. Wahl, J.B. Allen, K.L. Hines, T. Imamichi, A.M. Wahl, L.T. Furcht and J.B. McCarthy, J. Clin. Invest., 94, 655 (1994).
18. M. Isobe, J. Suzuki, H. Yagita, K. Okumura and M. Sekiguchi, Transplantation Proc., 26, 867 (1994).
19. S. Molossi, M. Elices, T. Arrhenius, R. Diaz, C. Coulber and M.J. Rabinovitch, J. Clin. Invest., 95, 2601 (1995).
20. M.I. Cybulsky and M.A. Gimbrone, Science, 251, 788 (1991).
21. M.I. Cybulsky, K. Iiyama, H. Li, S. Zhu, M. Chen, M. Iiyama, V. Davis, J-C Gutierrez-Ramos, P.W. Connelly and D.S. Milstone, J. Clin. Invest., 107, 1255 (2001).
22. C. Joneckis, R. Ackley, E. Orringer, E. Wayner and L. Parise, Blood, 82, 3548 (1993).
23. R.A. Swerlick, J.R. Eckman, A. Kumar, M. Jeitler and T.M. Wick, Blood, 82, 1891 (1993).
24. B. Gee and O. Platt, Blood, 85, 268 (1995).
25. G.A. Lutty, M. Taomoto, J. Cao, D.S. McLeod, P. Vanderslice, B.W. McIntyre, M. E. Fabry and R. Nagel, Invest. Opthalmol. Vis. Sci., 42, 1349 (2001).
26. P. Vanderslice, C.L. Munsch, E. Rachal, D. Erichsen, K.M. Sughrue, A.N. Truong, J.N. Wygant, B.W. McIntyre, S.G. Eskin, R.G. Tilton and P.J. Polverini, Angiogenesis, 2, 265 (1998).
27. J.T. Yang, H. Rayburn and R.O. Hynes, Development, 121, 549 (1995).
28. L. Kwee, H.S. Baldwin, H.M. Shen, C.L. Stewart, C. Buck, C.A. Buck and M.A. Labow, Development, 121, 489 (1995).
29. G.C. Gurtner, V. Davis, H. Li, M.J. McCoy, A. Sharpe and M. I. Cybulsky, Genes & Development, 9, 1 (1995).
30. S. Spence, C. Vetter, W.K. Hagmann, G. Van Riper, H. Williams, R.A. Mumford, T.J. Lanza, L.S. Lin, and J.A. Schmidt, Teratology, 65, 26 (2002).
31. D.M. Nowlin, F. Gorcsan, M. Moscinski, S. Chang, T.J. Lobl, and P.M. Cardarelli, J. Biol. Chem., 268, 20352 (1993).
32. P. Vanderslice, K. Ren, J.K. Revelle, D.C. Kim, D. Scott, R.J. Bjercke, E.T.H. Yeh, P.J. Beck, and T.P. Kogan, J. Immunol., 158, 1710 (1997).
33. G.W. Holland, J.M. Kassir, R.J. Biediger, A. Bourgoyne, Q. Chen, E.R. Decker, V.O. Grabbe, K.M. Keller, T.P. Kogan, G. Krudy, S. Lin, R.V. Market, D. Maxwell, C. Munsch, B.G. Raju, K. Ren, I.L. Scott, R. Tilton, H. West, P. Vanderslice, T. You and R.A. Dixon, Book of Abstracts, 219th Amer. Chem. Soc. Natl. Mtg., San Francisco, CA, MEDI 10 (2000).
34. J-P Xiong, T. Stehle, R. Zhang, A. Joachimiak, M. Frech, S.L. Goodman, and M.A. Arnaout, Science, 296, 151 (2002).

35. T.J. You, D.S. Maxwell, T.P. Kogan, Q. Chen, J. Li, J. Kassir, G.W. Holland, and R.A.F. Dixon, Biophysical J., 82, 447 (2002).
36. P. Lockey, A. Stoppard, E. White, M. White and W. Wong, Am. J. Respir. Crit. Care Med., 161, A201 (2000).
37. P. Bahra, K. Ebsworth, C.E. Lawrence, S.E. Weber and R.J. Williams, Am. J. Respir. Crit. Care Med., 161, A201 (2000).
38. J.A. Cairns, K. Kaik, J. Lloyd, S.E. Weber and R.J. Williams, Am. J. Respir. Crit. Care Med., 161, A202 (2000).
39. K. Ebsworth, C.E. Lawrence, S.E. Weber and R.J. Williams, Am. J. Respir. Crit. Care Med., 161, A201 (2000).
40. S.L. Underwood, L. Prince, K. Page, S.E. Weber and M.L. Foster, Am. J. Respir. Crit. Care Med., 161, A201 (2000).
41. J,J, Baldwin, E. McDonald, K.J. Moriarty, C.R. Sarko, N. Machinaga, A. Nakayama, J. Chiba, S. Hmura, and Y. Yoneda, Patent Application WO 01/00206 A1 (2001).
42. S. Wattanasin, and B. Weidmann, Patent Application WO 01/42192 A2 (2001).
43. J-D. Bourzat, A. Commercon, C. Filoche, V.N. Harris, and C. McCarthy, Patent Application WO 00/15612 (2000).
44. D.R. Brittain, M.S. Large, and G.M. Davies, Patent Application WO 01/53295 A1 (2001).
45. A. Okuyama, S. Ikegami, T. Maruyama, Y. Matsumura, N. Nagata, H. Fukui, and K. Fujimoto, Patent Application WO 01/32610 A1 (2001).
46. J.W. Tilley, G. Kaplan, K. Rowan, V. Schwinge, and B. Wolitzky, Bioorg. Med. Chem. Lett., 11, 1 (2001).
47. S.C. Archibald, J.C. Head, N. Gozzard, D. w. Howat, T.A.H. Parton, J.R. Porter, M. K. Robinson, A. Shock, G.J. Warrellow and W.A. Abraham, Bioorg. Med. Chem. Lett., 10, 997 (2000).
48. I.E. Kopka, D.N. Young, L.S. Lin, R.A. Mumford, P.A. Magriotis, M. MacCoss, S.G. Mills, G. Van Riper, E. McCauley, L.E. Egger, U. Kidambi, J.A. Schmidt, K. Lyons, R. Stearns, S. Vincent, A. Colletti, Z. Wang, S. Tong, J. Wang, S. Zheng, K. Owens, D. Levorse, and W.K. Hagmann, Bioorg. Med. Chem. Lett., 12, 637 (2002).
49. D. Sarantakis, R.B. Baudy, J.J. Bicksler et.al., Abstracts 220th Amer. Chem. Soc. Natl. Mtg., MEDI 136 (2000).
50. D. Sarantakis, J.J. Bicksler, C.Cannon, et.al., Abstracts 220th Amer. Chem. Soc. Natl. Mtg., MEDI 137 (2000).
51. I. Sircar, K. Gudmundsson, R. Martin, et.al., Abstracts 218th Amer. Chem. Soc. Natl. Mtg., MEDI 59 (1999).
52. I. Sircar, K. Gudmundsson, R. Martin, et.al., Abstracts 218th Amer. Chem. Soc. Natl. Mtg., MEDI 60 (1999).
53. W.K. Hagmann, S.E. Delaszlo, G. Doherty, L.L. Chang, and G.X. Yang, Patent Application WO 01/12183 A1 (2001).
54. S.E. Delaszlo, W.K. Hagmann, and T.M. Kamenecka, Patent Application WO 01/14328 A2 (2001).
55. A. Sidduri, and J.W. Tilley, Patent Application WO 01/42225 A2 (2001).
56. T. Takahashi, T. Ishigaki, M. Funahashi, K. Taniguchi, M. Kaneko, M. Kainoh, and H. Meguro, Patent Application WO 02/22563 A1 (2002).
57. S.E. de Laszlo, Patent Application WO 99/20272 A1 (1999).
58. H. Yusuf-Makagiansar, M.E. Anderson, T.V. Yakovleva, J.S. Murray, and T.J. Siahaan, Med. Res. Rev., 22, 146 (2002).

Chapter 8. Purine and Pyrimidine Nucleotide (P2) Receptors

Kenneth A. Jacobson

Molecular Recognition Section, Laboratory of Bioorganic Chemistry, National Institute of Diabetes and Digestive and Kidney Diseases, Bethesda Maryland 20892-0810, USA

Introduction – The existence of P2 nucleotide receptors was proposed originally by Burnstock, based on studies of the actions of ATP (1) on smooth muscles and blood vessels (1). Further evidence for a role for ATP as a transmitter was obtained more recently in the central nervous system, where P2 receptors are now known to be plentiful. The observation that ATP served as a transmitter substance was initially met with skepticism, since: 1) ATP was known to have an important intracellular role as the "energy currency" of the cell, and an extracellular role was considered doubtful. 2) ATP is readily hydrolyzed to adenosine in biological systems and the effects of ATP were difficult to distinguish from those of adenosine. The field of P2 receptor medicinal chemistry was slow to develop due to a variety of limitations in the study of nucleotide pharmacology. Such limitations include: Lack of control over the endogenous pool of nucleosides and nucleotides, lability of the known endogenous and synthetic ligands, rapid desensitization of some effects, and finally the difficulty of synthesizing pure nucleotide analogues for SAR studies. A further difficulty of studying these receptors made apparent through their cloning is uncertainty in the correspondence between native and expressed receptors.

Two large families of receptors, P2X (ligand-gated ion channels) and P2Y (G protein-coupled) receptors, were cloned beginning in 1993. The range of putative endogenous ligands acting at P2Y receptors has now expanded to include ADP (2), UTP (3), and UDP (4), as well as ATP. A large family of ectonucleotidases has been characterized and is closely associated with the production and degradation of P2 receptor agonists *in vivo* and generation of adenosine, which acts at a separate class of receptors (2). Intense medicinal chemical efforts are now directed towards designing selective agonists and antagonists for some of the 14 (or more) P2 receptor subtypes. Many of the existing P2 ligands, so far, are limited in studies of their pharmacological application by low affinity and selectivity, low bioavailability due to polyanionic groups and large molecular weights and low stability as a result of hydrolysis of phosphate groups. All of the known P2 agonists are nucleotide analogues. A major challenge to the medicinal chemist is to find uncharged, low molecular weight P2 antagonists. Both rational, structure-based approaches and high throughput screening techniques have been adopted towards this goal.

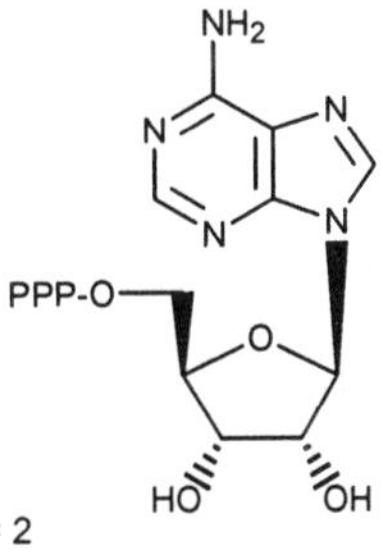

1, n = 2
2, n = 1

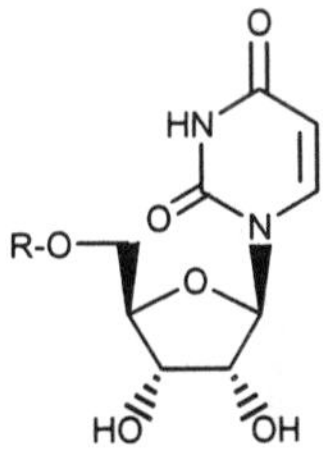

3, R = PPP
4, R = PP

PP =

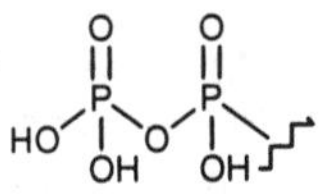

PPP =

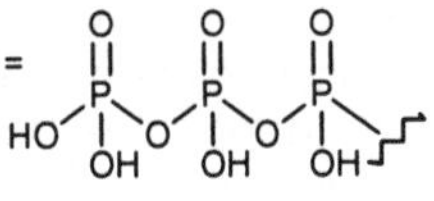

P2X RECEPTORS: STRUCTURE AND LIGANDS

There are seven subtypes of subunits of P2X ligand-gated ion channels (denoted P2X$_{1-7}$), and each functional ion channel consists of oligomers, probably trimers. Both homo- and heterooligomerization seem to occur commonly, however the P2X$_7$ subtype exists exclusively as a homomer and the P2X$_6$ subtype seems to never occur as a homomer (3). The cationic selectivity of the P2X receptors is Na$^+$, K$^+$ > Ca^{2+} (4). The distribution is broad for many of the subtypes, and many cell types express multiple P2X subunits. The P2X$_3$ subunit is of special interest in pain control, since it occurs mainly on the pain pathways, thus selective antagonists may be useful for this envisioned application (5). Other therapeutic interests related to P2X receptors are: inflammation (P2X$_7$), neuroprotectin (P2X$_7$), bladder control (P2X$_1$), cardiac function (P2X$_4$), and the central nervous system.

<u>P2X Agonists</u> – α,β–meATP (**5**) distinguishes Group I receptors, i.e. P2X$_1$ and P2X$_3$ receptors, which are activated by **5**, from the other subtypes, which are insensitive. Another distinguishing feature is the ability of some subtypes (mainly Group I) to desensitize rapidly. Thus, agonist pre-activation of P2X$_1$ receptors leading to desensitization, for example by **5**, has been used pharmacologically in place of antagonists.

The 5'-di- vs. triphosphate derivatives generally diverge in activity at P2X subtypes. Only at the P2X$_1$ receptor does ADP elicit agonist action. None of the P2X receptors are activated by AMP or adenosine, although 2-thioether substitution of the 5'-monophosphate produces measurable activity at some P2X subtypes (6). Both α,β–methylene, **5**, and β,γ–methylene modifications, as well as γ–thiophosphate analogues generally increase stability of the triphosphate group to enzymatic hydrolysis while maintaining potency at P2X receptors.

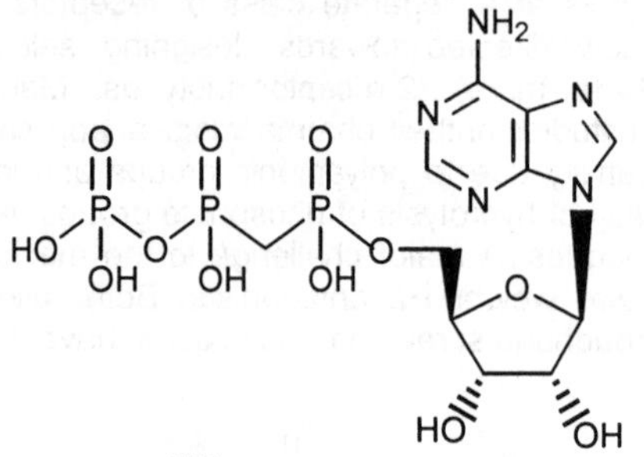

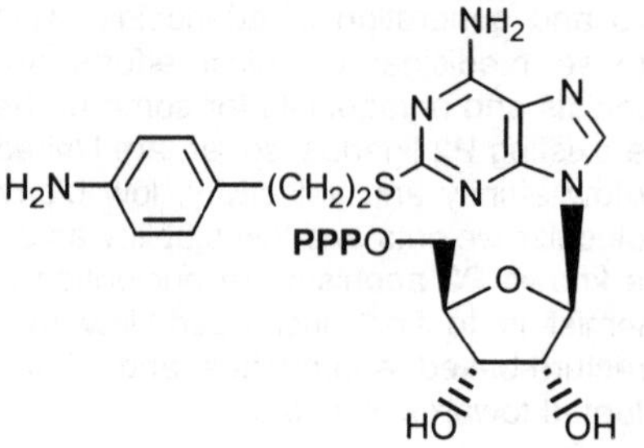

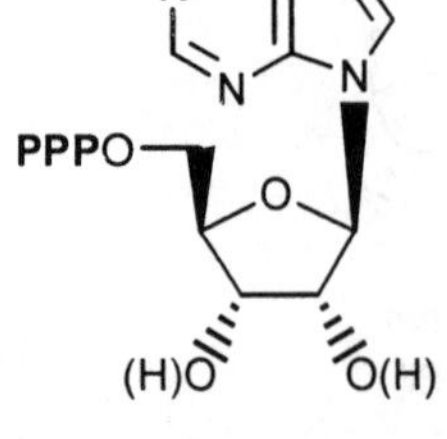

Substitutions of adenine moiety of ATP known to enhance potency are 2-thioethers. Thus, the *p*-aminophenylethyllthio analogue (PAPET) (**6**) is the most potent agonist yet reported (EC$_{50}$ 17 nM) of the rat P2X$_3$ receptor (6). Ribose substitution has not been fully explored at P2X receptors. The ester derivative 2′-&3′-O-(4-benzoyl-benzoyl)-ATP

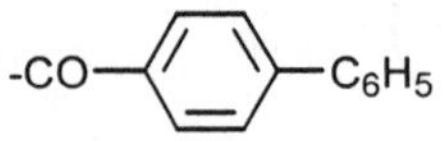

(BzATP) (**7**) is of nanomolar potency in activating the P2X$_1$ receptor and, although less potent, is nevertheless the most potent reported activator of the P2X$_7$ receptor (7). Dinucleotides of varying numbers of linked phosphate groups have been studied as agonists and partial agonists of P2X receptors (8).

P2X Antagonists – Derivatives of the trypanocidal drug suramin have been introduced as P2X antagonists, an action unrelated to the clinical use of suramin (9). For example, NF279 (**8**) and NF449 (**9**) have roughly nanomolar IC$_{50}$ values at the rat P2X$_1$ receptor (10). Antagonists derived from dyes, such as Reactive-blue 2, antagonize both P2X and P2Y receptors (11). The SAR of pyridoxal-phosphate derived antagonists, introduced by Lambrecht and coworkers have been explored at P2X receptors (9).

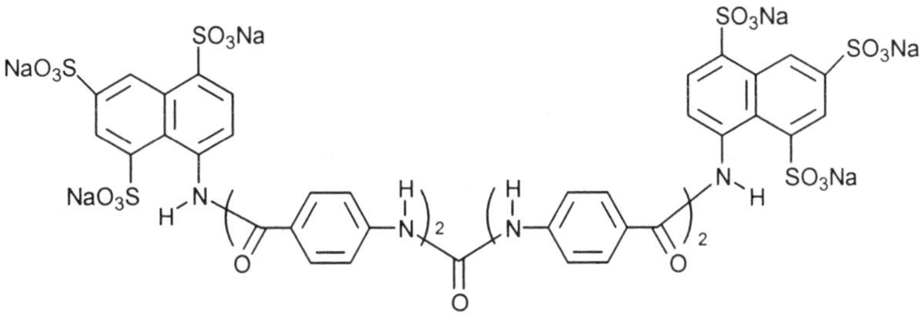

8

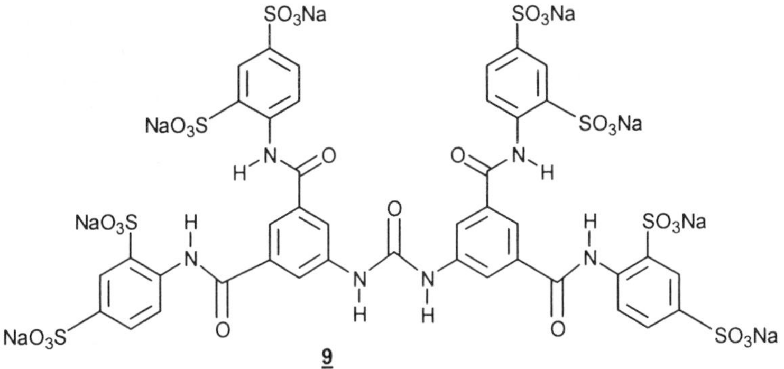

9

PPADS (**10a**) and its 2',5'-sulfonate isomer "isoPPADS" (**10b**) are somewhat more potent at P2X than at P2Y receptors. The phosphonate analogue MRS 2257 (**11**) is a highly potent antagonist at both rat P2X$_1$ and P2X$_3$ receptors (IC$_{50}$ 11 nM in an electrophysiological model) (12). PPNDS (**12**) is a highly potent antagonist at the P2X$_1$ receptor (13). Replacement of the diazo linkage of PPADS and its congeners with a carbon bridge has been accomplished, with MRS 2335 (**13**), which is roughly equipotent to the corresponding diazo derivative at the P2X$_1$ receptor. While high potency at P2X receptors has been achieved for PPADS derivatives, a disadvantage is the noncompetitive binding they display, accompanied by slow on- and off-rates (14).

Nucleotide derivatives, which have more favorable binding kinetic properties, have also been found to antagonize P2X receptors. Faster on- and off-rates of binding than for PPADS have been demonstrated for the antagonist TNP-ATP (**14**), which binds potently to P2X$_1$, P2X$_3$, and P2X$_{2/3}$ (heteromeric) receptors (14). A dinucleotide derivative, Ip$_5$I (**15**), potently antagonizes the P2X$_1$ receptor (15).

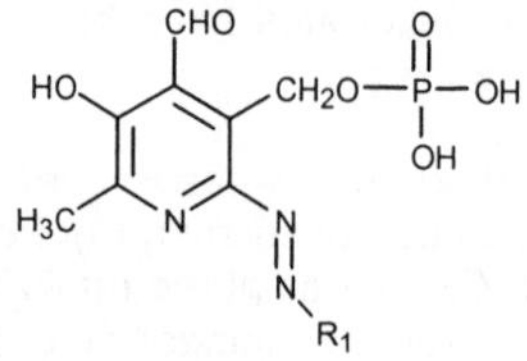

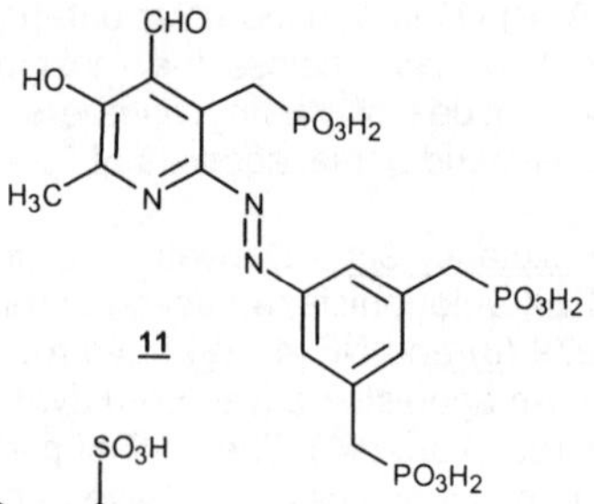

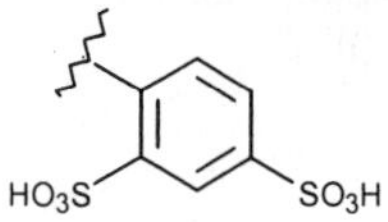

10a 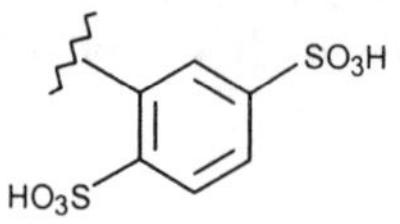   10b 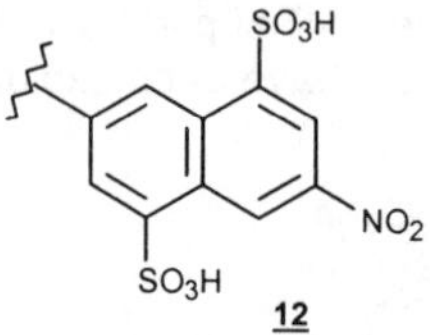   12

At the P2X$_4$ receptor, none of the commonly used antagonists bind appreciably. At P2X$_5$ and P2X$_6$ receptors, antagonists have not yet been identified, partly due to the difficulty of expressing these receptors in Xenopus oocytes. At the P2X$_7$ receptor, which occurs naturally in many mast cell cultures, antagonists have been identified among a variety of tyrosine isoquinoline derivatives, e.g. KN-62 (**16**), which were originally introduced as inhibitors of CaM kinase II (16). KN-62 antagonizes human P2X$_7$ receptors more potently than mouse or rat homologues. The tyrosine derivative MRS 2409 (**17**) and derivatives act as potent antagonists of both mouse and human P2X$_7$ receptors (17). A family of adamantane derivatives such as **18** have been shown to be antagonists of the P2X$_7$ receptor (18-20). Piperidine and piperazine derivatives such as **19** and **20** have been proposed as non-nucleotide P2X$_7$ receptor antagonists (21,22).

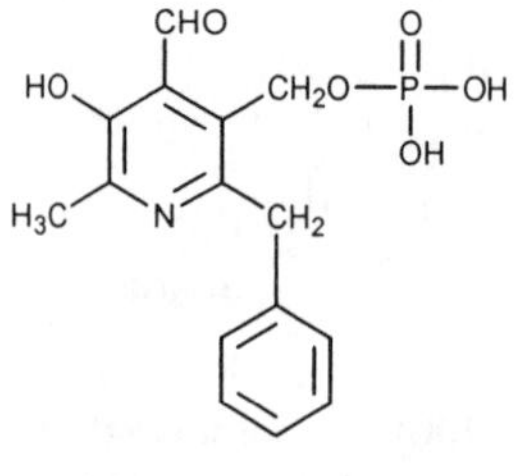

13 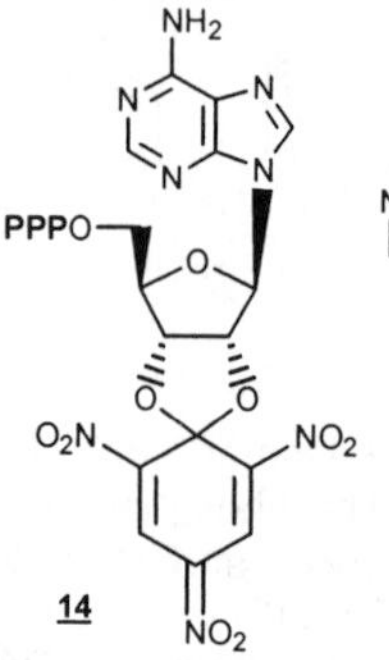   14 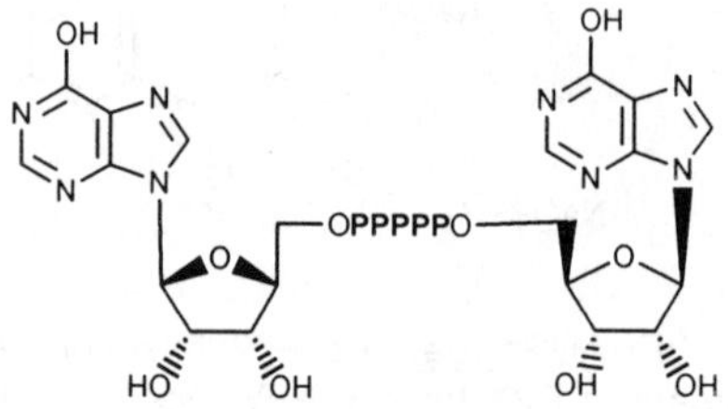   15

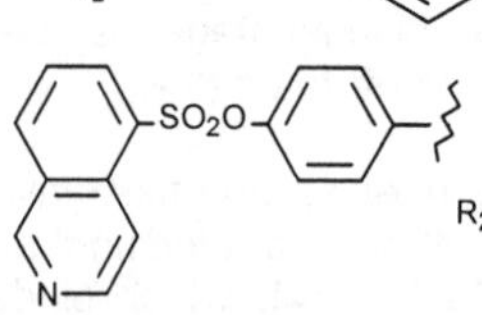

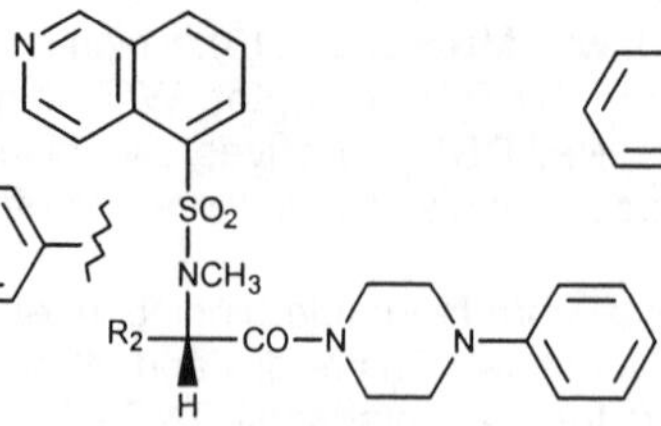

16 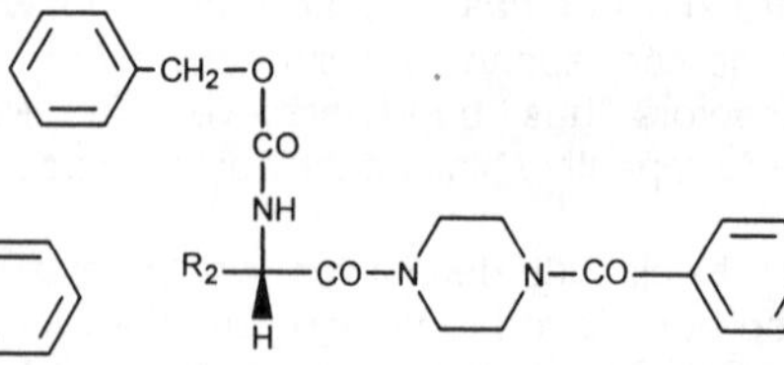   17

P2Y RECEPTOR STRUCTURE AND FUNCTION

P2Y receptors are 7TM receptors, which typically couple preferentially via Gq to phospholipase C, and thus lead to a rise in intracellular calcium. Subclasses of P2Y receptors have been defined based on: clustering of sequence (in general low homology among sybtypes), ligand preference, second messengers, and receptor sequence analysis. The numbering system for P2Y subtypes is currently discontinuous, i.e. $P2Y_{1-13}$ receptors have been defined, however six of the intermediate numbered cloned receptor sequences (e.g. $P2Y_3$, $P2Y_5$, $P2Y_7$, $P2Y_8$, $P2Y_9$, $P2Y_{10}$) do not correspond to distinct, functional mammalian nucleotide receptors (1). $P2Y_{12}$ and $P2Y_{13}$ receptors couple via Gi. Structural insights have been gained using molecular modeling based on a rhodopsin template in conjunction with mutagenesis to suggest recognition elements important for nucleotide binding in TMs 3, 5, 6, and 7 and in the 2nd extracellular loop (23).

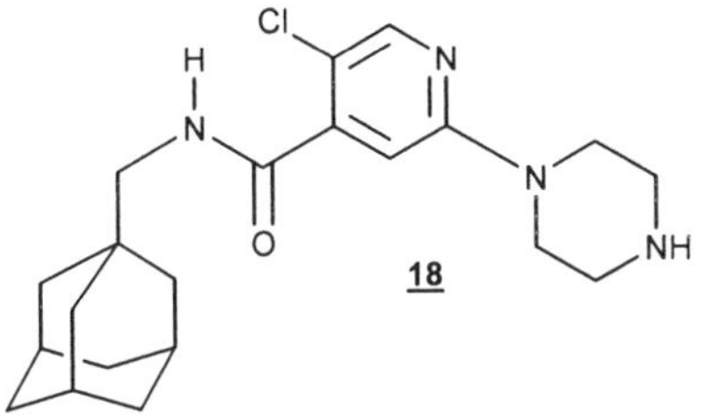

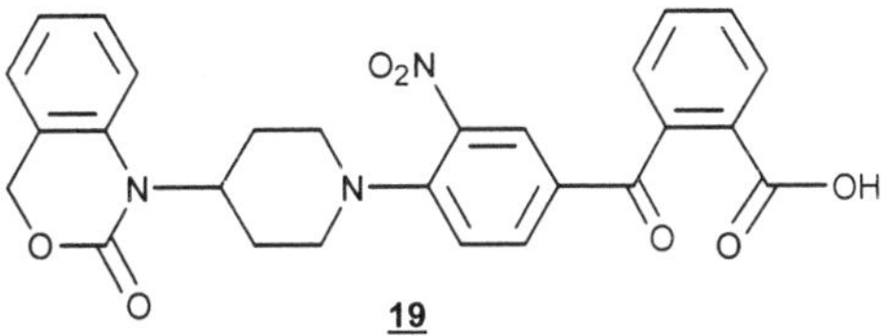

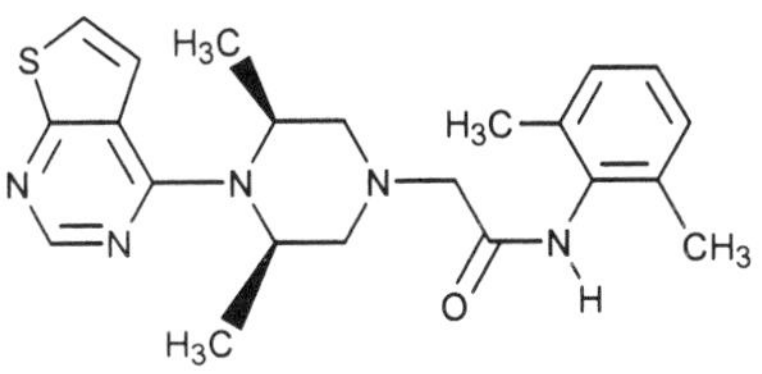

The distribution of P2Y receptors is broad, and the therapeutic interests include antithrombotic therapy, modulation of the immune system, diabetes, and treatment of cystic fibrosis and other pulmonary diseases. Other receptors classified as P2Y receptors without sequence information include a putative dinucleotide receptor (24). $P2Y_{12}$ stimulation of which is an important pro-aggregatory signal in platelets, and $P2Y_{13}$ receptors belong to a structurally distinct cluster of P2Y receptor sequences (25-27). A UDP-glucose receptor has been cloned and found to have a sequence more similar to the $P2Y_{12}$ and $P2Y_{13}$ receptors than other subtypes (26). Some recently reported orphan receptors have been noted to contain ligand recognition elements previously identified for $P2Y_1$ receptors (23,29).

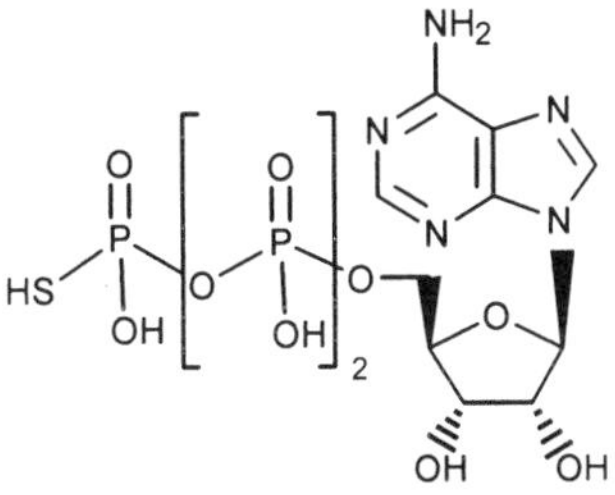

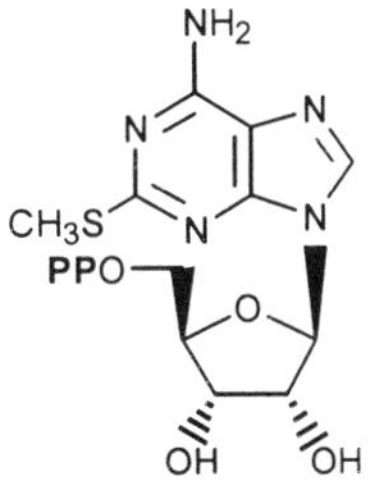

ADENINE-PREFERRING P2Y RECEPTORS: $P2Y_1$, $P2Y_{11}$, $P2Y_{12}$, AND $P2Y_{13}$

Agonists – $P2Y_1$ and $P2Y_{13}$ receptors are more potently activated by ADP than by ATP, while at $P2Y_{11}$ receptors ATP is preferred. At $P2Y_{12}$ receptors, ADP activates and ATP antagonizes. At $P2Y_{11}$ receptors, ATP-γ-S **21** is more potent than ATP. The 2-methylthio derivative of ADP **22** is a potent and selective agonist of the $P2Y_1$ receptor, while the corresponding triphosphate 2-MeS-ATP is less potent and selective (30).

A number of adenosine 5'-monophosphate analogues, such as **23** (EC$_{50}$ 70 nM at the turkey P2Y$_1$ receptor) have been found to potently activate P2Y$_1$ receptors (31). The α-thio modification of AMP analogues, i.e. **24**, increases potency at the P2Y$_1$ receptor (32). Such monophosphate derivatives have also been reported to inhibit ectonucleotidases, which complicates their use as P2Y receptor agonists. Among substitutions of the adenine moiety in nucleotides, the 2-alkylthio ethers appear to provide high P2Y$_1$ potency (33). Non-glycosidic ribose substitution in the form of anhydrohexitol **25** and the rigid (N)-methanocarba ring systems **26** have led to P2Y$_1$ receptor agonists, with EC$_{50}$ values of 1 µM and 0.4 nM, respectively (34). The binding site of the P2Y$_1$ receptor displays a preference for the "Northern" (N) conformation of the ribose-like ring over the corresponding "Southern" (S) conformation, as represented in an isomeric (S)-methanocarba series of ATP derivatives. The enhancement of P2Y$_1$ agonist potency upon freezing the preferred conformation in a pseudoribose ring may approach 300-fold.

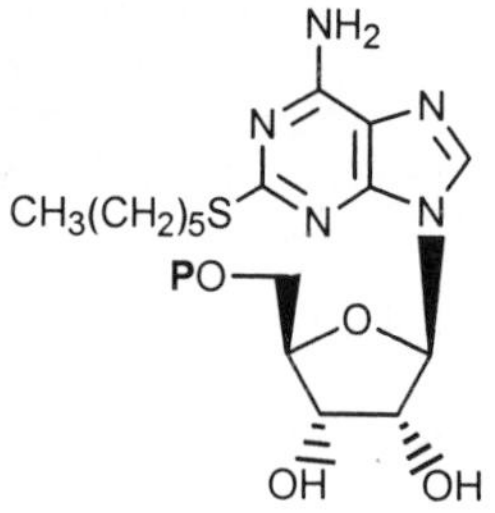

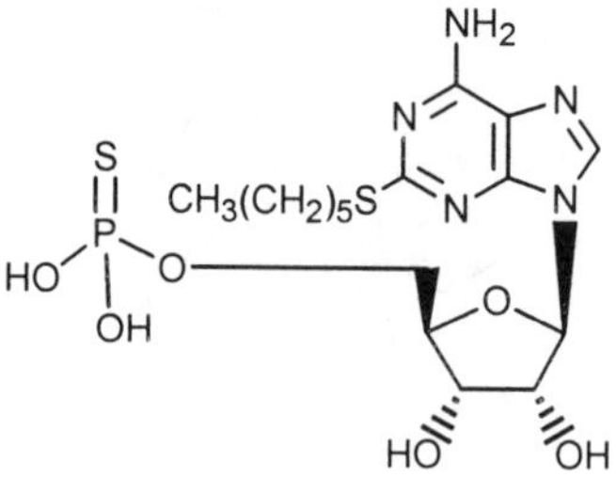

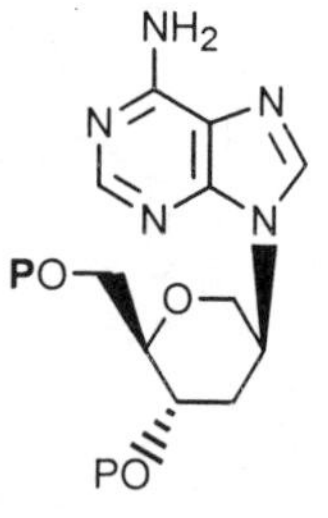

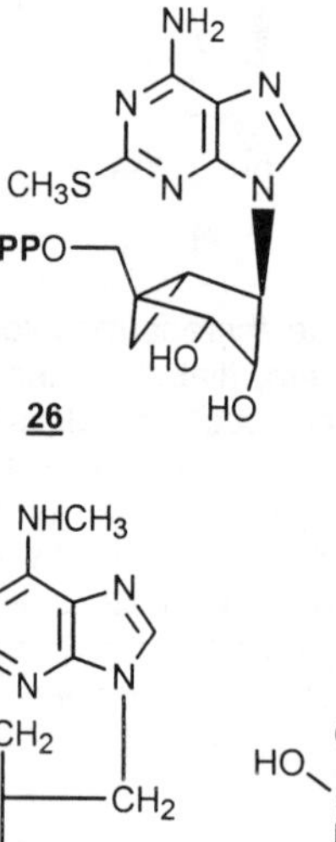

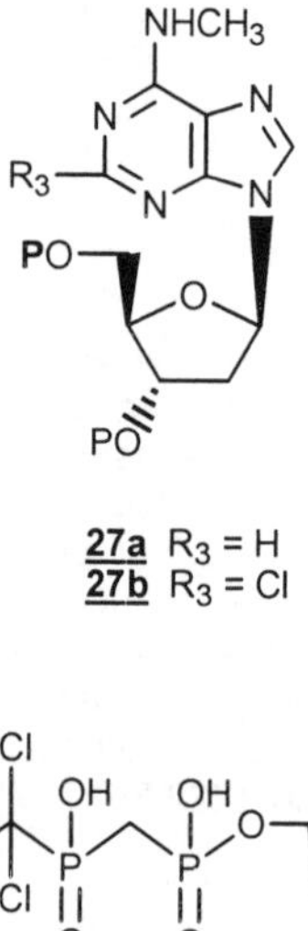

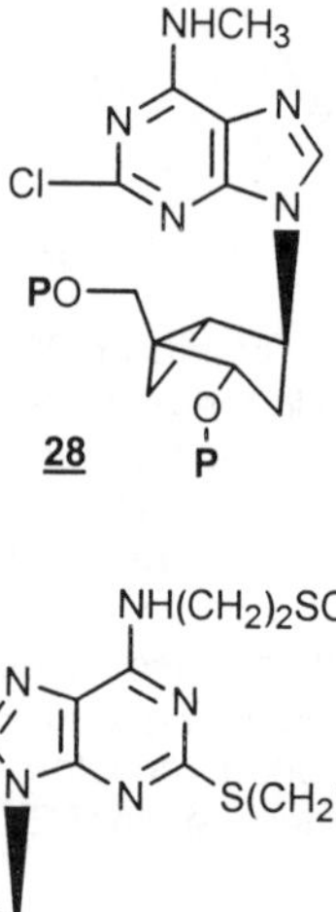

"P" = monophosphate

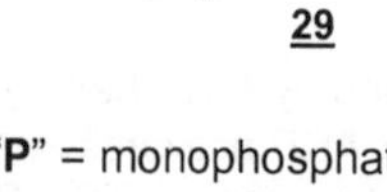

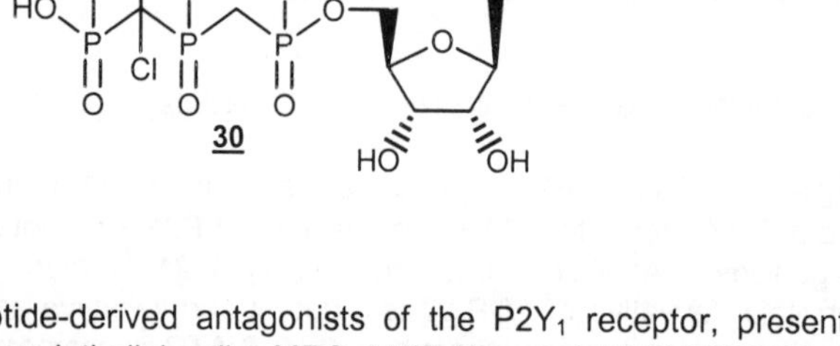

<u>Antagonists</u> – The nucleotide-derived antagonists of the P2Y$_1$ receptor, present on platelets and on vascular endothelial cells, MRS 2179 **27a** and MRS 2279 **28** were demonstrated to be high affinity competitive and selective antagonists at this subtype (27, 35). Unlike **27a**, the corresponding 2-chloro analogue MRS 2216 **27b** is inactive at

Group 1 P2X receptors (6). [^{33}P]MRS 2179 has been introduced as a radioligand for the P2Y$_1$ receptor in platelets (36). Acyclic nucleotide analogues such as MRS 2298 **29** were found to be moderately potent P2Y$_1$ receptor antagonists without residual agonism (37).

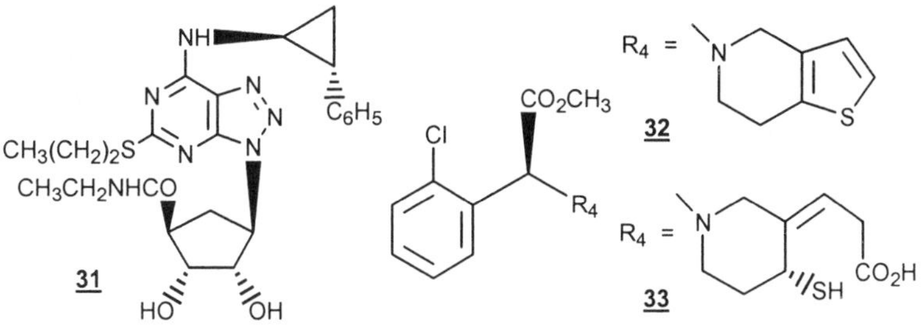

Potent and selective antagonists of the P2Y$_{11}$ or P2Y$_{13}$ receptors are still unknown. Nucleotide antagonists of high affinity for the platelet P2Y$_{12}$ receptor were reported, including ARL 69931MX **30** (38), which is in clinical trials as an antithrombotic agent. A nucleotide in this series (AR-C67085) has been shown to potently activate the P2Y$_{12}$ receptor (39). In this series has been possible to substitute the unwieldy triphosphate group in this series with short alcohols, esters, etc., thus proving that a highly anionic moiety is not needed for recognition by the P2Y$_{12}$ receptor leading to compounds such as **31** (40-44). The successful antithrombotic drug clopidogrel (**32**) produces a metabolite **33**, which acts as a P2Y$_{12}$ platelet receptor antagonist (45). Other non-nucleotide, non-highly charged P2Y$_{12}$ antagonists, such as CS-747 (**34**), which also acts through a metabolite, and the sulfonamide analog CT50547 (**35**) have been reported (46, 47).

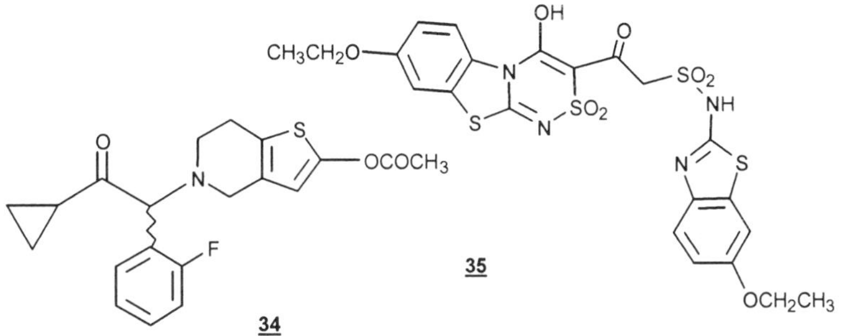

URACIL-PREFERRING P2Y RECEPTORS: P2Y$_2$, P2Y$_4$, AND P2Y$_6$

<u>Agonists</u> — The P2Y$_2$ receptor is activated by UTP or ATP, but not by the corresponding diphosphates. The P2Y$_6$ receptor is activated by di- but not triphosphates, while the human P2Y$_4$ receptor is activated by uracil nucleotide triphosphates and inhibited by adenine nucleotide triphosphates. Uridine β-thiodiphosphate (UDP-β-S) (**36**) and the γ-thiophophate (UTP-γ-S) (**37**) are selective agonists for P2Y$_6$ and P2Y$_4$ receptors, respectively (48). Substitutions of uracil moiety have been reported to reduce potency at the P2Y$_2$ receptor, while some uracil dinucleotides, such as INS 365 (**38**), potently activate the receptor (49). P2Y$_2$ receptor agonists are of clinical interest for the treatment of pulmonary and ophthalmic diseases, and possibly cancer. Ribose substitution with the (N)-methanocarba ring system has been shown to preserve the potency of both adenine and uracil nucleotides at the P2Y$_2$ receptor, and UTP (i.e. MRS2341 **39**) at the P2Y$_4$ receptor (50). However, inclusion of the same (N)-

methanocarba ring system an analogue of UDP prevents activation of the P2Y$_6$ receptor.

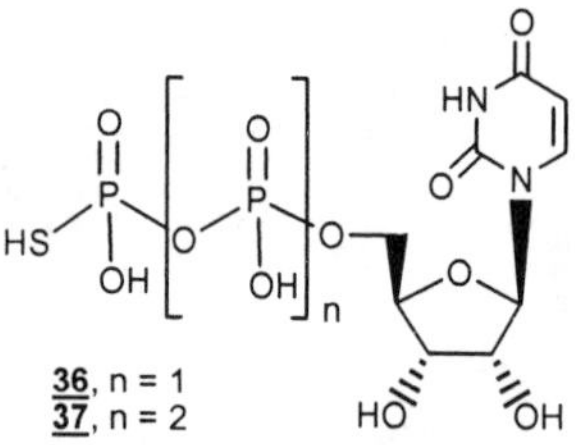

36, n = 1
37, n = 2

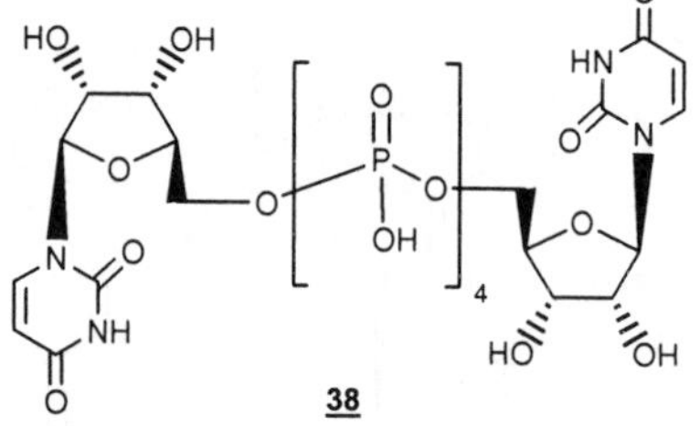

38

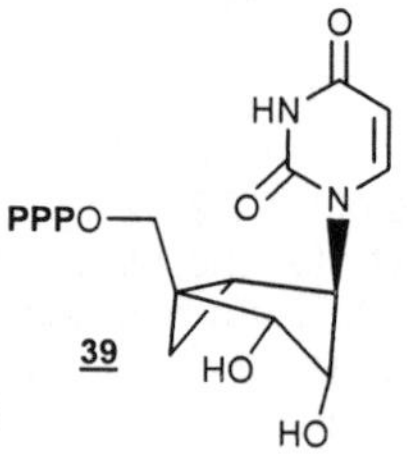

39

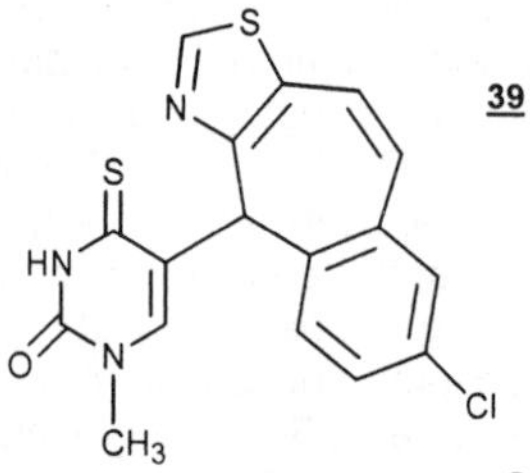

40

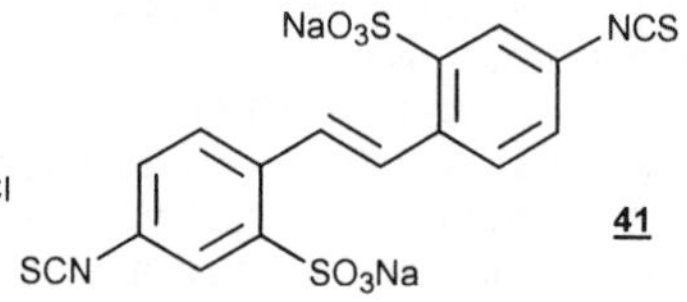

41

<u>Antagonists</u> – ATP antagonizes the human but not rat P2Y$_4$ receptor (51). Suramin is a weak antagonist at the P2Y$_2$ receptor. A family of heterocyclic antagonists of the P2Y$_2$ receptor, such as AR-C126313XX (**40**), has been reported. The isothiocyanate DIDS (**41**) has been shown to weakly antagonize the P2Y$_6$ receptor, although it is not selective since it also acts at P2X receptors (52).

<u>Summary</u> – There is rapidly growing interest in receptors for ATP, ADP, UTP, UDP and other nucleotides. The P2 receptor family comprises both metabotropic (P2Y) and ionotropic (P2X) receptors, at which ATP acts as a fast neurotransmitter, and at least one subtype of P2 receptors is found on nearly every cell. The future challenge for medicinal chemists lies in how to develop subtype-selective agonists and antagonists that are biovailable and stable in vivo. The discovery of such agents is relevant to treatment of disorders of the autonomic nervous system, central nervous system, respiratory system and the cardiovascular and renal systems.

References

1. G. Burnstock, and M. Williams, J. Pharmacol. Exp. Ther., 295, 862 (2000).
2. H. Zimmermann and N. Braun, Prog. Brain Res., 120, 371 (1999).
3. B.S. Khakh, G. Burnstock, C. Kennedy, B.F. King, R.A. North, P. Seguela, M. Voigt and P.P. Humphrey, Pharmacol. Rev., 53, 107 (2001).
4. B.F. King. Molecular biology of P2X purinoceptors, in Cardiovascular Biology of Purines, G. Burnstock, J.G. Dobson, B.T. Liang and J. Linden, eds., Kluwer, 1998, pp. 159-186.
5. M.F. Jarvis, C.T. Wismer, E. Schweitzer, H. Yu, T. van Biesen, K.J. Lynch, E.C. Burgard and E.A. Kowaluk, Br. J. Pharmacol., 132, 259 (2001).
6. S.G. Brown, B.F. King, B.F., Y.C. Kim, Y.-C., G. Burnstock and K.A. Jacobson, Drug Dev. Res., 49, 253 (2000).

7. B.R. Bianchi, K.J. Lynch, E. Touma, W. Niforatos, E.C. Burgard, K.M. Alexander, H.S. Park, H. Yu, R. Metzger, E. Kowaluk, M.F. Jarvis and T. van Biesen, Eur. J. Pharmacol., 376, 127 (1999).
8. O. Cinkilic, B.F. King, M. van der Giet, H. Schluter, W. Zidek and G. Burnstock, J Pharmaco.l Exp. Ther., 299, 131 (2001).
9. K. Braun, J. Rettinger M. Ganso, M. Kassack, C. Hildebrandt, H. Ullmann, P. Nickel, G. Schmalzing and G. Lambrecht, G. Naunyn-Schmiedeberg's Arch. Pharmacol., 34, 285 (2001).
10. K. Braun, J. Rettinger, M. Ganso, M. Kassack, C. Hildebrandt, H. Ullmann, P. Nickel, G. Schmalzing and G. Lambrecht, Naunyn-Schmiedeberg's Arch. Pharmacol. 34, 285 (2001).
11. M. Glänzel, R. Bültmann, K. Starke and A.W. Frahm, Proceedings, XVIth International Symposium on Medicinal Chemistry, Bologna, Italy (2000).
12. Y.C. Kim, S.G. Brown, T.K. Harden, J.L. Boyer, G. Dubyak, B.F. King, G. Burnstock and K.A. Jacobson, J. Med. Chem., 44, 340 (2001).
13. S.G. Brown, Y.C. Kim, S.A. Kim, K.A. Jacobson, G. Burnstock and B.F. King, Drug Devel. Res., 53, 281 (2001).
14. V. Spelta, L.-H. Jiang, A. Surprenant and R.A. North, Br. J. Pharmacol. 135, 1524 (2002).
15. B.F. King, M. Liu, J. Pintor, J. Gualix, M.T. Miras-Portugal and G. Burnstock, Br. J. Pharmacol., 128, 981 (1999).
16. B.D. Humphreys, C. Virginio, A. Surprenant, J. Rice and G.R. Dubyak, Mol. Pharmacol., 54, 22 (1998).
17. R.G. Ravi, S.B. Kertesy, G.R. Dubyak and K.A. Jacobson, Drug Devel. Res., 54, 75 (2001).
18. L. Alcaraz, M. Furber and M. Mortimore, WO Patent 0061569, (2000).
19. L. Alcaraz, M. Furber, T. Luker, M. Mortimore and P. Thorne, WO Patent 0142194, (2001).
20. L. Alcaraz and M. Furber, WO Patent 0194338, (2001).
21. A. Baxter, N. Kindon, Pairaudeau, B. Roberts and S. Thom, WO Patent 0144213, (2001).
22. P. Meghani and C. Bennion, WO Patent 0146200, (2001).
23. S. Moro, C. Hoffmann and K.A. Jacobson, Biochemistry, 38, 3498 (1999).
24. M. Diaz-Hernandez, J. Pintor and M.T. Miras-Portugal, Br. J. Pharmacol., 130, 434 (2000).
25. G. Hollopeter, H.-M. Jantzen, D. Vincent, G. Li, L. England, V. Ramakrishnan, R.B. Yang, P. Nurden, D. Julius and P.B. Conley, Nature, 409, 202 (2001).
26. D. Communi, N.S. Gonzalez, M. Detheux, S. Brézillon, V. Lannoy, M. Parmentier and J.M. Boeynaems, J. Biol. Chem., 276, 41479 (2001).
27. F.L. Zhang, L. Luo, E. Gustafson, K. Palmer, X. Qiao, X. Fan, S. Yang, T.M. Laz, M. Bayne and F. Monsma, Jr., J. Pharm. Exp. Ther., 301, 705 (2002).
28. J.K. Chambers, L.E. Macdonald, H.M. Sarau, R.S. Ames, K. Freeman, J.J. Foley, Y. Zhu, M.M. McLaughlin, P. Murdock, L. McMillan, J. Trill, A. Swift, N. Aiyar, P. Taylor, L. Vawter, S. Naheed, P. Szekeres, G. Hervieu, C. Scott, J.M. Watson, A.J. Murphy, E. Duzic, C. Klein, D.J. Bergsma, S. Wilson and G.P. Livi, J. Biol. Chem., 275, 10767 (2000).
29. D.K. Lee, T. Nguyen, K.R. Lynch, R. Cheng, W.B. Vanti, O. Arkhitko, T. Lewis, J.F. Evans, S.R. George and B.F. O'Dowd, Gene, 275, 83 (2001).
30. R.K. Palmer, J.L. Boyer, J.B. Schacter. R.A. Nicolas and T.K. Harden, Mol. Pharmacol. 54, 1118 (1998).
31. J.L. Boyer, S. Siddiqi, B. Fischer, T. Romera-Avila, K.A. Jacobson and T.K. Harden, Brit. J. Pharmacol., 118, 1959 (1996).
32. B. Fischer, A. Chulkin, J.L. Boyer, T.K. Harden, F.P. Gendron, A.R. Beaudoin, J. Chapal, D. Hillaire-Buys and P. Petit, J. Med. Chem., 42, 3636 (1999).
33. B. Fischer, J.L. Boyer, C.H.V. Hoyle, A.U. Ziganshin, A.L.Brizzolara, G.E. Knight, J. Zimmet, G. Burnstock, T.K. Harden and K.A. Jacobson, J. Med. Chem., 36, 3937 (1993).
34. E. Nandanan, S.Y. Jang, S. Moro, H. Kim, M.A. Siddiqui, P. Russ, V.E. Marquez, R. Busson, P. Herdewijn, T.K. Harden, J.L. Boyer and K.A. Jacobson, J. Med. Chem., 43, 829 (2000).
35. J. Boyer, M. Adams, R.G. Ravi, K.A. Jacobson and T.K. Harden, Br. J. Pharmacol., 135, 2004 (2002).
36. A. Baurand, P. Raboisson M. Freund, C. Leon, J.P. Cazenave, J.J. Bourguignon and C. Gachet, Eur. J. Pharmacol., 412, 213 (2001).
37. Kim, H.S., Barak, D., Harden, T.K., Boyer, J.L. and Jacobson, K.A., J. Med. Chem., 44, 3092 (2001).
38. A. H. Ingall, J. Dixon, A. Bailey, M.E. Coombs, J.I. McInally, S.F. Hunt, N.D. Kindon, B.J. Theobald, P. A. Willis, R.G. Humphries, P. Leff, J.A. Clegg, J.A. Smith and W. Tomlinson, J. Med. Chem. 42, 213 (1999).
39. D. Communi, B. Robaye and J.M. Boeynaems, Br. J. Pharmacol., 128, 1199 (1999).
40. R. Bonnert, A. Ingall, B. Springthorpe and P. Willis, WO Patent 9828300, (1998).
41. D. Hardern and B. Springthorpe, WO Patent 9905142, (1999).
42. S. Guile, A. Ingall, B. Springthorpe and P. Willis, WO Patent 9905143, (1999).

43. R. Brown and G. Pairaudeau, WO Patent 9905144, (1999).
44. R. Brown, G. Pairaudeau, B. Springthorpe, S. Thom and P. Willis, WO Patent 9941254, (1999).
45. P. Savi, J.M. Pereillo, M.F. Uzabiaga, J. Combalbert, C. Picard, J.P. Maffrand, M. Pascal and J.M. Herbert, Thromb. Haemost. 84, 891 (2000).
46. A. Sugidachi, F. Asai, K. Yoneda, R. Iwamura, T. Ogawa, K. Otsuguro and H. Koike, Br. J. Pharmacol., 132, 47 (2001).
47. R.M. Scarborough, A.M. Laibelman, L.A. Clizbe, L.J. Fretto, P.B.Conley, E.E. Reynolds, D.M. Sedlock and H.-M. Jantzen, Bioorg. Med. Chem. Lett., 11, 1805 (2001).
48. M. Malmsjö, M. Adner, T. K. Harden, W. Pendergast, L. Edvinsson and D. Erlinge, Br. J. Pharmacol., 131, 51 (2000).
49. W. Pendergast, B.R. Yerxa, J.G. Douglass III, S.R. Shaver, R.W. Dougherty, C.C. Redick, I.F. Sims and J. Rideout, Bioorg. Med. Chem. Lett., 11, 157 (2001).
50. Ravi, R.G., Kim, H.S., Servos, J., Zimmermann, H., Lee, K., Maddileti, S., Boyer, J.L., Harden, T.K. and Jacobson, K.A. J. Med. Chem., 45, 2090 (2002).
51. C. Kennedy, A.D. Qi, C.L. Herold, T.K. Harden and R.A. Nicholas, Mol. Pharmacol., 57, 926 (2000).
52. I. von Kügelgen and K.A. Jacobson, Abstr. German Pharmacol. Soc., Mainz, Germany (2001).

Chapter 9. Anticoagulants: Inhibitors of the Factor VIIa/ Tissue Factor Pathway

Leslie A. Robinson and Eddine M. K. Saiah
Deltagen Research Laboratories
San Diego, CA 92121

Introduction – The coagulation cascade has been the subject of intensive investigation over the last two decades and has been extensively reviewed (1). While tremendous progress has been made in the discovery of potent and selective inhibitors of thrombin and Factor Xa, key serine proteases on the intrinsic/common pathway, relatively less effort has been focused on the discovery of inhibitors of the extrinsic pathway (figure 1). Over the last few years, a broader understanding of the role of tissue factor (TF) in thrombotic disorders, as well as the discovery of potential roles in inflammation and cancer (2) has fueled interest in the identification of selective inhibitors of the factor VIIa/tissue factor (fVIIa/TF) pathway.

The fVIIa/TF pathway is recognized as the primary initiator of normal hemostasis (3). When exposed to plasma at the site of vascular injury, TF in the vessel wall binds to circulating fVII to form the activated fVIIa/TF complex. This complex triggers the coagulation cascade by activating both factor IX and factor X, ultimately leading to the production of thrombin at the platelet surface. The fVIIa/TF complex is inhibited on the TF-bearing cell by an endogenous protein, tissue factor pathway inhibitor (TFPI) complexed to factor Xa.

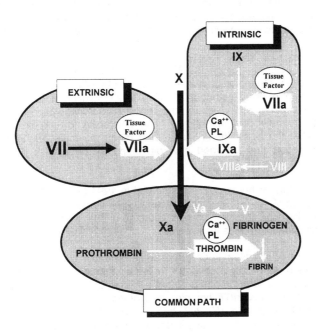

Figure 1. Coagulation pathway.

Several recent studies have sought to differentiate inhibition of the fVIIa/TF pathway from other anticoagulant strategies. Himber *et al*. (4) compared the effects of hTFAA, a soluble double mutant form of TF with decreased cofactor function for fX, with active site inhibited fIXa and fXa (fIXai and fXai, respectively) and heparin in a guinea pig model of recurrent arterial thrombosis (5). A bolus injection of hTFAA (2.5 mg/kg) inhibited thrombosis by about 90% without prolonging the cuticle bleeding time. At equally efficacious doses, fIXai, fXai and heparin prolonged bleeding time by 20%, 50% and 100% respectively, though the results for fIXai were not statistically significant (4). The ability of active site inhibited fVIIa (fVIIai) to prevent thrombosis was evaluated in a baboon surgical model. Bolus injection of fVIIai (1 mg/kg) prior to restoring flow prevented vascular thrombosis at the site of carotid endarterectomy without prolonging bleeding time or increasing surgical bleeding (6). Taken together, these and similar results reported for antibodies to fVIIa and for nematode anticoagulant protein C2 (NAPc2) (7), suggest that targeting the fVIIa/TF pathway may lead to effective antithrombotic agents with only minimal disturbance to normal hemostasis.

A number of protein-based inhibitors to the VIIa/TF pathway have been reported in the literature. These include inhibitors of TF (hTFAA) (8), inhibitors of fVIIa (fVIIa antibodies, fVIIai) (9-11) and naturally occurring inhibitors of the fVIIa/TF complex (TFPI, NAPc2) (12-13). These inhibitors have demonstrated efficacy in rodent and/or primate models of thrombosis, restenosis or sepsis. Fab antibodies to fVIIa as well as TFPI and rNAPc2 have entered clinical development. Protein-based approaches have been reviewed previously (14) and will not be described in detail here.

Over the past two years, a number of novel small molecule inhibitors of fVIIa have appeared in the patent and primary literature. Alternative strategies, targeting peptide exosites on fVIIa have also been reported. This review will summarize recent progress made towards discovering and developing selective inhibitors of the fVIIa/TF pathway.

Factor VIIa/Tissue Factor-Complex – In 1996, Banner et al. solved the X-ray crystal structure of an active site-inhibited human fVIIa (D-FFR-VIIa) complexed with a protease-cleaved form of human soluble TF (sTF), to a resolution of 2.0 Å (15). This was the first three-dimensional structure solved of a blood coagulation protease in complex with its cognate protein cofactor. The availability of protein structural information provided the basis for understanding many molecular aspects of the initiation of coagulation and provided new directions for developing potent and selective inhibitors of fVIIa (16-17).

The protein structure of human sTF was solved in 1994 (18), and subsequently refined to a resolution of 1.7 Å (19). Tissue factor, a member of the class 2 cytokine-receptor superfamily, was the first protein cofactor of the blood clotting system whose three-dimensional structure was solved.

Peptidic inhibitors of fVIIa - Nematode anticoagulant protein c2 (NAPc2) is an 85 amino acid protein originally isolated from a hematophagous hookworm Ancylostoma caninum. The protein is a potent anti-coagulant (Ki = 10 pmol/L) that inhibits the function of the fVIIa/TF complex by binding to a fXa exosite. A stable recombinant form of the protein, rNAPc2, has a human half-life of >50 hours and bioavailability of 90% after subcutaneous injection. In phase II clinical trials, rNAPc2 provided the first evidence that inhibition of the fVIIa/TF complex effectively reduces the development of postoperative venous thromboembolism. A dosage of 3 µg/kg administered within 1 hour after surgery provided optimal results, with an overall deep vein thrombosis rate of 12.3% and major bleeding rate of 2.3% (20).

A novel class of peptidic fVIIa inhibitors was discovered by phage display of naïve and partially randomized peptide libraries (21). Using this approach, E-76

(Ac-ALCDDPRVDRWYCQFVEG-NH$_2$) was identified as a high-affinity, calcium dependent binder to both fVIIa and the fVIIa/TF complex (K$_d$ 8.5 nM), but not TF. The binding site was determined through a combination of mutation, NMR and crystallographic studies to be a distinct exosite on the fVIIa protease domain close to the fVIIa active site. E-76 is a non-competitive inhibitor of fX activation by fVIIa/TF (IC$_{50}$ ~1 nM), most likely functioning through an allosteric mechanism involving structural changes in the 140's activation loop of fVIIa. The peptide is a potent anti-coagulant, inhibiting the TF-dependent clotting pathway, as evidenced by concentration dependent prolongation of prothrombin time (PT) without affecting the TF-independent activated partial thromboplastin time (APTT) (21).

A second series of peptidic exosite inhibitors of fVIIa, exemplified by peptide A-183 (EEWEVLCWTWETCER), was recently identified by phage display, analogously to the approach used to identify E-76 (22). A-183 is a potent and selective inhibitor of fVIIa/TF, inhibiting activation of both fX and fIX with IC$_{50}$'s of 1.6 and 3.5 nM, respectively. The peptide also potently inhibits the TF-dependent clotting pathway as measured in a PT clotting assay without effecting APTT. These peptides bind to a newly identified exosite on the protease domain of fVIIa, distinct from both the active site and the exosite occupied by E-76 (23-24).

Peptidomimetic inhibitors of fVIIa - Tetrapeptide mimetic **1** was reported to have a fVIIa Ki of 12 nM (25). No *in vivo* or pharmacokinetic data was disclosed for this compound. The related dipeptidomimetic **2** was reported to have a fVIIa Ki of 200 nM (26). Sulfonamide capped dipeptidomimetics such as **3** have also been claimed as fVIIa inhibitors (27). No biological data was disclosed for this series, but aliphatic replacements for the naphthyl and the thiomethyl functionalities were claimed.

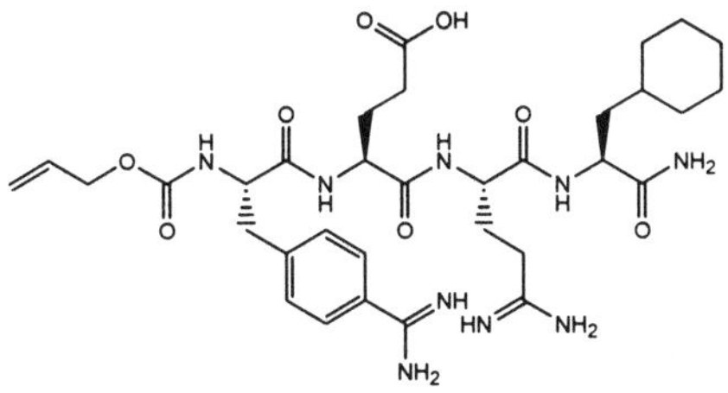

1

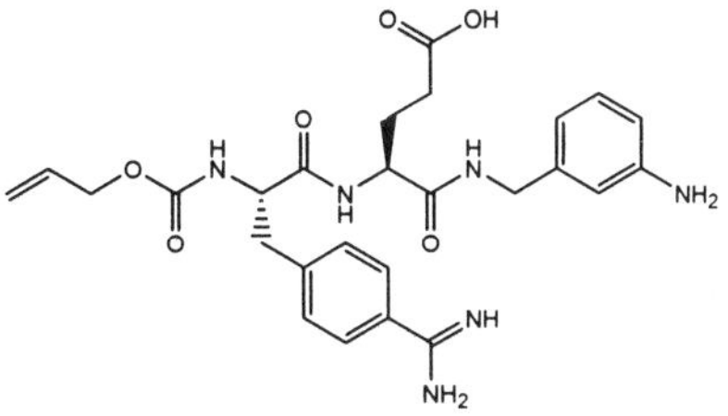

2

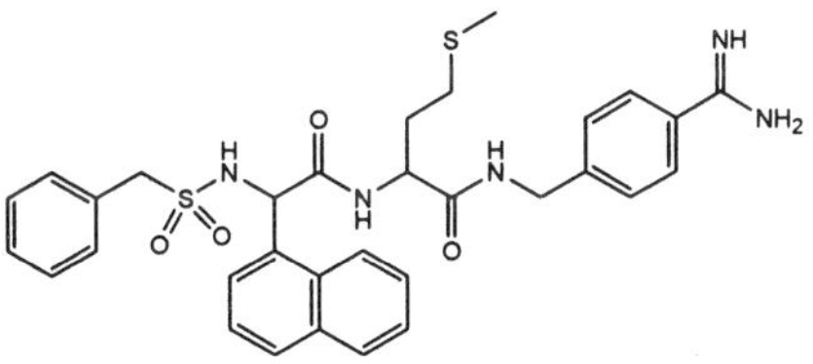

3

Small Molecule Non-Peptidic Inhibitors of fVIIa – A series of amidino-N-phenyl phenylglycine analogs (**4-9**) was recently disclosed as inhibitors of fVIIa (28). It appears that alkoxy groups on the C2-phenyl ring are preferred for activity. For example, the 4-benzyloxy analog **5** had a fVIIa Ki of 57 nM. The corresponding dimethoxy analog **4** was 5-fold less active (fVIIa Ki = 283 nM). Replacing the benzyl substituent with a more polar group, such as 4-picolyl derivative **6**, resulted in a 3-fold loss in activity (fVIIa Ki = 190 nM). Incorporation of larger substituents, such as phenethyl analog **7**, likewise resulted in a decrease in activity (fVIIa Ki = 210 nM).

Comparable activities were reported for 3-ethoxy analogs **8** and **9**, with fVIIa Ki's of 116 nM and 44 nM, respectively.

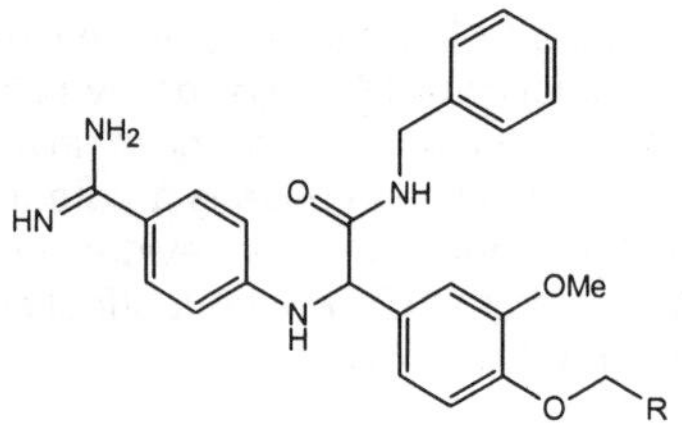

4 R = H
5 R = Ph
6 R = 4-pyridyl
7 R = CH2Ph

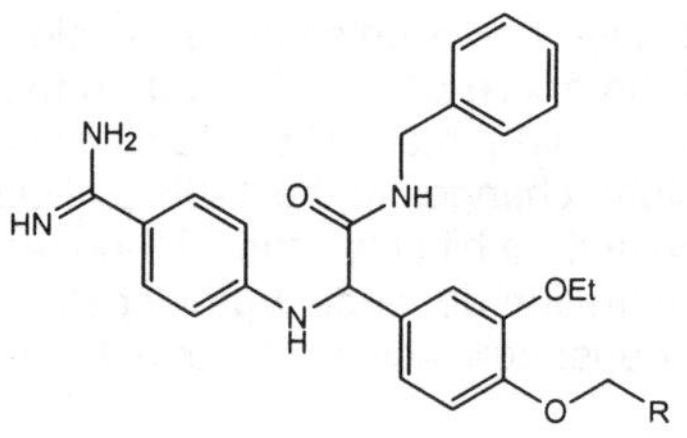

8 R = Me
9 R = Ph

A compound reported by the same group, Ro 67-8698, (structure undisclosed) was compared to the direct thrombin inhibitor napsagatran (29) by administration in a guinea pig carotid artery thrombosis model. In animals treated with Ro 67-8698 (30 mg/kg/day), prothrombin time (PT) was prolonged 1.25-fold. All animals survived and intestinal microbleeding was observed in only 1 of 8 animals. By contrast, 2 of 14 animals treated with napsagatran (4.3 mg/kg/day) died due to excess bleeding, with a 1.7-fold prolongation of PT. It was concluded that effective doses of the fVIIa inhibitor are associated with minimal bleeding risk (30).

Structurally related sulfonamide and acylsulfonamide derivatives have also been claimed as potent inhibitors of fVIIa (31). Acylsulfonamide derivative **10** was reported to have a fVIIa Ki of 3 nM, with good selectivity *vs.* thrombin (Ki = 113 nM) and fXa (Ki = 898 nM). Diethoxy analogs **11** and **12** had fVIIa Ki's of 5 nM and 7 nM, respectively, though **11** was also reported to be a potent inhibitor of kallikrein (Ki = 2 nM). No data for thrombin or fXa was reported for either analog. Sulfonamide **13** was reported to have a fVIIa Ki of 7 nM. No selectivity data was given for this compound, but a closely related analog **14** was a potent inhibitor of kallikrein (Ki = 1 nM).

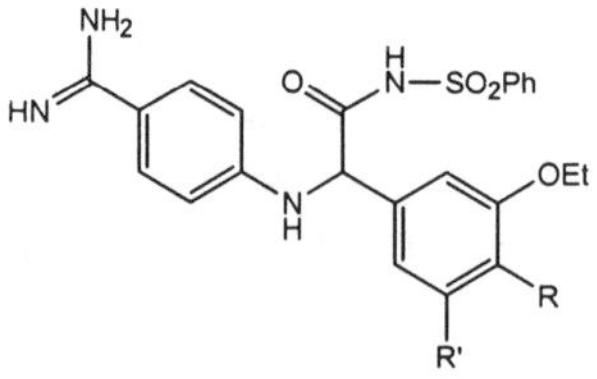

10 R = OiPr, R' = H
11 R = OEt, R' = H
12 R = H, R' = OEt

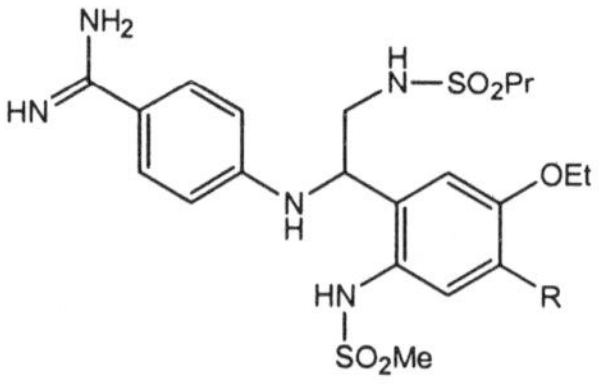

13 R = OBn
14 R = OiPr

Another report described a series of novel biaryl amides as potent inhibitors of fVIIa. The naphthyl analog **15** was reported to have a fVIIa IC$_{50}$ of 12 nM while the methoxy pyridyl analog **16** was reported to have an IC$_{50}$ of 13 nM (32). The 2'-carboxylic acid group appeared to be optimal for activity, while a variety of modifications at the 4'-carboxamide site were claimed. No SAR data and only limited synthetic information was reported for this series. Subsequently, a convergent synthesis of **15** was published by a different group (33), which confirmed that **15** was a potent inhibitor of fVIIa (Ki = 6.4 nM).

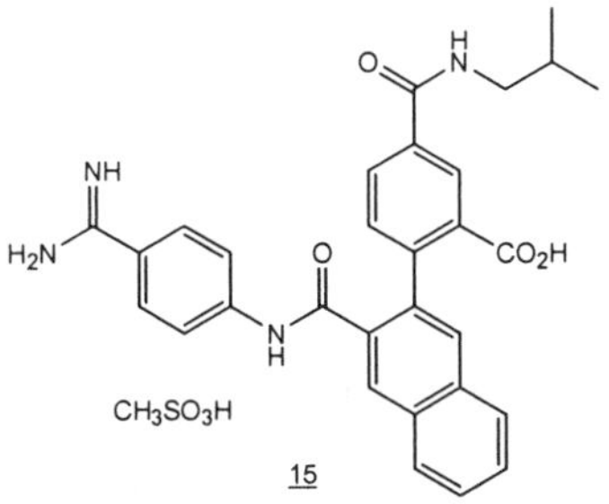

15

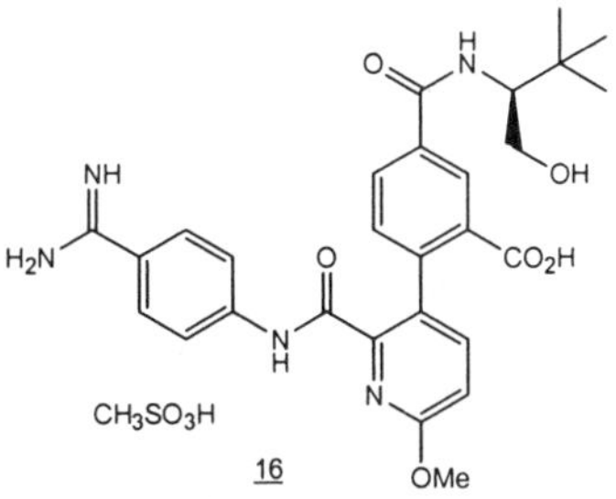

16

A series of biphenyl acetic acid derivatives was recently reported as selective inhibitors of fVIIa (34). The N-benzyl derivative **17** was reported to have a fVIIa IC_{50} of 670 nM, a thrombin IC_{50} of 9 μM, and was inactive on fXa.

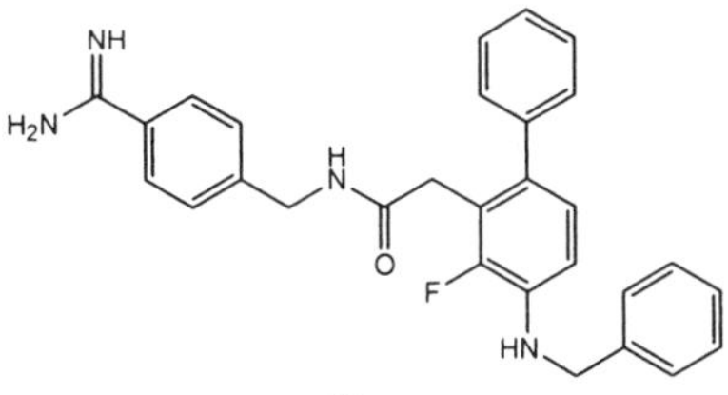

17

A recent patent application disclosed a series of naphthyl amidines as fVIIa inhibitors (35). Compound **18** was exemplified. However, no biological data was reported for this series.

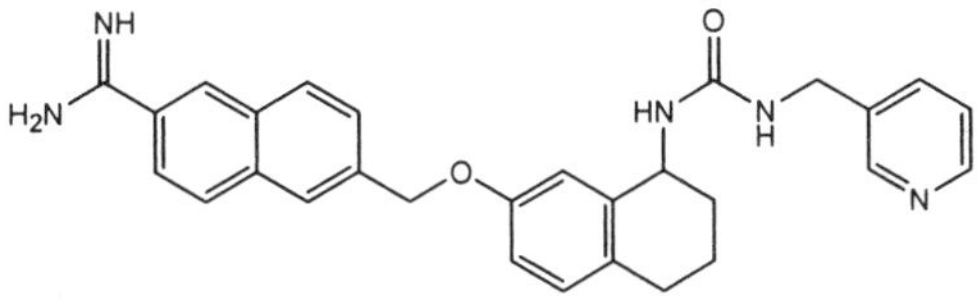

18

The amidinonaphthol derivative **19** was identified as a weak inhibitor of fVIIa/TF (IC_{50} = 60 μM) through random screening. A solid-phase library approach was applied to optimize this hit, leading to the identification of compounds **20** and **21** with IC_{50}'s of ca. 4 μM (36).

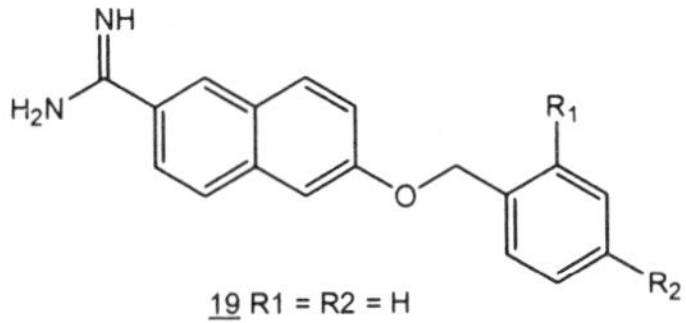

19 R1 = R2 = H

20 R1=H, R2 =

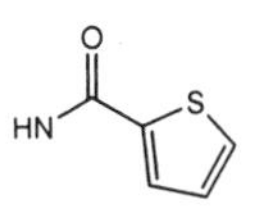

21 R1 = 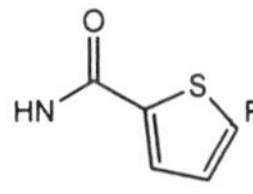　　　　　　R2 = Me

Benzimidazole **22** was identified as a potent inhibitor of fVIIa (Ki = 78 nM) by screening a targeted collection of serine protease inhibitors. While **22** showed good potency towards fVIIa, the compound was relatively non-selective towards fXa and uPa (Ki = 0.5 and 0.46 µM, respectively). Compound **22** also inhibited thrombin, trypsin and plasmin with Ki's less than 5 µM. The binding mode was proposed to involve a network hydrogen bonds between the phenol group, Ser195, His57 and a water molecule bound in the oxyanion hole. Replacement of the benzimidazole ring with an indole and installation of an acetic acid moiety to potentially interact with Lys 192 led to compound **23**. This compound had increased affinity for fVIIa (fVIIa Ki= 12 nM) and improved selectivity vs. thrombin, selectivity of **23** towards fXa, uPa and plasmin remained modest. Additional modifications aimed at replacing the aniline group led for the most part to weaker inhibitors with decreased selectivity. An exception was compound **24**, which demonstrated increased potency (fVIIa Ki = 3 nM) and selectivity vs. fXa (20 fold). Compound **24** doubled the PT time at a concentration of 1.7 µM. However, compound **25**, a somewhat weaker and less selective fVIIa inhibitor (fVIIa Ki = 27 nM, fXa selectivity 2x) doubled PT time at a similar concentration (2.2 µM). These results indicate the selectivity profile likely contributes to the overall anti-coagulant activity of these inhibitors (37).

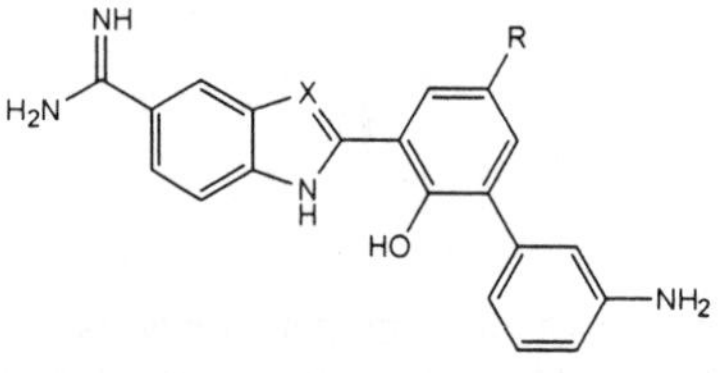

22 X = N, R = Cl
23 X = C, R = CH$_2$COOH

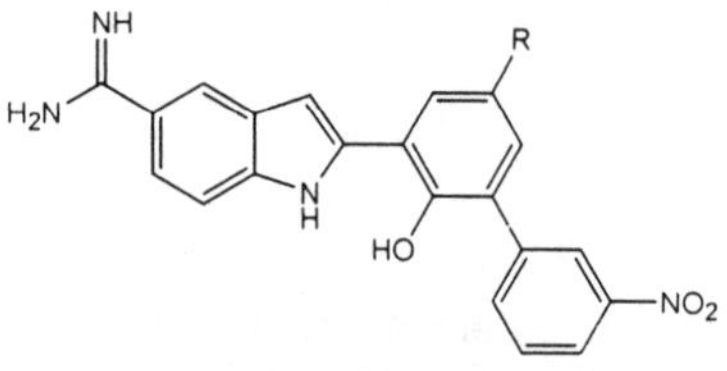

24 R = CH$_2$COOH
25 R = CH$_2$CH$_2$COOH

The alkoxybenzamidine derivatives, exemplified by compound **26**, were claimed in a recent patent application (38). A compound reported by the same group, AXC-4845, (structure undisclosed) is reported to be a potent and selective fVIIa inhibitor (fVIIa pIC$_{50}$= 8.2, fXa pIC$_{50}$= 5.7, thrombin pIC$_{50}$ =4). AXC-4845 doubled the PT time in human and rat plasma at concentrations of 3.5 µM and 30 µM respectively. AXC-4845 inhibited thrombus formation in a rat model of TF and stasis-induced venous thrombosis with an ID$_{50}$ of 0.4 mg/kg i.v. The compound also inhibited total occlusion time in an electrically induced carotid arterial thrombosis model in rats with an ID$_{50}$ of 0.15 mg/kg/h, suggesting potential for preventing both venous and arterial thrombosis (39).

Amino biphenylsulfonamide derivatives such as **27** have been claimed as inhibitors of fVIIa, fIXa, fXa and thrombin (40). No biological data was disclosed for these compounds, which are similar to compounds previously claimed as fXa inhibitors. The level of selectivity for **27** vs. fVIIa was not reported.

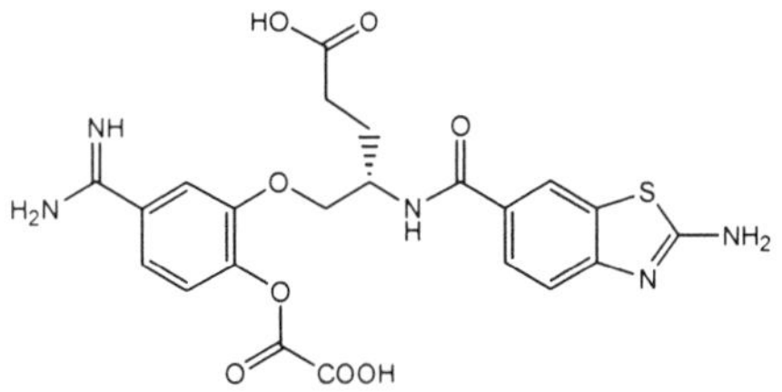

26 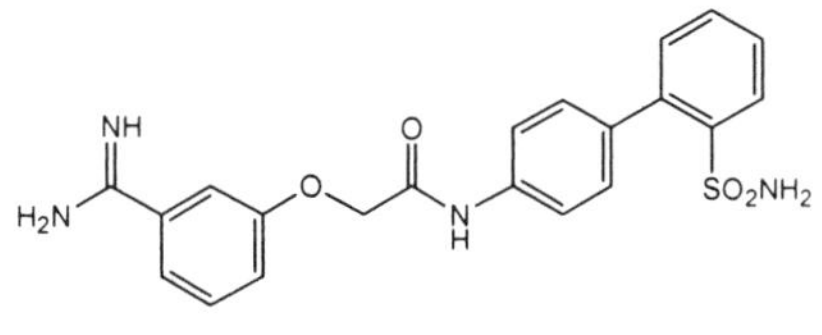   27

Aminobenzamidine derived ureas and thioureas have been claimed as fVIIa inhibitors (41). Compound **28** below was reported to have a fVIIa Ki of 26 nM. These compounds are claimed to be useful for the treatment of blood clotting, inflammation and cardiovascular disorders.

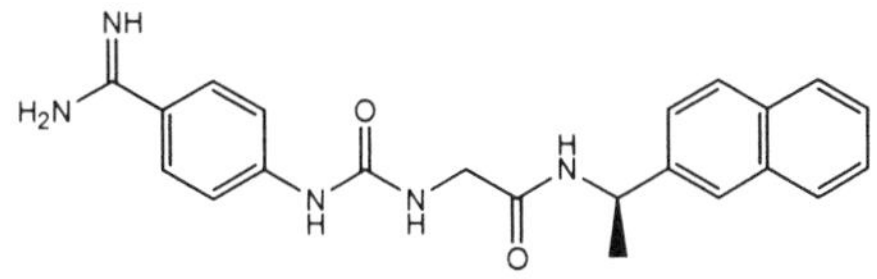

28

A use patent has been filed on YM-60828 **29** as a fVIIa inhibitor when given in combination with melagatran, a selective thrombin inhibitor currently in phase III clinical trials (42). When used in combination with melagatran, YM-60828 was reported to provide a synergistic anticoagulant effect. YM-60828, previously claimed as a fXa inhibitor with a Ki of 2.3 nM (43), reduced the incidence of occlusion and improved carotid arterial potency at 30 mg/kg po in an electrically induced arterial thrombosis model in rats.

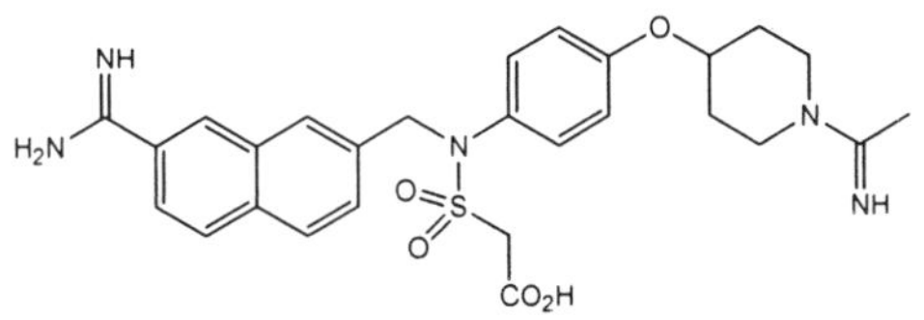

29

2-Aryl substituted 4H-3,1-benzoxazin-4-ones and related heterocyclic-fused oxazin-4-ones have been reported as specific inhibitors of the fVIIa/TF pathway (44-45). The proposed mechanism of action for this class of inhibitors is deactivation by acylation of the active site serine by the benzoxazinone lactone. Compounds were evaluated for inhibitory activity on the fVIIa/TF-induced activation of fX using a two-step amidolytic assay. Preferred compounds such as **30** and **31** inhibited fX activation with IC$_{50}$ values of ~1 μM and are selective vs. fXa and thrombin. Electronegative substituents were preferred on the 5, 6 or 7 positions of the benzoxazinone ring, while 2,6-difluoro substitution was preferred on the 2-phenyl group. These compounds prolonged PT time by 1.31 and 1.51 fold, respectively. In general, the heterocyclic fused compounds were weaker and less selective inhibitors of fX activation than the benzoxazin-4-ones. Pyridine analog **32** inhibited fX activation with an IC$_{50}$ value of 1.3 μM and prolonged PT time by 1.24-fold.

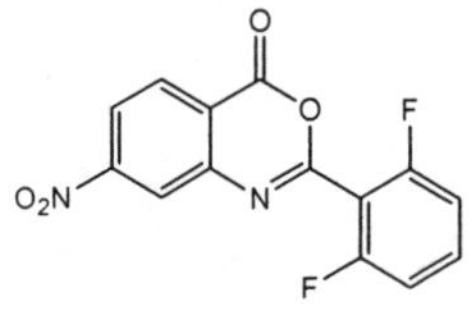

30 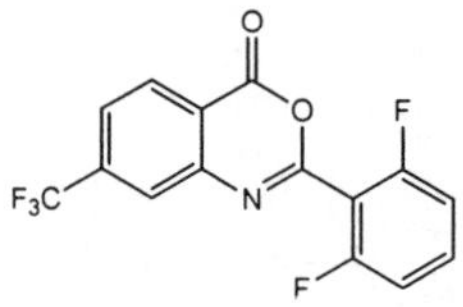   31 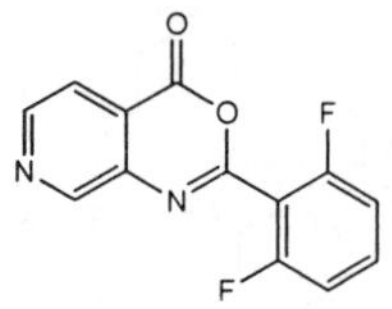   32

Unsaturated fatty acid derivatives of salicylic acid isolated from Chinese medicinal plants were identified as inhibitors of fVIIa/TF amidolytic activity at a concentration of ~30 µM (46). The compounds are proposed to bind to fVII and prevent its interaction with soluble TF. At least one cis double bond was required for activity. Representative structures **33** and **34** are shown below.

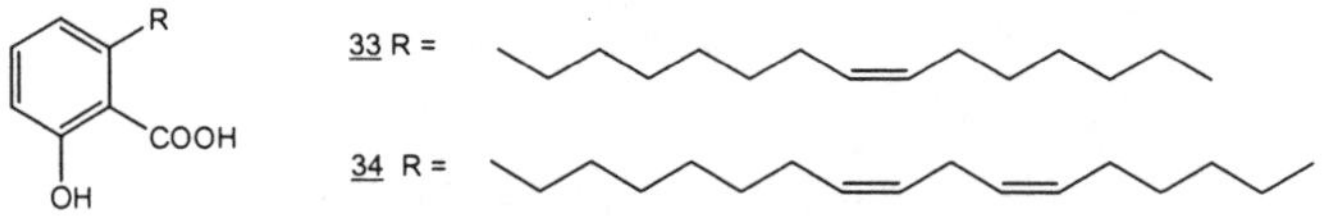

Dibenzothiophene sulfone **35** has been claimed as a TF blocking agent (47). The compound was reported to inhibit fX binding to the fVIIa/TF complex with an IC_{50} of 1 µM.

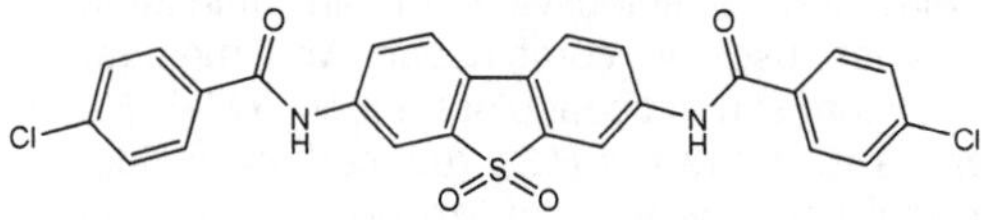

35

Conclusion - Over the last several years, selective inhibition of the fVIIa/TF pathway has emerged as an increasingly attractive therapeutic approach for the treatment of thrombotic disorders. Interest has been spurred by initial evidence suggesting that inhibition of this pathway may result in significantly less bleeding liability than currently available therapeutics. Significant advances have been made by a number of groups in identifying potent and selective small molecule inhibitors of fVIIa. The chemical diversity of these newly discovered compounds indicates the likelihood of multiple binding modes in the active site. However, most of these compounds contain a basic amidine moiety which could adversely affect their absorption and pharmacokinetic properties. The discovery of new, less basic or non-basic fVIIa inhibitors could lead to orally active compounds with appropriate plasma half lives. Advancement of such compounds to the clinic should help clarify the potential advantages of small molecule fVIIa inhibitors vs. fXa or thrombin inhibitors in the treatment of thrombotic disorders. An assessment of the success of this approach will await further development of these inhibitors.

References

1. E.W. Davie, K. Fujikawa and W. Kisiel, Biochemistry, 30, 10363 (1991).
2. B. Osterud, Thromb. Haemost., 78, 755 (1997).
3. L.C. Petersen, S. Valentin and U. Hedner, Thrombosis Res., 79, 1 (1995).
4. J. Himber, C.J. Refino, L. Burcklen, S. Roux and D. Kirchhofer, Thromb. Haemost., 85, 475 (2001).
5. S. Roux, J.P. Carteaux, P. Hess, L. Falivene and J.P. Clozel, Thromb. Haemost., 71, 252 (1994).
6. L.A. Harker, S.R. Hanson and A.B. Kelly, Thromb. Haemost., 78, 736 (1997).
7. W.E. Rote, G.L. Oldeschulte, E.M. Dempsey and G.P. Vlasuk, Circulation, 94 Suppl 1:I-695 (abstract) (1996).
8. R.F. Kelly, C.J. Refino. M.P. O'Connell, N. Modi, P. Sehl, D. Lowe, et al., Blood, 89, 3219 (1997).
9. B.J. Biemond, M. Levi, H. ten Cate, H.R. Soule, L.D. Morris and D.L. Foster, Thromb. Haemost., 73, 223 (1995).
10. B. Arnljots, M. Ezban and U. Hedner, J Vasc. Surg., 25, 341(1997).
11. B.B. Sorensen, E. Persson, P.O. Freskgard, M. Kjalke, M. Ezban, T. Williams, et al., J. Biol. Chem., 272, 11863 (1997).
12. G.J. Broze, Thromb. Haemost., 74, 90 (1995).
13. S.L. Maki, W. Ruf, S. Huang, C. Kelly and G.P. Vlasuk, Circulation, 94 Suppl 1: I-694 (abstract) (1996).
14. K. Johnson and D. Hung, Coronary Artery Disease, 9, 83 (1998).
15. D.W. Banner, A. D'Arcy, C. Chene, F.K. Winkler, A. Guha, W.H. Konigsberg, Y. Nemerson and D. Kirchhofer, Nature, 380, 41 (1996).
16. D.W. Banner, Thromb. Haemost., 78, 512 (1997).
17. D. Kirchhofer and Y. Nemerson, Curr. Opin. Biotech., 7, 386 (1996).
18. K. Harlos, D.M.A. Martin, D.P. O'Brien, E.Y. Jones, D.I. Stuart, I. Polikarpov, A. Miller, E.G.D. Tuddenham and C.W.G. Boys, Nature, 370, 662 (1994).
19. Y.A. Muller, M.H. Ultsch and A.M. de Vos, J. Mol. Biol., 256, 144 (1996).
20. A. Lee, G. Agnelli, H. Buller, J. Ginsberg, J. Heit, W. Rote, G. Vlasuk, L. Costantini, J. Julian, P. Comp, J. van Der Meer, F. Piovella, G. Raskob, M. Gent, Circulation, 104, 74 (2001)
21. M.S. Dennis, C. Eigenbrot, N.J. Skelton, M.H. Ultsch, L. Santell, M.A. Dwyer, M.P. O'Connell and R.A. Lazarus, Nature, 404, 465 (2000).
22. M.S. Dennis, M. Roberge, C. Quan and R.A. Lazarus, Biochemistry, 40, 9513 (2001).
23. M. Roberge, L. Santell, M.S. Dennis, C. Eigenbrot, M.A. Dwyer and R.A. Lazarus, Biochemistry, 40, 9522 (2001).
24. C. Eigenbrot, D. Kirchhofer, M.S. Dennis, L. Santell, R.A. Lazarus, J. Stamos and M.H. Ultsch, Structure, 9, 627 (2001).
25. P. Wildgoose, P. Safar, A. Safarova, WO Patent 0015658 (2000).
26. E. Defossa, U. Heinelt, H. Matter, O. Klingler, M. Schudok, G. Zoller, P. Safar, WO Patent 0075172 (2000).
27. C. Alcouffe, P. Bellevergue, G. Dellac, C. Latham, V. Martin, C. Masson, G. McCort, WO Patent 0058346 (2000).
28. K. Grobke, Y-H. Ji, S. Wallbaum, J. Mahe, Eur. Patent EP 0921116-A1 (1999).
29. H. Bounameaux, H. Ehringer, A. Gast, J. Hulting, H. Rasche, H.J. Rapold, G. Reber and T.B. Tschopp, , Thromb. Haemost., 81, 498 (1999).
30. Fingerle, J. et al. Thromb. Haemostasis, 18th Cong. Int. Soc. Thromb. Haemostasis, July 6-12 2001, Paris, Suppl.: Abst. OC1766
31. I. Aliagas-Martin, D.R. Artis, M.S. Dina, J.A. Flygare, R.A. Goldsmith, R.A. Munroe, A.G. Olivero, R. Pastor, T.E. Rawson, K.D. Roberge, D.P. Sutherlin, R.J. Weese, A. Zhou, Y. Zhu, WO Patent 0041531 (2000).
32. (a) K. Senokuchi and K. Ogawa, WO Patent 9941231 (1999); (b) K. Senokuchi, K. Ogawa, Eur. Patent EP 1078917-A1 (2001).
33. J.T. Kohrt, K.J. Filipski, S.T. Rapundalo, W.L. Cody and J.J. Edmunds, Tetrahedron Lett., 41, 6041 (2000).
34. J. Parlow, M.S. South, WO Patent 0168605 (2001).
35. C. Alcouffe, P. Bellevergue, C. Latham, G. Lassalle, S. Mallart, V. Martin, WO Patent 0066545 (2000).
36. P. Roussel, M. Bradley, P. Kane, C. Bailey, R. Arnold and A. Cross, Tetrahedron, 55, 6219 (1999).

37. (a) W.B. Young, A. Kolesnikov, R. Rai, P.A. Sprengeler, E.M. Leahy, W.D. Shrader, J. Sangalang, J. Burgess-Henry, J. Spencer, K. Elrod and L. Cregar, Bioorg. Med. Chem. Lett. 11, 2253, (2001); (b) D. A. Allen, J. M. Hataye, W. N. Hruzewicz, A. Kolesnikov, R. L. Mackman, R. Rai, W.D. Shrader, J. R. Spencer, E.J. Verner and W.B. Young, WO Patent 0035886 (2000).

38. K. Sagi, K. Fujita, K. Sakurai, M. Sugiki, S. Takehana, K. Tashiro, T. Kayahara, WO Patent 0142199 (2001).

39. S. Takehana, T. Kayahara, K. Tashiro, K. Sagi, M. Sugiki, A. Okajima, A. Chiba, K. Fujita, A. Kodaira, M. Yamanashi, K. Sakurai, M. Shoji, S. Iwata, Thromb. Haemostasis. 18th Cong. Int. Soc. Thromb. Haemostasis, July 6-12 2001, Paris, Suppl.: Abst p1393.

40. D. Dorsh, J. Gleitz, H. Juraszyk, W. Mederski, C. Barnes, C. Tsaklakidis, WO Patent 0170678 (2001)

41. (a) O. Kingler, H-P. Nestler, H. Schreuder, H. Matter, M. Schudok, WO Patent 0194301 (2001); (b) O. Kingler, H-P. Nestler, H. Schreuder, H. Matter, M. Schudok, Eur. Patent Appl. EP 1162194 (2001).

42. M. Christer and B. Ruth, WO Patent 0195932(2001).

43. F. Hirayama, H. Koshio, Y. Matsumoto, T. Kawasaki, S. Kaku, I. Yanagisawa, Eur. Patent Appl. EP 0798295 (1997).

44. P. Jakobsen, B.R. Pedersen and E. Persson, Bioorg. Med. Chem., 8, 2095 (2000).

45. P. Jakobsen, A.M. Horneman and E. Persson, Bioorg. Med. Chem., 8, 2803 (2000).

46. D. Wang, T.J. Girard, T.P. Kasten, R.M. LaChance, M.A. Miller-Wideman and R.C. Durley, J. Nat. Prod., 61, 1352 (1998).

47. J.A. Jiao, L.K. Leupschen, E. Nieves, H.C. Wong and D.P. Taylor, WO Patent 0130333 (2001).

SECTION III. CANCER AND INFECTIOUS DISEASES

Editor: Jacob J. Plattner, Chiron Corp
Emeryville, CA

Chapter 10. The Prospects for Microbial Genomics Providing Novel, Exploitable, Antibacterial Targets

Thomas J. Dougherty[a] and John F. Barrett[b]
[a]Pfizer Global Research & Development, Groton, CT, 06340
[b]Merck Research Laboratories, Rahway, NJ

Introduction – Antibiotics have been a major triumph in applied medical science since their introduction in the last century. The rapid improvement of patients afflicted with heretofore deadly infections led to these compounds being termed "miracle drugs". Unfortunately, it has become increasingly clear that bacterial resistance to these compounds is rising at a rate that threatens to undermine their future utility (1,2). In this sense, antimicrobials stand in a somewhat unique position among drug classes. Compounds that affect human physiology, such as lipid-lowering agents, ACE inhibitors, and anti-inflammatory agents, do not engender resistance to their pharmacological targets in the human population. However, in the case of infectious agents we are dealing with rapidly evolving populations of organisms placed under selective pressure by antibiotics (3). Exacerbating the situation is the fact that many of the existing antimicrobial classes are derived from natural products, and pre-existing resistance genes were present in the producer organisms (4). These genes have been mobilized on genetic elements such as transposons and plasmids, and disseminated widely among many pathogens. There is a clear need for new classes of antibiotics as part of the strategy to combat the emerging antibiotic resistance problem.

Genomics Development - In the mid-1990s, technologies were developed that permitted the rapid sequencing of the entire genetic complement of an organism (5,6). These involved random ("shotgun") sequencing of a gene library of an organism by the use of automated, high-throughput DNA sequencers. This process was coupled with sophisticated sequence assembly programs running on high-speed computers to link the multiple, redundant, relatively short sequences into a coherent genome. These methods were initially applied to microorganisms, and *Haemophilus influenzae* became the first organism whose genome was completed using what was termed "shotgun" sequencing and assembly in 1995 (7). This marked the beginning of the genomics era, and multiple microorganism genomes were rapidly sequenced, as were landmark organisms such as *Saccharomyces cerevisiae, Caenorhabditis elegans, Drosophilia melanogaster* and the human genome (8-11). The avalanche of gene sequence information created opportunities to understand organisms in fundamentally new ways that are still very much in their nascent stages. In the human genome, the initial mining of the genomic information has revealed new members of important gene families, such as fibroblast growth factors, keratin, and TGFβs (11). The extensive syntectic relationships between human and mouse genome permits the application of mouse mutants and genetics to solving physiological roles of human genes.

<u>Microbial Genomics</u> - With regard to microbial genomes, a fundamentally different approach is applied in mining genome information for the discovery of new antibacterial targets. In microbes, a small handful of essential, physiological processes are the targets of the existing classes of antibiotics. These include terminal stages in the synthesis of the bacterial cell wall peptidoglycan (e.g., β-lactams, glycopeptides), protein synthesis (macrolides, aminoglycosides, tetracyclines, oxazolidinones), and DNA topoisomerases (fluoroquinolones) (12). The advent of microbial genomics has suddenly thrust the complete "parts list" for multiple pathogens into the hands of biologists. By identifying additional essential gene products, it is believed that targets will become available to exploit in the identification of novel inhibitors. Development of some of these inhibitors into drugs will expand the classes of antimicrobial compounds. Most importantly, novel antimicrobials should circumvent resistance mechanisms to current antimicrobials.

Currently, the number of microbial genomes that have been sequenced since the initial *H. influenzae* genome is approximately fifty-nine, with many additional projects in progress. Limiting the examples to pathogens alone, organisms such as *Streptococcus pyogenes*, several strains of *Streptococcus pneumoniae*, *Chlamydia trachomatis*, and *Chlamydia pneumoniae*, *Borrelia burgdorferi*, *Treponema pallidum*, *Mycobacterium tuberculosis* and an enterohaemorrhagic strain of *Escherichia coli* (13-21) are among many additional pathogen genome sequences in the public domain. Two important organisms in which the bulk of microbial physiology has been performed, *E. coli* K-12 and *Bacillus subtilis*, are also available. These are key, as many of the putative function annotations of pathogen genomes are based on sequence similarities to gene products in these two organisms. With individual microbial genomes running between roughly 400,000 to over 4 million base pairs, the sheer volume of this information has made it imperative to use high-speed computer workstations to handle, sort, compare, and analyze the databases. In order to rationalize the information, programs have been devised to identify probable genes, compare multiple genomes, gather functional annotation evidence, and sort the gene products into biochemical pathways (22-26).

<u>Applications of Genomics to Antimicrobial Target Identification</u> - With the complete genome sequences of a large number of pathogenic bacteria now available, strategies to identify gene products that may be targets for inhibitors have been developed and deployed in a number of laboratories. These strategies can be based on a number of different suppositions, and the differences in approach reflect different models of the types of targets sought, as well as the methods to identify such targets. Figure 1 illustrates the general pathways and outcomes of these approaches to the common problem of novel target identification in bacteria. It is possible to take one of two differing general gene disruption approaches to identify novel antibacterial targets. It is also possible to test the disrupted genes on synthetic growth media (*in vitro*) or in the context of the infected host (*in vivo* animal infection model). In either case, the failure to recover a gene that has been disrupted on a plate, or in an infected animal, is considered evidence for the essential nature of the gene under the testing circumstances.

<u>Identification of Novel Genomic Targets: *Targeted Knockouts*</u> - As indicated above, the identification of novel antimicrobial targets can be approached by inactivation of genes to determine if their protein products are essential for bacterial survival. One way to identify these potential targets is by first examining and comparing genomes computationally to generate a list of likely target knockout candidates. Computational methods are available for comparing multiple microbial genomes to one another to identify, by sequence similarity, probable conserved gene functions (23,24,27-29). It is also possible to subtract eukaryotic sequences at different levels of similarity to reduce the likelihood of adverse events due to cross-inhibition of host functions.

This conservation of sequence similarity is used as a surrogate for spectrum of a potential antimicrobial agent against the target. Once a list of gene knockout candidates has been finalized, a decision on which bacterial species will

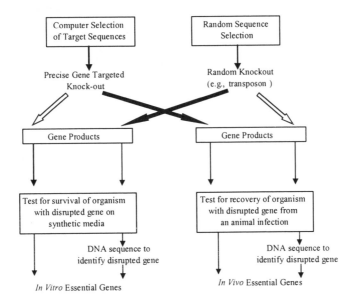

Figure 1. Strategies for random and directed gene knockouts, and subsequent test procedures to establish gene essentiality either on synthetic growth media or in an infected animal. Targeted knockouts test directly whether a gene is essential (not recovered after knockout) or not (recovered). In the case of the random knockouts, those genes with disruptions that are recovered and identified are non-essentials, and essential genes are deduced by their lack of recovery from large knockout campaigns.

be used as test cases for the gene knockouts must be made. Several targeted knockout systems have been described in the literature. Among these is the pEVP-3 plasmid system for use with the naturally transformable *S. pneumoniae* (30).

This plasmid, which carries only a gram-negative specific replicon, also contains a chloramphenicol resistance gene, which is expressed in both pneumococci and in *E. coli*, where the plasmid is replicated. The disruption system uses a small internal fragment lacking the coding for the N- and C-terminal regions of the target gene. This gene fragment, generated by PCR, is inserted into the multiple cloning site of pEVP-3. When introduced into the pneumococcus by transformation, the plasmid recombines into the chromosome via homologous recombination between the gene fragment and the target gene. The insertion duplication event leads to the generation of two partial copies (nonfunctional) of the target gene flanking the inserted plasmid. Insertion into a nonessential gene will lead to recovery of a chloramphenicol resistant recombinant. Failure to recover a chloramphenicol-resistant colony is presumptive evidence for the essential nature of the target gene. A similar system has been developed for the readily transformable *B. subtilis*, termed the pMUTIN vectors (31). These include a promoter, pSpac, to regulate downstream genes to address operon polarity issues. Yet another similar system generates gene knockouts in *S. aureus* using the plasmid, pSA3182 (32).

Other methods include the use of linear DNA generated by PCR with overlapping primers to construct antibiotic resistance cartridges with flanking ends of homologous DNA that match the sequence of genes on either side of a knockout target gene (33). This leads to deletion of the target gene and replacement with the antibiotic resistance gene in the amplified construct. Introduction of the construct via transformation or electroporation, followed by recombination into the chromosome mediated by the flanking homologies will lead to recovery of resistance only if the deleted target gene is not essential. The recent introduction of an *E. coli* strain with the λ phage *exo, bet,* and *gam* genes under temperature-sensitive repressor control can generate a hyper-recombinogenic strain that integrates linear DNA which is electroporated into the cell (34). Using PCR methods as described above, resistance cartridges can be constructed that have DNA homologous to the flanking regions of a target gene for deletion. Selection and recovery of resistance indicates lack of target gene essentiality.

Identification of Novel Genomic Targets: *Random Knockouts* - In contrast to the above methods in which precise targeting of knockouts occurs, methods for random mutagenesis have been developed. The use of transposons as gene disruption tools has been practiced for some time prior to the advent of genomics. In one general case, transposons can be utilized to knock out genes by random insertion into the total genome. The transposons usually possess some selective marker (e.g., antibiotic resistance), and cells with an integrated transposon are identified by their resistance. Used in this way, essential genes are deduced by virtue of the absence of isolation of transposons in these genes (35,36). Unfortunately, saturation of a chromosome and mapping of many transposon inserts is laborious, and polarity effects (downstream disruption of transcription by the transposon) can lead to equivocal results. A method to resolve such issues is the use of balanced lethals, in which plasmids with the wild type allele are employed with the transposon knockout (37-39).

To address some of these problems, several variations of transposon mutagenesis have been implemented. One system is termed GAMBIT (Genomic Analysis and Mapping By *In vivo* Transposition). In this technique, a variant of the eukaryotic *mariner* transposon, which is essentially a random insertion transposon, has been modified with antibiotic resistance genes (40). A somewhat related procedure is genome scanning (41). Another transposon-based method is the use of an element termed TnAraOut, which is a derivative of the *mariner* transposon with an antibiotic resistance marker and an arabinose inducible promoter (42).

Identification of Novel Genomic Targets *In Vivo* - A previously untapped source of antibacterial targets may lie in the genes that are expressed only in the infected host. These genes would only be expressed within the environs of an infected cell or animal, activated by signals only present under these conditions, such as pH, calcium ions, nutrient starvation, iron limitation, and stress (43-46). Microbial genes regulated by host factors may include adhesins, toxins, invasins, protein secretion systems (e.g., Type III secretion) and molecules that influence the host immune system (47,48). By identifying novel compounds that would interfere with the expression or function of one or more of these factors, it may be possible to abort an infection (49,50). Described below are several additional samples of experimental systems that have been devised to specifically identify such genes for the identification of novel antimicrobial and vaccine targets.

Signature-tagged mutagenesis - This was one of the first large-scale (comprehensive genome-wide) systems to search for genes that were specifically involved in virulence (51,52). The method incorporates a "label" of a random oligonucleotide sequence of 40 bp, flanked by constant DNA sequencing primers. These random 40 bp DNA tags are ligated into the transposon used for mutagenesis giving each transposon a unique

DNA tag. The transposons are used to mutate a bacterial species and individual antibiotic resistant colonies, each with a unique tagged transposon inserted into a different gene, are recovered (to date, species include *Salmonella typhimurium, V. cholerae, Yersinia enterocolitica, S. aureus,* and *Legionella pneumophila,* among others) (53). Pools of transposon-mutated cells are used to infect individual animals, and bacteria are recovered from the infected animals. The tag regions are amplified by PCR, and used as probes against colony blots of the original input organisms in the pool. Loss of a signal indicates that the transposon had inserted into a gene that was essential for survival in the infected animal. The pathogen carrying that transposon was eliminated, removing that transposon's unique tag from the pool.

In Vivo Expression Technology (IVET) - This system uses an ampicillin-resistant suicide delivery plasmid that contains inserted random small DNA fragments in a cloning site upstream of a promoterless essential nutrient bacterial gene transcriptionally-fused to a β-galactosidase gene (54,55). The plasmid is delivered into a recipient bacterial strain that was either lacking the essential nutritional gene, and recombines via a Campbell-type insertion into the chromosome mediated by the inserted DNA fragment. This results in a gene insertion duplication, leaving an intact promoter and copy of the target gene downstream of the insertion. After the cells are grown in media supplemented with the missing essential nutrient, the bacteria are used to infect an animal. During *in vivo* growth, only those bacteria with promoters in the inserted DNA that are turned on in the animal host survive, as the promoter drives transcription of the nutritional gene, essential for survival. A significant proportion of genes identified with this system have no known function nor similarity to other genes (54). Several additional systems to identify genes that are expressed *in vivo* have been devised (56). These include subtractive and differential hybridization, in which m-RNA levels are compared between *in vitro* and *in vivo* grown bacteria. It was possible, for example, to subtract the "housekeeping" genes of *Mycobacterium avium* grown in broth from the genes specifically induced upon growth of the organisms in macrophages (57). Chip-based hybridization methods are also used to monitor *in vitro* versus *in vivo* gene expression (58).

Identification of Novel Inhibitors by Genomic-based Target Screens - Other technologies, such as transcriptional analysis of whole genomes have been employed in mechanism of action studies for antimicrobial compounds, as well as pathway analysis of unknown gene function (59). These types of experiments point to future directions in which transcriptional profiles of antibiotic-treated cells, using antibiotics known to target different cell processes, could be used to study the effects of novel antimicrobial compounds. New antimicrobial targets identified by gene knockouts as essential, but of unknown function, could be placed behind controllable promoters (60). Down-regulation of these genes followed by transcriptional profiling could indicate the biochemical pathway involved and possible function of the novel gene. Likewise, proteomics, which has been defined as the measurement of the total expressed protein complement of a cell under defined conditions (61), has already been deployed to study protein profile changes upon drug treatment of microbes (59,62).

Early Results From Genomics-based Targets - The use of microbial genomics represents a fundamentally new way of identifying antimicrobial targets. In many ways, the information and tools necessary for successful utilization are still under development. For example, despite many years of combined genetic and physiologic studies, the function of a significant number of the over 4000 genes in *E. coli* K-12 remain unknown at present (63). Progress has been made in characterizing some of these gene functions since the release of the genome, but much experimental work remains to be performed. This work is essential to the progress of microbial genomics, as much of the sequence similarity and tentative functional assignments rest on the foundation of *E. coli* studies. Nonetheless, early discovery efforts focused on identifying novel compounds from genomics derived targets have been performed.

A recent article detailed studies performed on enzymes involved in bacterial fatty acid biosynthesis, and the use of key enzymes in the pathway as targets for novel drug development (64). The comparison of the genomes of *S. aureus, S. pneumoniae, E. faecalis,* and *H. influenzae* identified those fatty acid biosynthesis enzymes which were conserved among this group of organisms. An important finding was that the pneumococcus did not possess a *fabI* gene. The FabI protein had been thought to be the sole enzyme that could carry out enoyl-ACP reductase activity. The presence of such an alternative form of the enzyme, FabK, which bears no sequence similarity to *fabI*, had been deduced on the basis of biochemical studies (65). Bioinformatics revealed that some pathogens such as *E. faecalis* and *E. faecium* have both forms of the enzyme. Thus the advisability of using FabI as a target to search for novel inhibitors is questionable. Likewise, FabA and FabB were missing from the genomes of several key pathogens. In contrast, FabD, FabG, FabH, and FabZ were found among all four pathogens examined.

Another genomics-derived target that has been exploited is peptide deformylase (PDF). Although this essential metalloenzyme had been described in *E. coli* prior to the arrival of genomics (66), the microbial genomics databases indicated how widespread this enzyme was, it's importance in cell physiology, and enzyme studies in organisms other than *E. coli* (67). Screening of the enzyme yielded weakly binding compounds, which were N-hydroxy-acetamide derivatives (68). Structural information gathered from the purified enzyme complexed with the inhibitors suggested that related hydroxamic acid derivatives might be better inhibitors. Two compounds from two different hydroxamic acid series exhibited good activity against the PDF over other related metalloenzymes, but unfortunately antibacterial activity was weak. The genes encoding peptide deformylase were isolated from several different microbes such as *S. pneumoniae* and *H. influenzae*, and overexpression of these heterologous proteins was shown to restore growth of *E. coli* in the presence of the inhibitor. The PDF gene of *S. pneumoniae* was placed behind a regulated promoter, and decreased expression of PDF led to increased susceptibility to the inhibitors. Another group also reported the isolation of an inhibitor using isolated enzyme (69). In this case, an N-formyl-hydroxylamine derivative, BB-3497 (**1**), was isolated. BB-3497 is related to the natural product PDF inhibitor, actinonin. The inhibitor was co-crystallized with the *E. coli* PDF enzyme, and the structure of the inhibitor bound in the enzyme at 2.1 angstroms was obtained. Antibacterial activity against methicillin-resistant *S. aureus* and vancomycin-resistant *E. faecalis* was observed;

1

activity against gram-negatives was also obtained. Interestingly, this compound was shown to be orally bioavailable, and a single oral dose protected mice against intraperitoneal injections of methicillin-resistant *S. aureus* infections at a PD_{50} of 14 mg/kg.

Another genomics-derived set of targets is the amino-acyl-tRNA synthetases. The case for these enzymes as antibacterial targets has been recently reviewed (70). In a recent report, aminoalkyl adenylates and aminoacyl sulfamates were tested against *S. aureus* aminoacyl-tRNA synthetases (71). Arginyl derivatives were identified that inhibited the class I arginyl-tRNA synthetases. For the class II histidyl and threonyl t-RNA synthetases, the acyl sulfamates were found to be potent inhibitors. A natural product antibiotic, SB-219383 (**2**), which inhibits tyrosyl-t-RNA synthetase was also identified. Derivatives were prepared, and improved activity in the sub-nanomolar range of some ester- and amide-derivatives was noted (72). A follow-on study removed the cyclic hydroxylamine portion of SB-219383 and replaced it with several C-pyranosyl derivatives (73). Several potent analogs that inhibited tyrosyl-tRNA were obtained. Further carbocyclic derivatives of SB-219383 were also synthesized, with a nanomolar inhibitor identified (74). Several of these inhibitors were studied in X-ray crystals of the *S. aureus* tyrosyl-tRNA synthetase. The inhibitors were found to occupy known substrate binding pockets in the enzyme, as well as a binding pocket for butyl-moieties (75).

In addition to the above specific reports of genomics-based discovery programs, several general discussions of approaches and issues in the field have been published. For example, a recent review has covered the role of genomics in discovery of new antimycobacterial drugs (76). This organism is somewhat unique in terms of the structure of the outer cell wall and in the inhibitors that act to arrest growth. Additional overviews of genomic strategies in the hunt for novel antibiotics have also been recently published (77-79). While this is by no means an exhaustive list of all the activity in the field, it does give an overall snapshot of key activities. Given the very new nature not only of the genomic information, but the necessity to develop novel ways to analyze, identify and screen for inhibitors, the progress to date in identifying new inhibitors is most encouraging.

From Inhibitor to Drug: *The Next Set of Challenges* - Early indications are that it will be possible to find inhibitors for at least some of the essential functional antibacterial enzymes. Pragmatically, however, it is a long road from an *in vitro* inhibitor to a potent, safe, and efficacious antibiotic. In this section, we briefly sketch some of the considerations that must be addressed along the road from genomic target to effective drug.

One prime consideration is the rapidity with which bacteria can mutate or exchange genetic information. At least two of the major classes of existing drugs, β-lactams and fluoroquinolones, have multiple targets in most bacterial pathogens (80,81). As a result of multiple drug targets within a single organism, resistance tends to occur in incremental steps. In contrast, single target inhibitors such as rifampin, which inhibits the bacterial RNA polymerase, are noted for single step, high rate of mutational resistance development. It may be important to identify essential gene paralogs within cells as high value targets in order to delay resistance development. Early studies with lead compounds and other cognate targets to monitor mutational resistance will be an important aspect of future antibacterial drug development.

Equally important will be an improved understanding of all the factors involved in the penetration and retention of compounds in bacterial cells. Many inhibitors of *in vitro* enzyme based screens are found to be inactive in whole cell antibacterial assays. Since the bulk of targets will be intracellular, it will be increasingly important to develop guidelines for the chemical properties necessary to promote drug transport, either passive or active, into the bacterial cell. The bacterial membrane differs from the eukaryotic cell membrane in several components. For example, with the exception of a few mycoplasmal species, it does not contain cholesterol (82). Optimal charge, size, and $\log_P$ values for membrane penetration will guide the optimization of leads from enzyme based screens to "build-in" whole cell activity. Even our fundamental understanding of whether drug class uptake is active or passive through these membranes is incomplete. These desirable "drug-like" properties are enhanced, optimized with the SAR developed by the medicinal chemist, while checking that the original antimicrobial activity is maintained.

Just as important is an understanding of the numerous drug efflux pumps that have been discovered, many through the use of genomics-based searches. A recent study comparing the genomes of thirty-six microorganisms focused on the membrane transport systems and the predicted drug efflux systems, which varied among the bacterial species studied (83). Examination of the highly drug-resistant *E. faecalis* revealed the presence of thirty-four potential drug transporter systems in the genome. A gene knockout campaign revealed that the loss of several individual transporters led to marked increases in drug sensitivity (84). It has also been suggested, as reviewed recently, that the drug efflux transporters themselves may constitute excellent targets for antimicrobial intervention (85). The design of molecules that inhibit the function of these drug pumps could restore sensitivity to entire classes of antimicrobial agents already in use and currently limited by efflux-mediated resistance (85).

Turning from the challenges presented by the microbes to the problems associated with converting an enzyme inhibitor into a drug, many good antibacterial target inhibitors have been reported over the years, but many of these with poor or limited spectrum antimicrobial activity, and even fewer as acceptable drug candidates that ever reach the clinic, let-alone the market. The key remains to build-in the desirable "drug-like" properties, while carefully checking that the original biological effect - antimicrobial activity - is maintained. Therefore the question with any novel antimicrobial will be: *can the desired pharmacological properties be attained while retaining the ability to effectively inhibit the essential function target in bacteria?* To answer this question, the merging of both eukaryotic and microbial genomic information may provide clues to drug optimization. While the task ahead in developing these new compounds is daunting, the situation with the continued growth of antibiotic-resistant pathogens makes it imperative that the search for novel antimicrobials continues. It is clear that there is a large and growing role for many new technologies in this endeavor, most notably microbial genomics.

References

1.   H.C. Neu, Science, 257, 1064 (1992).
2.   S.B. Levy, N. Engl. J. Med., 338, 1376 (1998).
3.   R.C. Moellering, Clin. Inf. Dis., 27 (Suppl.1), S135 (1998).
4.   L. Silver, K. Bostian, Eur. J. Clin. Microbiol. Infect. Dis., 9, 455 (1990).
5.   G.S. Sutton, O. White, M.D. Adams, A.R. Kerlavage, Genome Sci. Technol., 1, 9 (1995).
6.   C.M. Fraser, R.D. Fleischmann, Electrophoresis, 18, 1207 (1997).
7.   R.D. Fleischmann, M.D. Adams, O. White, R.A. Clayton, E.F. Kirkness, A. R. Kerlavage, C.J. Bult, J.-F. Tomb, B.A. Dougherty, J.M. Merrick, *et al.*, Science, 269, 496 (1995).
8.   A. Goffeau, B.G. Barrell, H. Bussey, R.W. Davis, B. Dujon, H. Feldmann, F. Galibert, J.D. Hoheisel, C. Jacq, M. Johnston, E.J. Louis, Y. Murakami, P. Philippsen, S.G. Oliver, Science, 274, 546 (1996).
9.   The *C. elegans* Sequencing Consortium, Science, 282, 2012 (1998).
10. M.D. Adams, S.E. Celniker, R.A. Holt, C.A. Evans, J.D. Gocayne, P.G. Amanatides, S.E. Scherer, P.W. Li, R.A. Hoskins, R.F. Galle, R.A. George etal., Science, 287, 2185. (2000).
11. The International Human Genome Sequencing Consortium, Nature, 409, 860 (2001).
12. D. Greenwood, Antimicrobial Chemotherapy (4th ed), Oxford Univ. Press (2000).
13. J.J. Ferreti, W.M. McShan, D. Ajdic, D.J. Savic, G. Savic, K. Lyon, C. Primeaux, S. Sezate, A.N. Suvorov, S. Kenton, *et al.*, Proc. Natl. Acad. Sci. USA, 98, 4658 (2001).
14. R.H. Baltz, F.H. Norris, P. Matsushima, B.S. DeHoff, P. Rockey, G. Porter, S. Burgett, R. Peery, J. Hoskins, L. Braverman, *et al.*, Microb. Drug Resist., 4, 1 (1998).
15. J. Dopazo, A. Mendoza, J. Herrero, F. Caldara, Y. Humbert, L. Friedli, M. Guerrier, E. Grand-Schenk, C. Gandin, M. De Francesco, *et al.* Microb. Drug Resist., 7, 99 (2001).
16. H. Tetelin, K.E. Nelson, I.T. Paulsen, J.A. Eisen, T.D. Read, S. Peterson, J. Heidelberg, R.T. DeBoy, D. H. Haft, R.J. Dodson, *et al.*, Science, 293, 498 (2001).

17. T.D. Read, R.C. Brunham, C. Shen, S.R. Gill, J.F. Heidelberg, O. White, E.K. Hickey, J.Peterson, T. Utterback, K. Berry, *et al.*, Nucleic Acids Res., 28, 1397 (2000).
18. C.M. Fraser, S. Casjens, W.M. Huang, G.G. Sutton, R. Clayton, R. Lathigra, O. White, K. A. Ketchum, R. Dodson, E. K. Hickey, *et al.*, Nature, 390, 580 (1997).
19. C.M. Fraser, S.J. Norris, G.M. Weinstock, O. White, G.G. Sutton, R. Dodson, M. Gwinn, E.K. Hickey, R. Clayton, K.A. Ketchum, *et al.*, Science, 281, 375 (1998).
20. S.T. Cole, R. Brosch, J. Parkhill, T. Garnier, C. Churcher, D. Harris, S.V. Gordon, K. Eiglmeier, S. Gas, C. E. Barry, *et al.*, Nature, 393, 537 (1998).
21. N.T. Perna, G. Plunkett, V. Burland, B. Mau, J.D. Glausner, D.J. Rose, G.F. Mayhew, P.S. Evans, J. Gregor, H.A. Kirkpatrick, T.S., *et al.*, Nature, 409, 529 (2001).
22. A.L. Delcher, D. Harmon, S. Kasif, O. White, S.L. Salzburg, Nucleic Acids Res., 27, 4636 (1999).
23. R.E. Brucolleri, T.J. Dougherty, D.B. Davison, Nucleic Acids Res., 26, 4482 (1998).
24. G. Perrière, L. Duret, M. Gouy, Genome Res., 10, 379 (2000).
25. T. Gaasterland, M.A. Ragan, Microb. Compar. Genomics, 3, 177 (1998).
26. P.D. Karp, M. Krummernacker, S. Paley, J. Wagg, Trends Biotech., 17, 275 (1999).
27. F. Chetouani, P. Glaser, F. Kunst, Microbiol., 147, 2643 (2001).
28. S.F. Altschul, W. Gish, W. Miller, E.W. Meyers, D.J. Lipman, J. Mol. Biol., 215, 403 (1990).
29. J.R. Pearson, D.J. Lipman, Proc. Natl. Acad. Sci. USA, 85, 2444 (1988).
30. J.P. Claverys, A. Dintilhac, E.V. Pestova, B. Martin, D.A. Morrison, Gene 164, 123 (1995).
31. V. Vagner, E. Dervyn, S. Dusko-Ehrlich, Microbiol., 144, 3097 (1998).
32. M. Xia, R.D. Lunsford, D. McDevitt, S. Lordanescu, Plasmid, 42, 144 (1999).
33. S.N. Ho, H.D. Hunt, R.M. Horton, J.K. Pullen, L.R. Pease, Gene, 77, 51 (1989).
34. D. Yu, H.M. Ellis, E.C. Lee, N.A. Jenkins, N.G. Copeland, D.L. Court, Proc. Natl. Acad. Sci. USA, 97, 5978 (2000).
35. C.A. Hutchison, S.N. Peterson, S.R. Gill, R.T. Cline, O. White, C.M. Fraser, H.O. Smith, J.C. Venter, Science, 286, 2165 (1999).
36. H.E. Takiff, T. Baker, T. Copeland, S.M. Chen, D.L. Court, J. Bacteriol., 174, 1544 (1992).
37. C.K. Murphy, E.J. Stewart, J. Beckwith, Gene, 155, 1 (1995).
38. N. Gaiano, A. Amsterdam, K. Kawakami, M. Allende, T. Becker, N. Hopkins, Nature, 383, 829 (1996).
39. J. Bender, N. Kleckner, Proc. Natl. Acad. Sci. USA., 89, 7996 (1992).
40. B.J. Akerley, E.J. Rubin, A. Camilli, D.J. Lampe, H.M. Robertson, J.J. Mekalanos, Proc. Natl. Acad. Sci. USA, 95, 8927 (1998).
41. K.A. Reich, L. Chovan, P. Hessler, J. Bacteriol., 181, 4961 (1999).
42. N. Judson, J.J. Mekalanos, Nature Biotech., 18, 740 (2000).
43. J. Deiwick , M. Hensel, Electrophoresis, 20, 813 (1999).
44. J.L. Rakeman, S.I. Miller, Trends Microbiol., 7, 22 (1999).
45. E. Garcia-Vescovi, F.C. Soncini, E.A. Groisman, Res. Microbiol., 145, 473 (1994).
46. J.W. Foster, Y.K. Park, I.S. Bang, K. Karem, H. Betts, H.K. Hall, E. Shaw, Microbiology, 140, 341 (1994).
47. P.A. Cotter, V.J. DiRita, Ann. Rev. Microbiol., 54, 519 (2000).
48. A. Sukhan, Cell Mol. Life Sci., 57, 1033 (2000).
49. L.E. Alksne, S.J. Projan, Curr. Opin. Biotechnol., 11, 625 (2000).
50. D.M.Heithoff, R.L. Sinsheimer, D.A. Low, M.J. Mahan, Philos. Trans. R. Soc. Lond. B Biol. Sci., 355, 633 (2000).
51. J.M. Mei, F. Nourbakhsh, C.W. Ford, D.W. Holden, Mol. Microbiol., 26, 399 (1997).
52. K.E. Unsworth, D.W. Holden, Philos. Trans. R. Soc. Lond. B Biol. Sci. 355, 613 (2000).
53. J.E. Shea, J.D. Santangelo, R.G. Feldman, Curr. Opin. Microbiol., 3, 451 (2000).
54. M.J. Mahan, J.M. Slauch, J.J. Mekalanos, Science, 259, 686 (1993).
55. D.M. Heithoff, C.P. Conner, P.C. Hanna, S.M. Julio, U. Hentschel, M.J. Mahan, M.J., Proc. Natl. Acad. Sci. USA, 94, 934 (1997).
56. M. Handfield, R.C. Levesque, FEMS Microbiol. Rev., 23, 69 (1999).
57. G. Plum, J.E. Clark-Curtiss, Infect. Immunity, 62, 476 (1994).
58. M.B. Eisen, P.O. Brown, Methods Enzymol., 303, 179 (1999).
59. H. Gmuender, K. Kuralti, K. Di Padova, C.P. Gray, W. Keck, S. Evers, Genome Res. 11, 28 (2001).
60. L. Zhang, F. Fan, L.M. Palmer, M.A. Lonetto, C. Petit, L.L. Voelker, A. St. John, B. Bankosky, M. Rosenberg, D. McDevitt, Gene, 255, 297 (2000).
61. A. Pandley, M. Mann, Nature, 405, 837 (2000).

62. C.P. McAtee, P.S. Hoffman, D.E. Berg, Proteomics, 1, 516 (2001).
63. M.H. Serres, S. Gopal, L.A. Nahum, P. Liang, T. Gaasterland, M. Riley, Genome Biology, 2, 1 (2001).
64. D.J. Payne, P.V. Warren, D.J. Holmes, Y. Ji, J.T. Lonsdale, Drug Discov. Today, 6, 537 (2001).
65. R.J. Heath, C.O. Rock, Nature, 206, 145 (2000).
66. T. Meinnel, S. Blanquet, J. Bacteriol., 175, 7737 (1993).
67. C.M. Apfel, H. Locher, S. Evers, B. Takacs, C. Hubschwerlen, W. Pirson, M.G. Page, W. Keck, Antimicrob. Agents Chemother., 45, 1058 (2001).
68. C.M. Apfel, D.W. Banner, D. Bur, M. Dietz, C. Hubschwerlen, H. Locher, F. Marlin, R. Masciadri, W. Pirson, H. Stalder, J. Med. Chem., 44, 1847 (2001).
69. J.M. Clements, R.P. Beckett, A. Brown, G. Catlin, M. Lobell, S. Palan, W. Thomas, M. Whittaker, S. Wood, S. Salama, et al., Antimicrob. Agents Chemother., 45, 563 (2001).
70. P. Gallant, J. Finn, D. Keith, P. Wendler, Emerg. Ther. Targets, 4, 1 (2000).
71. A.K. Forrest, R.L. Jarvest, L.M. Mensah, P.J. O'Hanlon, A.J. Pope, R.J. Sheppard, Bioorg. Med. Chem. Lett., 10, 1871 (2000).
72. J.M. Berge, N.J. Broom, C.S. Houge-Frydrych, R.L. Jarvest, L. Mensah, D.J. McNair, P.J. O'Hanlon, A.J. Pope, S. Rittenhouse, J. Antibiot., 53, 1282 (2000).
73. P. Brown, D.S. Eggleston, R.C Haltwanger, R.L. Jarvest, L. Mensah, P.J. O'Hanlon, A.J. Pope, Bioorg. Med. Chem Lett., 11, 711 (2001).
74. R.L. Jarvest, J.M. Berge, C.S. Houge-Frydrych, L.M. Mensah, P.J. O'Hanlon, A.J. Pope, Bioorg. Med. Chem. Lett., 11, 2499 (2001).
75. X. Qiu, C.A. Janson, W.W. Smith, S.M. Green, P. McDevitt, K. Johanson, P. Carter, M. Hibbs, C. Lewis, A. Chalker, et al., Prot. Sci., 10, 2008 (2001).
76. C.E. Barry, R.A. Slayden, A.E. Sampson, R.E. Lee, Biochem. Pharmacol., 59, 221 (2000).
77. L. P. Kotra, S. Vakulenko, S. Mobashery, Microbes Infect., 2, 651 (2000).
78. H. Loferer, A. Jacobi, A. Posch, C. Gauss, S. Meier-Ewert, B. Seizinger, Drug Discov. Today, 5, 107 (2000).
79. D. McDevitt, M. Rosenberg, Trends Microbiol., 9, 611 (2001).
80. C. Goffin, J.-M. Ghuysen, Microbiol. Mol. Biol. Rev., 62, 1079 (1998).
81. K. Drlica, X. Zhao, Microbiol. Mol. Biol. Rev., 61, 377 (1997).
82. K. Magnuson, S. Jackowski, C.O. Rock, J.E.Cronan, Microbiol. Rev., 57, 522 (1993).
83. I.T. Paulsen, J. Chen, K.E. Nelson, M.H. Saier, J. Mol. Microbiol. Biotechnol., 3, 145 (2001).
84. D.R. Davis, J.B. McAlpine, C.J. Pazoles, M.K. Talbot, E.A. Alder, C. White, B.M. Jonas, B.E. Murray, G.M. Weinstock, B.L. Rogers, J. Mol. Microbiol. Biotechnol., 3, 179 (2001).
85. K. Poole, J. Pharm., Pharmacol., 53, 283 (2001).

Chapter 11. The Chemistry and Biology of the Tetracyclines

Mark L. Nelson
Paratek Pharmaceuticals, Inc.
75 Kneeland Street, Boston, MA 02111
mnelson@paratekpharm.com

Introduction – The chemistry and biology of the tetracyclines is currently undergoing a "renaissance" whereby new chemistries are being applied to basic fermentation-derived tetracycline scaffolds and semisynthetic tetracycline products, affording compounds with activity against both tetracycline sensitive and resistant bacteria. Tetracyclines also exhibit activity against parasites, fungi, and, as discovered within the last 15 years, activity against mammalian targets and disease processes. This review will highlight some of the new developments against prokaryotic and eukaryotic cells and cellular processes.

Tetracyclines were discovered in 1947, by Benjamin M. Duggar, bioprospecting for new antibiotics (1). Naturally produced by soil bacteria of the *Actinomycetales* order, *Streptomyces* genus, three tetracyclines **1** (aureomycin®, chlortetracycline), **2** (terramycin®, oxytetracycline), and **3** (tetracycn®, tetracycline), were found to be produced and had activity against a wide range of microorganisms (2-4). For the first time, the words *broad spectrum* were used to describe their bioactivity against both Gram-positive and Gram-negative microorganisms.

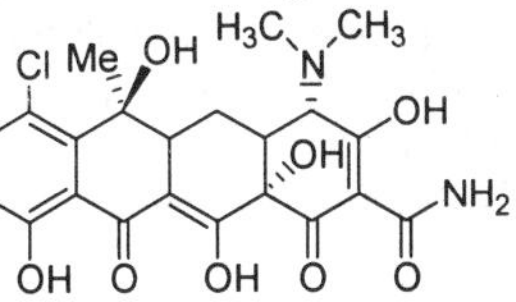

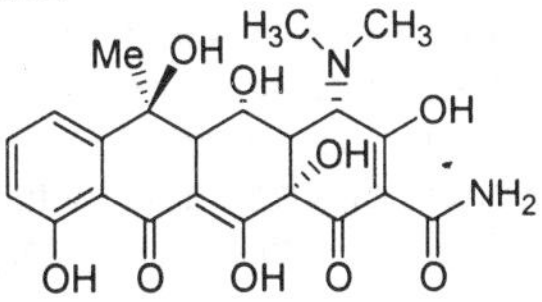

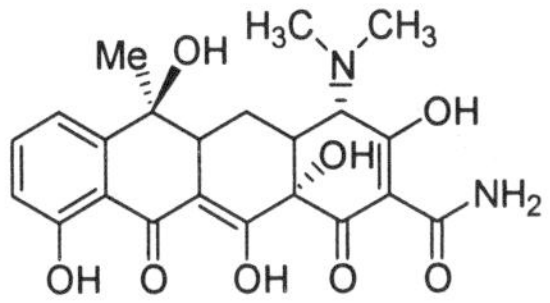

Tetracyclines are a diverse family of compounds produced by the sequential enzymatic synthesis and biotransformation of polyketide precursors, primarily studied within the order *Actinomycetales* (5,6). Other compounds modified within the naphthacene nucleus can be obtained from these producers, whose members number over 3,000 distinct species in over 40 genera (7).

Early antibacterial compounds discovered in *Streptomyces* fermentation broths include demecycline **4**, demeclocycline **5**, terramycin X **6**, and chelocardin **7**, while more recently, the tetracycline glycosides dactylocycline A **8**, and B **9**, were described as having activity against both sensitive and resistant *Staphylococcus* species with a MIC range of 1.3 to 6.1 □g/mL (8-13). Unfortunately, both compounds were produced in low yield via fermentation from *Dactylosporangium* species, are prone to acid hydrolysis and have not entered into further clinical development. All of the compounds (**4-9**) were found to possess Gram-positive activity, while being totally devoid of Gram-negative activity.

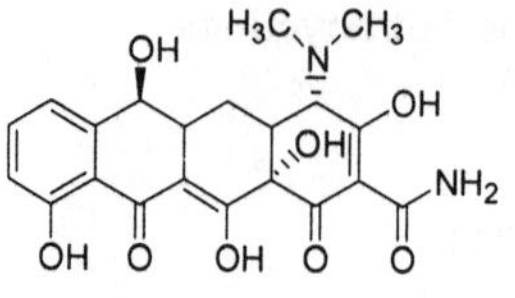

4 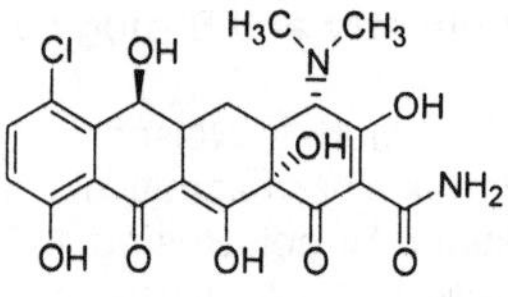 **5** 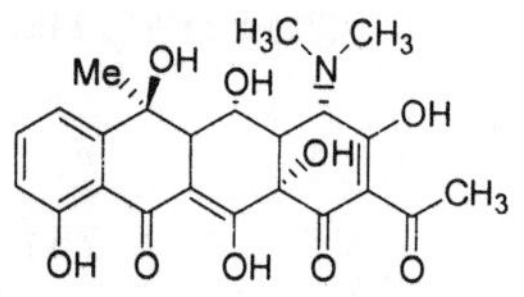 **6**

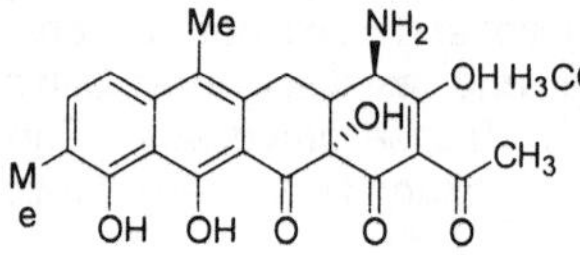

7 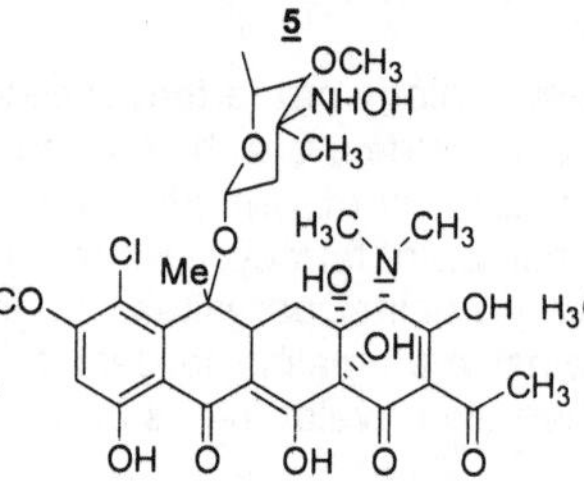 **8** 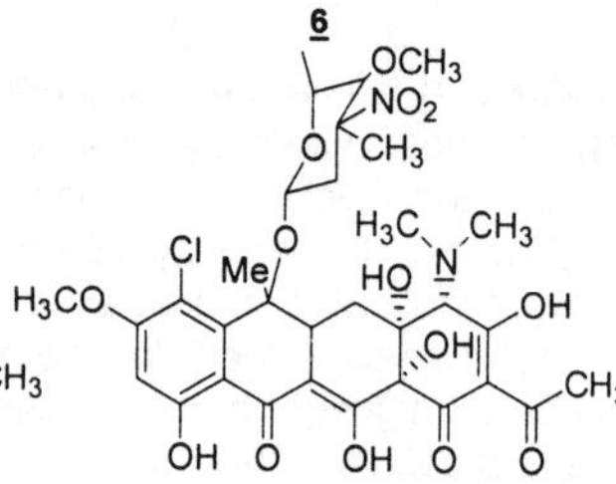 **9**

Streptomyces also produce a vast array of polyketide-derived natural products that have been found to be active against eukaryotic cells. The anticancer agent daunorubicin **10** and its congener doxorubicin (Adriamycin®) **11** and other anthracyclines are produced by several different *Streptomyces* species and possess a naphthacene ring system as a substructure. As anticancer agents they are effective and multifactorial in action, producing reactive species that intercalate with DNA, acting as DNA synthesis inhibitors in actively dividing cells, and they inhibit topoisomerase II and chelate metal ions such as Fe^{3+} and Cu^{2+}, initiating Fenton-type chemical redox reactions (14-16). This leads to a pleiotropic component or mechanism of action of these compounds, due presumably to their ability to form free radical species and non-specifically modify macromolecular targets.

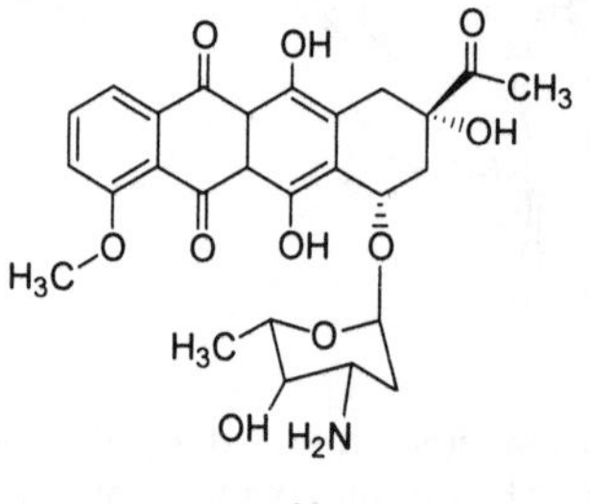

10 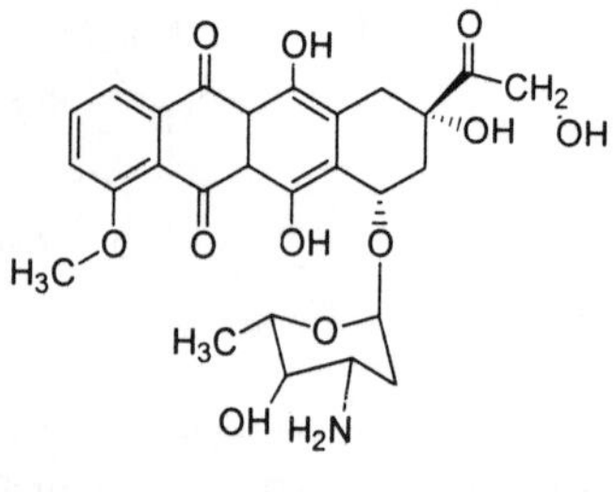 **11**

The antitumor antibiotic SF2575 (**12**) was isolated from *Streptomyces* sp. SF2575 and has been deposited at the Fermentation Research Institute, Agency of Industrial Science and Technology, Japan (17, 18). This naphthacene derivative was found to be cytotoxic against several different cell lines with an undetermined mechanism of action. Other antibiotics with similar substructures have been isolated from fermentation and described, e.g., the TAN series of anticancer agents. TAN tetracyclines were isolated from Streptomyces species AL-16012, producing three different structural variants designated TAN A **13**, TAN B **14** and TAN X, the latter of which had the same chemical composition as SF-2575 (**12**)(19).

Studies of the cellular action of the tetracyclines against both prokaryotes and eukaryotes show that they have a diverse range of biological activities. The antibacterial tetracyclines show dual activity depending upon the compound, where "typical" tetracyclines, those used clinically to treat infections, primarily inhibit protein synthesis via ribosomal binding to both the 30S and 50S subunits, with a delineation of a high affinity binding site on the 30S subunit (20).   Once bound, they interfere with binding of the aminoacyl-tRNA-EF-Tu-GTP complex to the ribosomal A-site by either non-competitive or allosteric effects and effectively stop protein synthesis (21).

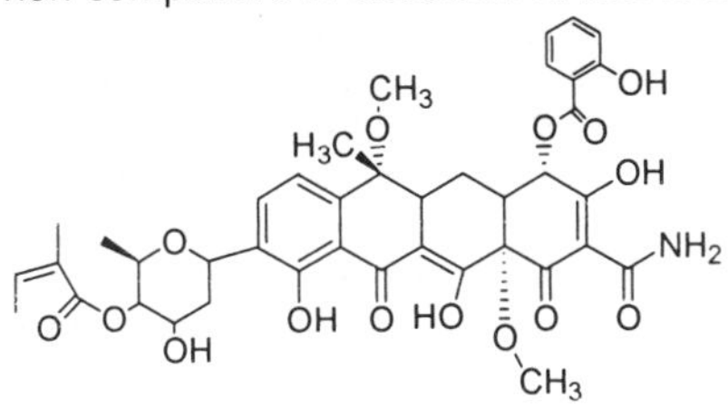

12

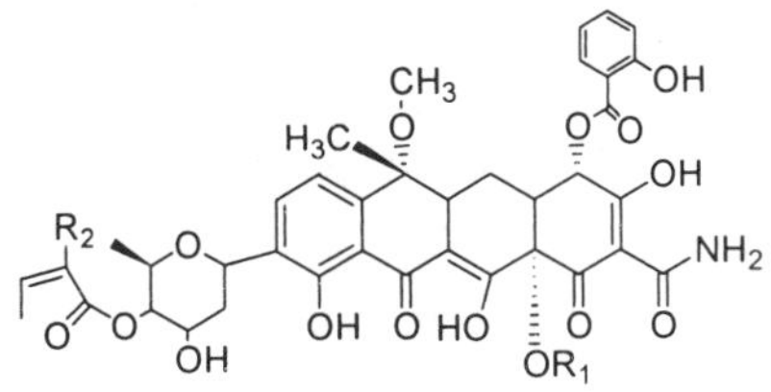

13   R1=H, R2=CH₃
14   R1=CH₃ R2=CH₂CH₃

Other, "atypical" and non-clinically used tetracyclines such as anhydrotetracycline **15**, a metabolite of tetracycline, and sancycline **16**, a semi-synthetic tetracycline, can exert deleterious effects at other bacterial cellular processes and macromolecular targets ranging from membrane perturbation, cellular disruption and impairment of membrane function via de-energization and autolysis (22-24).   Although both compounds have excellent activity, particularly against tetracycline resistant Gram-positive bacteria, they have never found use as antibacterial agents because of non-selective activity against cellular membranes.

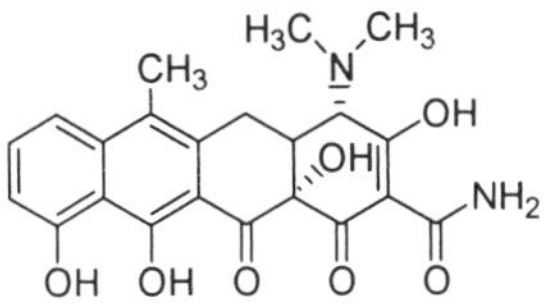

15 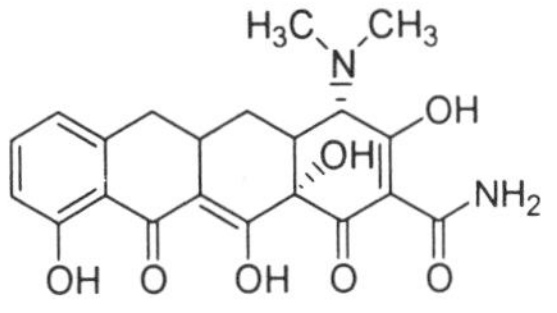   **16**

Chlortetracycline, **1** was found to have an effect on leucine incorporation in rat liver ribosomes while modulating protein synthesis in several different mammalian cell lines (25). Tetracyclines can be covalently incorporated into liver ribosomes and at clinically relevant levels, affect the production of migration inhibition factor in human lymphocytes stimulated with concavalin A, indicating even mammalian protein synthesis may be affected (26, 27). Tetracyclines also have profound activity against mammalian cell mitochondria, inhibiting mitochondrial protein synthesis and shifting membrane potential and respiratory control (28, 29). Tetracyclines also may have immunosuppressive activity via lymphocyte mitochondrial suppression, impairing mitochondrial biogenesis through blast cell modulation (30).

Tetracyclines have activity against parasites for perhaps the same reason, i.e., inhibitory effects against mitochondria. Oxytetracycline **2** is a potent inhibitor of mitochondrial protein synthesis in *Thelia parva* (East Coast fever in cattle), and of lymphoblast mitochondrial protein synthesis (31). Furthermore, oxytetracycline has macrofilaricidal activity against the filarial nematode *Onchocerca ochengi*, the chemotherapeutic model for *O. volvulus*, the causative agent of "river blindness" in humans.   It is believed that eradication of the nematode occurs by first eliminating *Wolbachia* symbiotic bacteria which are from the order Rickettsiales, which then is lethal to adult filariae (32).   The fact that tetracyclines affect mitochondria, Rickettsial-like symbionts such as *Wolbachia*, and are effective against Rickettsial infections alone is of no coincidence, as phylogenetic data support that all three have similar evolutionary origins (33).

BACTERIOLOGICAL USES OF TETRACYCLINES

The use of tetracyclines in medicine is limited to those that have been isolated by fermentation broths or by chemical modification through semisynthesis to produce chemically stable and effective compounds.   Two clinically used tetracyclines are presented below, along with their primary and secondary usages against bacterial pathogens (34). Natural products oxytetracycline **2,** tetracycline **3,** and demeclocycline **4**, are used to a lesser extent, while the semisynthetic tetracyclines, minocycline **17** and doxycycline **18**, are some of the most widely used antibiotics in medicine, animal health and agriculture, although antibiotic resistance has curtailed the use of these agents.

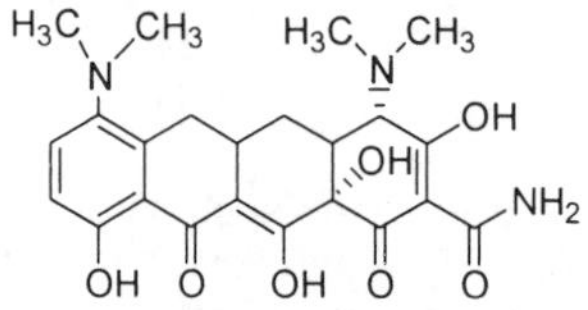

17

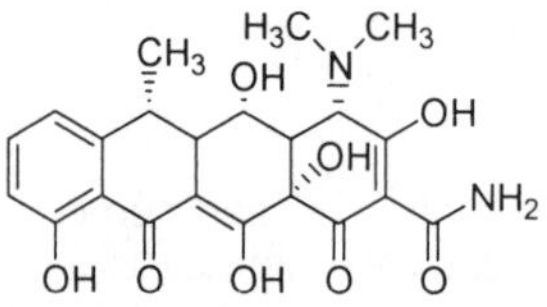

18

Susceptible Organisms and Indications[1]

Rickettsia

Mycoplasma pneumoniae

Psittacosis and Ornithosis

Borellia recurrentis and other species

Hemophilius ducreyi

Yersinia pestis

Bacteroides species

Vibrio cholera

Neisseria species

Treponema species

Clostridium species

Bacillus anthracis

Chlamydia species

Malaria prophylaxis

Enterococcus species[2]

Amoebiasis[3]

Severe acne[4]

E. coli[2]

Enterobacter aerogenes[2]

Shigella species[2]

Klebsiella species[2]

[1]When other antibiotics such as penicillin are contraindicated

[2] Susceptible strains from bacteriological testing

[3] Adjunct to amoebicides

[4] Adjunct to standard therapy

BACTERIAL RESISTANCE TO THE TETRACYCLINES AND RESISTANCE MODULATION

Both commensal and pathogenic bacteria have acquired resistance to the tetracyclines by two biochemical mechanisms that have been fairly well characterized in Gram-positive and Gram-negative bacteria, drug efflux and ribosomal protection (35), while inactivation of tetracyclines via enzymatic reactions are rare (36). Furthermore, the genes encoding both processes are found readily in the environment and distributed in bacteria as mobile transposons, plasmids and other genetic determinants that are readily transferable between species and genera. To date, over eighteen different *tet* genes encode for drug efflux, producing membrane-associated Tet proteins from the major facilitator superfamily (MFS) of export proteins, while ribosomal protection is represented by eight or more genes, encoding cytoplasmic proteins that protect ribosomes from the action of clinically used tetracyclines (37).

The Tet efflux proteins are readily expressed in *Eschericia coli,* where Class B proteins represent the most common resistant determinant found in many bacterial genera and has the highest transport activity of all the Tet proteins studied so far (38). When *E. coli* possessing the plasmid R222 with the Class B determinant on transposon Tn10 are lysed under high-pressure conditions, the inner cell membranes re-orient themselves inside-out, producing Tet(B) proteins trapped in an everted membrane vesicle--now pumping tetracyclines into the vesicle instead of pumping them out of the cell (39). Once isolated, the vesicles serve as an efficient assay for finding molecules that can block tetracycline efflux and modulate drug efflux-dependent resistance. This assay has been used with Michelis-Menten type kinetics to screen tetracyclines, tetracycline substructures and positional variants of tetracyclines to define the minimal structural pharmacophore(s) operative and/or synthetically modifiable regions maximal for Tet(B) efflux protein inhibition. Everted membrane assays would also provide data for future development of a Tet(B) efflux pump inhibitor as synergists with an antibacterial tetracycline. Inhibiting tetracycline efflux proteins may constitute a new approach to reversing tetracycline resistance, changing a resistant cell to a susceptible one.

Based on inhibitory data generated, it was found that an intact ABCD naphthacene ring is required for maximum inhibition of efflux, while the presence of the phenol keto-enol substructures along the lower peripheral region spanning C10, C11, C12 and C1 are required for activity (Figure 1) (40). The C4 dimethylamino group was not needed for inhibitory activity. Synthetic modifications at C2, C4, C5, C6, C7, C8 and C9 produced compounds of variable activity directed synthesis of compounds at the C13 exocyclic carbon to increase inhibitory activity.

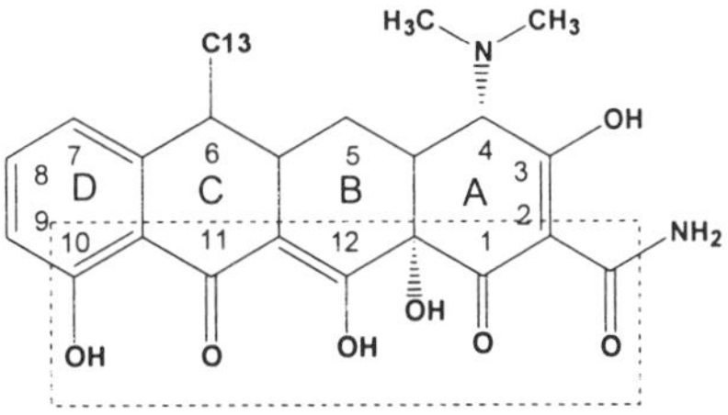

Figure 1. Molecular requirements for inhibition of the Tet(B) efflux protein. Boxed region represent areas most sensitive to chemical modification.

Position 13-alkylthio derivatives of tetracyclines, produced through the reaction of methacycline with mercaptans, produced the most potent inhibitors of efflux, 13-propylthio-5-OH tetracycline **19** with an IC_{50} of 0.7 $\square$M, and 13-cyclopentylthio-5-OH tetracycline (13-CPTC) **20** with an IC_{50} of 0.4 $\square$M, in contrast to the Km for tetracycline (15-20 $\square$M) and doxycycline (IC_{50} of 0.9 $\square$M) (41).

Studies of both compounds against susceptible and resistant bacteria showed that they were both effective alone against tetracycline susceptible Gram-positive S. aureus and E. faecalis, and effective against strains possessing Tet(K) and Tet(L) efflux proteins and were more effective than doxycycline against E. faecalis. They were not effective against susceptible E. coli, but did exhibit synergy with doxycycline **18** against E. coli possessing Tet(A), Tet(B) efflux proteins and E. faecalis possessing the Tet(M) ribosomal protection mechanism. A more detailed SAR study of the C13 position showed that the compounds exhibited activity that was dependent on the size, lipophilicity and polarity of the C13 substituent. Verloop-Hoogenstraten STERIMOL values suggested that the more potent inhibitors had substituent length values of between 4.4-6.2 Å, while optimal width was 3.0-4.2 Å. Lipophilicity also correlated well with bioactivity, where polar groups within a homologous series caused a decrease in inhibitory activity. Most of the active compounds fell within a range of lipophilicity values of 1.0 –2.5.

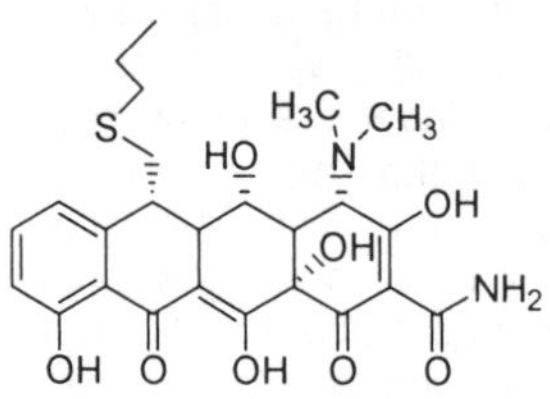

19 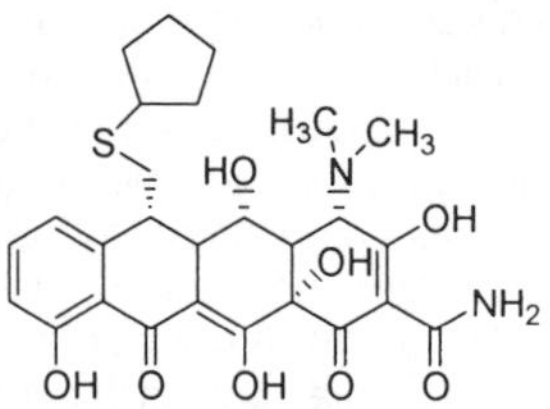   20

The effect of 13-CPTC on tetracycline accumulation kinetics in bacteria indicated that it had a greater affinity and was competing with tetracycline for the binding site on the Tet(B) protein, changing the distribution of Tet proteins of one of high affinity and translocation activity to one of decreased affinity for tetracycline and low translocation activity (42). In whole cells against Tet(B) efflux resistant E. coli, 13-CPTC caused a rapid uptake of tetracycline, indicating that it acted as a potent inhibitor of the efflux pump, binding competitively with tetracycline at the active site of the Tet protein. By combining subinhibitory amounts of 13-CPTC and doxycycline, it is possible to demonstrate pronounced inhibitory effects on both molecular processes and bacterial growth. These findings suggest that the tetracycline molecule may be synthetically modified toward the development of an effective efflux-blocking agent, modulating antibiotic transport and tetracycline resistance phenotypes, changing a resistant cell to a state of susceptibility, reversing antibiotic resistance.

CLINICAL DEVELOPMENT OF NOVEL TETRACYCLINES: TIGECYCLINE

Minocycline **17**, was the last tetracycline to be approved for use against bacterial infections in the United States. It is marketed under the tradename Minocin, and is produced from the natural product demeclocycline **5** by multi-step reactions via a minocycline intermediate. Researchers have produced semisynthetic derivatives of 9-aminominocycline **21** at C9, named the glycylcyclines, which are promising third-generation tetracyclines and effective against antibiotic resistant bacteria, in vitro and in vivo, and are the first new tetracyclines to enter clinical trials is over 30 years (43). Several compounds in this series were active against tetracycline resistant strains possessing efflux and ribosomal protection resistance determinants and harbored potent activity against methicillin-susceptible and resistant S. aureus, E. faecalis and E. faecium and other enterococcal species, including those that were vancomycin resistant. One compound, designated tigecycline **22**, formerly known as GAR-936, has shown broad spectrum activity against strains resistant to the tetracyclines, while exhibiting efficacy in animal models of infection using MRSA, vancomycin resistant E. faecalis or E. faecium or E. coli (44-46). Tigecycline also showed activity against Pseudomonas aeruginosa in a murine model of pneumonia, similar to gentamicin, decreasing bacteria lung counts when administered alone and in combination with gentamicin, although the compound lacked activity in vitro (47). Pharmacokinetic studies showed that **22** was widely distributed in the tissues, exhibited a dose-dependent half-life of 36 hours and was 59% serum-protein bound (48). The safety profile in humans showed effects similar to minocycline, namely, nausea, emesis and headaches (49).

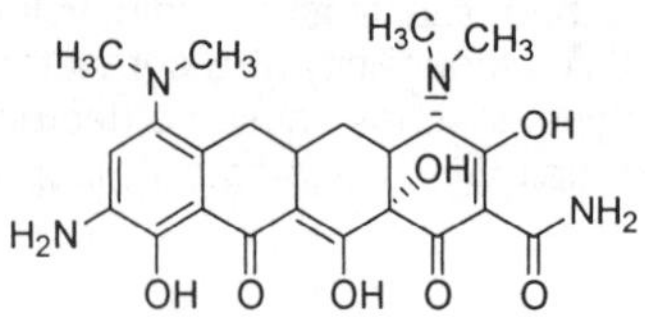

21 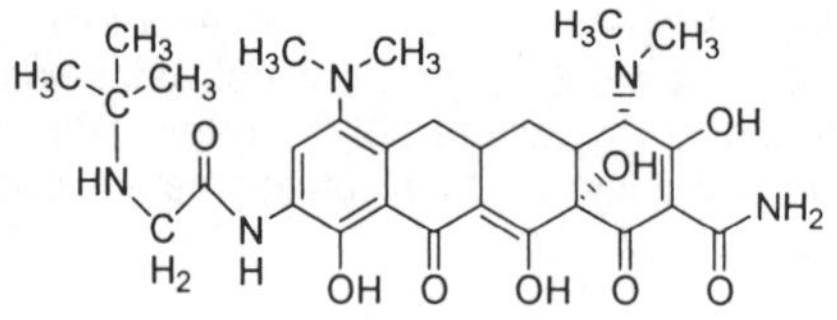   22

Other tetracycline scaffolds possessing the 9-t-butylglycylamido substructure have been described since this work began, including, 9-glycylamindo doxycycline **23** derivatives and L-amino acid derivatives of 9-aminodoxycycline **24,** but there was no improvement in either spectrum or potency of action (50).

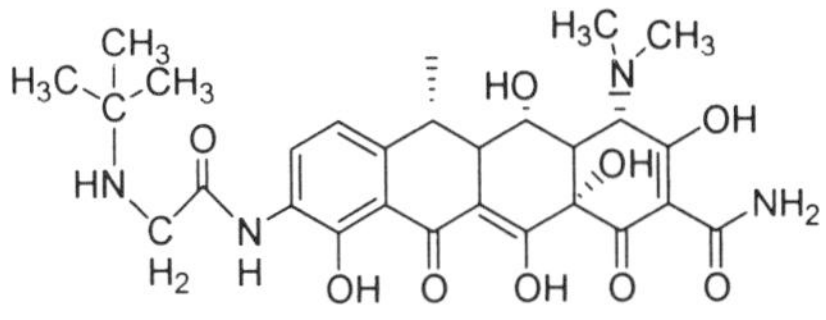

Tigecycline **22** has undergone three Phase II clinical trials in complicated skin, abdominal and urinary tract infections (58).

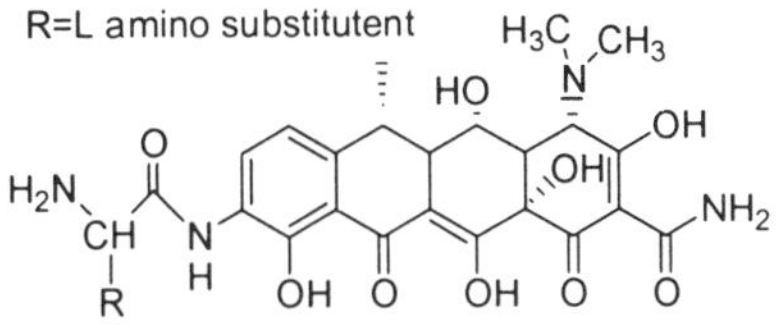

CHEMICAL MODIFICATIONS OF TETRACYCLINES

The semisynthesis of novel tetracyclines is plagued by numerous problems due inherently to the numerous functional groups and their related properties within the tetracycline nucleus. They are unstable to basic conditions, in some cases to strong acidity, and are prone to tautomerization between the A-ring keto-enolate functional groups, BC-ring keto-enolates and the C4 dimethylamino group (59-63). Synthesis research has centered on derivatization of C2, C5, C6-C13, C7, C8 and C9 positions, using chemistry that has been developed in the past 30 years and is now applied to the tetracycline family (Figure 2).

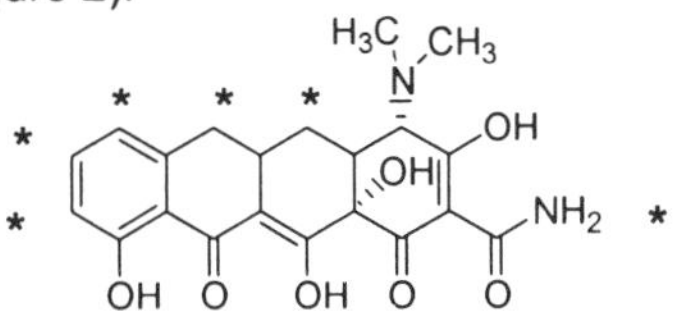

Figure 2. Positions modified to produce structurally diverse tetracyclines. Asterisks signify substructure series.

The patent literature recently describes transition metal-catalyzed reactions applied to reactive intermediates of tetracyclines, producing derivatives in high yield with remarkable chemical diversity (Figure 3)(52, 53). The method is applicable to all tetracyclines, where derivatives of sancycline **16**, minocycline **17** and doxycycline **18** possessing C-C linkages have been produced with few competing side reactions and tetracycline degradation. Transition metal couplings employ Suzuki, Heck, Stille and carbonylation reaction conditions for all of the aromatic positions within the tetracycline nucleus producing a variety of novel tetracycline derivatives. By using C8 reactive tetracycline intermediates possessing a halogen or reactive functional group, it is also possible to substitute the C8 position with phenyl, heteroaromatic groups, alkenes, alkynes and other C-C bond-forming substituents for the first time.

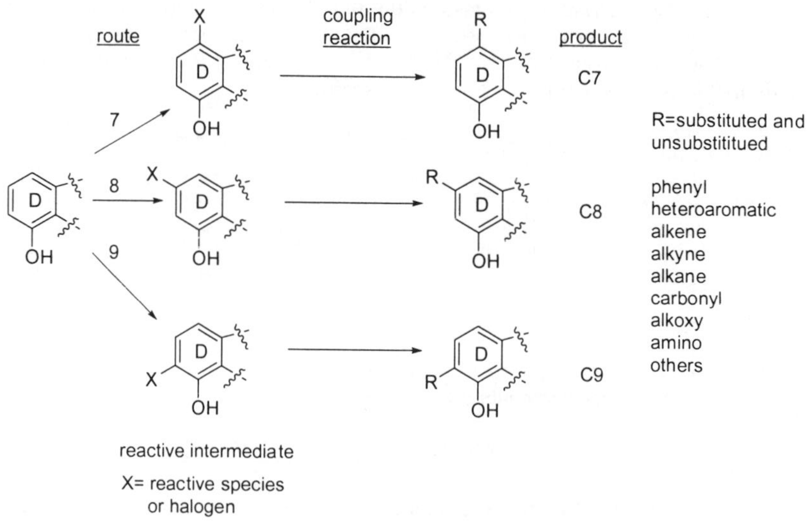

Figure 3. General reaction scheme for the production of C7, C8 and C9 tetracycline derivatives via transition metal catalyzed coupling reactions

Very few reactions have been successfully aimed at modifying the exocyclic double bond of methacycline **25**. Using transition metal-catalyzed reactions and unsubstituted or substituted phenylboronic acids, the C13 position undergoes facile derivatization, producing 13-phenyl derivatives of methacycline **26** in high yield (Figure 4) (54).

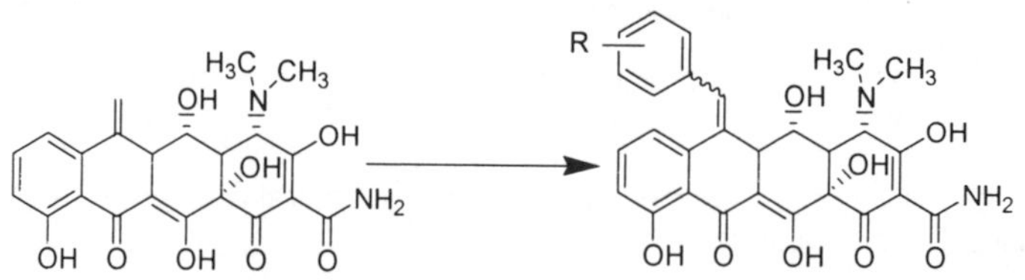

25   **26**

Figure 4. Reaction scheme for the production of C13 methacycline derivatives via transition metal catalyzed coupling reactions

Conclusion – The tetracyclines have a long history as antibacterials through the activities of only a handful of clinically used compounds, namely doxycycline and minocycline. Novel tetracyclines have been found to inhibit or evade mechanisms related to antibacterial resistance, namely, efflux and ribosomal protection, and are currently being studied towards the development of more potent and effective antibacterial agents. Only one compound, tigecycline, has been advanced to clinical studies, where Phase II trials indicate its potential. Tetracyclines, by virtue of their chemical uniqueness, can affect both microbial and mammalian targets and processes, ensuring continued interest as potential pharmacological agents.

References

1. B. M. Duggar, Ann. N.Y. Acad. Sci., 51, 177 (1948).
2. C. R. Stevens, L.H. Conover, R. Pasternak, F. A. Hochstein, W. T. Moreland, P. P. Regna, F. J. Pilgrim, K. J. Brunings and R. B. Woodward, J. Am. Chem. Soc., 76, 3568 (1954).
3. J. H. Kane and A. C. Finlay, U. S. Patent 2,482,055 (1949).
4. L. H. Conover, W. T. Moreland, A. R. English, C. R. Stephens and R. J. Pilgrim, J. Amer. Chem. Soc., 75, 462 (1953).
5. J. R. D. McCormick, E. R.Jensen, N. Arnold, H.S. Corey, U. H. Joachim, S. Johnson and N. O. Solander, J. Amer. Chem. Soc., 90, 7127 (1968).
6. Z. Hostalek, M. Tintertov. V. Jechova, M. Blumauerova, J. Suchy and Z. Vanek, Biotechnol. Bioeng, Biotechnol. Bioeng, 11, 539 (1969).
7. J. M. Brown, M. M. McNeil and E.P. Desmond, Man. Clin. Micro., 7th Ed. ASM Press, Washington, DC, 370 (1999).
8. D. Perlman and L. Hauser, U. S. Patent 3, 466,093 (1969).
9. L. H. Conover, W. T. Moreland, A. R. English, C. R. Stephens and R. J. Pilgrim, J. Amer. Chem. Soc., 75, 462 (1953).
10. L. A. Mitscher, J. V. Juvarkar, W. Rosenbrook, W. W. Andres, J. Schenk and R. S. Egan, J. Amer. Chem. Soc., 92, 6073 (1970).
11. J. S. Wells, J. O'Sullivan, C. Aklonis, H. A. Ax, A. A. Tymiak, D. R. Kirsch, W. H. Trejo and P. Principe, J. Antibiotics, 45, 1892 (1992).
12. A. A. Tymiak, H. A. Ax, M. S. Bolger, A. D. Kahle, M. A. Porubcan and N. H. Andersen, J. Antibio, 45, 1899 (1992).
13. C. Aklonis, H. A. Ax, J. O'Sullivan, A. A. Adrienne, J. S. Wells, J. Scott and D. R. Kirsch, Eur. Pat. Appl. 322717A2 (1990).
14. F. Acramone, Med. Res. Rev., 4, 153 (1984).
15. J. H. Peters, D. G. Streeter, J. S. Johl, G. R. Gordon and M. Tracy, Anticancer Res., 7, 1189 (1987).
16. G. Minotti, R. Ronchi, E. Salvatorelli, P. Menna and G. Cairo, Can. Res., 61, 8422 (2001).
17. M. Hatsu, T, Sasaki, H. Watabe, S. Miyadoh, M. Nagasawa, T. Shomura, M. Sezaki, S. Inouye and S. Kondo, J. Antibiot., 45, 320 (1992).
18. M. Hatsu, T. Sasaki, S. Gomi, Y. Kodama, M. Sezaki, S. Inouye and S. Kondo, J. Antibiot., 45, 325 (1992).
19. T. Horiguchi, K. Hayashi, S. Tsubotani, S. Iunuma, S. Harada and S. Tanida, J. Antibiot., 47, 545 (1994).
20. G. R. Craven, R. Gavin and T. Fanning, Cold Spring Harbor Symp. Quant. Bio., 34, 129 (1969).
21. E. F. Gale, E. Cundliffe, P. E. Reynolds, M. H. Richmond and M. J. Waring, The molecular basis of antibiotic action, Wiley, London (1981).
22. B. Rasmussen, H. F. Noller, G. Daubresse, B. Oliva, Z. Misulovin, D. M. Rothstein, G. A. Ellestad, Y. Gluzman, F. P. Tally and I. Chopra, Antimicrob. Agents Chemother., 35, 2306 (1991).
23. 23. B. Oliva, G. Guay, P. McNicholas, G. Ellestad and I. Chopra, Antimicrob. Agents Chemother., 36, 913 (1992).
24. B. Oliva and I. Chopra, Antimicrob. Agents Chemother., 36, 876 (1992).
25. S. D. Yeh and M. E. Shils, Proc. Soc. Exp. Biol. Med., 121, 729 (1966)
26. A. M. Reboud, S. Dubost and J. P. Reboud, Eur. J. Biochem, 124, 389 (1982).
27. R. Ganguly, D. G. Pennock and R. M. Kluge, Allerg. Immunol., 30, 104 (1984).
28. C. van den Bogert, G. van Kernebreek, L. de Leij and A. M. Kroon, Cancer Lett., 32, 41 (1986).
29. H. A. Pershadsingh, A. P. Martin, M. L. Vorbeck, J. W. Long and E. B. Stubbs, J. Biol. Chem., 257, 12481 (1982).
30. C. van den Bogert, T. E. Melis and A. M. Kroon, J. Leukocyte Biol., 46, 128 (1989).
31. P. R. Spooner, Parasitology, 101, 387 (1990).
32. N. G. Langworthy, A. Renz, U. Mackenstedt, K. Henkle-Duhrsen, M. B. Bronsvoort, V. N. Tanya, M. J. Donnely and A. J. Trees, Proc. R. Soc. Lond. B, 267, 1063 (2000).
33. V. V. Emelyanov, Biosci. Rep., 21, 1 (2001).
34. Physicians' Desk Reference, 54th Edition, Medical Economics Co. Montvale, NJ, 2371, 1543, 1537, 1528, 2371, 2384, 3164 (2000).
35. S. B. Levy, Antimicrob. Agents Chemother., 36, 695 (1992).
36. B. S. Speer, L. Bedzyk and A. A. Salyers, J. Bacteriol. 173, 176 (1991).
37. I. Chopra and M. Roberts, Micro. Mol. Bio. Rev., 65, 232 (2001).

38. L. M. McMurry, R. E. Petrucci and S. B. Levy, Proc. Natl. Acad. Sci. USA, 77, 3974 (1980).
39. B. P. Rosen and J. S. Clees, Proc. Natl. Acad. Sci. USA, 71, 5942 (1974).
40. M. L. Nelson, B. H. Park and S. B. Levy, J. Med. Chem., 37, 1355 (1994).
41. 41. M. L. Nelson, B. H. Park, J. S. Andrews, V. A. Georgian, R. C. Thomas and S. B. Levy, J. Med. Chem., 36, 370 (1993).
42. M. L. Nelson and S. B. Levy, Antimicrob. Agents Chemother., 43, 1719 (1999).
43. P. E. Sum, V. Lee, R. T. Testa, J. J. Hlavka, G. Ellestad, J. D. Bloom, Y. Gluzman and F. Tally, J. Med. Chem., 37, 184 (1994).
44. F. A. Fraise, N. Brenwald, J. M. Andrews and R. Wise, J. Antimicrob. Chemother., 35, 877 (1995).
45. A. C. Gales and R. N. Jones, Diagn. Microbiol. Infect. Dis., 36, 19 (2000).
46. P. J. Petersen, N. V. Jacobus, W. J. Weiss, P. E. Sum and R. T. Testa, Antimicrob. Agents Chemother., 43, 738 (1999).
47. A. P. Johnson, Curr. Opin. Anti-infect. Investig. Drugs, 2, 164 (2000).
48. M. L. Van Ogtrop, D. Andes, W. A. Craig, W. J. Weiss and O. Vesga, 38th ICAAC, San Diego, CA, US (1998). Abstract F-134.
49. G. Muralidharan, J. Getsy, P. Mayer, I. Paty, M. Micalizzi, J. Speth, B. Webster and P. Mojaverian, , 39th ICAAC, San Francisco, CA, US (1999). Abstract 416.
50. T. C. Barden, B. L. Buckwalter, R. T. Testa, P. J. Petersen and V. J. Lee, J. Med. Chem., 37, 3205 (1994).
51. D. C. Hooper, 41st ICAAC, Chicago, IL, US (2001). Abstract 602.
52. M. L. Nelson, G. Rennie and D. Koza, PCT Int. Appl. WO0119784 (2001).
53. M. L. Nelson and D. Koza, PCT Int. Appl. WO0204404 (2002).
54. M. L. Nelson, L. McIntyre, G. Rennie and B. Bhatia, PCT Int. Appl. WO0204405 (2001).

Chapter 12. Intracellular Signaling Targets for Cancer Chemosensitization

Thomas G. Gesner and Stephen D. Harrison

Chiron Corporation, Emeryville, CA 94608

Introduction - Chemotherapy and radiation exposure are the major options for the treatment of cancer. The application of both of these therapies is limited by severe adverse effects on normal tissues and the frequent development of tumor cell resistance. It is therefore highly desirable to improve the efficacy of these treatments without increasing the toxic side effects and to counteract the resistance mechanisms that can render tumors insensitive to therapy.

In this review we will focus on the sensitization of tumors to chemotherapy. Most chemotherapeutic agents are cytotoxic drugs that kill cancer cells by interfering with the cell cycle and inducing cell death, either by disrupting nucleotide precursor synthesis, inducing DNA damage or impairing the function of microtubules. Intrinsic and acquired resistance to such chemotherapy may result from a number of mechanisms that fall into broad classes. These are: decreased accumulation of the drug (e.g. by efflux pump overexpression); limitation of drug-induced injury (e.g. by glutathione overproduction); mutations in the target protein (e.g. topoisomerase I); repair of DNA damage or alterations in the cellular signaling pathways that regulate the cell cycle and cell death. Targeting of the former mechanisms has been extensively discussed (1,2). However as our understanding of the cell cycle has improved we have been able to identify a number of cell signaling proteins that may be effectively targeted to sensitize tumor cells to cytotoxic drugs. It is these latter targets that form the basis for this review.

CELL CYCLE REGULATION AND CHECKPOINT ARREST

Ultraviolet light, ionizing radiation, environmental agents and cytotoxic drugs can result in damage to cellular DNA integrity. When such damage occurs during DNA replication or cell division it is potentially catastrophic and may result in cell death.

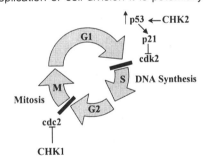

The cellular response is to arrest the cell cycle at one of two checkpoints (G1/S or G2/M) to either permit DNA repair or initiate apoptosis.

The G1/S checkpoint is regulated by the p53 transcriptional activator protein and the absence of this critical protein is often an important step in tumorigenesis, thus defining p53 as a tumor suppressor. In fact, nearly 50% of all cancers are p53 defective due to mutation (3). In response to DNA damage, checkpoint kinase 2 (CHK2) phosphorylates p53 and this results in stabilization of the protein and an elevation in p53 levels (4). Consequently, negative cell cycle regulators, such as p21$^{Waf1/Cip1}$, are activated and halt the cell cycle at the G1/S checkpoint (5).

The G2/S checkpoint is monitored by the serine/threonine checkpoint kinase 1 (CHK1). Upon DNA damage the protein kinase ATR (ataxia-telangiectasia mutated - rad53 related kinase) is activated (6,7). ATR-dependent phosphorylation of CHK1 promotes its phosphorylation of cdc25 and Wee1 and ultimately inactivation of cdc2.

Thus, CHK1 phosphorylation of cdc25c targets it for nuclear export to the cytoplasm and as a result cdc25c phosphatase is rendered unavailable to activate cdc2 by dephosphorylation (8-12). In addition, CHK1 activates the protein kinase Wee1, which phosphorylates and inactivates cdc2 (13,14). These dual pathways thus converge to result in cell cycle arrest.

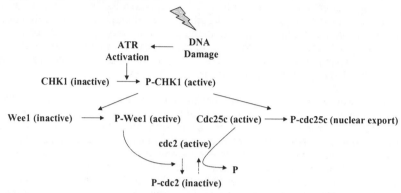

<u>CHK1</u> - As cell cycle arrest is a potential mechanism by which tumor cells can overcome the damage induced by cytotoxic agents, abrogation of these checkpoints with novel therapeutic agents should increase the sensitivity of tumors to chemotherapy. The presence of two checkpoints, coupled with the tumor specific abrogation of one of these by p53 mutations in 50% of cancers, can be exploited to design tumor-selective agents. Thus, in p53 minus tumors, therapeutic inhibition of G2/M arrest leaves cancerous cells no options for DNA damage repair and results in apoptosis. Normal cells have wild type p53 and retain an intact G1/S checkpoint. Thus these cells have an opportunity to correct DNA damage and survive. One approach to the design of chemosensitizers that abrogate the G2/M checkpoint is to identify inhibitors of the key G2/M regulatory kinase, CHK1.

Proof of concept for the use of CHK1 inhibition as a means of sensitizing cells to the effects of DNA-damaging agents comes from the use of a variety of approaches. In one recent study, HeLa cells were arrested at G2/M in the cell cycle by exposure to the topoisomerase II inhibitor etoposide (15). HeLa cells expressing a dominant negative CHK1 gene (dnHeLa) were shown to progress through mitosis in the presence of etoposide. Irradiation of dnHeLa cells also caused increased mitotic entry over that seen in untransfected cells. This resulted in a profound reduction in clonogenic survival in dnHeLa cells (15). Additionally, transfection of H1299 cells with either antisense or ribozymes that target CHK1 RNA abrogated adriamycin-induced G2/M arrest and rendered the cells more susceptible to killing by that agent (16).

Low molecular weight organic inhibitors of CHK1 have also been identified. 7-OH-staurosporin (UCN-01), <u>**1a**</u>, a natural isolate from Streptomyces, has been identified as an inhibitor of CHK1 and abrogates G2/M cell cycle arrest (17-21). UCN-01 is a close relative of another microbial alkaloid, staurosporin, <u>**1b**</u>, which was first discovered as a potent protein kinase C inhibitor (IC_{50} = 2.7 nM) with a broad spectrum of activity on other kinases (22). UCN-01 also inhibits many protein kinases, but has greater potential to overcome the cell-cycle arrest, as it does not inhibit cdc2 as potently (23,24). The kinase inhibition profile of UCN-01 has lead to its investigation as a anticancer agent and clinical trials are in progress in the United States and Japan for UCN-01 as a single-agent. However, as outlined below, there is considerable potential for this compound as a chemosensitizer.

Recent studies have been conducted investigating the ability of UCN-01 to potentiate the sensitization of existing cancer therapeutics on p53 wild type and p53 deficient cells. In p53 null MDA-MB-231 and G101A breast cancer cells, UCN-01 increased the sensitivity to camptothecin, a topoisomerase I inhibitor. The synergistic interaction of the two compounds achieved a potency increase of 30- and 40-fold respectively, on the cells compared to camptothecin alone. p53 containing endothelial cells displayed no such interaction, suggesting selectivity for p53 status (25). Other authors have demonstrated sensitization of additional cytotoxic agents. Sugiyama and coworkers have observed enhanced mitomycin C sensitization with subtoxic concentrations of UCN-01 (26). In a clonogenic assay, the p53 deficient epidermoid A432 and pancreatic PSN-1 carcinoma cells showed potentiation by the combined drug exposure. The p53 wild type HCT116 colon and MCF breast cancer cells showed no enhancement. In this paper it was suggested that the increased mitomycin C-induced cytoxicity and apoptosis by UCN-1 was related to p53 function (26).

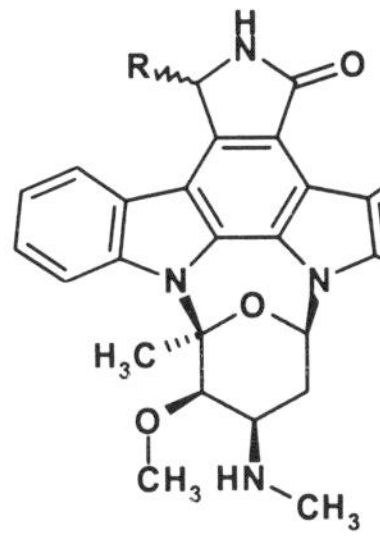

1a: R=OH   **2**   **3**
1b: R=H

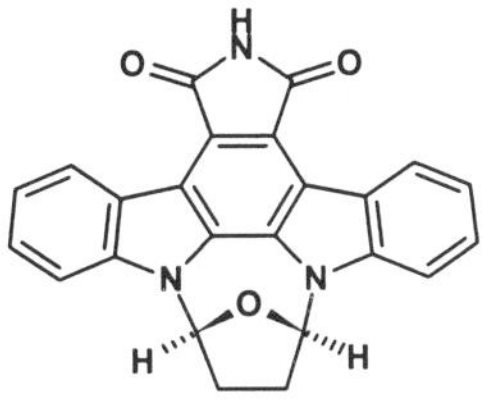

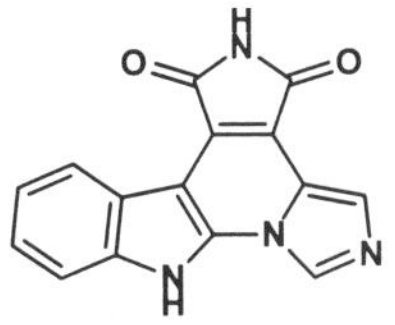

Despite the rationale outlined above, p53 status may not be a universal determinant of chemosensitization by UCN-01 for all cell types or chemotherapeutic agents. For example, temozolomide-induced G2/M arrest in p53 mutant and wild type glioma cell lines is abrogated by UCN-01 and nontoxic doses of UCN-01 inhibit temozolomide-induced cdc2 and cdc25 phosphorylation in both cases. However, UCN-01 also mediated a five-fold decrease in the temozolomide IC_{50} on both glioma cell lines in a colony formation assay, apparently independent of the p53 status (27).

In addition to its effect on the G2/M checkpoint, UCN-01 appears to chemosensitize cells through additional mechanisms. The nucleoside analogue gemcitabine promotes S-phase arrest in MC-1 cells, an acute myelogenous leukemic cell line. This S-phase arrest could not be released by addition of UCN-01. However, upon UCN-01 treatment, the S-phase population of cells was shown to undergo rapid apoptosis without cell cycle progression (28). This observation may be related to the recent report by the authors suggesting that UCN-01 may be inhibiting the phosphatidylinositol-3-kinase (PI 3-kinase)/Akt-bad cell survival pathway (29). Another report suggests that CHK1 is involved in S phase arrest in response to stalled replication. This function is also sensitive to UCN-01 (30).

Recently other structural analogues of staurosporin have been reported. The indolocarbozole, SB-218078, **2**, has improved selectivity over staurosporin and is a relatively less potent inhibitor of cdc2 (23). SB-218078 sensitized HeLa and HR29 cells to topotecan, a camptothecin analogue, in a cytotoxicity assay. As for UCN-01, SB218078 was shown to initiate cell cycle progression through a G2 checkpoint arrest that was induced by irradiation or topotecan in HeLa cells (23,31). Another related compound, isogranulatamide, **3**, isolated from *Didemnum granulatum,* also promotes release from the G2 checkpoint (32).

Wee1 - One of the functions of CHK1 is to phosphorylate and activate the Wee1 kinase. As Wee1 promotes the inactivation of cdc2 and consequent cell cycle arrest, inhibition of Wee1, like antagonism of CHK1 activity, might be expected to sensitize cells to DNA damaging agents. A low molecular weight Wee1 inhibitor (IC_{50} = 24 nM) has recently been described. The pyridopyrimidine, PD0166285, **4**, inhibits Wee1 and the related kinase Myt1 and blocks threonine 14 and tyrosine 15 phosphorylation of cdc2 (33,34). Radiation induced G2 arrest in HT29 cells was abrogated by PD0166285 and radiosensitization caused by this agent was specific to p53 null cells and did not occur in p53 wild type cells (34). Thus, PD0166285 has potential as a chemosensitizer.

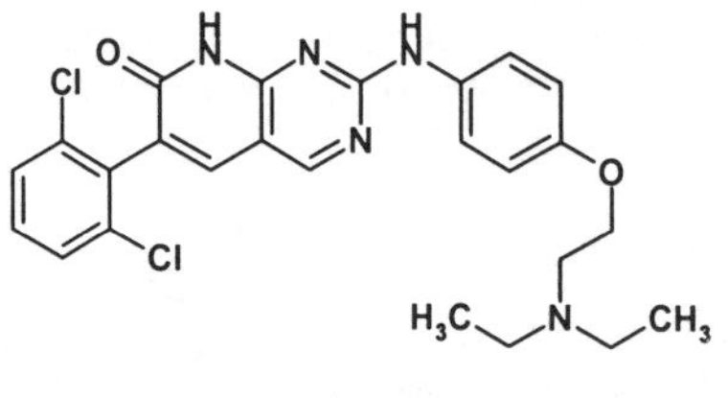

4 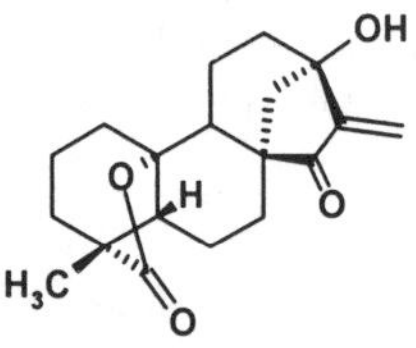   **5** 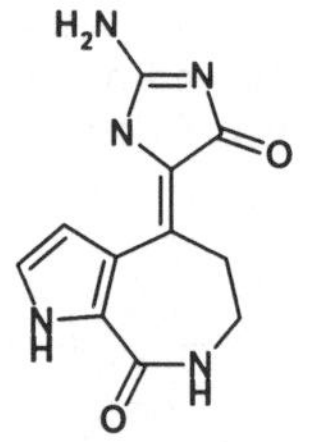   **6**

Other G2/M checkpoint regulators - The *ent*-kaurene diterpenoid, 13-hydroxy-15-oxoapatlin (OZ), **5**, has recently been identified as an antagonist of G2 arrest. An extract from the bark of the African tree *Parinari curatelliflora* possessed an activty that abrogates the radiation-induced G2 arrest and this is attributable to OZ (11). However, OZ did not inhibit a select panel of checkpoint kinases and its mechanism action remains to be elucidated. MCF7 and HCT116 cells that had been engineered to be p53 negative (MCF7 p53mp and HCT116p53$^{-/-}$ respectively) were modestly released from G2 checkpoint arrest by OZ. However, in combination with other inhibitors of the G2 checkpoint, particularly isogranulatamide and debromohymenialdesine (DBH), **6**, OZ additively increased the percentage of p53mp in mitosis (32,35). OZ and DBH represent new structural classes of agents that could sensitize cells to G2 arresting drugs.

APOPTOSIS REGULATION

All major classes of chemotherapeutic drug induce tumor cell death through the programmed response known as apoptosis. This process involves the activation of a signaling cascade and can be blocked by the upregulation of proteins that negatively regulate the pathway. Thus cancer cells can potentially adapt to chemotherapy induced stress by overexpressing key anti-apoptotic proteins and indeed this has been demonstrated in the case of some of the proteins discussed below. These anti-apoptotic proteins are thus potential targets for drugs that sensitize tumors to chemotherapy.

The bcl-2 family of proteins - One of the best-known classes of cytoprotective genes are those that encode the bcl-2-related proteins. The prototypic member of this class, bcl-2, was originally identified as being overexpressed in B cell lymphomas as a result of t(14;18) chromosomal translocations. Since that time it has been found to be upregulated in approximately half of all human tumors (36). Of particular relevance to this discussion, high levels of bcl-2 expression have been linked to chemotherapy resistance in lymphoma, leukemia and prostate cancer (37,38). Thus targeting of bcl-2 activity might be expected to have disproportionately more effect on tumor cells than on normal cells, which express less of the target protein.

Evidence for a causal relationship between elevated levels of bcl-2 and resistance to anti-cancer drugs has been provided by a variety of experiments. Experimental overexpression of bcl-2 in tumor cell lines increases their resistance to many chemotherapy agents, including cisplatin, mitoxantrone, adriamycin, as well as radiation (39-44). Furthermore antisense oligonucleotides and transgenically expressed antisense molecules increase the sensitivity of tumor cells to chemotherapy (45).

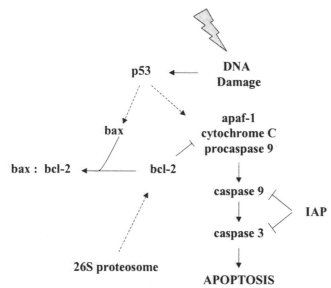

The validity of bcl-2 as a chemosensitization target has now been confirmed *in vivo*. G3139, a 18-mer phosphorothioate oligonucleotide designed to hybridize to the first six codons of bcl-2 mRNA, has demonstrated clinical responses in a phase I trial as a single agent in non-Hodgkin's lymphoma (46). This agent sensitizes human tumor xenografts in SCID mice to the chemotherapy agents dacarbazine and cisplatin and this combination efficacy has now been confirmed in the clinic (47,48). In a clinical trial, 14 patients with advanced malignant melanoma were treated with intravenous or subcutaneous doses of increasing concentrations of G3139 along with a standard regimen of dacarbazine (49). Six of the patients showed anti-tumor responses. Reductions in bcl-2 expression were measured and correlated with increased tumor cell apoptosis. The combination of G3139 and dacarbazine is now being evaluated in a phase III clinical trial and combinations of G3139 with other cytotoxic drugs (docetaxol, mitoxantrone, fludarabine, cytarabine, irinotecan) are ongoing in Phase I/II trials (50). In particular, it was recently reported that intravenous infusion of G3139 with docetaxel in patients with progressive androgen-independent prostate cancer resulted in stable disease after four cycles for four out of six patients (51). Synergistic chemosensitization by G3139 in combination with another taxane, paclitaxel had previously been demonstrated in the mouse LNCAP prostate cancer model of progression to androgen independence (52).

In addition to bcl-2, there are more than 15 closely homologous proteins in the human genome. Like bcl-2, some members of the class, such as bcl-X$_L$ function to inhibit apoptosis. Other members of the class, such as bad, bak and bax are pro-apoptotic (36,53). Targeting other anti-apoptotic members of the family may also have the potential to sensitize cells to chemotherapy. Lebedeva and colleagues have recently demonstrated that antisense downregulation of bcl-X$_L$ protein sensitizes bladder carcinoma cells to various cytotoxic compounds (54).

Thus, interfering with the activity of bcl-2 and related proteins seems well validated as an approach to chemosensitization. However, the clinical development of low molecular weight organic antagonists has lagged behind the discovery of antisense therapeutics. This is most likely attributable to the lack of any inhibitable enzymatic activity associated with bcl-2 function. Indeed it is most likely that bcl-2 family members exert their antiapoptotic effects through binding to other proteins, in particular the pro-apoptotic members of the same class. Crystal structures of bcl-X_L from human and rat and homology modeling with bcl-2 have identified a hydrophobic surface pocket formed by domains conserved across the family of proteins (BH1, BH2 and BH3 domains) (55,56). Mutations in this site block the anti-apoptotic function of the protein, presumably by interfering with binding of associated proteins, such as bad, bak and bax (57). The nature of this interaction has been elucidated by the generation of the co-crystal structure between bcl-X_L and the BH3 domain of bak. These data and NMR structures of the family members have facilitated the design of the low molecular weight compounds described below that bind within the pocket, interfere with protein-protein interactions and thus promote apoptosis (53).

Some of the first identified antagonists of the bcl-2 protein were small peptides derived from the BH3 domain of Bad. When these peptides are made cell permeable through the covalent addition of a fatty acid moiety, they selectively killed tumor cells in culture and slowed the growth of human myeloid leukemia cells in SCID mice (58). The authors of the previous study have subsequently reported the discovery of a non-peptidic molecule, HA14-1, **7**, designed using a structure-based approach. HA14-1 induces apoptosis in leukemic cells (59). This agent has now been used to confirm the potential of low molecular weight bcl-2 antagonists in chemosensitization. A synergistic interaction of HA14-1 has been demonstrated in conjunction with etoposide and cytarabine in leukemia cell lines (60).

The generality of sensitization to DNA-damaging stimuli has been confirmed using another agent that antagonizes the effect of bcl-2 related molecules, in this case bcl-X_L and Mcl-1. The quinoline PK11195, **8**, sensitized cholangiocarcinoma cells to UV light and radiotherapy (61).

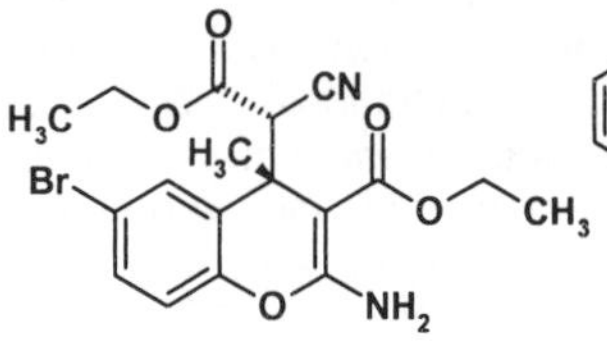

7 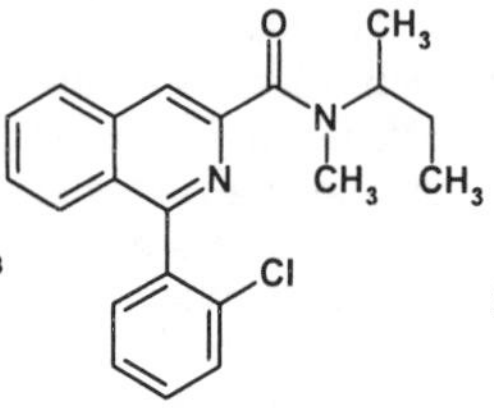   **8** 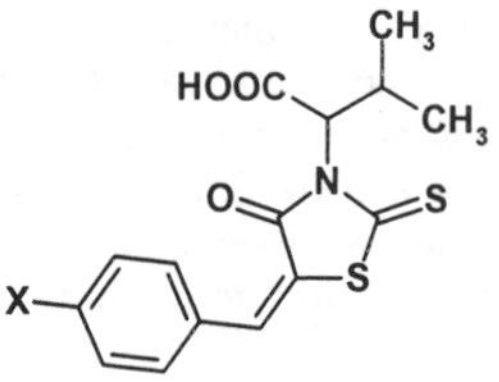   **9**

More recently, high-throughput screening and structure-based drug design have been instrumental in the discovery of additional classes of bcl-2 family member antagonists. Degterev and coworkers have described two classes of antagonist (62). The BH3I-1 series of thiooxothiozolidines, **9**, exhibit Ki values for the inhibition of the bak/ bcl-X_L interaction ranging from 2.4µM (X=Br) to 12.5µM (X=H). The BH3I-2 series of 2-hydroxybenzamides, **10**, have similar potency with Ki values from 3.3µM (Y=Cl, Z=I) to 6.4µM (Y=I, Z=I). Both classes of molecule induce apoptosis in JK cells. In another study, computational modeling was used to predict that the natural product antimycin A, **11**, would bind to the bcl-2 interaction domain. This has been confirmed by showing that antimycin A and a BH3 peptide from bak compete for binding to bcl-2 (63). The antimycin A is toxic to cells and inhibits several of the cellular functions of bcl-2. Additional compounds were identified in a related approach in which the NCI 3D compound database was searched for compounds predicted to bind to the bcl-2 interaction site (64). One of these compounds, "compound 6", **12**, has a binding affinity (IC_{50}) of 10μM and induces apoptosis selectively in cancer cell lines, while having little effect on normal cells. Most recently another bcl-2 binding compound, GX-01, has been reported. GX-01 induces apoptosis in tumor cells in culture and improves the survival of mice bearing ovarian tumors (65).

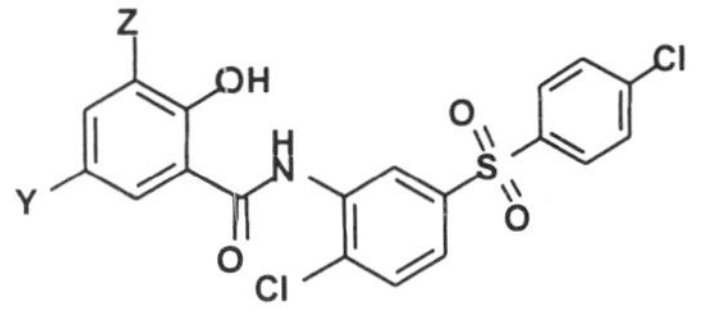

10 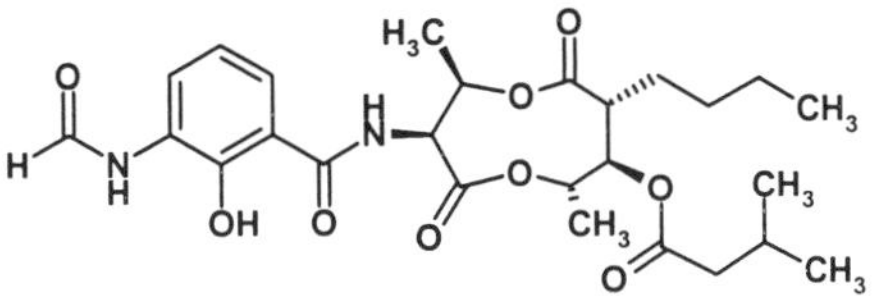   **11**

The recent proliferation of novel low molecular weight bcl-2 antagonists suggests that the development of such agents as chemosensitizers may well be feasible.

Heat Shock Proteins - Heat shock proteins (HSPs) are induced under conditions of cell stress and protect against cellular damage by acting as protein chaperones. Overexpression of certain HSPs can protect cells from apoptosis, including that induced by cytotoxic drugs (50). HSP protein expression has also been associated with resistance to treatment and HSP70 expression inversely correlates with the response of breast cancer to chemotherapy (50). Thus HSPs and other cytoprotective protein chaperones may provide good targets for chemosensitization and a number of approaches are being taken to develop anti-cancer agents that target HSPs (66,67). Perhaps the most advanced of these compounds are the ansamycins, geldanamycin, **13**, and 17-allylaminogeldanamycin (17-AAG), **14**. These compounds bind to the ATP/ADP binding site of HSP90 and, like ADP, block the ability of the HSP90 chaperone to protect intracellular proteins from ubiquitin-dependent degradation (68,69). Some of the proteins degraded as a result of ansamycin treatment include important anti-apoptotic proteins, including 3-phosphoinositide dependent kinase 1 (PDK1) (70). This fact may explain the observation that the ansamycins show chemosensitization *in vitro* and *in vivo* (71-73). 17-AAG has now completed phase I clinical trials and phase II trials are being planned in combination with a variety of cytotoxic agents. The crystal structure of HSP90 has been published and reveals the ATP/ADP binding site to be a very attractive pocket for drug binding, raising hopes that additional small molecule inhibitors against this target may soon be developed using structure based design (69).

Other protein chaperones are also being targeted in the development of chemosensitizers. Clusterin is a secreted glycoprotein that is thought to act in a manner functionally analogous to HSPs to protect cells from stress and death. Antisense oligonucleotides that inhibit the expression of clusterin are currently being developed as chemosensitizers. Clusterin is overexpressed in 69% of renal cell cancers, as well as in prostate, bladder, lung, ovarian and urothelial cancers (74). Antisense oligonucleotides significantly enhanced chemosensitivity to paclitaxel *in vitro* and acted synergistically *in vivo* to increase apoptosis and delay tumor growth (74).

26S Proteosome - The proteosome is a large multicatalytic protease complex that is responsible for the majority of non-lysosomal protein degradation within cells. Inhibition of this complex with small molecular weight inhibitors results in the stabilization of proteins and in proliferating cells leads to cell death (75). A number of mechanisms for the anti-apoptotic effect of proteasome inhibitors have been proposed, and may involve inhibition of NF-kB signaling and the downregulation of bcl-2 (76). The proteosome inhibitor MLN-341, **15**, has demonstrated the potential to sensitize tumor cells to gemcitabine *in vivo* and it is the first agent of this class to enter clinical trials (76). A number of phase I trials have been initiated with various chemotherapeutic agents, including docetaxel, gemcitabine and irenotecan. MLN-341 is also showing promise in phase II clinical trials as a single agent in patients with hematological malignancies (77).

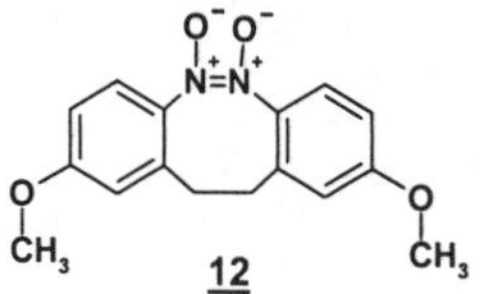

12

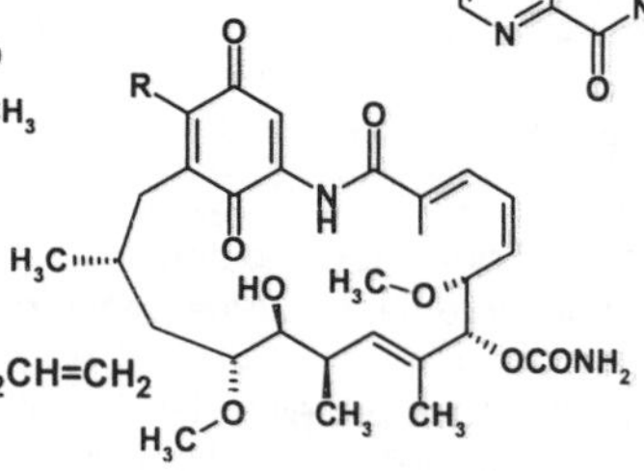

13: R=OCH₃
14: R=NHCH₂CH=CH₂

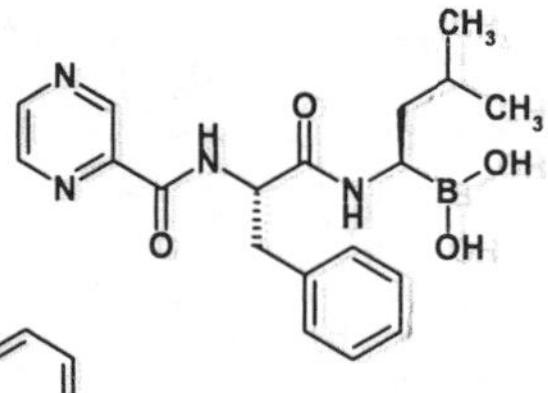

15

Clearly the principle of inhibiting anti-apoptotic proteins extends beyond the targets outlined above. Other potential targets include the components of the PI3-kinase/ PDK1/ akt pathway, IAP family members and potentially many more (50,78).

Conclusions - Chemosensitizers are being developed to increase the therapeutic index of known cytotoxic drugs. The potential to achieve this goal depends on the tumor selectivity of these agents, a property that is made possible by the defective cell cycle regulation of tumor cells and their increased reliance on the upregulation of particular anti-apoptosis pathways. The data reviewed above show that increased tumor selectivity is a realizable goal. The strategy of improving the therapeutic index is a highly pragmatic approach to extending the utility of cytotoxic drugs with a proven record of efficacy. It may even allow the continued development of chemotherapeutic drugs that had previously demonstrated unacceptable toxicity.

As stated above, the development of chemosensitizers that target intracellular signaling pathways has been stimulated by our increased knowledge of the human genome. In parallel with the sequencing of the normal genome, there has been an extensive effort to characterize the mutations that lead to disease and which cause therapy to fail. As these efforts identify additional signaling mechanisms that cause chemotherapy resistance in tumors, the number of possibilities for the development of chemosensitizers will increase.

While this is an exciting area of drug development, like all new therapeutic approaches, many details of the clinical application of chemosensitizers need to be worked out. The relative dose schedule of the cytotoxic drug and the sensitizer can have profound effects on outcome and clinical trials will have to be designed to evaluate this. In addition, as few precedents exist for this therapeutic approach, a constant dialog will have to be maintained with clinical oncologists and with regulatory authorities in order to ensure that all potential concerns are addressed and that this promising therapeutic alternative will receive wide acceptance.

References

1.   D.M. Bradshaw and R.J. Arceci, J.Clin.Oncol., 16, 3674 (1998).
2.   D.J. Harrison, J.Pathol., 175, 7 (1995).
3.   T. Soussi, Ann.N.Y.Acad.Sci., 910, 121 (2001).

4. A. Hirao, Y.Y. Kong, S. Matsuoka, A. Wakeham, J. Ruland, H. Yoshida, D. Liu, S.J. Elledge and T.W. Mak, Science, 287, 1824 (2000).
5. B. Vogelstein, D. Lane and A.J. Levine, Nature, 408, 307 (2000).
6. H. Zhao and H. Piwnica-Worms, Mol.Cell Biol., 21, 4129 (2001).
7. Q. Liu, S. Guntuku, X.S. Cui, S. Matsuoka, D. Cortez, K. Tamai, G. Luo, S. Carattini-Rivera, B. DeMayo F, A. , L.A. Donehower and S.J. Elledge, Genes Dev., 14, 1448 (2000).
8. Y. Sanchez, C. Wong, R.S. Thoma, R. Richman, Z. Wu, H. Piwnica-Worms and S.J. Elledge, Science, 277, 1497 (1997).
9. C.Y. Peng, P.R. Graves, R.S. Thoma, Z. Wu, A.S. Shaw and H. Piwnica-Worms, Science, 277, 1501 (1997).
10. T.A. Chan, H. Hermeking, C. Lengauer, K.W. Kinzler and B. Vogelstein, Nature, 401, 616 (1999).
11. N.T. Rundle, L. Xu, R.J. Andersen and M. Roberge, J.Biol.Chem., 276, 48321 (2001).
12. A. Lopez-Girona, B. Furnari, M. O. and P. Russell, Nature, 397, 172 (1999).
13. J. Lee, A. Kumagai and W.G. Dunphy, Mol.Biol.Cell, 12, 551 (2001).
14. L.L. Parker and H. Piwnica-Worms, Science, 257, 1955 (1992).
15. K. Koniaras, K.R. Cuddihy, H. Christopoulos, A. Hogg and M.J. O'Connell, Oncogene, 20, 7453 (2001).
16. Y. Luo, S.K. Rockow-Magnone, P.E. Kroeger, L. Frost, Z. Chen, E.K. Han, S.C. Ng, R.L. Simmer and V.L. Giranda, Neoplasia, 3, 411 (2001).
17. I. Takahashi, E. Kobayashi, M. Toshida and H. Nakano, J.Antibiot., 40, 1782 (1987).
18. P.R. Graves, L. Yu, J.K. Swartz, J. Gales and E.A. Sausville, J.Biol.Chem., 275, 5600 (2000).
19. R.T. Bunch and A. Eastman, Clin.Cancer.Res., 2, 791 (1996).
20. E.C. Busby, D.F. Leistritz, R.T. Abraham, L.M. Karnitz and J.N. Sarkaria, Cancer Res., 60, 2108 (2000).
21. Q. Wang, S. Fan, A. Eastman, P.J. Worland, E.A. Sausville and P.M. O'Connor, J.Natl.Cancer Inst., 88, 956 (1996).
22. T. Tamaoki, H. Nomoto, I. Takahashi, M. Kato and F. Tomita, Biochem.Biophys.Res.Commun., 135, 397 (1986).
23. J.R. Jackson, A. Gilmartin, C. Imburgia, J.D. Winkler, L.A. Marshal and A. Rushak, Cancer Res., 60, 566 (2000).
24. T. Tamaoki, Methods Enzymol., 201, 340 (1991).
25. C.B. Jones, M.K. Clements, S. West and S.S. Daoud, Cancer Chemother.Pharmacol., 45, 252 (2000).
26. K. Sugiyama, M. Shimizu, T. Akiyama, K. Yamaguchi, R. Takahashi, A. Eastman and S. Akinaga, Int.J.Cancer, 85, 703 (2000).
27. Y. Hirose, M.S. Berger and R.O. Pieper, Cancer Res., 61, 5843 (2001).
28. Z. Shi, A. Azuma, D. Sampath, Y. Li, P. Huang and W. Plunkett, Cancer Res., 61, 1065 (2001).
29. Z. Shi and W. Plunkett, Proc.Am.Assoc Cancer Res., 41, 313 (2000).
30. C. Feijoo, C. Hall-Jackson, R. Wu, D. Jenkins, J. Leitch, D.M. Gilbert and C. Smythe, J.Cell Biol., 154, 913 (2001).
31. G.A. Gilmartin, M.L. Ho, C.S. Imburgia, A.K. Roshak and M. Lago, WO Patent 0016781 (2000).
32. R.G.S. Berlinck, R. Britton, E. Piers, L. Lim, M. Roberge, R.M.d. M. da Rocha and R.J. Andersen, J.Org.Chem., 63, 9850 (1998).
33. J. Li, Y. Wang, Y. Sun and T.S. Lawrence, Radiat.Res., 157, 320 (2002).
34. Y. Wang, J. Li, R.N. Booher, A. Kraker, T. Lawrence, W.R. Leopold and Y. Sun, Cancer Res., 61, 8211 (2001).
35. D. Curman, B. Cinel, D.E. Williams, N. Natalie Rundle, W.D. Block, A.A. Goodarz, J.R. Hutchins, P.R. Clarke, B.-B. Zhou, S.P. Lees-Miller, R.J. Andersen and M. Roberge, J.Biol.Chem., 276, 17919 (2001).
36. J.C. Reed, Adv.Pharm., 41, 501 (1997).
37. L. Campos, J.-P. Roualult, O. Sabido, N. Roubi, C. Vasselon, F. Archimbaud, M. J.-P. and D. Guyotat, Blood, 81, 3091 (1993).
38. Z.T. Maung, F.R. MacLean, M.M. Reid, A.D.J. Pearson, S.J. Proctor, P.J. Hamilton and A.G. Hall, Brit.J.Haem., 88 105 (1994).
39. T. Miyashita and J.C. Reed, Cancer Res., 52, 5407 (1992).
40. T. Miyashita and J.C. Reed, Blood, 81, 151 (1993).
41. M.I. Walton, D. Whysong, P.M. O'Connor, D. Hockenbery, S.J. Korsmeyer and K.W. Kohn, Cancer Res., 53, 1853 (1993).

42.   S. Kamesaki, H. Kamesaki, T.J. Jorgensen, A. Tanizawa, Y. Pommier and J. Cossman, Cancer Res., 53, 4251 (1993).

43.   T.C. Fisher, A.E. Milner, C.D. Gregory, A.L. Jackman, G.W. Aherne, J.A. Hartley, D. C. and J.A. Hickman, Cancer Res., 53, 3321 (1993).

44.   C. Tang, M.C. Willingham, J.C. Reed, T. Miyashita, S. Ray, V. Ponnathpur, Y. Huang, M.E. Mahoney, G. Bullock and K. Bhalla, Leukemia, 8, 1960 (1994).

45.   S. Kitada, S. Takayama, K. De Riel, S. Tanaka and J.C. Reed, Antisense Res.Dev., 4, 71 (1994).

46.   J.S. Waters, A. Webb, D. Cunningham, P.A. Clarke, F. Raynaud, F. di Stefano and F.E. Cotter, J.Clin.Oncol., 18, 1812 (2000).

47.   B. Jansen, H. Schlagbauer-Wadl, B.D. Brown, R.N. Bryan, A. van Elsas, M. Muller, K. Wolff, H.G. Eichler and H. Pehamberger, Nat.Med., 4, 232 (1998).

48.   V. Wacheck, E. Heere-Ress, J. Halaschek-Wiener, T. Lucas, H. Meyer, H.G. Eichler and B. Jansen, J.Mol.Med., 79, 587 (2001).

49.   B. Jansen, V. Wacheck, E. Heere-Ress, H. Schlagbauer-Wadl, C. Hoeller, T. Lucas, M. Hoermann, U. Hollenstein, K. Wolff and H. Pehamberger, Lancet, 356, 1728 (2000).

50.   I. Tamm, B. Dorken and G. Hartmann, Lancet, 358, 489 (2001).

51.   M.J. Morris, W.P. Tong, C. Cordon-Cardo, M. Drobnjak, W.K. Kelly, S.F. Slovin and N. Rosen, Euro.J.Cancer, 37, Abs 801 (2001).

52.   S. Leung, H. Miyake, T. Zellweger, A. Tolcher and M.E. Gleave, Int.J.Cancer, 91, 846 (2001).

53.   Z. Huang, Oncogene, 19, 6627 (2000).

54.   I. Lebedeva, A. Raffo, R. Rando, J. Ojwang, P. Cossum and C.A. Stein, J.Urol., 166, 461 (2001).

55.   S. Muchmore, W.M. Sattler, H. Liang, R.P. Meadows, J.E. Harlan, H. Yoon, S. , D. Nettesheim, B.S. Chang, C.B. Thompson, S.L. Wong, S.L. Ng and S.W. Fesik, Nature, 381, 335 (1996).

56.   M. Aritomi, N. Kunishima, N. Inohara, Y. Ishibashi, S. Ohta and K. Morikawa, J.Biol.Chem, 57.   X.M. Yin, Z.N. Oltvai and K. SJ., Nature, 369, 321 (1994).

58.   J.L. Wang, Z.J. Zhang, S. Choksi, S. Shan, Z. Lu, C.M. Croce, E.S. Alnemri, R. Korngold and Z. Huang, Cancer Res., 60, 1498 (2000).

59.   J.L. Wang, D. Liu, Z.J. Zhang, S. Shan, X. Han, S.M. Srinivasula, C.M. Croce, E.S. Alnemri and Z. Huang, Proc.Natl.Acad.Sci.U.S.A., 97, 7124 (2000).

60.   D.A. Fennell, M. Corbo, A. Pallaska and F.E. Cotter, Blood, 98, Abst 3489 (2001).

61.   A. Okaro and D.A. Fennell, Unit.Eur.Gast.Week, , Abst 453 (2000).

62.   A. Degterev, A. Lugovskoy, M. Cardone, B. Mulley, G. Wagner, T. Mitchison and J. Yuan, Nat.Cell Biol., 3, 173 (2001).

63.   S.P. Tzung, K.M. Kim, G. Basanez, C.D. Giedt, J. Simon, J. Zimmerberg, K.Y. Zhang and D.M. Hockenbery, Nat.Cell Biol ., 3, 183 (2001).

64.   I.J. Enyedy, Y. Ling, K. Nacro, Y. Tomita, X. Wu, Y. Cao, R. Guo, B. Li, X. Zhu, Y. Huang, Y.Q. Long, P.P. Roller, D. Yang and S. Wang, J.Med.Chem., 44, 4313 (2001).

65.   M.S. Murthy, Intl.Conf.Mol.Targets Cancer Ther., , Abst 313 (2001).

66.   P.W. Piper, Curr.Opin.Investig.Drugs, 2, 1606 (2001).

67.   P. Workman and A. Maloney, Expert Opin.Biol.Ther., 2, 3 (2002).

68.   L. Whitesell, E.G. Mimnaugh, B. De Costa, C.E. Myers and L.M. Neckers, Proc.Natl.Acad.Sci.U.S.A., 91, 8324 (1994).

69.   C.E. Stebbins, A.A. Russo, C. Schneider, N. Rosen, F.U. Hartl and N.P. Pavletich, Cell, 89, 239 (1997).

70.   N. Fujita, S. Sato, A. Ishida and T. Tsuruo, J.Biol Chem., (2002).

71.   P.N. Munster, A. Basso, D. Solit, L. Norton and N. Rosen, Clin.Cancer Res., 7, 2228 (2001).

72.   D.M. Nguyen, D. Lorang, G.A. Chen, J.H.t. Stewart, E. Tabibi and D.S. Schrump, Ann.Thorac.Surg., 72, 371 (2001).

73.   M.V. Blagosklonny, T. Fojo, K.N. Bhalla, J.S. Kim, J.B. Trepel, W.D. Figg, R. Y. and L.M. Neckers, Leukemia, 15, 1537 (2001).

74.   T. Zellweger, H. Miyake, L.V. July, M. Akbari, S. Kiyama and M.E. Gleave, Neoplasia, 3, 360 (2001).

75.   H.C. Drexler, Apoptosis, 3, 1 (1998).

76.   R.J. Bold, V. S. and D.J. McConkey, J.Surg.Res., 100, 11 (2001).

77.   C. Papandreo, D. Daliani, R. Millikan, S. Tu, L. Pagliaro, J. Adams, P. Elliott, P. Dieringer and C. Logothetis, Eur.J.Cancer, 37, Abst. 532 (2001).

78.   J. Downward, Curr.Opin.Cell Biol., 10, 262 (1998).

Chapter 13. Emerging Microtubule Stabilizing Agents for Cancer Chemotherapy

David C. Myles
Kosan Biosciences, Inc.
3832 Baycenter Place, Hayward, CA 94545

Introduction - The discovery of the potent anti-cancer properties and subsequent elucidation of the mechanism of action of paclitaxel firmly established this compound as a leader in the treatment of difficult cancers. In addition, the unusual mechanism of action of paclitaxel, the stabilization of microtubules, distinguished it from other microtubule-binding agents, such as the vinca alkaloids, that act by destabilizing microtubules. In this paper, recent advances in the identification of non-taxane microtubule stabilizing agents for cancer chemotherapy will be discussed with an emphasis on work published in 2000-02.

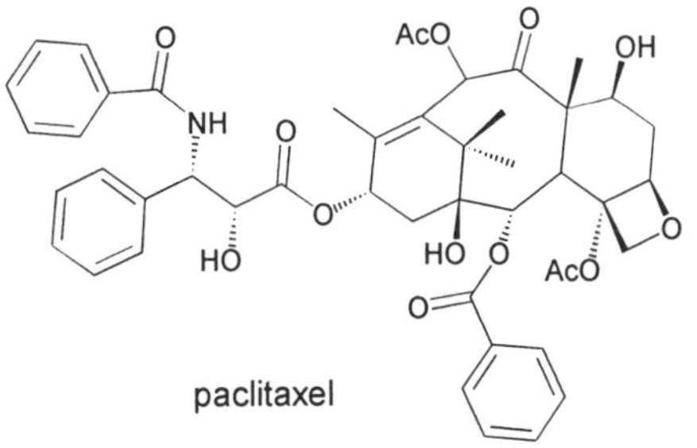

paclitaxel

Microtubule (MT) stabilizing agents target the dimer of α- and β-tubulin, binding in a hydrophobic cleft on the β form, and promote the formation of and impart unusual stability to the microtubule (1). The chemistry and structural biology of this interaction and the interaction of other families of drugs that bind to microtubules has been the object of considerable study using a range of techniques, particularly photo affinity labeling (2). Not surprisingly, these experiments show that the site targeted by MT stabilizing drugs is distinct and well separated from the site at which other mechanistic classes of MT interacting agents bind to the target. More recently, an electron diffraction structure of oligomeric α,β-tubulin prepared in the presence of paclitaxel confirmed the presence of the drug in the site suggested by the earlier labeling experiments (3). The low resolution (ca. 4 A) of the structure makes structure-based drug design based on it difficult.

During mitosis, the interaction of drugs with the microtubule interferes with the proper formation of the mitotic spindle and causes arrest of the cell cycle at the G2/M transition. This effect ultimately results in apoptosis. The MT stabilizing effect of paclitaxel and other MT stabilizing agents can also be observed in vitro using mammalian α- and β-tubulin. Thus, data from *in vitro* cytotoxicity, microtubule bundling, and flow cytometry can all be used to drive medicinal chemistry efforts directed toward identifying novel potent analogs of existing microtubule stabilizing agents.

The last decade has seen paclitaxel and related taxanes emerge as aggressive front line therapies for advanced tumors including breast, lung and ovarian carcinomas. Although the use of taxanes for these indications is considerable, commanding the lion's share of the market and generating combined yearly sales of nearly 2 billion dollars, the high toxicity and poor solubility of these drugs may ultimately limit their utility. Furthermore, these compounds are substrates for P-glycoprotein, a transmembrane efflux pump that serves to limit the intracellular concentration of drug substrates in cells. Thus, tumor cells that express P-glycoprotein are resistant to taxanes and other hydrophobic drugs that are substrates for the efflux pump. These issues combine to create an opportunity for new drugs, free of these liabilities, to enter the market and gain acceptance.

Recently, several other new structurally distinct classes of MT stabilizing natural products have been discovered and join paclitaxel in this still rare class of molecules (4). Not surprisingly, these new compounds have attracted significant attention in both the synthetic organic and the medical communities due to their complex structure and potential for use in cancer chemotherapy. Although there are few examples of non-natural product related MT stabilizing agents, the existence of even a small number indicates that this mode of action is not the exclusive realm of natural products and related drugs. Several of these new compounds or classes have entered human clinical trials, others are well advanced in preclinical evaluation. Table 1 summarizes *in vitro* cyctotoxicity data on a number of the more promising MT stabilizing agents (5).

Table 1. Inhibition of Growth of Human Cancer Cell Lines by Microtubule Stabilizing Agents, IC$_{50}$ (nM).

Compound[h]

Cell Line	Pacl	Epo A	Epo B	Epo D	Epo F	Disco	Eleu	Lau	Dict	BMS
MCF-7	1.8[g]	1.5[g]	0.18[g]	–	–	8.1[b]	–	–	2.7[c]	–
MCF-7/ADR	9105[g]	27.1[g]	2.92[g]	–	–	–	–	–	20[c]	–
A549	1.4[a]	–	–	–	1.0[d]	3.8[a]	14[a]	20-100[f]	0.95[c]	–
KB-31	2.3[g]	2.1[g]	0.19[g]	2.7[g]	0.77[d]	–	–	30[e]	–	–
HCT 116	2.3[d]	3.2[d]	0.42[d]	6.5[d]	–	–	–	–	–	3.6[d]
MES-SA/DX5	1645[c]	–	–	–	–	–	–	–	11[c]	–

a[20], b[5a], c[30], d[5b], e[5c], f[5d], g[5e], h: pacl = paclitaxel, Epo A = epothilone A, Epo B = epothilone B, Epo D = epothilone D, Epo F = epothilone F, Disco = discodermolide, Eleu = eluthrobin, Lau = laulimalide, Dict = dictyostatin, BMS = BMS-247550.

Epothilones-Of the new MT stabilizers, the epothilone class is, by far, the most advanced clinically. Epothilones A (**1**), and B (**2**) were isolated from the myxobacterium *sorangium cellulosum* and originally identified due to their anti fungal activity (6). The potent anti tumor properties and the elucidation of the MT stabilizing effect of these compounds heightened interest in the epothilones (7). As well as epothilones A, and B, the culture of the producing organism also yields a wide variety of other epothilone related structures including deoxy analogs epothilones C D (**3**) and F (**4**) (8). In addition to these major natural products, a number of totally synthetic and semisynthetic analogs have been prepared and studied *in vitro* (9,10).

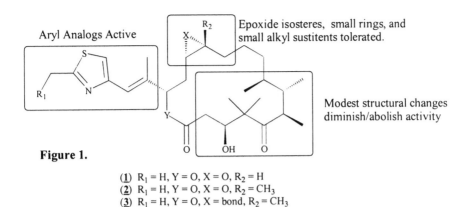

Figure 1.

(**1**) $R_1 = H, Y = O, X = O, R_2 = H$
(**2**) $R_1 = H, Y = O, X = O, R_2 = CH_3$
(**3**) $R_1 = H, Y = O, X = bond, R_2 = CH_3$
(**4**) $R_1 = OH, Y = O, X = O, R_2 = CH_3$
(**5**) $R_1 = H, Y = NH, X = O, R_2 = CH_3$

The structure activity relationships that emerge from this body of work are described in Figure 1. In general, the macrocycle can be divided into two halves for the purposes of discussing the SAR. The C1 to C8 region of the molecule is highly sensitive to modification, with even modest changes abrogating activity. In contrast the C9 to C17 portion affords a greater degree of flexibility. In various forms, epoxide isosteres, small heterocycles, as well as alkyl substituents may be substituted in the C12-C13 region with modest or no reduction in activity. Likewise a range of side chains emanating from C15 have been found to maintain varying degrees of activity.

To date, four epothilone analogs have entered human clinical trials. Compound **2** will enter Phase II clinical trials in 2002. Likewise, BMS-247550 (**5**), the semisynthetic lactam analog of epothilone B, is advancing to Phase II in 2002. The synthesis of this material, prepared in a three step, one pot procedure from **2**, was presumably motivated by its more secure patent position and the possibility for it to show reduced hydrolytic instability as compared to the natural product. In late 2001 KOS-862 (**3**) entered Phase I human clinical trials. In a head to head comparison with **2**, the *in vivo* toxicity of **3** in mice was found to be as much as 40 X lower (4). Although these results await confirmation in human clinical trials, they do shed light on the possible role of the epoxide moiety in both potency and toxicity. The most recent epothilone analog to enter the clinic is BMS-310705 (11). Although the precise structure of this compound has not yet been disclosed, it is described as having potency similar to other MT stabilizers and a toxicity profile similar to BMS-247550. It is clear from this early clinical data that activity in the area of epothilone chemistry and clinical work will continue at a brisk pace for some time to come.

Discodermolide-Discodermolide (**6**) was isolated from the deep-water sponge *Discodermia dissoluta* by Gunasekera and coworkers (12). Initial biological tests indicated that **6** has immunosuppressive activity, while recent interest has focused on its potent antimitotic activity. Interestingly, **6** has been reported to show potent activity against multi-drug resistant carcinoma cell lines [13] It is thought to be ca. 100-fold more soluble in water than paclitaxel and thus is likely to be more easily formulated. Additionally, **6** has a reduced affinity for p-glycoprotein, a known mediator of paclitaxel resistance.

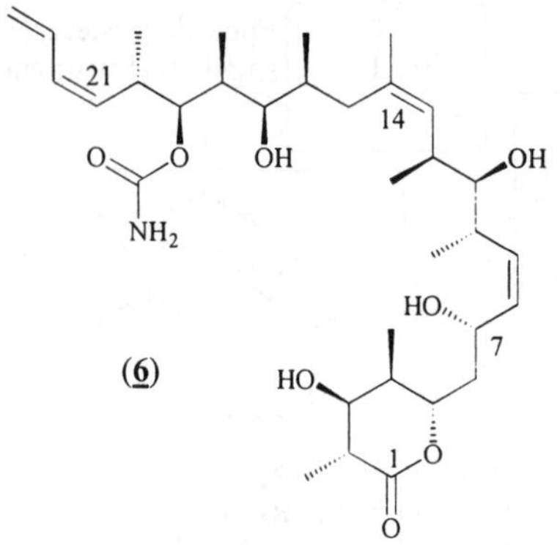

The natural source of **6** provides only minute quantities of material, with the reported isolated yield of 7 mg from 434 gm of frozen sponge. At this time, the only reliable source of meaningful quantities of the natural product is total chemical synthesis. Total syntheses of **6** have been reported by several groups (14, 15, 16, 17, 18). These syntheses break the **6** into three fragments and begin with an enantiomer of methyl 3-hydroxy-2-methylpropionate, but differ significantly in the manner in which the fragments are prepared and coupled. A detailed discussion of these syntheses has been published (19). From analogs obtained by synthesis and semisynthesis, a faint *in vitro* SAR for **6** is now emerging. Poly-acetylated versions of the compound show a range of activity (20). Totally synthetic analogs in which methyl groups at either C16 (21) or C14 have been deleted retain activity as does the 2,3 dehydro analog (22). Analogs of **6** in which the C.1 lactone cabonyl has been reduced and stored as the thiophenyl acetal show activity similar to the natural product. The retention of activity by analogs in which the lactone ring has been modified suggest that this region of the molecule may be a fruitful area for further analog work. One group has prepared in excess of 50 analogs of **6** by total synthesis and has indicated that certain isosteres of the C21-C24 diene maintain potency while substitution of the C13-C15 carbon atoms with an N-Methyl amide abolish activity (23). This synthetic effort has also served as the foundation for a Phase I clinical trial of **6** in 2002.

Eleuthrobin-Eleuthrobin (**7**) and the structurally related sarcodictyins (**8**) and (**9**) are marine natural products that were identified as cyctotoxins (24,25). The unique structure of these compounds and their antiproliferative activity initially made them quite attractive to the synthetic and medicinal chemistry community. These compounds have been synthesized and a number of analogs have been prepared and studied *in vitro* as potential anticancer agents (4). Like the taxanes, these compounds were found to be substrates for P-glycoprotein and hence show dramatically reduced cytotoxicity against MDR expressing cell lines. The SAR of the eleuthrobin/sarcodictyin class of microtubule stabilizing agents has been explored in a combinatorial fashion (26). This effort led to the conclusion that the conserved urocanic acid side chain was required for activity and that acetal and ester variants (R$_1$ and R$_2$ in **8** and **9**) were more easily tolerated. Although these compounds showed early promise as novel anti-proliferative agents, the failure of this class to address adequately resistant tumor lines suggests that it is unlikely that it will yield clinically useful cytotoxins.

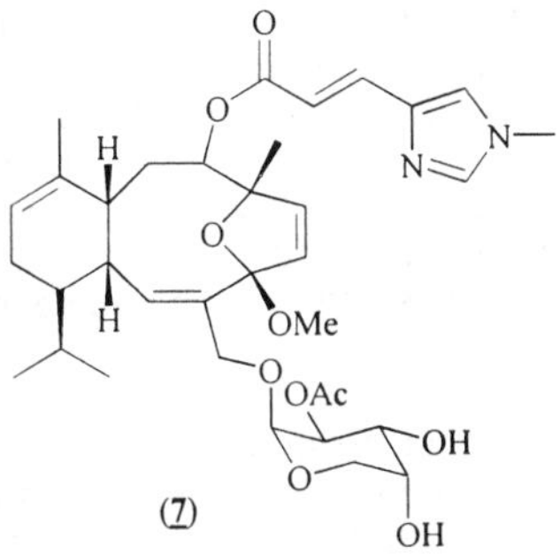

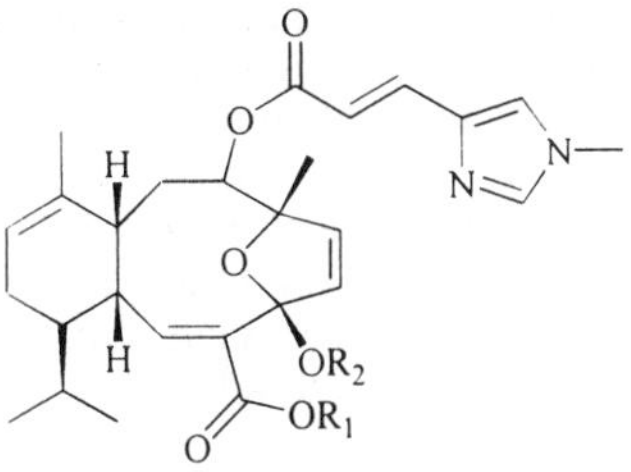

(8) R_1 = Me, R_2 = H
(9) R_1 = Et, R_2 = H

Laulimalide-The MT stabilizing macrocyclic lactone laulimalide (**10**), also known as fijianolide, has been reported in small quantities from a number of difference sponge species (27). This compound is quite potent (see Table) against susceptible tumor cell lines. Perhaps even more remarkable is the fact that this compound is ca. 100X more potent against a P-glycoprotein-expressing cell line (SKVLB) when compared to paclitaxel (28). This compound is quite unstable at low pH and undergoes a rapid conversion to isolaulimalide (**11**) via displacement of the epoxide by the C20 alcohol with concomitant loss in potency. Despite its promise as an antiproliferative agent, the *in vivo* evaluation of **10** has not yet been described, presumably due to lack of material for such experiments. Not surprisingly, **10** has been the subject of considerable synthetic effort. Three total syntheses of the compound have recently appeared (29, 30, 31). These and later total syntheses will create additional material with which to study this interesting molecule and its analogs.

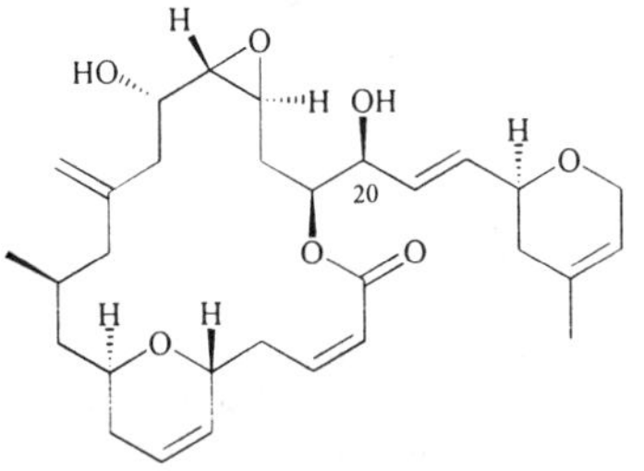

(10) 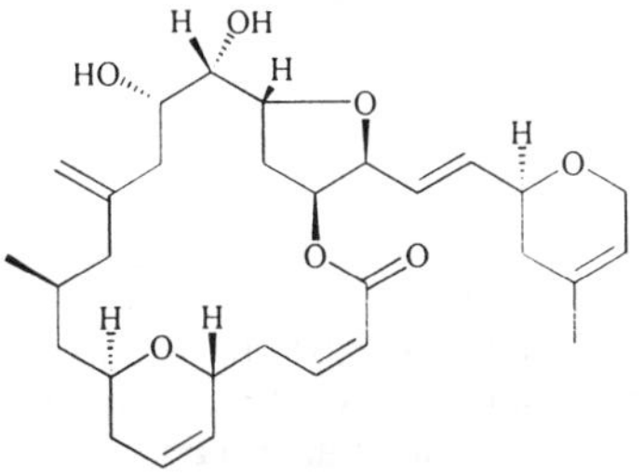   (11)

Dictyostatin-A recent and potentially quite interesting entrant into the class of microtubule stabilizing agents is dictyostatin (**12**). This 22-membered macrocyclic lactone was isolated several years ago in extremely small quantity from *spongia* sp. and found to be potently cytotoxic against P388 lymphocytic leukemia (32). This

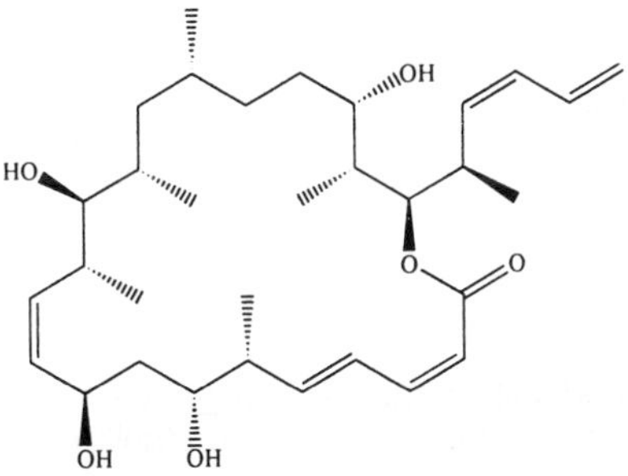

(12)

compound was recently re-isolated from a deep dwelling (2300 ft.) sponge *theonellidae* sp. and re-examined for cytotoxicity and mechanism (33). Compound **12** was found to function as a MT stabilizer with nM potency against susceptible and resistant cell lines (see Table). It is interesting to note the high degree of homology between the structure of **12** and **6**. Although little else is know about **12**, its potency, including against P-glycoprotein-expressing lines, makes **12** a worthy target of further study. Full evaluation of the antitumor properties of **12** may be hampered by the limited supply of material.

Other Microtubule Stabilizing Molecules   There exist several other natural and synthetic materials that are inhibitors of microtubule depolymerization. The isolation and characterization of WS9885B (**13**) has been reported (34). This material was found to possess cyctotoxicity similar to paclitaxel (35). A family of polyisoprenyl benzophenones (eg. **14**) has been isolated from the fruit of the Malasian plant *Garcinia pyrifera* have shown low uM cytotoxicities (36). Although **14** is only weakly cyctotoxic and structurally complex, it is isolated in relatively large quantity (27 g of **14** from 1.8 kg of dried fruit) and has served as the starting point for an analog program based on semisynthesis.

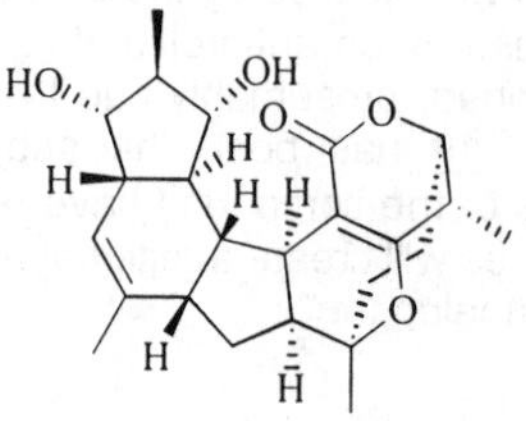

(**13**) 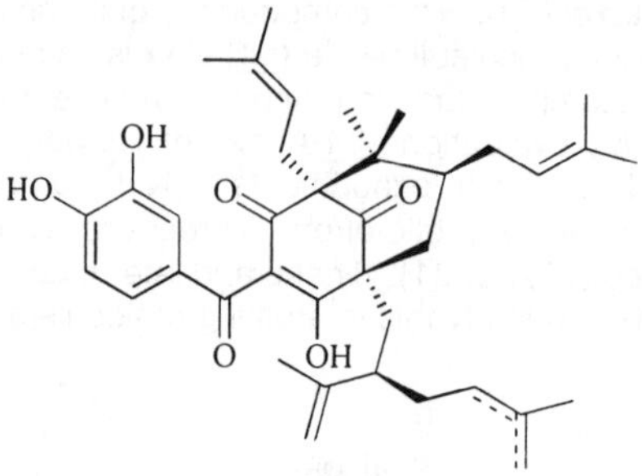   (**14**)

Several synthetic MT stabilizers have also been identified. Although the best of these compounds are significantly less potent than paclitaxel and other clinically useful microtubule stabilizing agents, they are as a group considerably less complex than their natural product counterparts. Thus, it is likely that these compounds can serve as more amenable lead structures for medicinal chemistry. GS-64 (**15**), a small heterobicycle that shows all the mechanistic hallmarks expected of an MT stabilizer has recently been described (37). This compound was found to possess uM cytotoxicity, roughly 1000X less potent than paclitaxel. Homologated estradiol **16** was likewise found to be a promoter of MT assembly at the uM level (38)

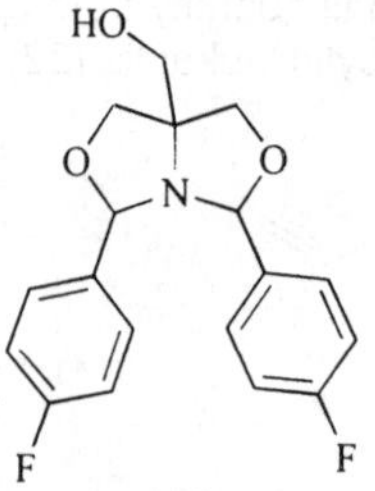

(**15**) 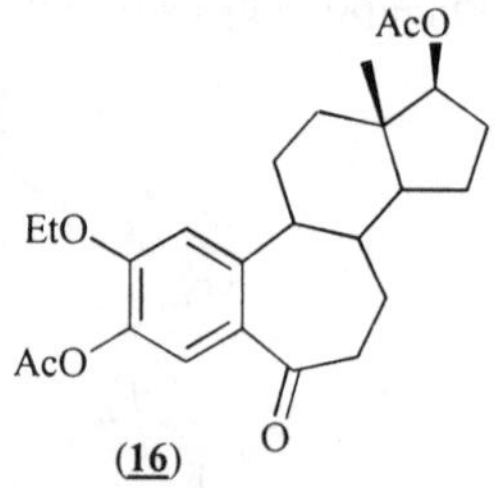   (**16**)

Other workers have identified sulfonamide **17** via a high throughput screening strategy. This compound has been reported to stabilize microtubules *in vitro* with potency similar to paclitaxel, however, it is far less potent as a cytotoxin (39). A similar divergence between *in vitro* and cell based potency was observed for synthetic borneol esters such as **18** (40).

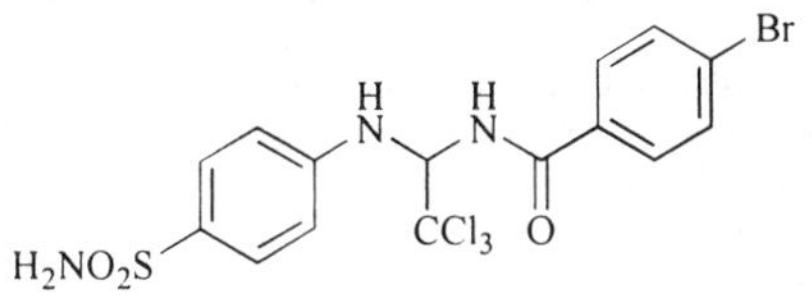

(17) 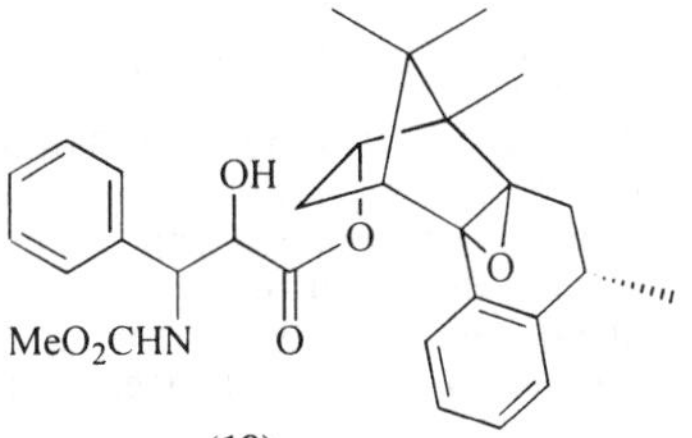   (18)

Conclusion – The clinical effectiveness of the taxanes has paved the way for the more recently discovered non-taxane microtubule stabilizing agents. These new compounds, most discovered in the last 10 years, have provided new avenues for research in cancer chemotherapy, as well as significant challenges to the synthetic community to supply sufficient quantities of these frequently complex structures. The early clinical success of the epothilones is an encouraging first step towed the introduction of a non-taxane MT stabilizer into wide spread clinical use. The advancement of discodermolide toward clinical evaluation is further cause for optimism. Total synthesis has played a key role in the advancement of dicodermolide through preclinical evaluation. Furthermore, totally synthetic material will be used in clinical trials. The complexity of this undertaking illustrates what may be the single biggest challenge for further evaluation of other microtubule stabilizing natural products, that being the lack of adequate supplies of material. Biotechnology and heterologous expression of biosynthetic machinery of these natural products can alleviate this issue, as is the case with the epothilones. The discoveries of a small number of simple, synthetic materials that show cytotoxiciy and function as microtubule stabilizers may serve to promote additional medicinal chemistry efforts on these structures.

References

1. L. He, G.A. Orr, S. B. Horowitz, Drug Discovery Today, $\underline{6}$(22), 1153 (2001).
2. J. Jimenez-Barbero, F. Amat-Guerri, J.P. Snyder, Curr. Med. Chem. $\underline{2}$, 91 (2002).
3. E. Nogales, S. G. Wolf, K. H. Downing, Nature, $\underline{391}$, 199 (1998)
4. S.J. Stachel, K. Biswas, S.J. Danishefsky, Current Pharm. Design $\underline{7}$, 1277 (2001)
5. (a) E. ter Haar, R. J. Kowalski, E. Hamel, C. M. Lin, R. E. Longley, S. P. Gunasekera, H. S. Rosenkranz, B. W. Day, Biochem. $\underline{35}$, 243 (1996); (b) G. Hoefle, N. Glaser, M. Kiffle, H. J. Hecht, F. Sasse, H. Reichenbach, Angew. Chem. Int. Ed. Engl. $\underline{38}$, 1971 (1999).; (c) P.G. Corley, R. Herb, R.E. Moore, P.J. Scheuer, V.J. Paul, J. Org. Chem. $\underline{53}$, 3644 (1988); (d) J.-I. Tanaka, T. Higa, G. Bernadelli, C.W. Jefford, Chem. Lett. 255 (1996); (e) K.H. Altman, M. Wartmann, T. O'Reilly, Biochem. Biophys. Acta, M79 (2000).
6. G. Hofle, N. Bedorf, H. Gerth, H. Reichenbach, Angew. Chem. Int. Ed. Engl. $\underline{35}$, 1567 (1996).
7. D.M. Bolag, P.A. McQueney, J. Zhu, O. Hensens, L. Koupal, J. Liesch, M. Goetz, E. Lazarides, C.M. Woods, Cancer Research, $\underline{55}$, 2325 (1995).
8. I.H. Hardt, H. Steinmetz, K. Gerth, F. Sasse, H. Reichenbach, G. Hofle, J. Nat. Prod. 847 (2001).
9. A. Florsheimer and K.-H. Altmann, Expert Opin. Ther. Patent $\underline{11}$(6), 951 (2001).
10. K.C. Nicolaou, F. Roschanger, D. Voulourmis, Angew. Chem. Int. Ed. Engl. 2014 (1998).
11. F.Y. Lee, G. Vite, G. Hofle, S.-H. Kim, J Clark, K. Fager, K. Kennedy, R. Smykla, M.-L. Wen, K. Leavitt, K. Johnson, R. Patterson, A. Kamath, M. Franchini, G. Schulze, C. Fairchild, K. Raghaven, B. Long, R. Kramer, AACR National Meting, Abstract number 3928, 2002.

12. S.P. Gunasekera, M. Gunasekera, R.E. Longley, G.K. Shulte. J. Org. Chem. 55, 4912 (1991) (corrigendum: ibid. J. Org. Chem. 56, 1346 (1991))
13. J.R.Kowalski, P. Giannakakou, S.P. Gunasekera, R.E. Longley, B.W. Day, E. Hamel Mol. Pharmacol. 52, 613 -622 (1997).
14. J. B. Nerenberg, D. T. Hung, P. K. Somers, S. L. Schreiber, J. Am Chem. Soc. 115(26), 12621 (1993).
15. S. S. Harried, G. Yang, M. A. Strawn, D. C. Myles, J. Org. Chem. 62(18), 6098 (1997).
16. A. B. Smith III, T. J. Beauchamp, M. J. LaMarche, H. Arimoto, Org. Letts. 1, 1823 (1999).
17. J. A. Marshall & B. A. J. Org. Chem. 63, 7885 (1999).
18. I. Paterson, G.J. Florence, K. Gerlach, J.P. Scott, Angew. Chem. Int. Ed. 39, 377 (2000).
19. M. Kalesse Chem. Bio. Chem. 1, 171 (2000).
20. R.A. Insbruker, S. P. Gunasekera, R.E. Longley, Cancer-Chemother. Pharmacol. 48(1) 29 (2001).
21. D. T. Hung, J. B. Nerenberg, S. L. Schreiber, J. Am. Chem. Soc. 118(45) 11054 (1996).
22. L.A. Martello, M.J. LeMarche, L. He, T. J. Beuachamp, A. B. Smith III, S. B. Horowitz, Chem. Bio. 8, 843 (2001).
23. F. R. Kinder Jr., K. W. Bair, W. Chen, G. Florence, C. Francavill, P. Geng, S. Gunasekera, P. T. Lassota, R. E. Longley, M. Palermo, I. Patterson, S. Pomponi, T. M. Ramsey, L. Rogers, M. Sabio, N. Sereinig, E. Sorensen, R. Wang, A. Wright, Abstract number 3650 AACR (2002).
24. B.H. Long, J. M Carboni, A. J. Wasserman, L. A> Cornell, A. M. Casazza, P. R. Jensen, T. Lindel, W. H. Fenical, C. R. Fairchild Cancer Research, 58, 1111 (1998).
25. M. D'Ambrosio, A. Guerriero, F. Pietra, Helv. Chim. Acta 70(8), 2019 (1987).
26. K. C. Nicolaou, D. Wissinger, D. Voulourmis, T. Oshima, S. Kim, J. Pfefferkorn, J.-Y. Xu, T. Li, J. Am. Chem. Soc. 120, 10814 (1998).
27. (a) E. Quinoa, Y. Kakou, Crews, P, J. Org. Chem. 53, 3642 (1988); (b) D.G. Corley, R. Herb, R.E. Moore, P.J. Scheuer, V. Paul, J. Org. Chem, 53, 3644, (1988); (c) J.I. Tanaka, T. Higa, G. Bernadelli, C.W. Jefford, Chem. Lett, 255 (1996).
28. S.L. Mooberry, G. Tien, A.H. Hernandez, A. Plubrukarn, B.S. Davidson, Cancer Research, 59, 653 (1999).
29. J. Mulzer, E. Ohler, Angew. Chem, Int. Ed. Engl. 40(20) 3842 (2001).
30. I. Paterson, C. De Savi, M. Tudge, Org. Lett. 3(20), 3149 (2001).
31. A.K. Ghosh and Y. Wang, J. Am. Chem. Soc. 122, 11027 (2000).
32. G.R. Pettit, Z.A, Cichacz, F. Gao, M. R. Boyd, J. M. Schmidt, J. Chem. Soc. Chem. Commun. 1111 (1994).
33. A.E. Wright, R.A. Insbruker, R. E. Longley, J. Cummins, P. Linley, S.A. Pomponi, Abstract 107, American Society of Pharmacognasy, National Meeting, 2001.
34. H. Muramatsu, M. Miyauchi, B. Sato, S. Yoshimura, Abstract number 83, 40th Symposium on the Chemistry of Natural Products, Fukuoka Japan, 1998.
35. C.D. Vanderwal, D.A. Vosberg, S. Weiler, E. J. Sorensen, Org. Lett., 1, 645 (1999) and references therein.
36. D. Roux, H. A. Hadi, S. Thoret, D. Guenard, O. Thoison, M. Pais, T. Sevenet, J. Nat. Prod., 63, 1070 (2000).
37. Y. Shintani, T. Tanaka, Y. Nozaki, Cancer Chemother. Pharmacol. 40, 513 (1997).
38. Z. Wang, D. Yang, A. Mohanakrishnan, P. E. Fanwick, P. Nampoothiri, E. Hamel, M. Cushman, J. Med. Chem., 43, 2419 (2000).
39. S.J. Haggarty, T. U. Mayer, D.T. Miyamoto, R. Fathi, R.W. King, T.J. Mitcheson, S.L. Schreiber, Chem. Biol. 7, 275 (2000).
40. U. Klar, H. Graf, O. Schenk, B. Rohr, H. Schulz, Bioorg. Med. Chem. Lett. 8 1397 (1998).

Chapter 14. Antiviral Agents

Nicholas A. Meanwell, John F. Kadow and Paul M. Scola
The Bristol-Myers Squibb Pharmaceutical Research Institute
5, Research Parkway, Wallingford, CT 06492

Introduction – The discovery and development of antiviral agents has entered a new era of opportunity, driven by unmet medical needs. These include the continuing problem associated with the emergence of resistance to drugs used for the treatment of the chronic diseases human immunodeficiency virus (HIV) and hepatitis B virus (HBV) and the potential offered by the limited therapeutic options currently available to treat hepatitis C virus (HCV), where the pace of drug discovery has noticeably quickened (1,2). In addition, a fuller appreciation of the effects of the acutely infectious respiratory viruses influenza, human rhinovirus (HRV) and respiratory syncytial virus (RSV) has fostered the discovery of preventive and therapeutic agents. Recent significant developments in drug discovery initiatives for each of these viruses will be summarized in that order.

INHIBITORS OF HIV

Overview: Marketing authorization for the nucleoside reverse transcriptase inhibitor (NRTI) tenofovir disoproxil was granted by both the FDA (10-26-2001) and the European Medicines Evaluation Agency (EMEA) (2-7-2002). This NRTI exhibits excellent activity against a broad range of resistant HIV strains and brings the clinically approved HIV drugs for use in combination therapy to 7 NRTIs, 3 non-nucleoside reverse transcriptase inhibitors (NNRTIs) and 6 protease inhibitors (PIs).

HIV Entry Inhibitors – The inhibition of HIV entry is emerging as an important new target for therapeutic intervention with blockade of the HIV gp120 envelope protein-CD4 interaction, interference with the HIV gp-41 fusion protein and the antagonism of the host cell chemokine receptors CCR5 and CXCR4, used by the virus as co-receptors, established as viable opportunities (3,4). The clinical efficacy of the tetrameric immunoglobulin-CD4 fusion protein PRO-542 provides proof-of-principle for gp120 as a target whilst FP-21399 (**1**), a parenterally administered bis-azo dye that interferes with the V3 loop of gp120, offers potential as a small molecule inhibitor (5,6). The betulinic acid derivative IC-9564 (**2**) has been suggested to interfere with gp120 (7). A subcutaneous formulation of enfuvirtide (T-20), a synthetic 36 amino acid peptide derived from gp41 that interferes with the intramolecular association of gp41 during a crucial step in the fusion process, is currently in phase 3 (P3) trials with a more potent analogue T-1249 in development (8). Additional peptidic and non-peptidic inhibitors of gp41 have been described (9,10).

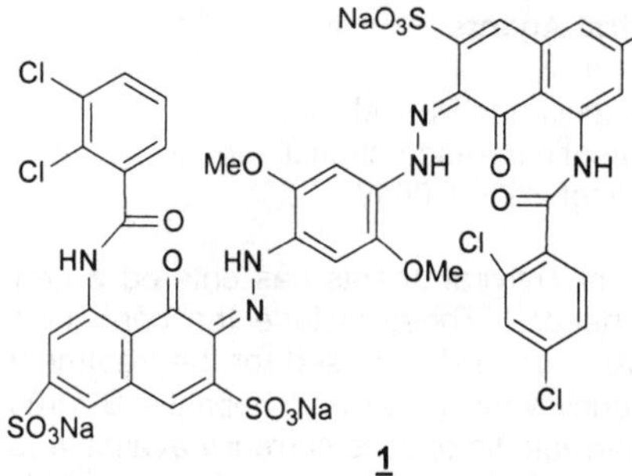

In vitro studies with PRO-140, an anti-CCR5 monoclonal antibody, have revealed potent inhibition against a broad range of major genetic subtypes of HIV, supporting the concept of CCR5 inhibition as an antiviral strategy (11). TAK-779 was the first small molecule CCR5 antagonist with antiviral activity to be described and SAR evolution has produced non-quaternary ammonium antiviral agents (12,13). Sch-351125 (**3**) is a potent and orally bioavailable CCR5 antagonist that has established proof-of-principle for small molecule CCR5 antagonists in a P2 clinical study (14-16). A 25 mg oral dose of **3** administered b.i.d as monotherapy to HIV-infected individuals reduced viral load by over 0.5 $\log_{10}$ (16). *In vitro* resistance studies indicate that repeated passaging of virus in the presence of **3** does not result in the selection of viruses preferring CXCR4 co-receptor use (17). The spiroketopiperazine E-913 (**4**) is both a potent CCR5 antagonist and antiviral agent, EC_{50} = 30-60 nM *in vitro*, that is a recent addition to this rapidly growing new class of HIV inhibitor (4,18).

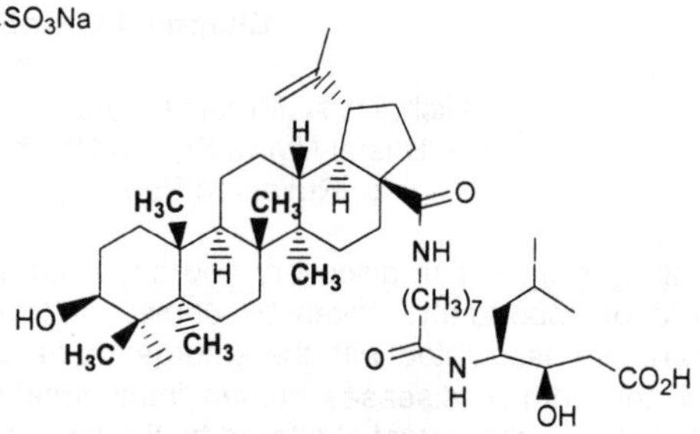

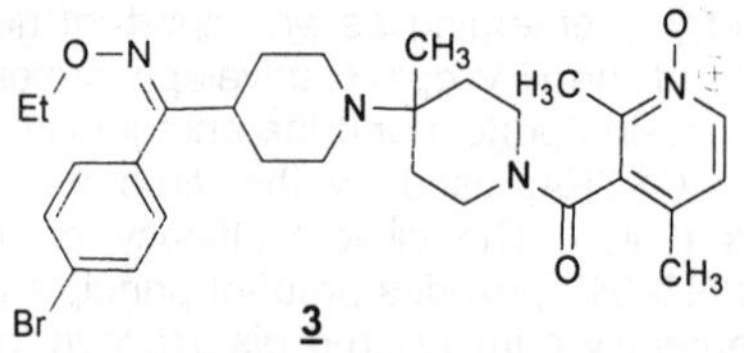

Clinical pharmacokinetic data on the parenterally-administered CXCR4-antagonizing bicyclam AMD-3100 (**5**) has been described and, whilst promising against X4 strains, development of this drug was terminated in May, 2001 due to cardiac irregularities in two patients and a failure to achieve robust efficacy end points (4,19-21). Current efforts in the cyclam area are directed towards preparing monocyclams similar to AMD-3465 (**6**), which exhibits antiviral activity and potency comparable to **5**, that are likely to display improved pharmaceutical properties (22,23).

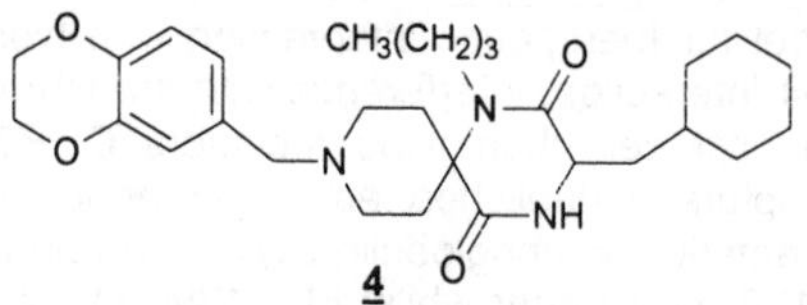

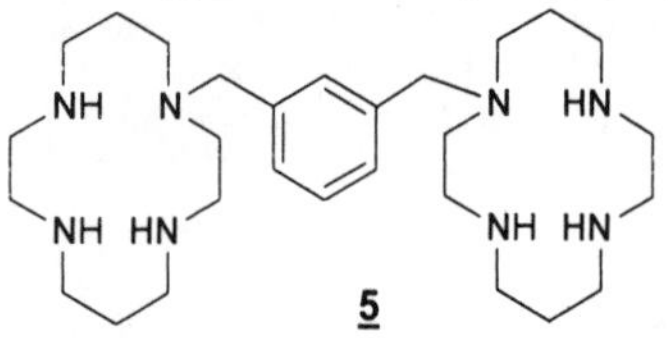

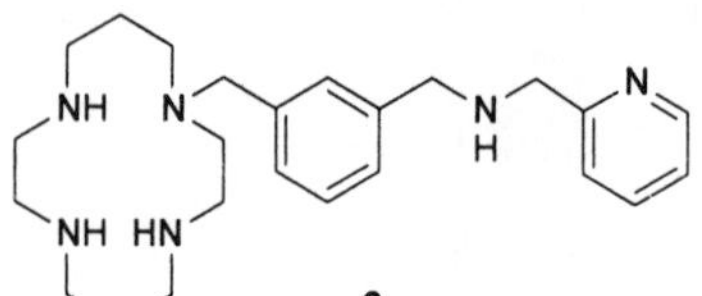

HIV Reverse Transcriptase Inhibitors – NRTIs are well established as clinically effective HIV inhibitors but the development of resistance remains problematic (24,25). Tenofovir disoproxil (**7**) is the most recent RTI to be approved and demonstrates activity towards AZT- and 3TC-resistant HIV strains (26). The dioxolane nucleoside amdoxivir (DAPD) (**8**), a prodrug of the guanosine derivative, is a promising inhibitor of both HIV and HBV that is active against NRTI-resistant HIV (27,28). P3 clinical studies with emtricitabine (**9**), a fluoro derivative of 3TC, demonstrate efficacy with a side effect profile similar to 3TC (29).

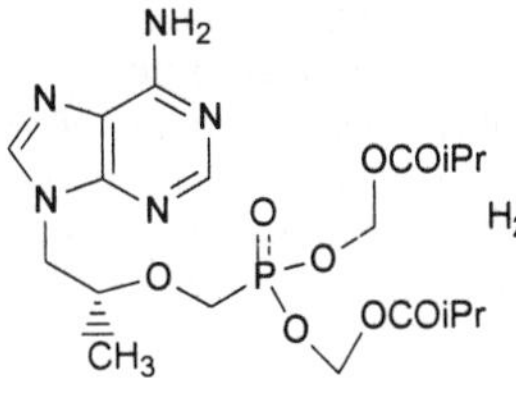

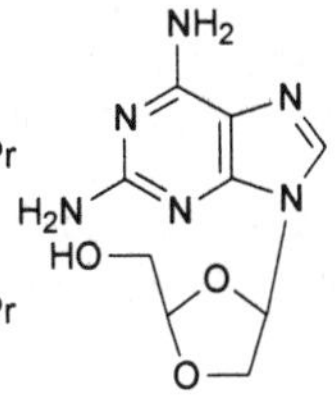

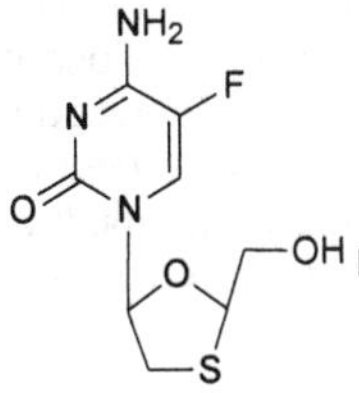

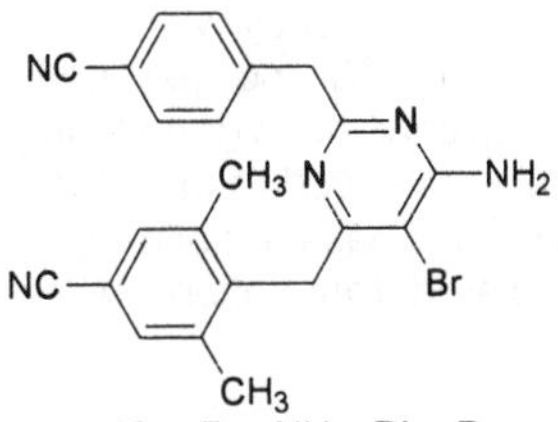

7　　　　　　　　**8**　　　　　　　　**9**　　　　　　　**10a**: R = NH$_2$; R' = Br
10b: R, R' = H

The three approved NNRTI's have assumed greater importance in clinical regimens based on their efficacy and pharmacokinetic properties although resistance remains an issue (30,31). More than 30 different classes of NNRTIs have been identified to date from which the pyrimidine derivative TMC-125 (**10a**) has emerged as an extremely potent NNRTI active against both wild-type and a panel of clinically significant single and double mutant strains of HIV-1 (32-35). Early clinical data for **10a** indicate that it is highly potent in lowering HIV viral loads in treatment-naïve patients as monotherapy (34,35). A course of 900 mg of **10a** b.i.d for 1 week administered to anti-retroviral naïve HIV-1 infected patients produced a 1.92 log$_{10}$ reduction in HIV RNA with a concomitant increase in CD4 cell count (35). The close analogue TMC-120 (R-147681) (**10b**) is more potent *in vivo*, diminishing viral load by 1.5 log$_{10}$ at a dose of 100 mg administered bid for 1 week (36). Early P2 clinical data for DPC-083 (**11**) indicate efficacy in patients infected with NNRTI-resistant HIV whilst development of capravirine (**12**) has resumed (37,38). SJ-3366 (**13**) is an HIV-1 NNRTI that also inhibits HIV-2, an effect believed to be a consequence of interference with the fusion mechanism (32,39).

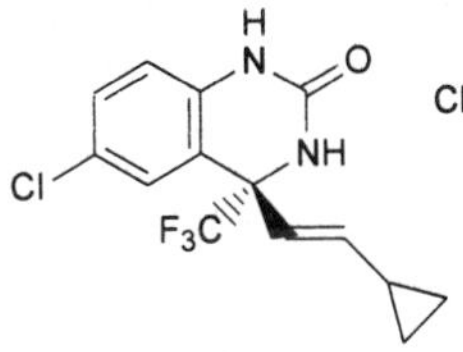

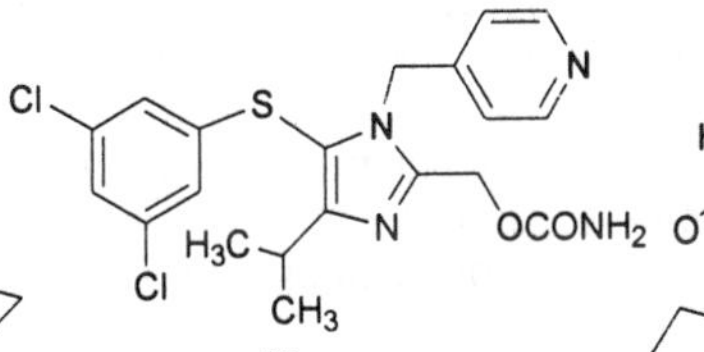

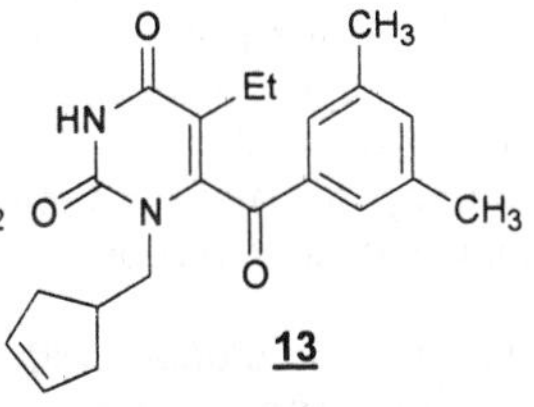

11　　　　　　　　　　**12**　　　　　　　　　　**13**

<u>HIV Integrase Inhibitors</u> – Inhibitors of HIV integrase that demonstrate antiviral activity are beginning to emerge (40,41). S-1360 (**14**), evolved from a carboxylic acid-containing prototype, is the first of this class to enter clinical evaluation (42). Carboxylic acids continue to be of interest, including the amide **15**, whilst the aza-naphthalenyl carboxamide **16** represents an interesting new integrase inhibitor motif (43,44).

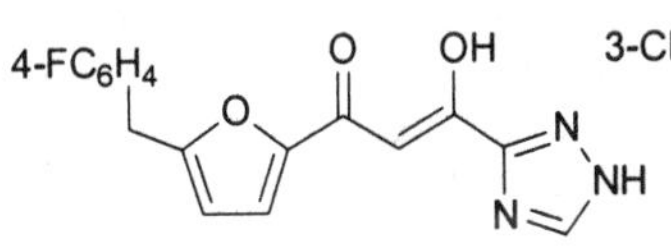

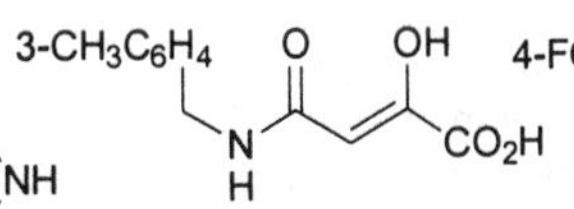

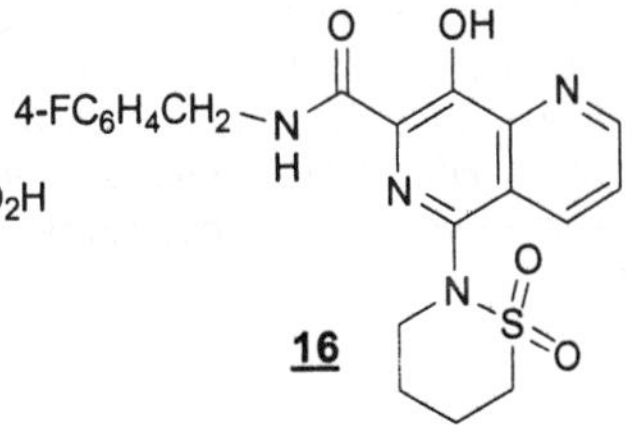

14　　　　　　　　　　　　**15**　　　　　　　　　　**16**

HIV Protease Inhibitors – HIV protease inhibitors are an important component of highly active antiretroviral therapy (HAART) for which the molecular basis for resistance has been examined (45,46). P3 clinical data for atazanavir (BMY-232632) demonstrate antiviral efficacy with a favorable lipid profile after 48 weeks of therapy at doses of 400 mg or 600 mg per day (47,48). Fosamprenavir (**17**), a phosphate prodrug of amprenavir designed to improve dissolution in gastric juices and, hence, bioavailability, has advanced to late stage clinical studies (49). UIC-94003 (TMC-126, **18**) is a potent HIV protease inhibitor, active against multi-protease inhibitor-resistant strains of HIV-1 isolated from patients who were unresponsive to currently available therapy (50).

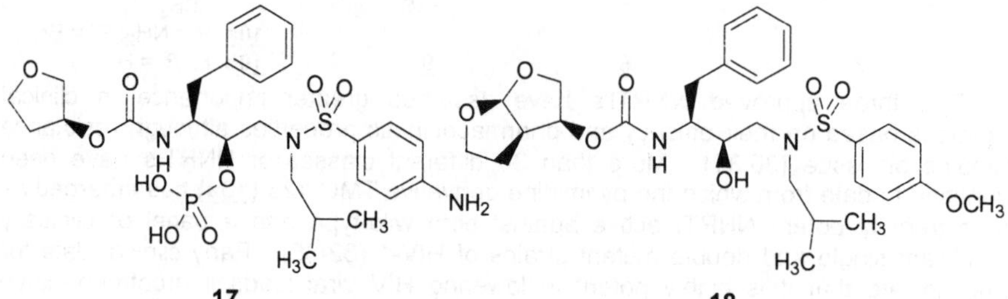

17   **18**

INHIBITORS OF HEPATITIS B AND C VIRUS

Inhibitors of HBV – Lamivudine (3TC) and IFNα are the only approved agents for treating HBV infection but the development of resistance to 3TC, mapped to mutations in the HBV polymerase gene (M550I or M550V + L526M), leads to poor durability (51,52). This is thought to be due to the limited potency and efficacy associated with 3TC, a deficiency being addressed by a series of newer, more potent nucleoside analogues currently in clinical development that retain demonstrable activity against 3TC-resistant HBV. Marketing approval for adefovir dipivoxil was sought on 3-21-2002. Entecavir (**19**) is currently in P3 whilst several other nucleoside analogues are in earlier stages of development (53). However, agents that target other proteins essential to HBV replication are beginning to emerge and hold promise for an opportunity to more effectively control HBV infection in the clinic using drug combinations that include novel nucleoside analogues.

Long term studies with (**19**) in the woodchuck hepatitis virus (WHV) model demonstrate effective viral suppression over 12 months of weekly oral therapy at a dose of 0.5 mpk (53,54). Very potent antiviral activity has been established for **19** in HBV-infected patients. Oral doses of **19** ranging from 50 µg to 1 mg per person, administered for 28 days, produced viral load reductions in excess of 2 $\log_{10}$ (55).

The recent emphasis on L-sugar-based nucleoside derivatives has provided a number of potentially useful inhibitors of HBV (56). LdT (**20**) and LdC (**21**) are potent and selective inhibitors of HBV replication in HepG2 cells with **20**, EC_{50} = 190 nM, currently in P1/2 clinical trials based on efficacy in the woodchuck model and safety in preclinical toxicology studies (57,58). Preliminary clinical data with **20** indicated efficacy at oral doses of 25-200 mg/day, which produced a 2-4 $\log_{10}$ reduction in plasma HBV DNA within 2-4 weeks (59). The 3',5'-bis-valine ester prodrug of **21** is being developed to overcome the limited oral bioavailability of the parent molecule (60).

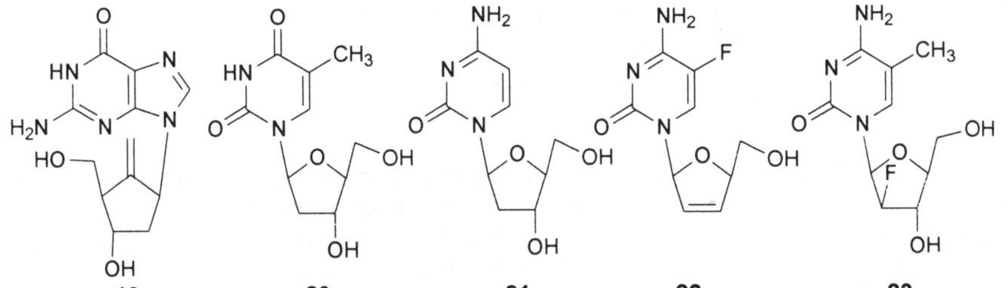

19   **20**   **21**   **22**   **23**

β-LFd4C (ACH-126443) (**22**) shows efficacy in the woodchuck model with 4 mpk/day providing a sustained antiviral response whilst P1 clinical data indicate good oral bioavailability that predicts once-daily dosing (61,62). Clevudine (L-FMAU) (**23**) shows good efficacy in the woodchuck model with daily doses of 0.1 mpk determined as the threshold for a significant antiviral effect (63). However, *in vitro* studies indicate that the L526M, M550I and dual M550V + L526M mutations that confer resistance to 3TC are also resistant to **23** whilst, in contrast, **8**, **19** and adefovir dipivoxil (**24**) retain activity against these clinically relevant variants (64).

The nucleoside phosphonate derivative **24** has completed pivotal P3 clinical trials with positive efficacy data disclosed recently (65). A 10 mg daily dose of this prodrug for 48 weeks reduced plasma HBV DNA by a median 3.52 $\log_{10}$, which compared to a reduction of 0.55 $\log_{10}$ in placebo controls (66). Liver histology was significantly improved in 53% of treated patients. Adefovir dipivoxil has also shown efficacy in patients infected with 3TC-resistant HBV and an NDA was filed on 3-21-2002 (65-67).

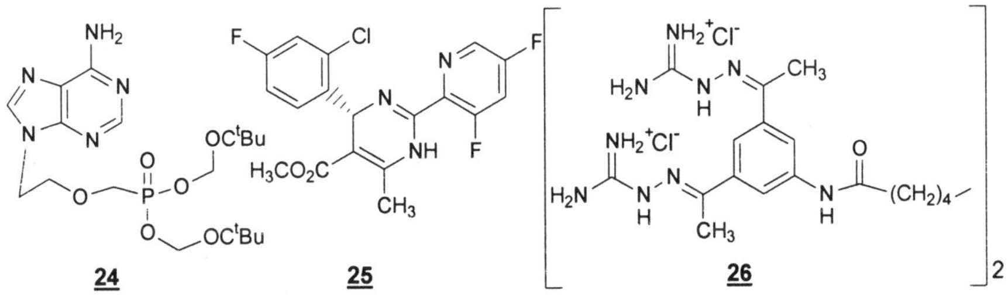

24   **25**   **26**

Bay-41-4109 (**25**) represents a new mechanistic class of HBV inhibitor that potently blocks replication in HepG 2.2.15 cells, EC_{50} = 53 nM, by binding to the core protein and interfering with virus assembly (68-71). This compound is 30% bioavailable in mice and oral administration at doses of 3-100 mpk bid to HBV transgenic mice for 28 days reduced plasma and liver HBV DNA levels with efficacy comparable to that of lamivudine (70). AXD-455 (**26**) inhibits HBV replication in cell culture with an EC_{50} = 3.2 µM and a CC_{50} > 20 µM and fully retains activity towards the replication of 3TC-resistant virus, EC_{50} = 2.8 µM (72). An additive interaction with 3TC in cell culture was observed but, since **26** is also active against M- and T-tropic HIV, it lacks virus specificity (73). This is consistent with the proposed mode of action based on inhibition of eIF-5A, a host cell initiation factor that facilitates the export of viral mRNA from the cell nucleus to the cytoplasm (73).

Inhibitors of HCV – HCV clinical therapy is currently restricted to interferon α (IFNα) in conjunction with ribavirin. The first pegylated form of IFNα, a chemical modification that allows for once weekly administration, was approved in the US last year (74-76). Two *in vivo* models of HCV infection, in which human liver cells transplanted into immunodeficient mice produce grafts that support HCV infection, have been developed

(77,78). These models offer considerable promise for the pre-clinical evaluation of HCV inhibitors, a timely development against the backdrop of the recent marked progress that has been made towards the identification of small drug-like inhibitors of HCV (79,80) and cell culture systems (81,82).

The HCV NS2/3 protease has been characterized as a Zn^{2+}-dependent enzyme essential for viral replication for which *in vitro* assays utilizing truncated peptide fragments that retain autocatalytic activity have been developed (83,84). This enzyme is inhibited by NS4a peptide fragments, N-terminal cleavage products and substrate-like peptides that span the active site region (83,85).

Potent peptide-based inhibitors of the HCV NS3 serine protease have been identified based on several structural motifs (86-94). α-Keto amide derivatives and structurally related peptides that present an electrophilic carbonyl to the active site serine hydroxyl form the basis of several series of potent and selective inhibitors of HCV NS3 (86-89,94). For example, the amide **27** inhibits HCV NS3 with an IC_{50} of 11 nM (91). Peptidic boronates have been identified as inhibitors of HCV NS3 with Ac-Asp-Glu-Val-Val-Pro-boroAlg-OH exhibiting a Ki of 13 nM, potent inhibition that is dependent on the presence of the NS4a co-factor (93). The dicarboxylic acid **28** is representative of a series of peptide-like inhibitors that inhibit HCV NS3 by binding to prime sites of the protease (95). GW-3112 (**29**) is a mechanism-based inhibitor of the enzyme, $k_{obs}/I = 7760$ $M^{-1}s^{-1}$ in a FRET assay, that exhibits an EC_{50} of 0.45 µM in a replicon assay (96). APC-6336 (**30**) is one of a series of Zn^{2+}-dependent inhibitors of HCV protease, $IC_{50} = 200$ nM that is 800-fold less potent in the absence of Zn^{2+} (97).

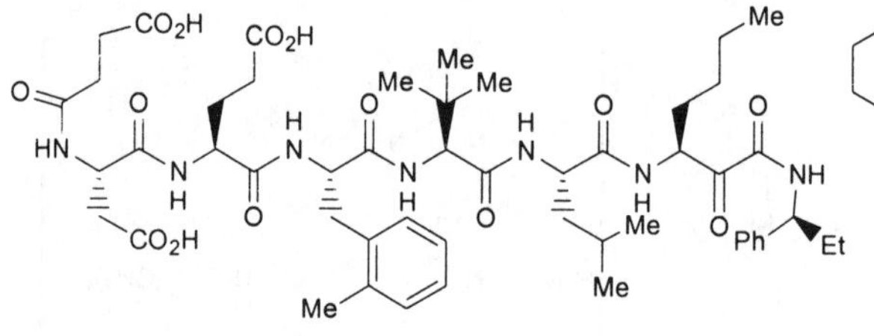

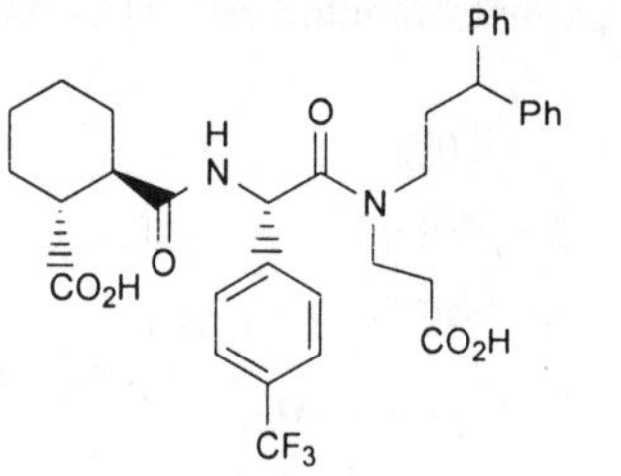

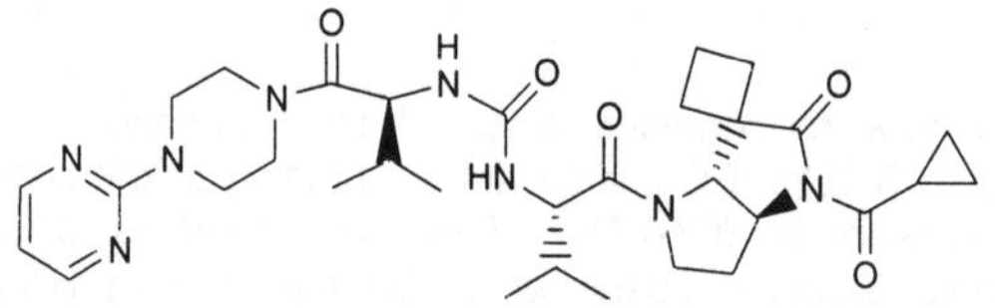

The HCV NS5b polymerase is a virus-specific target for which assays have been developed and crystallographic data are available (98,99). The polymerase is believed to function in multimeric form with oligomerization elements mapped to Glu-18 and His-502 (100). Amongst the several mechanisms of antiviral action attributed to ribavirin, inhibition of HCV polymerase may play a role (101,102). Levovirin, the L-enantiomer of ribavirin, retains immunomodulatory and antiviral activity whilst exhibiting reduced toxicity and is under clinical development (103). Two recent patent applications disclose inhibitors of HCV NS5b based on a structurally similar benzimidazole scaffold, represented by **31** and **32** (105,106).

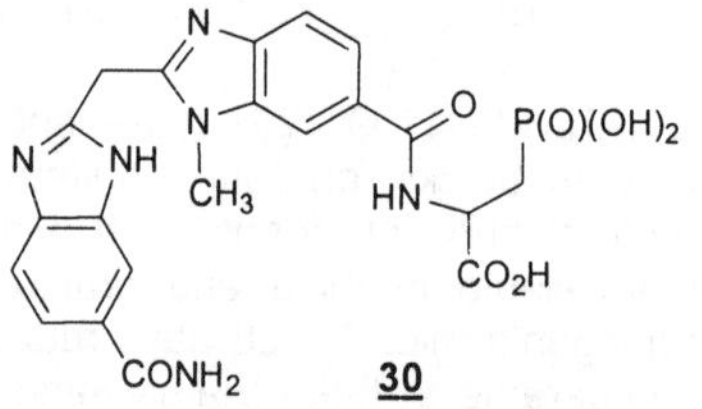

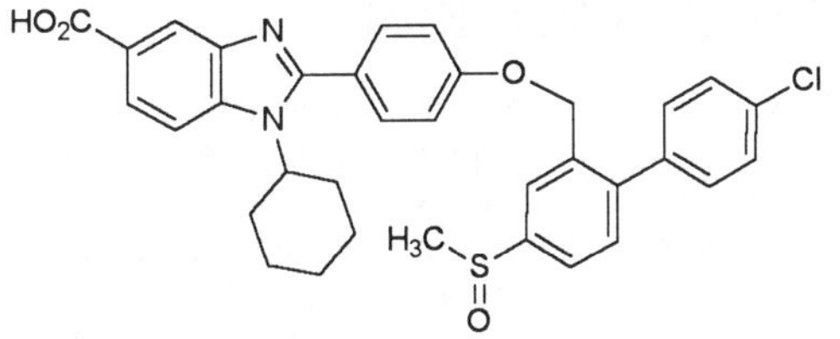

__31__ 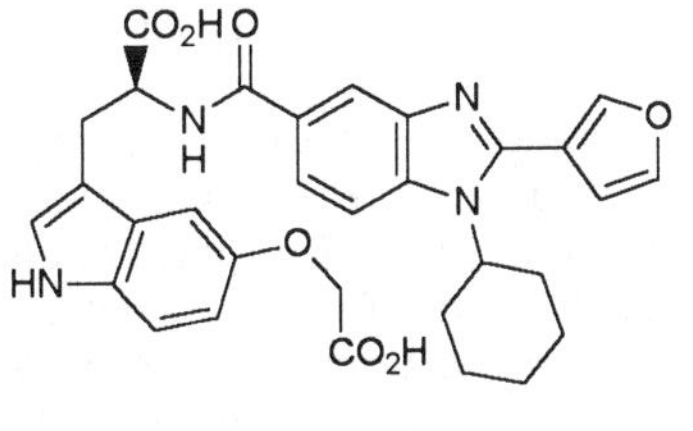   __32__

A series of pyrimidine derivatives, typified by __33__, have been claimed as HCV polymerase inhibitors (107).   Rhodanine derivative __34__ is an irreversible inhibitor of NS5b, $IC_{50} < 30$ μM, that functions as a Michael acceptor towards the thiol of Cys-366 that is found in the active site of the enzyme (108).

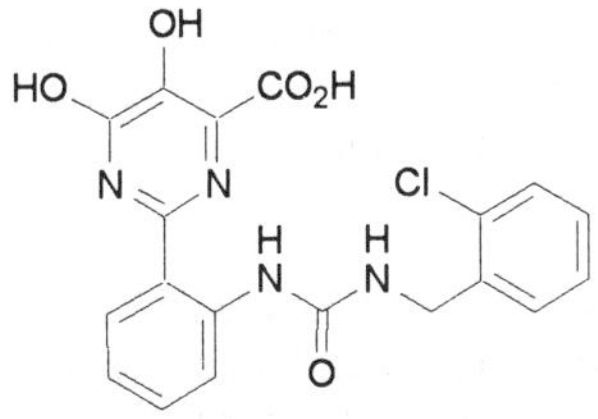

__33__ 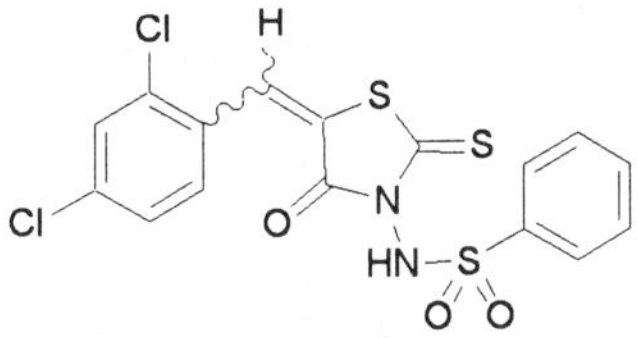   __34__

Other small molecule inhibitors of HCV polymerase include __35__ and __36__ whilst 3'-deoxycytidine-5'-triphosphate is claimed to inhibit NS5b with an IC_{50} of 30 nM (109-112).   Nucleoside analogues form the basis of patent applications from which __37__ exhibits an EC_{50} of 50 nM in a replicon assay (112-114).

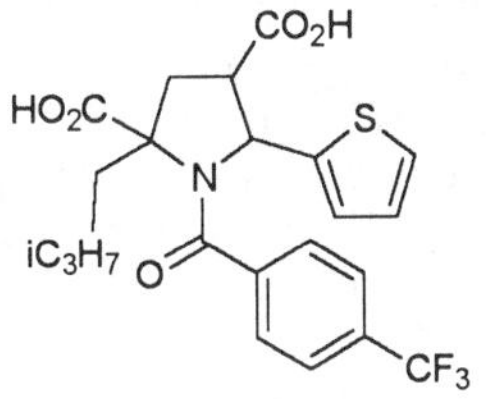

__35__ 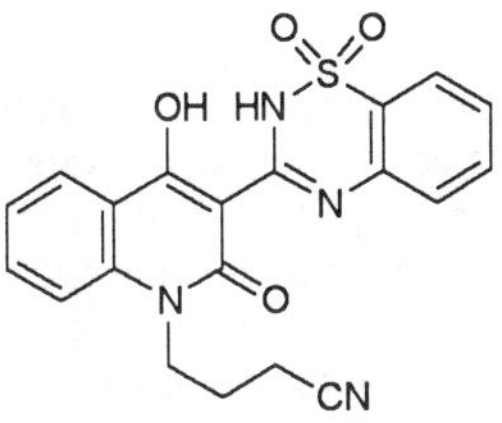   __36__ 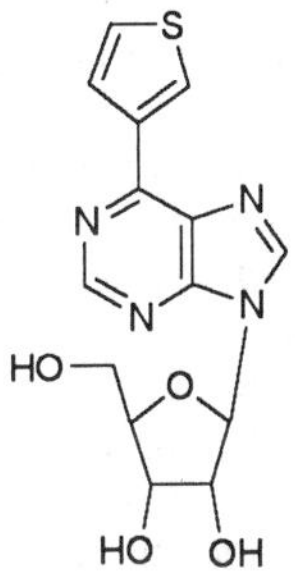   __37__

Structural information for the viral RNA internal ribosomal entry site (IRES) in the presence of bound 40S subunit has been elucidated by cryo-electron microscopy at 20 Å resolution (115). The antisense oligonucleotide ISIS-14803 was designed to interact with the HCV IRES and this agent is in early clinical trials as a potential treatment for HCV, both as a single-agent and in combination with ribavirin and PEG-IFN-α (116).

HCV entry offers a panoply of targets and assays have been developed (117). A crystal structure of human CD81, an HCV receptor recognized by the viral E2 envelope glycoprotein and a member of the tetraspanin family, has been solved at 1.6 Å resolution, providing a potential opportunity for structure-based drug design (118,119).

INHIBITORS OF RESPIRATORY VIRUSES

Overview - The three major respiratory viruses, human rhinovirus (HRV), respiratory syncytial virus (RSV) and influenza, continue to be significant sources of epidemic infection (120). The prevalence of RSV remains underestimated since this virus is frequently mistaken for influenza, a situation complicated further with the identification of a new pneumovirus that causes symptoms similar to RSV (121-123). This virus, designated human metapneumovirus (hMPV), has been circulating for at least 50 years in Europe and may be a significant cause of respiratory infections (123). The influenza inhibitor market is dominated by the orally active neuraminidase (NA) inhibitor oseltamivir, prescribed preferentially over the topical NA inhibitor zanamivir, whilst a third NA inhibitor, peramivir, is undergoing pivotal P3 trials (124,125). An NDA for the HRV capsid inhibitor pleconaril (picovir) was filed in July 2001, seeking approval to market this agent as the first antiviral treatment for the common cold, but was recently rejected by an advisory committee. The topically administered HRV 3C protease inhibitor ruprintavir is currently undergoing clinical evaluation whilst the RSV market is largely restricted to the humanized monoclonal antibody palavizumab (Synagis), licensed for prophylactic use. However, interest in small molecule RSV inhibitors is gaining momentum.

Inhibitors of Influenza Virus – Concerns about the considerable threat presented by unpredictable and potentially devastating influenza pandemics have lead to the suggestion that a surveillance network be established, compelling since the origin of virulence of the 1918 Spanish influenza outbreak remains obscure (126-128). Neuraminidase inhibitors remain of contemporary interest and updates on the clinical application of the influenza NA inhibitors oseltamivir and zanamivir have appeared (129-131). P3 clinical trials with peramivir (BCX-1812, RWJ-270201) (**38**) have been reinitiated for the 2001-2 influenza season. A full account of the design and discovery of **38** has appeared along with detailed descriptions of the *in vivo* antiviral properties of this potent NA inhibitor (132-134). A recent patent application claims an extensive range of peramavir prodrug derivatives modified at both the guanidine and carboxylic acid moieties (135). Several series of NA inhibitors based on novel structural scaffolds have been designed (136-140). The pyrrolidine derivative **39** has been profiled in some detail whilst the more potent NA inhibitor A-315675 (**40**), which has been prepared by enantioselective syntheses, shows good oral bioavailabilty when administered as an ester prodrug (141-144). Both the ethyl and isopropyl esters are rapidly cleaved in plasma and deliver the parent drug after oral administration to rats with compound readily detectable in lung tissue (141,143). Structural evolution of zanamivir has focussed on dimeric species, represented by **41**, that are able to simultaneously bind to two neuraminidase active sites and are markedly more potent as a consequence (145). Importantly, significant concentrations of **41** are retained in the lungs of rats for a week following a single topical administration (145).

38   **39**   **40**   **41**

Additional potential influenza targets have been reviewed (146). Parallel synthesis methodology has produced two topologically distinct but potent inhibitors of influenza, **42** and **43**, that function by preventing the conformational rearrangement of hemagglutinin, the key step in fusion of virus and host cell membranes (147).

However, these compounds lack broad spectrum activity since inhibition is restricted to influenza A H1 and H2 subtypes. T-705 (**44**) is a new influenza inhibitor that protects mice against a lethal infection (148).

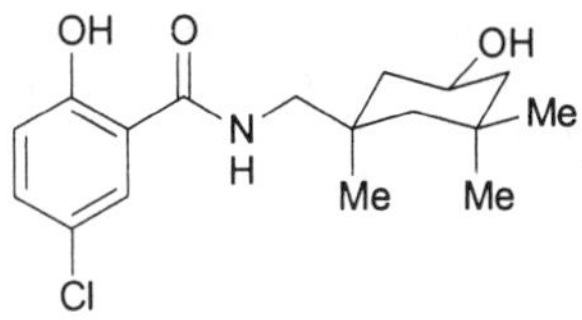

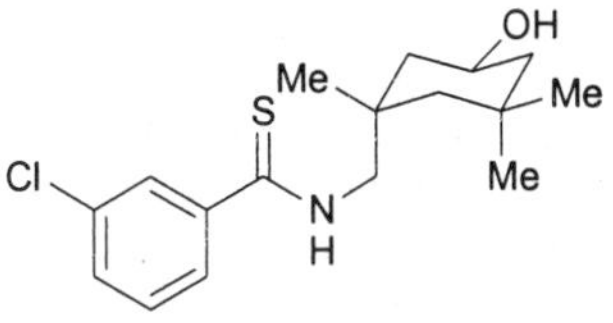

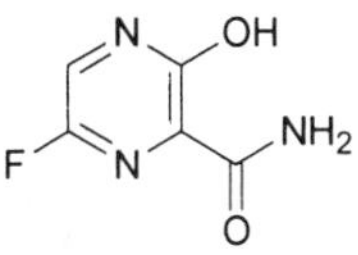

42   **43**   **44**

Inhibitors of Human Rhinovirus – Approaches to the treatment of rhinovirus infections and a summary profile of the preclinical and clinical properties of pleconaril have been published (149-151). The pleconaril NDA is in late stages of review by the FDA and HRV capsid inhibitors remain of contemporary interest. BTA-188 (**45**) shows potent *in vitro* activity against 87 of 100 rhinovirus serotypes with a median EC_{50} = 10 ng/mL, superior to pleconaril (152,153). In rats and dogs, **45** exhibits good oral bioavailability, with drug effectively accumulating in the nasal epithelium of dogs at concentrations significantly exceeding the median EC_{50} (154). Additional examples of this class of HRV inhibitor have been disclosed (155). The use of a novel, high-throughput mass spectrometry screening assay to identify HRV capsid binding agents has been described (156). Screening of a library of compounds synthesized from fragments selected after analysis of the binding pocket of VP1 identified benzothiazole **46** as an inhibitor of rhinovirus 14 in cell culture with an EC_{50} = 0.8 μM (156).

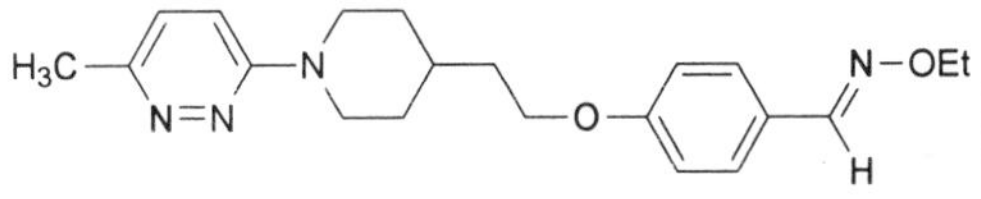

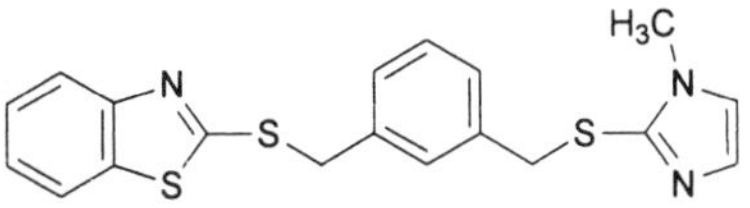

45   **46**

Ruprintivir (**47**) is the prototypical HRV 3C protease inhibitor, a topically applied compound currently undergoing clinical evaluation for the prevention and treatment of the common cold (157,158). Structural evolution of **47** has focussed on the identification of inhibitors with reduced peptidic character that would be more suitable for oral administration (159-161). Pyridone **48** retains broad spectrum HRV inhibitory activity with an average EC_{50} of 45 nM across 15 serotypes of HRV, but is 48% bioavailable in the dog, markedly superior to the 8% observed with ruprintavir (161).

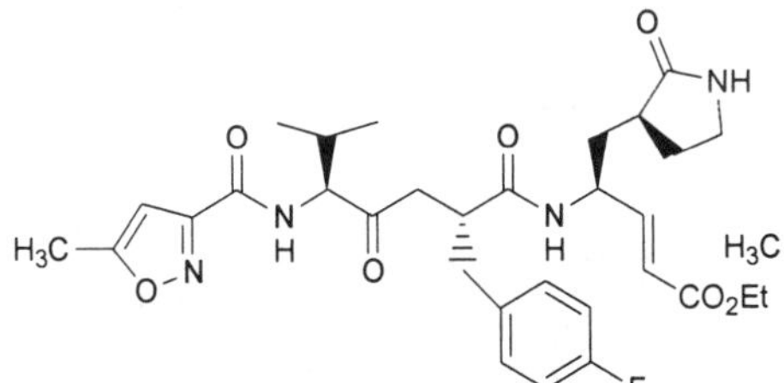

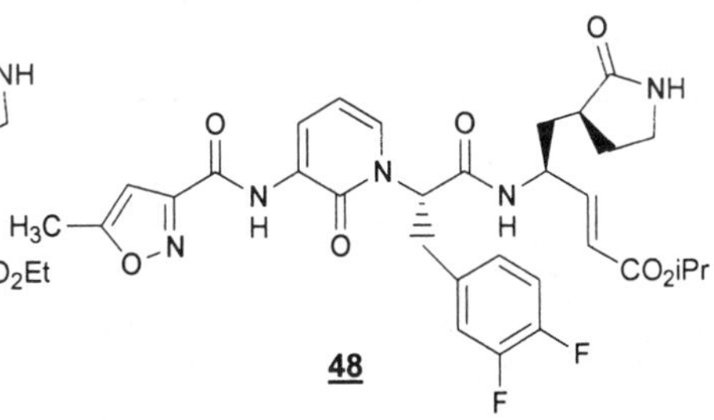

47   **48**

Inhibitors of Respiratory Syncytial Virus – The epidemiology and clinical significance of RSV have been reviewed whilst progress towards antiviral agents have been summarized (121,122,162-164). The RSV fusion (F) protein, critical to virus entry and replication and the target of the clinically successful prophylactic humanized monoclonal antibody palivuzumab, is beginning to emerge as a viable target for small molecule antiviral agents. It is now known that the RSV F protein requires 2 proteolytic cleavages in order to be fully activated (165). The first crystal structure of a paramyxovirus F protein, that from Newcastle disease virus, has been solved, revealing a new structural paradigm for virus fusion mechanisms (166). Inhibitors of RSV fusion reported to date are structurally disparate and include RF-641 (**49**), although this compound also appears to interfere with the attachment (G) protein, and R-170591 (**50**) (167-170). Development of RF-641 has been discontinued because of safety concerns and drug delivery problems (171). Mode of action studies with RD3-0028 (**51**), a cyclic disulfide with no direct virucidal effect, indicate an inhibition of events late in the virus replication cycle (172). Resistance mapped to a $Asn_{276}Tyr$ mutation in the RSV F protein, leading to the conclusion that **51** interferes with F protein processing rather than directly blocking fusion. The benzimidazole **52** is representative of three series of benzimidazole-based antiviral agents claimed recently as inhibitors of RSV (173-175).

References

1. W.A. Haseltine, Sci. Amer., Nov., 56 (2001).
2. E. De Clercq, Nature Rev. Drug Disc., 1, 13 (2002).
3. S. Redshaw and M. Westby, Expert Opin. Emerging Drugs, 6, 209 (2001).
4. A. Mastrolorenzo; A. Scozzafava and C.T. Supuran, Expert Opin. Ther. Patents, 11, 1245 (2001).
5. B.M. O'Hara, W.C. Olson, R.J. Israel, M. Barish, P.J. Maddon, I. Lowry and J.M. Jacobson, 15th Int. Conf. Antiviral Res., Prague, Czech Republic, Mar. 17-21, Abst. LB2 (2002).
6. G. Poli and E. Vicenzi, IDrugs, 4, 1293 (2001).
7. S. Holz-Smith, I.-C. Sun, L. Jin, T.J. Matthews, K.-H. Lee and C.H. Chen, Antimicrob. Agents Chemother., 45, 60 (2001).
8. J. Lalezari, E. DeJesus, D. Northfelt, G. Richmond, J. Delehanty, R. DeMasi and M. Salgo, 9th Conf. Retroviruses Opportunistic Infections, Seattle, WA, Feb. 24-28, Abst. 418W (2002).
9. C.A. Bewley, J.M. Louis, R. Ghirlando and G.M. Clore, J. Biol. Chem., 277, 14238 (2002).
10. M.J. Root, M.S. Kay and P.S. Kim, Science, 291, 884 (2001).
11. A. Trkola, T.J. Ketas, K.A. Nagashima, L. Zhao, T. Cilliers, L. Morris, J.P. Moore, P.J. Maddon and W.C. Olson, J. Virol., 75, 579 (2001).
12. J.A. Esté, Curr. Opin. Invest. Drugs, 2, 354 (2001).
13. S. Imamura, S. Hashiguchi, T. Hattori, O. Nishimura, N. Kanzaki, M. Baba and Y. Sugihara, PCT Patent Appl. WO 0125200-A1 (2001).
14. J. Esté, Curr. Opin. Invest. Drugs, 3, 379 (2002).
15. J.M. Strizki, S. Xu, N.E. Wagner, L. Wojcik, J. Liu, Y. Hou, M. Endres, A. Palani, S. Shapiro, J.W. Clader, W.J. Greenlee, J.R. Tagat, S. McCombie, K. Cox, A.B. Fawzi, C-C. Chou, C. Pugliese-Sivo, L. Davies, M.E. Moreno, D.D. Ho, A. Trkola, C.A. Stoddart, J.P. Moore, G.R. Reyes and B.M. Baroudy, Proc. Natl. Acad. Sci. USA, 98, 12718 (2001).
16. J. Reynes, R. Rouzier, T. Kanouni, V. Baillat, B. Baroudy, A.Keung, C. Hogan, M. Markowitz and M. Laughlin, 9th Conf. Retroviruses Opportunistic Infections, Seattle, WA, Feb. 24-28th, Abst. 1 (2002).
17. A. Trkola, S.E. Kuhmann, J.M. Strizki, E. Maxwell, T. Ketas, T. Morgan, P. Pugach, S. Xu, L. Wojcik, J. Tagat, A. Palani, S. Shapiro, J.W. Clader, S. McCombie, G.R. Reyes, B.M. Baroudy and J.P. Moore, Proc. Natl. Acad. Sci. USA, 99, 395 (2002).
18. K. Maeda, K. Yoshimura, S. Shibayama, H. Habashita, H. Tada, K. Sagawa, T. Miyakawa, M. Aoki, D. Fukushima and H. Mitsuya, J. Biol. Chem., 276, 35194 (2001).
19. B.J. Doranz, L.J. Filion, F. Diaz-Mitoma, D.S. Sitar, J. Sahai, F. Baribaud, M.J. Orsini, J.L. Benovic, W. Cameron and R.W. Doms, AIDS Res. Human Retrovirus., 17, 475 (2001).

20. D. Schols, S. Claes, E. De Clerq, C. Hendrix, G. Bridger, G. Calandra, G.W. Henson, S. Fransen, W. Huang, J.M. Whitcomb, C.J. Petropoulos and the AMD-3100 HIV Study Group, 9th Conf. Retroviruses Opportunistic Infections, Seattle, WA, Feb. 24-28th, Abst. 2 (2002).

21. C. Hendrix, A. Collier, M. Lederman, R. Pollard, S. Brown, M. Glesby, C. Flexner, G. Bridger, K. Badel, R. MacFarland, G. Henson and G. Calandra for the AMD-3100 HIV Study Group, 9th Conf. Retroviruses Opportunistic Infections, Seattle, WA, Feb. 24-28th, Abst. 391T (2002).

22. G.J. Bridger, E.M. Boehringe, Z. Wang; D. Schols, R.I. Skerlj and D.E. Bogucki, PCT Patent Appl. WO 0144229-A1 (2001).

23. S. Hatse, K. Princen, G. Bridger, R. Skerlj, G. Henson, E. De Clercq and D. Schols, 15th Int. Conf. Antiviral Res., Prague, Czech Republic, Mar. 17-21, Abst. 5 (2002).

24. L. K. Naeger and M.D. Miller, Curr. Opin. Invest. Drugs, 2, 335 (2001).

25. C.J. Loveday, Acq. Immune Defic. Syndr., 26 (Suppl. 1), S10 (2001).

26. B.G. Gazzard, Int. J. Clin. Practice, 55, 704 (2001).

27. A.H. Corbett and J.C. Rublein, Curr. Opin. Invest. Drugs, 2, 348 (2001).

28. P.A. Furman, J. Jeffrey, L.L. Kiefer, J.Y. Feng, K.S. Anderson, K. Borroto-Esoda, E. Hill, W.C. Copeland, C.K. Chu, J.P. Sommadossi, I. Liberman, R.F. Schinazi and G.R. Painter, Antimicrob. Agents Chemother., 45, 158 (2001).

29. D.D. Richman, Antiviral Ther., 6, 83 (2001).

30. G. Moyle, Drugs, 61, 19 (2001).

31. S.G. Deeks, J. Acq. Immune Defic. Syndr., 26 (Suppl. 1), S25 (2001).

32. R. W. Buckheit, Exp. Opin. Invest. Drugs, 10, 1423 (2001).

33. J. Balzarini, R. Esnouf and E. De Clercq, Antiretroviral Ther., 63 (2001).

34. D.W. Ludovici, B.L. De Corte, M.J. Kukla, H. Ye, C.Y. Ho, M.A. Lichtenstein, R.W. Kavash, K. Andries, M.-P. de Bethune, H. Azijn, R. Pauwels, P.J. Lewi, J. Heeres, L.M.H. Koymans, M.R. de Jonge, K.J.A. Van Aken, F.F.D. Daeyaert, K. Das, E. Arnold and P.A.J. Janssen. Bioorg. Med. Chem. Lett., 11, 2235 (2001).

35. S. Sankatsing, G. Weverling, G. van't Klooster, J. Prins and J. Lange, 9th Conf. Retroviruses Opportunistic Infections, Feb. 24-28, Seattle, WA, Abst. 5 (2002).

36. K. Das, E. Arnold, A.D. Clark Jr., D.W. Ludovici, M.J. Kukla, B. DeCorte, R.W. Kavash, K. Andries, R. Pauwels, M-P. de Béthune, G.A.E. van't Klooster, P. Lewi, S.H. Hughes and P.A.J. Janssen, 9th Conf. Retroviruses Opportunistic Infections, Feb. 24-28, Seattle, WA, Abst. 4 (2002).

37. L. Lim, Curr. Opin. Invest. Drugs, 2, 1209 (2001).

38. N. Ruiz, R. Nusrat, E. Lauenroth-Mai, D. Berger, C. Walworth, L.T. Bacheler, L. Ploughman, P. Tsang, D. Labriola, R. Echols, R. Levy and the DPC 083-203 Study Team, 9th Conf. Retroviruses Opportunistic Infections, Feb. 24-28, Seattle, WA, Abst. 6 (2002).

39. R.W. Buckheit Jr., K. Watson, V. Fliakas-Boltz, J. Russell, T.L Loftus, M.C. Osterling, J.A. Turpin, L.A. Pallansch, E.L. White, J.-W. Lee, S.-H. Lee, J.-W. Oh, H.-S. Kwon, S.-G Chung and E.-H. Cho, Antimicrob. Agents Chemother., 45, 393 (2001).

40. S. D. Young, Curr. Opin. Drug Discov. Devel., 4, 402 (2001).

41. N. Neamati, C. Marchand, H. E. Winslow and Y. Pommier, Antiretroviral Ther., 87 (2001).

42. T. Yoshinaga, A. Sato, T. Fujushita and T. Fujiwara, 9th Conf. Retroviruses Opportunistic Infections, Feb. 24-28, Seattle, WA, Abst. 8 (2002).

43. M.A. Walker, T.D. Johnson, N.A. Meanwell and J. Banville, PCT Patent Appl. WO 0196283-A2 (2001).

44. N.J. Anthony, R.P. Gomez, S.D. Young, M. Egbertson, J.S. Wai, L. Zhuang, M. Embrey, L. Tran, J. Melamed, H.M. Langford, J.P. Guare, T.E. Fisher, S.M. Jolly, M.S. Kuo, D. Perlow, J.J. Bennett and T.W. Funk, PCT Patent Appl. WO 0230930-A2 (2001).

45. J.R. Huff and J. Kahn, Adv. Protein Chem., 56, 213 (2001).

46. W. Wang and P.A. Kollman, Proc. Natl. Acad. Sci. USA, 98, 14937 (2001).

47. G. Witherell, Curr. Opin. Invest. Drugs, 2, 340 (2001).

48. P. Piliero, P. Cahn, G. Pantaleo, J. M. Gatell, K. Squires, L. Percival, I. Sanne, R. Wood, P. Phanuphak, S. Shelton, A. Lazzarin, A. Thiry, T. Kelleher, M. Giordano and S.M. Schnittman, 9th Conf. Retroviruses Opportunistic Infections, Feb. 24-28, Seattle, WA, Abst. 706T (2002).

49. A.H. Corbett and A.D.M. Kashuba, Curr. Opin. Invest. Drugs, 3, 384 (2002).

50. K. Yoshimura, R. Kato, M.F. Kavlick, A. Nguyen, V. Maroun, K. Maeda, K.A. Hussain, A.K. Ghosh, S.V. Gulnick, J.W. Erickson and H. Mitsuya, J. Virol., 76, 1349 (2002).

51. K.P. Fischer, K.S. Gutfreund and D.L. Tyrrell, Drug Resistance Updates, 4, 118 (2001).

52. W.E. Delaney, IV, S. Locarnini and T. Shaw, Antiviral Chem. Chemother., 12, 1 (2001).

53. A. Billich, Curr. Opin. Invest. Drugs, 2, 617 (2001).

54. R.J. Colonno, E.V. Genovesi, I. Medina, L. Lamb, S.K. Durham, M.-L. Huang, L. Corey, M. Littlejohn, S. Locarnini, B.C. Tennant, B. Rose and J.M. Clark, J. Infect. Dis., 184, 1236 (2001).

55. R.A. DeMan, L.M.M. Wolters, F. Nevens, D. Chua, M. Sherman, C.L. Lai, A. Gadano, Y. Lee, F. Mazzotta, N. Thomas and D. DeHertogh, Hepatology, 34, 578 (2001).

56. G. Gumina, G.-Y. Song and C.K. Chu, FEMS Microbiol. Lett., 201, 9 (2001).

57. M.L. Bryant, E.G. Bridges, L. Placidi, A. Faraj, A.-G. Loi, C. Pierra, D. Dukhan, G. Gosselin, J.-L. Imbach, B. Hernandez, A. Juodawlkis, B. Tennant, B. Korba, P. Cote, P. Marion, E. Cretton-Scott, R.F. Schinazi and J.-P. Sommadossi, Antimicrob. Agents Chemother., 45, 229 (2001).

58. D.N. Standring, E.G. Bridges, L. Placid, A. Faraj, A. Giulia Loi, C. Pierra, D. Dukhan, G. Gosselin, J.-L. Imbach, B. Hernandez, A. Juodawlkis, B. Tennant, B. Korba, P. Cote, E.

Cretton-Scott, R.F. Schinazi, M. Myers, M.L. Bryant and J.-P. Sommadossi, Antiviral Chem. Chemother., 12, 119 (2001).
59. X. Zhou, C. Lai, N. Brown, M. Myers, D. Pow and S. Lim, 41st ICAAC, Dec. 16-19, Chicago, IL, Abst. H-464 (2001).
60. M.L. Bryant, G. Gosselin and J.-L. Imbach, PCT Patent Appl. WO 0196353-A (2001).
61. F. Le Guerhier, C. Pichoud, C. Jamard, S. Guerret, M. Chevallier, S. Peyrol, O. Hantz, I. King, C. Trepo, Y.-C. Cheng and F. Zoulim, Antimicrob. Agents Chemother., 45, 1065 (2001).
62. L.M. Dunkle, S. Oshana, J. Dickson, Y.-C. Cheng, W.G. Rice, 41st ICAAC, Dec. 16-19, Chicago, IL, Abst. A-508 (2001).
63. S.F. Peek, P.J. Cote, J.R. Jacob, I.A. Toshkov, W.E. Hornbuckle, B.H. Baldwin, F.V. Wells, C.K. Chu, J.L. Gerin, B.C. Tennant and B.E. Korba, Hepatology, 33, 254 (2001).
64. R. Chin, T. Shaw, J. Torresi, V. Sozzi, C. Trautwein, T. Bock, M. Manns, H. Isom, P. Furman and S. Locarnini, Antimicrob. Agents Chemother., 45, 2495 (2001).
65. P. Marcellin, 52nd Ann. Mtg. Amer. Assoc. Study Liver Dis. (AASLD), Dallas, TX, Nov. 12. (2001).
66. Y. Benhamou, M. Bochet, V. Thibault, V. Calvez, M.H. Fievet, P. Vig, C.S. Gibbs, C. Brosgart, J. Fry, H. Namini, C. Katlama and T. Poynard, Lancet, 358, 718 (2001).
67. K.M. Walsh, T. Woodall, P. Lamy, D.G.D. Wright, S. Bloor and G.J.M. Alexander, Gut, 49, 436 (2001).
68. S. Goldmann, J. Stoltefuss, U. Niewöhner, T. Krämer, E. Graef, K. Deres, R. Masantschek, O. Weber and A. Paessens, 41st ICAAC, Dec. 16-19, Chicago, IL, Abst. F-1664 (2001).
69. A. Paessens, E. Graef, K. Deres, U. Niewoehner, J. Stoltefuss, D. Haebich, H. Ruebsamen-Waigmann, R.N.A. Masantschek, D. Koletzki, W.S. Mason, C.H. Schroeder, O. Weber and S. Goldmann, 41st ICAAC, Dec. 16-19, Chicago, IL, Abst. F-1665 (2001).
70. O. Weber, K.-H. Schlemmer, E. Hartmann, I. Hagelschuer, A. Paessens, E. Graef, K. Deres, S. Goldmann, U. Niewoehner, J. Stoltefuss, D. Haebich, H. Ruebsamen-Waigmann and S. Wohlfeil, Antiviral Res., 54, 69 (2002).
71. K. Deres, C.H. Schroeder, U. Pleiss, E. Graef, S. Goldman, U. Niewoehner, J. Stoltefuss, O. Weber, D. Koletzki, R. Masantschek, A. Paessens, D. Haebich and H. Ruebsamen-Waigmann, 41st ICAAC, Dec. 16-19, Chicago, IL, Abst. F-1667 (2001).
72. D. Bevec and S. Obert, PCT Patent Appl. WO 0200613-A (2002).
73. D. Bevec, 41st ICAAC, Dec. 16-19, Chicago, IL, Abst. LB-14 (2001).
74. G.M. Lauer and B.D. Walker, New Engl. J. Med., 345, 41 (2001).
75. L.J. Scott and C.M. Perry, Drugs, 62, 507 (2002).
76. S. Zeuzem, J.E. Heathcote, N. Martin, K. Nieforth and M. Modi, Expert Opin. Invest. Drugs, 10, 2201 (2001).
77. D.F. Mercer, D.E. Schiller, J.F. Elliott, D.N. Douglas, C. Hao, A. Rinfret, W.R. Addison, K.P. Fischer, T.A. Churchill, J.R.T. Lakey, D.L.J. Tyrrell and N.M. Kneteman, Nature Medicine, 7, 927 (2001).
78. E. Ilan, J. Arazi, O. Nussbaum, A. Zauberman, R. Eren, I. Lubin, L. Neville, O. Ben-Moshe, A. Kischitzky, A. Litchi, I. Margalit, J. Gopher, S. Mounir, W. Cai, N. Daudi, A. Eid, O. Jurim, A. Czerniak, E. Galun and S. Dagan, J. Infect. Dis., 185, 153 (2002).
79. B.W. Dymock, Emerging Drugs, 6, 13 (2001).
80. T. Wilkinson, Curr. Opin. Invest. Drugs, 2, 1516 (2001).
81. R. Bartenschlager and V. Lohmann, Antiviral Res., 52, 1 (2001).
82. R.E. Lanford and C. Bigger, Virology, 293, 1 (2002).
83. D. Thibeault, R. Maurice, L. Pilote, D. Lamarre and A. Pause, J. Biol. Chem., 276, 46678 (2001).
84. M. Pallaoro, A. Lahm, G. Biasiol, M. Brunetti, C. Nardella, L. Orsatti, F. Bonelli, S. Orru, F. Narjes and C. Steinkuhler, J. Virol., 75, 9939 (2001).
85. P.L. Darke, A.R. Jacobs and L.C. Kuo, PCT Patent Appl. WO 0116379 A1 (2001).
86. R. Perni, J. Court, E. O'Malley, G.R. Bhisetti, PCT Patent Appl. WO 0174768 A2 (2001).
87. A.K. Saksena, V.M. Girijavallabhan, S.L. Bogen, R.G. Lovey, E.E. Jao, F. Bennett, J.L. McCormick, H. Wang, R.E. Pike, Y.-T. Liu, Y. Chen, Z. Zhu, A. Arasappan, K.X. Chen, S. Venkatraman, T.N. Parekh, P.A. Pinto, B. Santhanam, F.G. Njoroge, A.K. Ganguly, H.A. Vaccaro, S.J. Kemp, O.E. Levy, M. Lim-Wilby and S.Y. Tamura, PCT Patent Appl. WO 0208187 A1 (2002).
88. A.K. Saksena, V.M. Girijavallabhan, R.G. Lovey, E.E. Jao, F. Bennett, J.L. McCormick, H. Wang, R.E. Pike, S.L. Bogen, T.-Y. Chen, Y.-T. Liu, Z. Zhu, F.G. Njoroge, A. Arasappan, T.N. Parekh, A.K. Ganguly, K.X. Chen, S. Venkatraman, H.A. Vaccaro, P.A. Pinto, B. Santhanam, W. Wu, S. Hendrata, Y. Huang, S.J. Kemp, O.E. Levy, M. Lim-Wilby and S.Y. Tamura, PCT Patent Appl. WO 0208244 A2 (2002).
89. A.K. Saksena, V.M. Girijavallabhan, R.G. Lovey, E.E. Jao, F. Bennett, J.L. McCormick, H. Wang, R.E. Pike, S.L. Bogen, Y.-T. Liu, A. Arasappan, T.N. Parekh, P. Pinto, F.G. Njoroge, A.K. Ganguly, T.K. Brunck, S.J. Kemp, O.E. Levy, M. Lim-Wilby and S.Y. Tamura, PCT Patent Appl. WO 0208256 A2 (2002).
90. W. Han, PCT Patent Appl. WO 0140262 A1 (2001).
91. R. Beevers, M.G. Carr, P.S. Jones, S. Jordan, P.B. Kay, R.C. Lazell and T.M. Raynham, Bioorg. Med. Chem. Lett., 12, 641 (2002).
92. S. Colarusso, B. Gerlach, U. Koch, E. Muraglia, I. Conte, I. Stansfield, V.G. Matassa and F. Narjes, Bioorg. Med. Chem. Lett., 12, 705 (2002).

93. S.J. Archer, D.M. Camac, Z.J. Wu, N.A. Farrow, P.J. Domaille, Z.R. Wasserman, M. Bukhtiyarova, C. Rizzo, S. Jagannathan, L. Mersinger and C.A. Kettner, Chem. Biol., <u>9</u>, 79 (2002).
94. R. Babine, S.-H. Chen, J.E. Lamar, N.J. Snyder, X.D. Sun, M.J. Tebbe, F. Victor, Q.M. Wang, Y.Y.M. Yip, I. Collado, C. Garcia-Paredes, R.S. Parker, III, L. Jin, D. Guo and J.I. Glass, PCT Patent Appl. WO 0218369 A2 (2002).
95. P. Ingallinella, D. Fattori, S. Altamura, C. Steinkühler, U. Koch, D. Cicero, R. Bazzao, R. Cortese, E. Bianchi and A. Pessi, Biochemistry, <u>41</u>, 5483 (2002).
96. M.J. Slater, D.M. Andrews, G.R. Baker, S. Bethell, S.J. Carey, H.M. Chaignot, B.E. Clarke, B.A. Coomber, M.R. Ellis, N.M. Gray, G.W. Hardy, M.R. Johnson, P.S. Jones, G. Mills, J.E. Robinson, T. Skarzynski and D.O. Somers, 15th Int. Conf. Antiviral Res., Mar. 17-21, Prague, Czech Republic, Abst. 11 (2002).
97. K.-S. Yeung, N.A. Meanwell, Z. Qiu, D. Hernandez, S. Zhang, F. McPhee, S. Weinheimer, J.M. Clark, and J.W. Janc, Bioorg. Med. Chem. Lett., <u>11</u>, 2355 (2001).
98. C. Lesburg, R. Radfar and P.C. Weber, Curr. Opin. Invest. Drugs, <u>1</u>, 296 (2000).
99. L. Goobar-Larsson, L. Wiklund and S. Schwartz, Arch. Virol., <u>146</u>, 1553 (2001).
100. W. Qin, H. Luo, T. Nomura, N. Hayashi, T. Yamashita and S. Murakami, J. Biol. Chem., <u>277</u>, 2123 (2002).
101. G.R. Reyes, Curr. Opin. Drug Discov. Devel., <u>4</u>, 651 (2001).
102. D. Maag, C. Castro, Z. Hong and C. Cameron, J. Biol. Chem., <u>276</u>, 46094 (2001).
103. R.C. Tam, K. Ramaswamy, J. Bard, B. Pai, C. Lim and D.R. Averett, Antimicrob. Agents Chemother., <u>44</u>, 1276 (2000).
104. R.C. Tam, J.Y.N. Lau and Z. Hong, Antiviral Chem. Chemother., <u>12</u>, 261 (2002).
105. H. Hashimoto, K. Mizutani and A Yoshida, Patent Appl. EP 1162196 A1 (2001).
106. P.L. Beaulieu, G. Fazal, J. Gillard, G. Kukolj and V. Austel, PCT Patent Appl. WO 0204425 A2 (2002).
107. C. Gardelli, C. Giuliano, S. Harper, U. Koch, F. Narjes, J.M. Ontoria Ontoria, M. Poma, S. Ponzi, I. Stansfield and V. Summa, PCT Patent Appl. WO 0206246 A1 (2002).
108. J.C. Jaen, D.E. Piper, J.P. Powers, N.P.C. Walker and Y. Li, PCT Patent Appl. WO 0177091 A2 (2001).
109. D. Dhanak and T. Carr, PCT Patent Appl. WO 0185720 A1 (2001).
110. D. Dhanak, A.C. Kaura and A. Shaw, PCT Patent Appl. WO 0185172 A1 (2001).
111. M. Hale, F. Maltais, C.Baker, J. Janetka, Y.C. Moon and J. Saunders, PCT Patent Appl. WO 0107027 A2 (2001).
112. H.M.A. Ismaili, Y.-X. Cheng, J.-F. Lavallee, A. Siddiqui and R. Storer, PCT Patent Appl. WO 0160315 A2 (2001).
113. J.-P. Sommadossi and P. Lacolla, PCT Patent Appl. WO 0190121 A2 (2001).
114. R. Devos, B.W. Dymock, C.J. Hobbs, W.-R. Jiang, J.A. Martin, J.H. Merrett, I. Najera, N. Shimma and T. Tsukuda, PCT Patent Appl. WO 0218404 A2 (2002).
115. C.M.T. Spahn, J.S. Kieft, R.A. Grassucci, P.A. Penczek, K. Zhou, J.A. Doudna and J. Frank, Science, <u>291</u>, 1959 (2001).
116. G. Witherell, Curr. Opin. Invest. Drugs, <u>2</u>, 1523 (2001).
117. Y. Matsuura, H. Tani, K. Suzuki, T. Kimura-Someya, R. Suzuki, H. Aizaki, K. Ishii, K. Moriishi, C.S. Robison, M.A. Whitt and T. Miyamura, Virology, <u>286</u>, 263 (2001).
118. K. Kitadokoro, D. Bordo, G. Galli, R. Petracca, F. Falugi, S. Abrignani, G. Grandi and M. Bolognesi, EMBO J., <u>20</u>, 12 (2001).
119. A. Op De Beeck, L. Cocquerel and J. Dubuisson, J. Gen. Virol., <u>82</u>, 2589 (2001).
120. A.C. Schmidt, R.B. Couch, G.J. Galasso, F.G. Hayden, J. Mills, B.R. Murphy and R.M. Chanock, Antiviral Res., <u>50</u>, 157 (2001).
121. M.C. Zambon, J.D. Stockton, J.P. Clewley and D.M. Fleming, Lancet, <u>358</u>, 1410 (2001).
122. E.A.F. Simoes, Lancet, <u>358</u>, 1382 (2001).
123. B.G. van den Hoogen, J.C. de Jong, J. Groen, T. Kuiken, R. de Groot, R.A.M. Fouchier and A.D.M.E. Osterhaus, Nature Medicine, <u>7</u>, 719 (2001).
124. R.B. Couch, New Engl. J. Med., <u>343</u>, 1778 (2000).
125. M.G. Ison and F.G. Hayden, Curr. Opin. Pharmacol., <u>1</u>, 482 (2001).
126. I.Gust, A.W. Hampson and D. Lavanchy, Rev. Med. Virol., <u>11</u>, 59 (2001).
127. S.P. Layne, T.J. Beugelsdijk, C. Kumar, N. Patel, J.K. Taubenberger, N.J. Cox, I.D. Gust, A.J. Hay, M. Tashiro and D. Lavanchy, Science, <u>293</u>, 1729 (2001).
128. M.J. Gibbs, J.S. Armstrong and A.J. Gibbs, Science, <u>293</u>, 1842 (2001).
129. A.F. Abdel-Magid, C.A. Maryanoff and S.J. Mehrman, Curr. Opin. Drug Discov. Devel., <u>4</u>, 776 (2001).
130. K.McClellan and C.M. Perry, Drugs, <u>61</u>, 2 (2001).
131. S.M. Cheer and A.J. Wagstaff, Drugs, <u>62</u>, 71 (2002).
132. P. Chand, P.L. Kotian, A. Dehgani, Y. El-Kattan, T.-H. Lin, T.L. Hutchinson, Y.S. Babu, S. Bantia, A.J. Elliott and J.A. Montgomery, J. Med. Chem., <u>44</u>, 4379 (2001).
133. R.W. Sidwell, D.F. Smee, J.H. Huffman, D.L. Barnard, K.W. Bailey, J.D. Morrey and Y.S. Babu, Antimicrob. Agents Chemother., <u>45</u>, 749 (2001).
134. D.F. Smee, J.H. Huffman, A.C. Morrison, D.L. Barnard and R.W. Sidwell, Antimicrob. Agents Chemother., <u>45</u>, 743 (2001).
135. Y.S. Babu, P. Chand, J.A. Montgomery, K.B. Watts, D.J. Hlasta and G.W. Caldwell, PCT Patent Appl. WO 0162242-A1 (2001).
136. G.T. Wang, Y. Chen, S. Wang, R. Gentles, T. Sowin, W. Kati, S. Muchmore, V. Giranda, K. Stewart, H. Sham, D. Kempf and W.G. Laver, J. Med. Chem., <u>44</u>, 1192 (2001).

137. W.M. Kati, D. Montgomery, C. Maring, V.S. Stoll, V. Giranda, X. Chen, W.G. Laver, W. Kohlbrenner and D.W. Norbeck, Antimicrob. Agents Chemother., 45, 2563 (2001).

138. C.J. Maring, V.L. Giranda, D.J. Kempf, V.S. Stoll, M. Sun, C. Zhao, Y.G. Gu, G.T. Wang, A.C. Krueger, Y. Chen, D.A. DeGoey, D.J. Grampovnik, W.M. Kati, A.L. Kennedy, Z. Lin, D.L. Madigan, S.W. Muchmore, H.L. Sham, K.D. Stewart, S. Wang and M.C. Yeung, PCT Patent Appl. WO 0128979-A2 (2001).

139. C.J. Maring, V.L. Giranda, D.J. Kempf, V.S. Stoll, M. Sun, C. Zhao, Y.G. Gu, S. Hanessian, G.T. Wang, A.C. Krueger, H.-j. Chen, Y. Chen, D.A. Degoey, W.J. Flosi, D.J. Grampovnik, W.M. Kati, A.L. Kennedy, L.L. Klein, Z. Lin, D.L. Madigan, K.F. McDaniel, S.W. Muchmore, H.L. Sham, K.D. Stewart, N.P. Tu, F.L. Wagenaar, S. Wang, P.E. Wiedeman, Y. Xu, M.C. Yeung, M. Bayrakdarian and X. Luo, PCT Patent Appl. WO 0128996-A2 (2001).

140. C.J. Maring, D.A. DeGoey, W.J. Flosi, V.L. Giranda, D.J. Kempf, A.L. Kennedy, L.L. Klein, A.C. Krueger, K.F. McDaniel, V.S. Stoll, M. Sun and C. Zhao, PCT Patent Appl. WO 0129050-A2 (2001).

141. S.N. Raja, K. St. George, L. Fan, G. Nequist, T. Reisch, C. Maring, K. McDaniel, D. GeGoey and J. Darbyshire, 41st ICAAC, Dec. 16-19, Chicago, IL, Abst. F-1683 (2001).

142. W.M. Kati, D. Montgomery, R. Carrick, L. Gubareva, C. Maring, K. McDaniel, K. Steffy, A. Molla, F. Hayden, D. Kempf and W. Kohlbrenner, Antimicrob. Agents Chemother., 46, 1014 (2002).

143. D.A. DeGoey, H.-J. Chen, J. Flosi, D.J. Grampovnik, C.M. Yeung, L.L. Klein and D.J. Kempf, J. Org. Chem., ASAP (2002).

144. S. Hanessian, M. Bayrakdarian and X. Luo, J. Am. Chem. Soc., 124, 4716 (2002).

145. S.P. Tucker, V.T.T. Nguyen, B. Jin, D.B. McConnell, K.G. Watson, R. Cameron, S. Hamilton, S. Macdonald, R. Fenton, P.A. Reece and W.-Y. Wu, 15th Int. Conf. Antiviral Res., Mar. 17-21, Prague, Czech Republic, Abst. 99 (2002).

146. S. Shigeta, Antiviral Chem. Chemother., 12 (Suppl. 1) 179 (2001).

147. M.S. Deshpande, J. Wei, G. Luo, C. Cianci, S. Danetz, A. Torri, L. Tiley, M. Krystal, K.-L. Yu, S. Huang, Q. Gao and N.A. Meanwell, Bioorg. Med. Chem. Lett., 11, 2393 (2001).

148. Y. Furuta, K. Takahashi, Y. Fukuda, M. Kuno, T. Kamiyama, K. Kozaki, N. Nomura, H. Egawa, S. Minami, Y. Watanabe, H. Narita and K. Shiraki, Antimicrob. Agents Chemother. 46, 977 (2002).

149. R.B. Turner, Antiviral Res., 49, 1 (2001).

150. M.A. McKinlay, Curr. Opin. Pharmacol., 1, 477 (2001).

151. J.R. Romero, Exp. Opin. Invest. Drugs, 10, 369 (2001).

152. K.G. Wilson, W.Y. Wu, G.Y. Krippner, D.B. McConnell, B. Jin, P.C. Stanislawski, R.N. Brown, D.K. Chalmers, S.P. Tucker, R. Cameron-Smith, S. Hamilton, A. Luttick, J. Ryan and P.A. Reece, Antiviral Res., 50, A126 (2001).

153. F.G. Hayden, C. Crump, P.A. Reece, K. Watson, J. Ryan, R. Cameron-Smith and S.P. Tucker, Antiviral Res., 50, A127 (2001).

154. J. Ryan, R.W. Sidwell, D. Barnard, S.P. Tucker and P.A. Reece, Antiviral Res., 50, A128 (2001).

155. W-Y. Wu, K. Watson, D. McConnell, B. Gin and G. Krippner, PCT Patent Appl. WO 0078746-A1 (2000).

156. S.K. Tsang, J. Cheh, L. Isaacs, D. Joseph-McCarthy, S-K. Choi, D.C. Pevear, G.M. Whitesides and J.M. Hogle, Chem. Biol., 8, 33 (2001).

157. P.S. Dragovich, Exp. Opin. Ther. Patents, 11, 177 (2001).

158. P-H. Hsyu, Y.K. Pithavala, M. Gersten, C.A. Penning and B.M. Kerr, Antimicrob. Agents Chemother., 46, 392 (2002).

159. P.S. Dragovich, T.J. Prins, R. Zhou, T.O. Johnson, E.L. Brown, F.C. Maldonado, S.A. Fuhrman, L.S. Zalman, A.K. Patick, D.A. Matthews, X. Hou, J.W. Meador, III, R.A. Ferre and S.T. Worland, Bioorg. Med. Chem. Lett., 12, 733 (2002).

160. S.E. Webber, J.T. Marakovits, P.S. Dragovich, T.J. Prins, R. Zhou, S.A. Fuhrman, A.K. Patick, D.A. Matthews, C.A. Lee, B. Srinivasan, T. Moran, C.E. Ford, M.A. Brothers, J.E.V. Harr, J.W. Meador, III, R.A. Ferre and S.T. Worland, Bioorg. Med. Chem. Lett., 11, 2683 (2001).

161. P.S. Dragovich, T.J. Prins, R. Zhou, E.L. Brown, F.C. Maldonado, S.A. Fuhrman, L.S. Zalman, T. Tuntland, C.A. Lee, A.K. Patick, D.A. Matthews, T.F. Hendrickson, M.B. Kosa, B. Liu, M.R. Batugo, J.-P.R. Gleeson, S.K. Sakata, L. Chen, M.C. Guzman, J.W. Meador, III, R.A. Ferre and S.T. Worland, J. Med. Chem., 45, 1607 (2002).

162. P.A. Cane, Rev. Med. Virol., 11, 103 (2001).

163. C. Breese-Hall, New Engl. J. Med., 344, 1917 (2001).

164. G. Prince, Exp. Opin. Invest. Drugs, 10, 297 (2001).

165. L. Gonzalez-Reyes, M. Begona Ruiz-Arguello, B. Garcia-Barreno, L. Calder, J.A. Lopez, J.P. Alba, J.J. Skehel, D.C. Wiley and J.A. Melero, Proc. Natl. Acad. Sci. USA, 58, 9859 (2001).

166. L. Chen, J.J. Gorman, J. McKimm-Breschkin, L.J. Lawrence, P.A. Tulloch, B.J. Smith, P.M. Colman and M.C. Lawrence, Structure, 9, 255 (2001).

167. A.A. Nikitenko, Y.E. Raifeld and T.Z. Wang, Bioorg. Med. Chem. Lett., 11, 1041 (2001).

168. V. Razinkov, A. Gazumyan, A. Nikitenko, G. Ellestead and G. Krishnamurthy, Chem. Biol., 8, 645 (2001).

169. C.C. Huntley, W.J. Weiss, A. Gazumyan, A. Buklan, B. Field, W. Hu, T.R. Jones, T. Murphy, A.A. Nikitenko, B. O'Hara, G. Prince, S. Quartuccio, Y.E. Raifeld, P. Wyde and J.F. O'Connell, Antimicrob. Agents Chemother., 46, 841 (2002).

170. K. Andries, M. Moeremans, T. Gevers, R. Willebrords, J. Lacrampe and F. Janssens, 40th ICAAC, Sept. 17-20, Toronto, CA, Abst. H-1160 (2000).
171. D. Holzman, ASM News, 68, 154 (2002).
172. K. Sudo, K. Konno, W. Watanabe, S. Shigeta and T. Yokota, Microbiol. Immunol., 45, 531 (2001).
173. K.-L. Yu, R.L. Civiello, M. R. Krystal, K.F. Kadow and N.A. Meanwell, PCT Patent Appl. WO 0004900 A1 (2000).
174. K.-L. Yu, R.L. Civiello, K.D. Combrink, H.B. Gulgeze, N. Sin, X. Wang, N.A Meanwell and B.L. Venables, PCT Patent Appl. WO 0195910 A1 (2001).
175. K.-L. Yu, R.L. Civiello, K.D. Combrink, H.B. Gulgeze, B.C. Pearce, X. Wang, N.A Meanwell and Y. Zhang, PCT Patent Appl. WO 0226228 A1 (2002).

Chapter 15. Aerosol Delivery of Antibiotics

Peter B. Challoner
Chiron Corporation, Emeryville CA

Introduction – In theory, using inhaled antibiotics for the treatment of pulmonary infections should have several advantages over the same compounds presented by more conventional systemic routes. For endobronchial infections in particular, the deposition of inhaled aerosol particles on the surface of the airways can result in peak concentrations at the infection site that are several orders of magnitude higher than can be achieved by oral or intravenous delivery (1,2). The topical administration of an inhaled aerosol to the airways can generate these high pulmonary concentrations while simultaneously minimizing both peak serum levels and the exposure of non-involved tissues to the drug (3). Thus, inhalation can provide a way to circumvent the dose-limiting systemic toxicity of an antibiotic. Finally, for compounds with low oral availability, inhalation is a cost-effective alternative to intravenous delivery for outpatient use.

Historically, the clinical and commercial success of inhalation antibiotics has been inextricably linked to, and limited by, the capabilities of the aerosol delivery technologies that were available at the time. Several novel methods for generating therapeutic aerosols are now under development, and are beginning to be applied to the problem of delivering antibiotics. The first section of this review provides an overview of aerosol delivery technologies in relation to the particular problems posed by inhalation antibiotics. The second section summarizes clinical results obtained from the use of inhalation antibiotics, organized by indication.

AEROSOL DELIVERY TECHNOLOGIES FOR INHALATION ANTIBIOTICS

General Aerosol Principles - Efficient and accurate dosing by inhalation presents unique challenges that are not shared by other routes of administration. The relationship between the particle size distribution of an aerosol and pulmonary deposition is well understood, and was reviewed recently (4). Particles above five microns in diameter are likely to be deposited by impaction in the oropharynx before they reach the lung, while particles below one micron in diameter are to small to be deposited in the airways by either impaction or sedimentation, and so are lost by exhalation. The fundamental problem faced by all aerosol delivery devices then, is how to produce particles in this very narrow size range at a usable rate of output. The newer technologies described below all represent attempts to improve on these two parameters. A second factor that affects deposition is the velocity of the inhaled aerosol particle. For example, particles in the one to five micron "respirable range" will still deposit by impaction in the mouth and throat if they are ejected from a pressurized nozzle at a high velocity. Conversely, executing a "breath-hold" maneuver following the inhalation of an aerosol will give suspended particles time to deposit by sedimentation.

Besides particle size distribution and velocity, a number of patient-related factors, such as breathing pattern, inspiratory airflow rate, respiration rate, airway anatomy, and underlying pulmonary disease status (for example, obstructive airway disease) also affect aerosol deposition. At least some of the variability in dosing caused by these patient-related factors may be mitigated by restricting aerosolization to the inhalation phase of the breath cycle. Such "breath-actuated" devices either use airflow data to turn aerosolization on and off during each cycle of normal tidal breathing, or are designed to deliver an aerosol bolus during a single inhalation, followed by a breath-hold maneuver.

TABLE 1. – Major Classes of Aerosol Delivery Technologies

Device Class	Aerosol Generation Methods	Supported Formulations	Considerations for the Delivery of Inhalation Antibiotics
Metered-Dose Inhalers (MDIs)	CFC propellant-driven spray from a pressurized canister	Liquids and Suspensions	Not generally used for inhalation antibiotics: payload capability, accuracy, and reproducibility are not adequate.
Dry Powder Inhalers (DPIs)	Passive devices (powder dispersed by inspiratory airflow)	Milled powders, blended with coarse carrier particles	
	Active devices (powder dispersed by energy supplied from the device)	Spray-dried, amorphous glassy particles	New particle engineering technologies under development that yield easily dispersible powders with high drug loading. Precise control of particle size distribution. Potential to deliver an inhalation antibiotic from a simple passive dry powder inhaler in a few breaths.
		Spray-dried, low density, porous particles	
		Supercritical fluid condensates	
Nebulizers	Ultrasonic	Liquids and Suspensions	Nebulizers represent a fully developed aerosol delivery technology, and are by far the most commonly used devices for inhalation antibiotics. The better breath-enhanced models can reliably deliver an inhalation antibiotic in less than 20 minutes. Disadvantages include long delivery times, low efficiencies, broad particle size distributions, and a lack of portability.
	Air-jet (continuous output, breath-enhanced, and breath-actuated models)		
Soft-Mist Inhalers	Vibrating porous membrane	Liquids and Suspensions	Several different methods for high output liquid aerosol generation under development. More precise control of particle size than nebulizers. Potential to improve on the delivery time, efficiency, and portability of nebulizers. Most soft-mist inhalers utilize some form of breath-actuation.
	Pressure-driven porous membrane		
	Electrohydrodynamic		

MDIs and DPIs - An overriding consideration for inhalation antibiotics is the need to deliver a deposited aerosol dose that results in concentrations that are well above the MIC_{90} of the target organism. This precludes the use of pressurized metered-dose inhalers (MDIs) and "first generation" milled powders that are mixed with coarse, lactose-based carrier particles and delivered by passive dry powder inhalers (DPIs). The pulmonary dose that these popular inhalers can support is typically several hundred micrograms: this is sufficient for the potent selective □2-adrenoceptor agonists and corticosteroids that are used to treat asthma, but is far below the tens of milligrams required for inhalation antibiotics. Furthermore, the accuracy and reproducibility of MDIs and DPIs are not adequate for inhalation antibiotics, where, unlike most asthma drugs, the patient does not receive immediate feedback to indicate that an efficacious dose was inhaled. The relative merits of pMDIs and DPIs have been reviewed recently (5).

Advances in Dry Powder Particle Engineering – Reducing the particle size distribution to the one to five micron range by milling tends to create substantial cohesive and electrostatic forces that make it difficult to disperse the particles again into the patient's inspired airflow. Mixing with coarse, inert carrier particles can help to disperse respirable particles containing the active compound, but the mass of the drug in the aerosol bolus then becomes to low for inhalation antibiotics. Several recent advances in powder processing technology have the potential to produce respirable particles with both high drug loading and low cohesive properties (6). For example, spray drying a solution containing an active compound and an excipient such as a simple carbohydrate or amino acid yields respirable particles that exist in a stable amorphous glass state (7,8). Spray drying more complex emulsions containing excipients such as poly(lactic-acid-co-glycolic acid) produces "large porous particles" (9,10). These particles may be as large as 30 microns, but their low density gives them the aerodynamic characteristics of particles in the respirable range. The large geometric diameter reduces the number of interparticle contacts, and therefore the amount of energy required to disperse the powder. Smaller hollow porous particles have also been produced by spray drying emulsions containing drug, phosphatidylcholine derivatives, and a fluorocarbon-based blowing agent (11). Finally, an alternative to spray drying that can generate dispersible powders with high drug loading is supercritical fluid condensation (12,13). These new particle engineering technologies have evolved to a point where a simple passive DPI could be used to deliver the pulmonary dose of an inhalation antibiotic in a few breaths.

Air-Jet and Ultrasonic Nebulizers – The term "nebulizer" actually describes two very different methods for aerosol generation. In air-jet nebulizers, liquid is sheared into a polydisperse aerosol by a high velocity jet of compressed air. In ultrasonic nebulizers, a piezoelectric element vibrating at an ultrasonic frequency creates standing waves in the liquid that release aerosol particles from the surface. Both types of devices require baffle systems to condense and recycle large particles into a reservoir. The common feature of air-jet and ultrasonic nebulizers that led to their widespread use for the delivery of inhalation antibiotics is quite simply a large liquid reservoir that can accommodate an aqueous solution containing several hundred milligrams of an antibiotic. Until recently, nebulizers were the only aerosol delivery devices with an adequate payload capacity for inhalation antibiotics.

Air-jet nebulizers are more commonly used for inhalation antibiotics than ultrasonic nebulizers because they are less expensive, simpler, easier to clean and decontaminate, and do not heat the drug solution. However, standard air-jet nebulizers generate aerosols continuously at a low output rate. Besides wasting the aerosol that is emitted during exhalation, these devices also retain a substantial residual volume at the end of a treatment that cannot be aerosolized. Thus, the pulmonary deposition obtained from basic air-jet nebulizers is often less than ten percent of the nominal dose, and treatment times can exceed thirty minutes. The performance of continuous air-jet nebulizers has been improved by the development of breath-assisted, open vent designs that boost aerosol output during inhalation (14). In addition, two breath-actuated nebulizers are now available that completely eliminate aerosolization during exhalation (15-17). The technical aspects of nebulization have been reviewed (18).

Soft-Mist Inhalers – The term "soft-mist" refers to several new technologies for generating low velocity liquid aerosols at higher output rates than air-jet nebulizers can achieve. Three types of soft-mist inhalers that may be useful for delivering inhalation antibiotics are currently under development. Vibrating porous membrane aerosol generators have a piezoelectric element bonded to a membrane containing an array of precisely sized holes (19-22). When liquid is applied to one side of the membrane, the vibration induced by the piezoelectric element creates a pumping action that forces liquid through the holes to produce an aerosol. A related approach also has an array of holes in a membrane, but instead uses pressure to drive the liquid from a unit-dose blister through the holes (23,24). The quality of the aerosol produced by either method depends on controlling hole geometry, driving force, and solution viscosity. In the third class of soft-mist technology, electrohydrodynamic aerosolization, a strong electric field is applied to a liquid meniscus to produce a cone-shaped tip that sprays a nearly monodisperse population of particles (25).

Portable, handheld, battery-operated inhalers are under development for all three of the soft-mist technologies. Most of the soft-mist inhaler models can aerosolize the volumes required for inhalation antibiotics faster than air-jet nebulizers, with additional efficiencies gained by having negligible residual volumes and some form of breath-actuation.

APPLICATIONS FOR INHALATION ANTIBIOTICS

Cystic Fibrosis – The primary source of morbidity and mortality in cystic fibrosis (CF) patients is a chronic endobronchial infection caused by *Pseudomonas aeruginosa*. Colonization by this pathogen initiates a cycle of inflammation and airway destruction that ultimately leads to a severe reduction in lung function. *P. aeruginosa* can grow to densities in excess of 10^9 per milliliter in CF sputum, an environment that is both antagonistic to, and difficult to access by, systemic antibiotics. The need for effective chronic suppressive therapy to manage *P. aeruginosa* in CF led to extensive experimentation with inhalation antibiotics in the 1980s. A number of the available anti-pseudomonal agents were tested, including carbenicillin (26), ceftazidime (27), cephaloridine (28), colistin (29,30), gentamycin (31,32), and tobramycin (33,34). These early studies employed several standard air-jet nebulizer models with different performance levels, charged with extemporaneous solutions compounded from the intravenous formulations of the antibiotics. These factors, coupled with small sample sizes, undoubtedly contributed to some variability in the outcomes of the clinical trials. Nevertheless, meta-analysis of these results indicates that, in general, the inhalation antibiotics improved lung function and reduced the frequency of hospitalizations required to treat acute exacerbations (35,36).

Aerosolized tobramycin, and to lesser extent colistin, are currently used for the chronic suppression of *P. aeruginosa* in CF (37,38). Aerosolized colistin is also one component of an aggressive chemoprophylaxis regimen designed to prevent or delay *P. aeruginosa* colonization in pediatric patients (39,40). However, colistin inhalation has been shown to cause bronchoconstriction in pediatric patients and adults (41-43). In contrast, aerosolized tobramycin is well tolerated, and the delivery system, pulmonary dose, and treatment regimen for aerosolized tobramycin have all undergone extensive refinements (44). Two studies tested the safety and efficacy of a 600 milligram dose of preservative-free tobramycin, administered three times daily for four or twelve weeks with an ultrasonic nebulizer (45,46). Although the increase in lung function after four weeks was greater than previously observed with 80 milligram doses, the time and effort required for ultrasonic nebulization three times a day was not practical for outpatient use. Ultimately, "tobramycin solution for inhalation" (TSI) was developed specifically for aerosolization; with osmolarity, pH, and chloride ion concentration all adjusted for optimal lung tolerance (47,48). The safety and efficacy of this 300 milligram dose, delivered twice daily by an efficient breath-enhanced nebulizer, was tested over three 28 day treatment periods, separated by 28 day off-drug intervals (3). On average, the forced expiratory volume in one second (FEV_1) increased by ten percent in the treatment group at week twenty, while FEV_1 declined by two percent in the control group. TSI also decreased the likelihood of hospitalization by 23 percent. While this study only evaluated inhaled tobramycin in patients aged six and above, the 300 milligram aerosol dose was recently shown to produce similar serum and endobronchial concentrations in pediatric patients between six months and six years of age (49).

Since inhaled tobramycin for CF is the best understood use of an aerosolized antibiotic, it is the logical first choice for assessing the value of emerging technologies for the delivery of inhalation antibiotics. TSI has been delivered by a soft-mist inhaler with a vibrating porous membrane aerosol generator to both healthy volunteers (50,51) and CF patients (52). The soft-mist inhaler required about one third of the nebulizer fill volume to deliver an equivalent pulmonary dose, and the inhaler treatment time was less than half of the nebulizer treatment time.

Increasingly sophisticated dry powder formulations and DPIs have been applied to the problem of delivering aminoglycoside aerosols to CF patients. In several small studies, micronized gentamycin particles were mixed with coarse lactose carrier particulates, and delivered by either a simple unit-dose DPI (53,54), or a multi-dose reservoir DPI (55). The pulmonary levels produced by these first generation dry powders and inhalers were more than ten-fold below what is possible with a concentrated tobramycin solution delivered by an efficient breath-enhanced nebulizer. More recently, emulsion-based spray drying has been used to generate hollow porous particles loaded with as much as 90 percent tobramycin sulfate (56). These powders, administered by a simple passive DPI, can equal the pulmonary tobramycin levels produced by TSI delivered with a breath-enhanced nebulizer (57). The dry powders were 3.5-fold more efficient, and less variable, than the nebulizer.

Bronchiectasis – The initiating events for bronchiectasis can be as diverse as an acute viral or bacterial infection that causes lung damage, or the aspiration of a foreign body. The primary injury ultimately leads to a chronic infection, and to a cycle of inflammation and airway destruction that is similar to CF, but usually more localized. Like CF, there has been extensive experimentation by bronchiectatics with the inhalation of intravenous formulations of antibiotics, delivered by standard air-jet nebulizers. Aerosolized amoxycillin reduced the secretion of purulent sputum in patients that had previously failed to improve with oral amoxycillin therapy (58,59). Similarly, aerosolized gentamycin decreased mucus hypersecretion, with a concomitant reduction in sputum myeloperoxidase (MPO) activity that is consistent with an effect on MPO-mediated airway destruction (60). The long-term inhalation of ceftazidime and tobramycin did not result in a change in FEV_1, but the frequency and duration of hospitalization both decreased (61).

The positive results obtained with several different inhalation antibiotics and nebulizers prompted a larger study of the safety and efficacy of TSI for the treatment of bronchiectasis (62). The microbiological response to TSI in this population was impressive, with an average 4.5 $\log_{10}$ reduction in *P. aeruginosa* colony-forming units per gram of sputum after four weeks of therapy. *P. aeruginosa* was completely eradicated in a third of the patients at two weeks following the discontinuation of TSI. In general, bronchiectatics did not tolerate TSI as well as CF patients had in previous studies (3). The incidence of adverse events such as cough, dyspnea, chest pain, and wheezing all increased with TSI administration, indicating that some adjustment in the quantity or form of the aerosol dose may be necessary.

Mechanical Ventilation – The decision to place a critically ill patient on mechanical ventilation is often confounded by the associated risks of tracheobronchitis and pneumonia. It has been particularly difficult to assess the efficacy of inhalation antibiotics for the chemoprophylaxis and treatment of these indications, because of additional complications for aerosol delivery that arise from the mechanics of ventilation. First, a variable amount of the respirable aerosol that is introduced into a ventilator circuit will deposit in the inspiratory limb, Y piece, and endotracheal tube before reaching the airways. Second, the overall efficiency of pulmonary deposition is generally lower during mechanical ventilation than during spontaneous breathing. Finally, the underlying pulmonary disease can cause occluded regions that are poorly ventilated. In early studies, these obstacles were reflected in pulmonary doses equal to one to three percent of the nebulizer charge (63,64), and stimulated a concerted effort to identify and optimize the variables that affect the efficiency of aerosol delivery during mechanical ventilation (64-68). This work ultimately led to recommendations for ventilator duty cycle settings, nebulizer location in the ventilator circuit, nebulizer fill volume, ventilator cycle-dependent actuation, the use of an aerosol holding chamber, and stopping humidification prior to aerosolization (69). If all of these factors are attended to, pulmonary deposition may exceed 20 percent of the nebulizer charge (70). Not surprisingly, given the differences in pulmonary deposition and the complexity of the patient population, the results obtained from several small clinical studies on the use of inhalation antibiotics during mechanical ventilation have been mixed (71). It may now be possible to use surrogate markers, such as increases in the volume of respiratory secretions and the levels of inflammatory cytokines, to define high-risk candidates for inhalation antibiotics to prevent progression to ventilator-associated pneumonia (72).

Pneumocystis carinii Pneumonia (PCP) – Patients suffering from acquired immunodeficiency syndrome (AIDS) are highly susceptible to pneumonias caused by the protozoan *Pneumocystis carinii*. Trimethoprim-sulfamethoxazole (TMP-SMX) and pentamidine are both active against *P. carinii*, but adverse events associated with the systemic delivery of those compounds can be difficult to manage in severely immunocompromised individuals. The need for well-tolerated and effective treatment and prophylaxis of PCP became acute prior to the availability of antiretroviral agents, and led to several studies exploring the use of inhaled pentamidine.

Given the inhalation technologies that were available in the late 1980s, the delivery of pentamidine aerosols presented several challenges. First, a filter to capture exhaled aerosol was necessary to protect health care workers and family members from environmental exposure to pentamidine. Second, unlike the bacterial pathogens of pneumonia, *P. carinii* colonizes alveolar spaces without invading interstitial tissue. While this places the organism at a site that is accessible to an inhaled aerosol, particles with a mass median aerodynamic diameter (MMAD) of approximately one micron are required to maximize deposition in the alveoli and peripheral airways, and to minimize loss from deposition in the conducting airways, tracheobronchial region, and oropharynx. The Respirgard II was the only air-jet nebulizer available at the time capable of generating particles in this size range, but it required 40 minutes to aerosolize six milliliters of a pentamidine isethionate solution. Fortunately, pentamidine that deposits in the alveoli is not well absorbed to the pulmonary circulation, and is located below the portion of the airways that are cleared by the mucociliary escalator. Consequently, aerosolized pentamidine is typically administered only once per month.

Aerosolized pentamidine delivered by the Respirgard II proved to be effective for treating the first episode of PCP in AIDS patients (73), but was subsequently found to be inferior to both intravenous pentamidine and TMP-SMX against more severe forms of PCP (74,75). Although a subset analysis showed that outcomes against mild PCP were similar for aerosolized pentamidine and TMP-SMX, it is not included in current PCP treatment regimens (76). Aerosolized pentamidine has also been used successfully for PCP chemoprophylaxis (77). Again, TMP-SMX was found to be more effective and less expensive (78), but aerosolized pentamidine is still recommended for patients that cannot tolerate the first line therapies (76).

Pulmonary Aspergillosis – Invasive aspergillosis is a frequent cause of mortality in neutropenic patients undergoing transplantation or chemotherapy. Intravenous amphotericin B is commonly used for the chemoprophylaxis and treatment of these infections, but its use is often limited by systemic toxicity. Chemoprophylaxis with aerosolized amphotericin B has shown promise in oncology patients, bone marrow transplant recipients, and heart-lung transplant recipients (79-82). However, in the largest study of this type, there was no significant difference in the incidence of invasive aspergillosis between the patients who received aerosolized amphotericin B and those who did not (83). Consequently, current guidelines for the use of antimicrobial agents in neutropenic patients do not recommend aerosolized amphotericin B (84). The devices that have been used for aerosolizing amphotericin B were standard air-jet nebulizers, charged with simple solutions of amphotericin B formulated for intravenous delivery. Given the continuing high incidence of invasive aspergillosis in neutropenic patients, it would be worthwhile to evaluate more efficient devices, liposomal formulations that may increase lung residence time, or even one of the newer antifungal agents, for this indication.

Respiratory Syncytial Virus Infection – Respiratory syncytial virus (RSV) is a common viral pathogen that causes serious bronchiolitis and pneumonia. It is particularly a concern for pediatric patients with complications of prematurity or impaired immune function. The use of aerosolized ribovirin, an antiviral nucleoside, against RSV is the subject of a recent review and consensus guideline document (85,86). Aerosolized ribovirin treatment is both complex and inefficient: particles generated by a small particle aerosol generator (SPAG) must be inhaled continuously from a face mask or humidification tent. Although the clinical value of aerosolized ribovirin has been questioned, it is effective for preventing RSV-infected bone marrow transplant patients from progressing to the lower respiratory tract infections that are associated with serious morbidity and mortality (87-89).

A new chemical entity, VP14637, has demonstrated potency against RSV clinical isolates at concentrations between 0.1 and 80 nanomoles (90,91). It does not effect related viral species, and specifically inhibits RSV replication by interfering with functions associated with the viral F (fusion) protein. VP14637 is currently being developed for inhalation with a device that uses electrohydrodynamic soft-mist aerosol generation technology (25).

Influenza – Zanamivir inhibits the activity of influenza virus type A and B neuraminidase glycoprotein, an enzyme that promotes viral budding by degrading sialic acid-containing receptors on the surfaces of both cell and virion membranes. Since Zanamivir has poor oral bioavailability, it was developed for inhalation. The use of inhaled Zanamivir in phase III clinical trials shortened the average duration of symptoms by one to 2.5 days, with the largest effect seen in patients defined as high risk (92,93). Efficacy and tolerability data for inhaled Zanamivir was reviewed recently (94).

Zanamivir is potent enough to be formulated as a first generation milled powder mixed with lactose-based carrier particles, and is delivered by the Diskhaler, a multi-unit dose DPI. Unlike pMDIs, the Diskhaler does not require the patient to coordinate the actuation of the device with inhalation. Unfortunately, many elderly patients, an at-risk population that would particularly benefit from inhaled Zanamivir, are unable to master the complexity of loading and priming the Diskhaler, even following a tutoring session on the operation of the device (95).

CONCLUSIONS

The utility of the topical application of an aerosolized antibiotic directly to the site of a pulmonary infection has now been demonstrated for a number of compounds and indications, but most extensively for aminoglycosides in CF. In parallel, several aerosol delivery technologies have advanced to a point where substantial improvements in treatment times, convenience, compliance, and dosing accuracy for inhalation antibiotics are possible.

References

1. J. Eisenberg, M. Pepe, J. Williams-Warren, M. Vasiliev, A.B. Montgomery, A.L. Smith, and B.W. Ramsey, Chest, 111, 955 (1997).
2. P.M. Mendelman, A.L. Smith, J. Levy, A. Weber, B. Ramsey, and R.L. Davis, Am. Rev. Respir. Dis., 132, 761 (1985).
3. B.W. Ramsey, M.S. Pepe, J.M. Quan, K.L. Otto, A.B. Montgomery, J. Williams-Warren, K.M. Vasiljev, D. Borowitz, C.M. Bowman, B.C. Marshall, S. Marshall, and A.L. Smith, N. Engl. J. Med., 340, 23 (1999).
4. C. O'Callaghan and P.W. Barry, Thorax, 52 Suppl 2, S31 (1997).
5. D. Ganderton, J. Aerosol. Med., 12 Suppl 1, S3 (1999).
6. J. Peart and M.J. Clarke, American Pharmaceutical Review, 4, 37 (2001).
7. M. Eljamal, J.S. Patton, L.C. Foster, and R.M. Platz, U.S. Patent 6 187 344 (2001).
8. L.C. Foster, M. Kuo, and S.R. Billingsley, U.S. Patent 6 258 341 (2001).
9. D.A. Edwards, J. Hanes, G. Caponetti, J. Hrkach, A. Ben-Jebria, M.L. Eskew, J. Mintzes, D. Deaver, N. Lotan, and R. Langer, Science, 276, 1868 (1997).
10. D.A. Edwards, G. Caponetti, J.S. Hrkach, N. Lotan, J. Hanes, R.S. Langer, and A. Ben-Jebria, U.S. Patent 6 254 854 (2001).
11. T.E. Tarara, J.G. Weers, and L.A. Dellamary, Respiratory Drug Delivery VII, Tarpon Springs, USA (2000).
12. M. Hanna and P. York, U.S. Patent 6 063 138 (2000).
13. R.E. Sievers and U. Karst, U.S. Patent 6 095 134 (2000).
14. S.L. Ho, W.T. Kwong, L. O'Drowsky, and A.L. Coates, J. Aerosol. Med., 14, 467 (2001).
15. Y.M. Christenson and C.J. Flanigan, 47th International Respiratory Congress Meeting, San Antonio, USA (2001).
16. D.E. Geller and B. Kesser, American Thoracic Society 97th International Conference, San Francisco, USA (2001).
17. J. Denyer, D. Pavia, and B. Zierenberg, Eur. Respir. Rev., 10, 187 (2000).
18. P.P. Le Brun, A.H. de Boer, H.G. Heijerman, and H.W. Frijlink, Pharm. World Sci., 22, 75 (2000).
19. Y. Ivri and C.H. Wu, U.S. Patent 6 085 740 (2000).
20. Y. Ivri and C.H. Wu, U.S. Patent 5 586 550 (1996).
21. M. Knoch, International Society for Aerosols in Medicine 13th International Congress, Interlaken, Switzerland (2001).
22. L. De Young, F. Chambers, S. Narayan, and C.H. Wu, Respiratory Drug Delivery VI, Hilton Head, USA (1998).
23. J. Schuster, R. Rubsamen, P. Lloyd, and J. Lloyd, Pharm. Res., 14, 354 (1997).
24. L.J. Lloyd, P.M. Lloyd, R.M. Rubsamen, and J.A. Schuster, U.S. Patent 6 014 969 (2000).
25. W.C. Zimlich, J.Y. Ding, D.R. Busick, R.R. Moutvic, M.E. Placke, P.H. Hirst, G.R. Pitcairn, S. Malik, S.P. Newman, F. Macintyre, P.R. Miller, M.T. Shepherd, and T.M. Lukas, Respiratory Drug Delivery VII, Tarpon Springs, USA (2000).
26. M.E. Hodson, A.R. Penketh, and J.C. Batten, Lancet, 2, 1137 (1981).
27. R.J. Stead, M.E. Hodson, and J.C. Batten, Br. J. Dis. Chest, 81, 272 (1987).
28. G. Nolan, P. Moivor, H. Levison, P.C. Fleming, M. Corey, and R. Gold, J. Pediatr., 101, 626 (1982).
29. J.M. Littlewood, M.G. Miller, A.T. Ghoneim, and C.H. Ramsden, Lancet, 1, 865 (1985).

30. T. Jensen, S.S. Pedersen, S. Garne, C. Heilmann, N. Hoiby, and C. Koch, J Antimicrob. Chemother., 19, 831 (1987).

31. P. Kun, L.I. Landau, and P.D. Phelan, Aust. Paediatr. J., 20, 43 (1984).

32. J.S. Ilowite, J.D. Gorvoy, and G.C. Smaldone, Am. Rev. Respir. Dis., 136, 1445 (1987).

33. G. Steinkamp, B. Tummler, M. Gappa, A. Albus, J. Potel, G. Doring, and H. von der Hardt, Pediatr. Pulmonol., 6, 91 (1989).

34. I.B. MacLusky, R. Gold, M. Corey, and H. Levison, Pediatr. Pulmonol., 7, 42 (1989).

35. D.J. Touw, R.W. Brimicombe, M.E. Hodson, H.G. Heijerman, and W. Bakker, Eur. Respir. J., 8, 1594 (1995).

36. S. Mukhopadhyay, M. Singh, J.I. Cater, S. Ogston, M. Franklin, and R.E. Olver, Thorax, 51, 364 (1996).

37. G. Doring, S.P. Conway, H.G. Heijerman, M.E. Hodson, N. Hoiby, A. Smyth, and D.J. Touw, Eur. Respir. J., 16, 749 (2000).

38. P.W. Campbell, 3rd and L. Saiman, Chest, 116, 775 (1999).

39. B. Frederiksen, C. Koch, and N. Hoiby, Pediatr. Pulmonol., 28, 159 (1999).

40. B. Frederiksen, C. Koch, and N. Hoiby, Pediatr. Pulmonol., 23, 330 (1997).

41. S. Cunningham, A. Prasad, L. Collyer, S. Carr, I.B. Lynn, and C. Wallis, Arch. Dis. Child., 84, 432 (2001).

42. J. Maddison, M. Dodd, and A.K. Webb, Respir. Med., 88, 145 (1994).

43. M.E. Dodd, J. Abbott, J. Maddison, A.J. Moorcroft, and A.K. Webb, Thorax, 52, 656 (1997).

44. V.B. Pai and M.C. Nahata, Pediatr. Pulmonol., 32, 314 (2001).

45. A.L. Smith, B.W. Ramsey, D.L. Hedges, B. Hack, J. Williams-Warren, A. Weber, E.J. Gore, and G.J. Redding, Pediatr. Pulmonol., 7, 265 (1989).

46. B.W. Ramsey, H.L. Dorkin, J.D. Eisenberg, R.L. Gibson, I.R. Harwood, R.M. Kravitz, D.V. Schidlow, R.W. Wilmott, S.J. Astley, M.A. McBurnie, and et al., N. Engl. J. Med., 328, 1740 (1993).

47. A.L. Smith, B.W. Ramsey, and A.B. Montgomery, U.S. Patent 5 508 269 (1996).

48. A. Weber, G. Morlin, M. Cohen, J. Williams-Warren, B. Ramsey, and A. Smith, Pediatr. Pulmonol., 23, 249 (1997).

49. M. Rosenfeld, R. Gibson, S. McNamara, J. Emerson, K.S. McCoyd, R. Shell, D. Borowitz, M.W. Konstan, G. Retsch-Bogart, R.W. Wilmott, J.L. Burns, P. Vicini, A.B. Montgomery, and B. Ramsey, J. Pediatr., 139, 572 (2001).

50. P.B. Challoner, M.G. Flora, P.H. Hirst, M.A. Klimowicz, S.P. Newman, B.A. Schaeffler, R.J. Speirs, and S.J. Wallis, American Thoracic Society 97th International Conference, San Francisco, USA (2001).

51. S.P. Newman, M. Flora, P.H. Hirst, M. Klimowicz, B. Schaeffler, R. Speirs, W. Sun, S.J. Wallis, and P. Challoner, International Society for Aerosols in Medicine 13th International Congress, Interlaken, Switzerland (2001).

52. R.W. Wilmott, B. Chatfield, M. Dyson, D. Geller, L. Milgram, P.G. Noone, D. Rodman, M. Rosenfeld, D.A. Waltz, B. Schaeffler, and P. Challoner, Pediatr. Pulmonol., 22(Suppl.), 292 (2001).

53. J.M. Goldman, S.M. Bayston, S. O'Connor, and R.E. Meigh, Thorax, 45, 939 (1990).

54. S.P. Conway, A. Watson, M.N. Pond, S.M. Bayston, R. Booth, and A. Ghonheim, Pediatr. Pulmonol., 9(Suppl.), 250 (1993).

55. N.R. Crowther Labiris, A.M. Holbrook, H. Chrystyn, S.M. Macleod, and M.T. Newhouse, Am. J. Respir. Crit. Care Med., 160, 1711 (1999).

56. J.G. Weers, T.E. Tarara, and A. Clark, U.S. Patent 20020017295 (2002).

57. S.A. Sisk, M.T. Newhouse, S.P. Duddu, Y. Walter, M. Eldon, T. Tarara, A.R. Clark, and J. Weers, International Society for Aerosols in Medicine 13th International Congress, Interlaken, Switzerland (2001).

58. S.L. Hill, H.M. Morrison, D. Burnett, and R.A. Stockley, Thorax, 41, 559 (1986).

59. R.A. Stockley, S.L. Hill, and D. Burnett, Clin. Ther., 7, 593 (1985).

60. H.C. Lin, H.F. Cheng, C.H. Wang, C.Y. Liu, C.T. Yu, and H.P. Kuo, Am. J. Respir. Crit. Care Med., 155, 2024 (1997).

61. R. Orriols, J. Roig, J. Ferrer, G. Sampol, A. Rosell, A. Ferrer, and A. Vallano, Respir. Med., 93, 476 (1999).

62. A.F. Barker, L. Couch, S.B. Fiel, M.H. Gotfried, J. Ilowite, K.C. Meyer, A. O'Donnell, S.A. Sahn, L.J. Smith, J.O. Stewart, T. Abuan, H. Tully, J. Van Dalfsen, C.D. Wells, and J. Quan, Am. J. Respir. Crit. Care Med., 162, 481 (2000).

63. H.D. Fuller, M.B. Dolovich, G. Posmituck, W.W. Pack, and M.T. Newhouse, Am. Rev. Respir. Dis., 141, 440 (1990).

64. C.J. Harvey, M.J. O'Doherty, C.J. Page, S.H. Thomas, T.O. Nunan, and D.F. Treacher, Thorax, 50, 50 (1995).
65. J.M. Hughes and B.S. Saez, Respir. Care, 32, 1131 (1987).
66. M.J. O'Doherty, S.H. Thomas, C.J. Page, D.F. Treacher, and T.O. Nunan, Am. Rev. Respir. Dis., 146, 383 (1992).
67. T.G. O'Riordan, M.J. Greco, R.J. Perry, and G.C. Smaldone, Am. Rev. Respir. Dis., 145, 1117 (1992).
68. T.G. O'Riordan, L.B. Palmer, and G.C. Smaldone, Am. J. Respir. Crit. Care Med., 149, 214 (1994).
69. M.J. O'Doherty and S.H. Thomas, Thorax, 52 Suppl 2, S56 (1997).
70. L.B. Palmer, G.C. Smaldone, S.R. Simon, T.G. O'Riordan, and A. Cuccia, Crit. Care Med., 26, 31 (1998).
71. G.C. Wood and B.A. Boucher, Pharmacotherapy, 20, 166 (2000).
72. G.C. Smaldone and L.B. Palmer, Respir. Care, 45, 667 (2000).
73. A.B. Montgomery, R.J. Debs, J.M. Luce, K.J. Corkery, J. Turner, E.N. Brunette, E.T. Lin, and P.C. Hopewell, Lancet, 2, 480 (1987).
74. G.W. Soo Hoo, Z. Mohsenifar, and R.D. Meyer, Ann. Intern. Med., 113, 195 (1990).
75. A.B. Montgomery, D.W. Feigal, Jr., F. Sattler, G.R. Mason, A. Catanzaro, R. Edison, N. Markowitz, E. Johnson, S. Ogawa, M. Rovzar, and et al., Am. J. Respir. Crit. Care Med., 151, 1068 (1995).
76. J.A. Kovacs, V.J. Gill, S. Meshnick, and H. Masur, J.A.M.A., 286, 2450 (2001).
77. G.S. Leoung, D.W. Feigal, Jr., A.B. Montgomery, K. Corkery, L. Wardlaw, M. Adams, D. Busch, S. Gordon, M.A. Jacobson, P.A. Volberding, and et al., N. Engl. J. Med., 323, 769 (1990).
78. S.A. Bozzette, D.M. Finkelstein, S.A. Spector, P. Frame, W.G. Powderly, W. He, L. Phillips, D. Craven, C. van der Horst, and J. Feinberg, N. Engl. J. Med., 332, 693 (1995).
79. E. Conneally, M.T. Cafferkey, P.A. Daly, C.T. Keane, and S.R. McCann, Bone Marrow Transplant, 5, 403 (1990).
80. B. Hertenstein, W.V. Kern, T. Schmeiser, M. Stefanic, D. Bunjes, M. Wiesneth, J. Novotny, H. Heimpel, and R. Arnold, Ann. Hematol., 68, 21 (1994).
81. S.E. Myers, S.M. Devine, R.L. Topper, M. Ondrey, C. Chandler, K. O'Toole, S.F. Williams, R.A. Larson, and R.B. Geller, Leuk. Lymphoma, 8, 229 (1992).
82. H. Reichenspurner, P. Gamberg, M. Nitschke, H. Valantine, S. Hunt, P.E. Oyer, and B.A. Reitz, Transplant Proc., 29, 627 (1997).
83. S. Schwartz, G. Behre, V. Heinemann, H. Wandt, E. Schilling, M. Arning, A. Trittin, W.V. Kern, O. Boenisch, D. Bosse, K. Lenz, W.D. Ludwig, W. Hiddemann, W. Siegert, and J. Beyer, Blood, 93, 3654 (1999).
84. W.T. Hughes, D. Armstrong, G.P. Bodey, A.E. Brown, J.E. Edwards, R. Feld, P. Pizzo, K.V. Rolston, J.L. Shenep, and L.S. Young, Clin. Infect. Dis., 25, 551 (1997).
85. M.G. Ottolini and V.G. Hemming, Drugs, 54, 867 (1997).
86. American Academy of Pediatrics Committee on Infectious Diseases, Pediatrics, 97, 137 (1996).
87. R.A. Bowden, Am. J. Med., 102, 27 (1997).
88. R. Adams, J. Christenson, F. Petersen, and P. Beatty, Bone Marrow Transplant, 24, 661 (1999).
89. R.H. Adams, Biol. Blood Marrow Transplant, 7 Suppl, 16S (2001).
90. T.J. Nitz, D.C. Peaver, T.R. Bailey, K.L. Bender, C.W. Blackledge, D. Cebzanov, T. Draper, S.B. Ellis, J. Hehman, M.A. McWherter, D.J. Rys, J. Swestock, T.M. Tull, and N. Ye, 40th Interscience Conference on Antimicrobial Agents and Chemotherapy, Toronto, Canada (2000).
91. D.C. Peaver, T.M. Tull, R. Direnzo, N. Ma, T.J. Nitz, and M.S. Collett, 40th Interscience Conference on Antimicrobial Agents and Chemotherapy, Toronto, Canada (2000).
92. Management of Influenza in the Southern Hemisphere Trialists Study Group, Lancet, 352, 1877 (1998).
93. F.G. Hayden, A.D. Osterhaus, J.J. Treanor, D.M. Fleming, F.Y. Aoki, K.G. Nicholson, A.M. Bohnen, H.M. Hirst, O. Keene, and K. Wightman, N. Engl. J. Med., 337, 874 (1997).
94. S.M. Cheer and A.J. Wagstaff, Drugs, 62, 71 (2002).
95. P. Diggory, C. Fernandez, A. Humphrey, V. Jones, and M. Murphy, BRIT. MED. J., 322, 577 (2001).

SECTION IV. IMMUNOLOGY, ENDOCRINOLOGY AND METABOLIC DISEASES

Editor: William K. Hagmann
Merck Research Laboratories, Rahway, NJ 07065

Chapter 16. Therapeutic Applications of Non-peptidic δ-Opioid Agonists

Graham N. Maw and Donald S. Middleton
Pfizer Global Research and Development, Department of Discovery Chemistry
Ramsgate Road, Sandwich, Kent, CT13 9NJ, U.K.

Introduction – Opiate pharmacology research has been the subject of significant academic and pharmaceutical activity over the last two decades. Three major subtypes, mu (μ), delta (δ), and kappa (κ) belonging to the G-protein coupled superfamily of receptors, have been pharmacologically characterised and cloned (1-3). The clinical application of morphine-based μ-opioids in pain management is well precedented and is driven by their powerful analgesic properties. However, centrally-mediated μ-agonism is also characterised by significant side-effects including toleration, respiratory depression, immunosuppression, and decreased resistance to infection (4-6). In addition, the strongly addictive properties of morphine-based μ-opioids make this class of agents far from ideal for medical use.

More recently, the targeting of δ-opioid receptor agonists to modulate antinociceptive activity has emerged as an attractive alternative (7). Cloned and expressed δ-opioid receptors are known to couple to both L-type Ca^{2+} and K^+ channels through Gi proteins (8). Two pharmacologically distinct subtypes of the δ-receptor (designated δ_1 and δ_2) have been identified, based upon both in vitro and in vivo studies. While the full pharmacological functions of these receptors remains to be elucidated, both are implicated to play a role in neuropathic pain regulation (9). In contrast to the morphine-like μ–opioids, δ-opioid agonist activity produces few adverse effects on respiratory function and has significantly lower potential for development of physical dependence (10, 11). In addition, alternative potential therapeutic applications of this class are beginning to emerge. For example, δ-agonists have been shown to possess immunostimulative effects, in contrast to the immunosuppressive effects observed with μ-agonism, thus potentially offering a further advantage over μ-agonists in pain treatment where patients may be immunocompromised (12). δ-Receptor stimulation has also been shown to mediate the cardioprotective effect of ischemic preconditioning (13, 14). δ-Agonist-mediated increases in synaptic dopamine levels may offer new therapeutic opportunities in the treatment of Parkinsons disease and as alternative non-addictive maintenance therapies in cocaine addiction (15, 16).

Despite these potential clinical applications, research in this area has been hampered by the lack of suitable selective, δ-opioid agonists which possess both high oral bioavailability and good CNS penetration following oral administration. Definitive clinical trials in these areas await the availability of suitable compounds (17).

Morphine-based derivatives - Work based on the "message-address" concept as a strategy to provide non-peptidic agonists in this area has provided potent compounds for pharmacological study (18, 19). The discovery that δ-antagonist **1** could be converted into a potent partial agonist, displaying 65% maximal response in the mouse vas deferens (MVD) agonist assay, by simple replacement of the cyclopropylmethyl substituent with a methyl group (**2**; δ K_i = 0.7nM, μ K_i = 468nM, κ K_i = 4467nM) stimulated further investigation in this structural class (20).

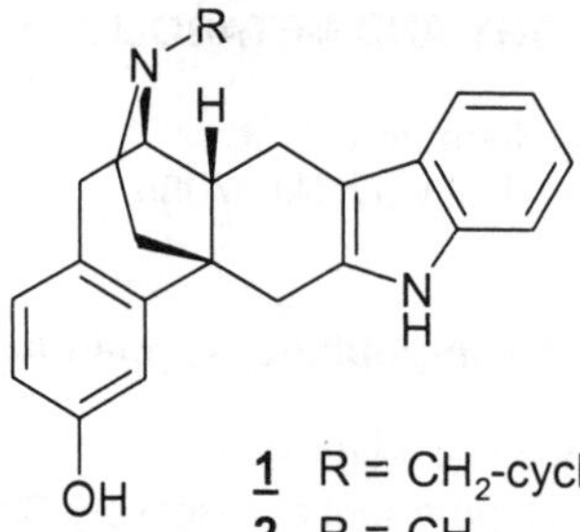

1 R = CH$_2$-cycloC$_3$H$_5$
2 R = CH$_3$

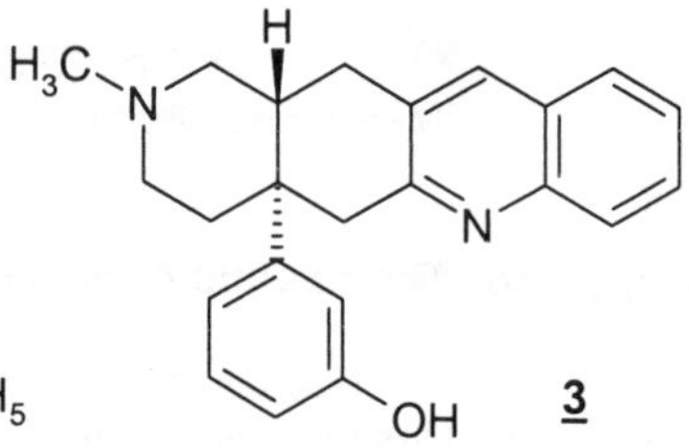

3

A novel class of octahydroisoquinolines **3** was identified which is formally derived from the fragmentation of **1** (21). The full agonist TAN-67 (**3**) is the most striking example, retaining excellent potency against the δ-receptor (δ K$_i$ = 1.1nM) and >1,000-fold selectivity over the related μ and κ receptors. More recently, the message-address concept has also been exploited in the reported study of a series of pyrrolomorphinans of general structure **4-6** as a novel class of selective δ-opioid agonists (22). The ester **4** shows high affinity for the δ-receptor (K$_i$ 2.3 nM) with >100-fold selectivity over μ- and κ-receptors (145- and 300-fold, respectively). Both amides **5** (δ K$_i$ = 4.8 nM) and ketones **6** (δ K$_i$ = 1.3 nM) also show high affinity for the δ-receptor.

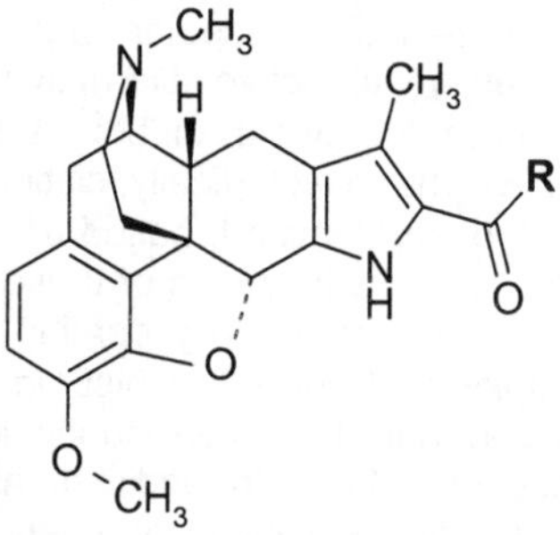

	R
4	OC$_2$H$_5$
5	NH-i-C$_3$H$_7$
6	i-C$_3$H$_7$

In rat *in vivo* models of inflammatory pain, **7** (δ K$_i$ = 6.2 nM; μ/δ = 110, κ/δ = 160) was shown to possess both good brain penetration (brain / plasma concentration = 2) and antihyperalgesic properties similar to morphine with no tolerance observed. These antihyperalgesic effects could be selectively reversed by the selective δ-antagonist naltrindole.

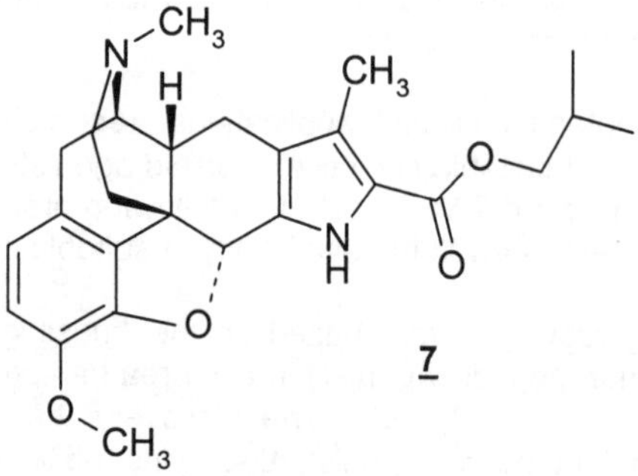

7

A novel series of substituted cycloheptenes has recently been described (23). Compound **8** is claimed to show potent δ-activity in the human binding assay (δ K$_i$ = 1.4 nM) and good *in vivo* activity in a mouse writhing model (ED$_{50}$ = 3.3 mg/kg i.v.), although no opioid sub-type selectivity data for this compound was reported.

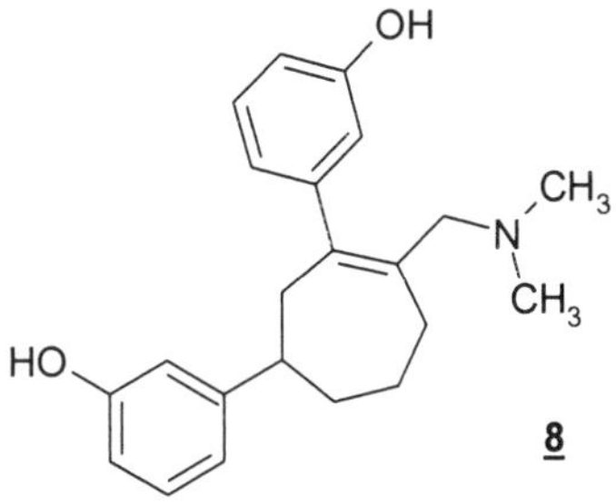

8

Benzhydrylpiperazine Derivatives - A further breakthrough in identifying non-peptide δ-opioid receptor agonists with drug-like potential was disclosed (24). These efforts resulted in the identification of BW373U86 (**9**) (Table 1). The SAR of this compound has been described and has formed the basis of much of the research activity in the area of δ-opioid receptor agonists. A systematic investigation of the SAR of BW373U86 (**9**) and SNC-80 (**11**) established the crucial importance of the α-benzhydryl configuration. In both the phenolic (**9**, **10**) and ether (**11**-**13**) sub-series, it has been established that the αR configuration leads to higher levels of potency and selectivity in binding assays, and that this outweighs the effect of the piperazine configuration. In addition, the methyl ether in this particular template confers the best combination of potency and selectivity relative to the phenols.

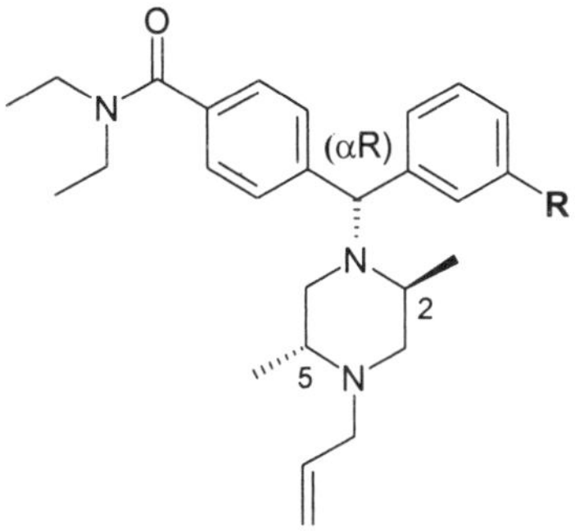

Table 1

	R	Configuration	δ binding IC$_{50}$ nM	μ/δ	Number
9	OH	αR, 2S, 5R	0.3	31	BW373U86
10	OH	αR, 2R, 5S	0.5	341	
11	OCH$_3$	αR, 2S, 5R	1.1	2327	SNC-80
12	OCH$_3$	αR, 2R, 5S	3.5	2611	
13	OCH$_3$	αS, 2S, 5R	56	101	

However, these compounds have also been associated with adverse *in vivo* effects such as convulsions (25). In addition, these analogues suffer from rapid metabolism and high *in vivo* clearance (26). It is for these reasons, as well as the desire to discover still more selective molecules with the biopharmaceutic properties required for an oral drug, that research has continued in the area. Early work to optimise the SNC-80 series (**11**) established the importance to binding and selectivity of the relative and absolute configuration of the piperazine moiety (27-29). Further efforts also established that the 4-substituted amides were greatly preferred over either the 2- or 3-substituted analogues. This optimisation succeeded in identifying compound **14** (δ IC$_{50}$ = 8 nM; μ/δ = 1514) which incorporated minor changes relative to the SNC-80 structure (30).

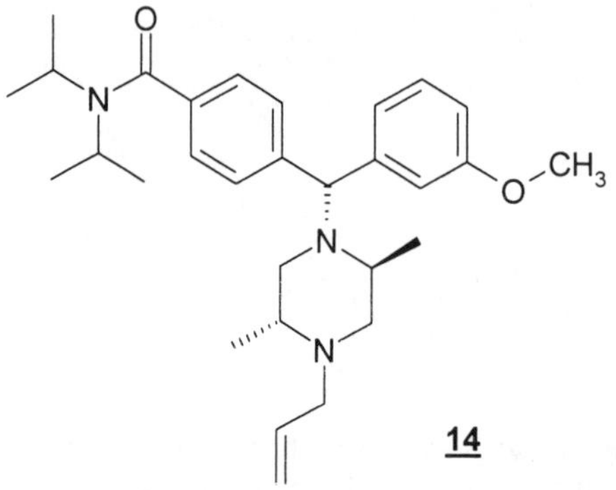

14

A related series of tetrazolyl compounds (**15-17**) targeting a peripheral site of action have been reported in a series of patents (31). Very high levels of functional activity have been reported for these analogues in the electrically stimulated MVD assay. In the parent tetrazole (**15**; MVD pIC_{50} = 9.7), replacement of N-allyl by benzyl (**16**; MVD pIC_{50} = 10.7) resulted in a 10-fold improvement in δ-opioid functional activity. Subnanomolar functional activity was retained in the 5-substituted pentanoic acid analogue (**17**) (MVD pIC_{50} = 9.5).

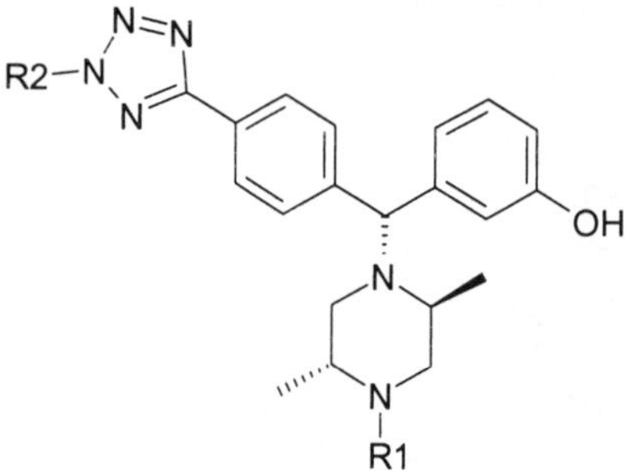

15 R1 = allyl, R2 = H

16 R1 = allyl, R2 = benzyl

17 R1 = allyl, R2 =

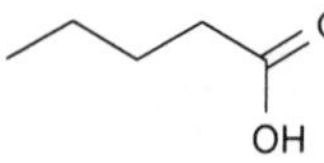

UK-321,130 (**17**) has been reported to show excellent oral bioavailability in the dog (F = 93%) (32). The compound shows peripheral δ-opioid activity and inhibits colonic motility in the dog after intravenous infusion (1µg/ml/kg). In addition, **17** demonstrates 17,000-fold selectivity over centrally mediated δ-opioid events. These tetrazoles (**15-17**) were the subject of a patent application disclosing a wide range of related zwitterionic compounds targeting opioid receptors found in the periphery. A range of heterocyclic tethers for the carboxylate moiety, including thiazole (**18**) and tetrahydroisoquinoline (**19**), are disclosed although their potencies are not reported. The relative stereochemistry found in **18** is different from that usually described for the most active examples of the 2,5-dimethylpiperazines suggesting that further efforts to optimise the central core may be of some benefit.

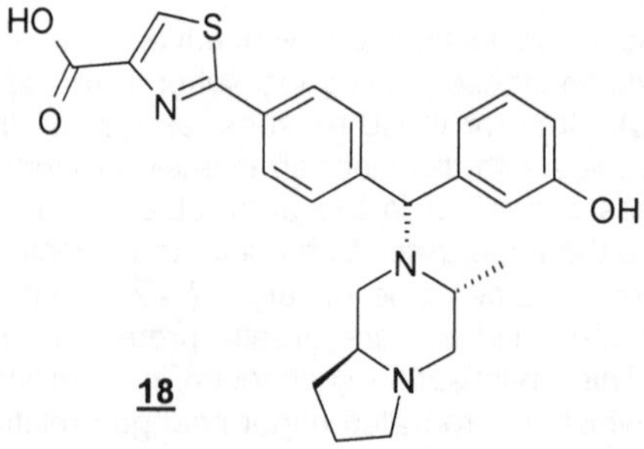

18

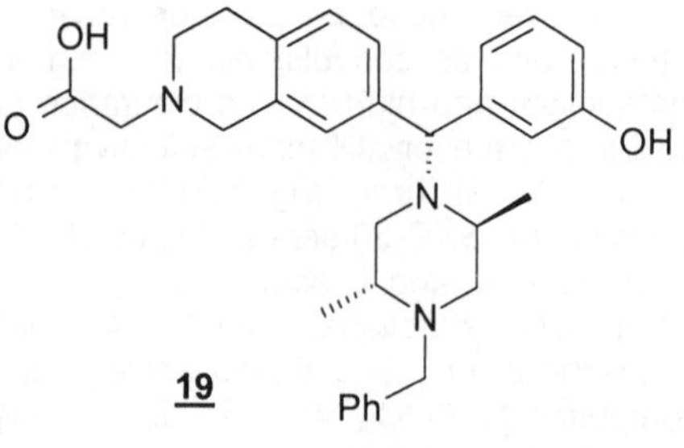

19

A complimentary series of compounds based on the SNC-80 **11** core template has been reported (33). The core could be simplified through the removal of chiral complexity and reduction of molecular weight as a means to reduce metabolic vulnerability. The removal of the methylether moiety was combined with both piperazine **20** and homopiperazine **21** replacements for 4-allyl-2,5-dimethylpiperazine, without significant loss of activity.

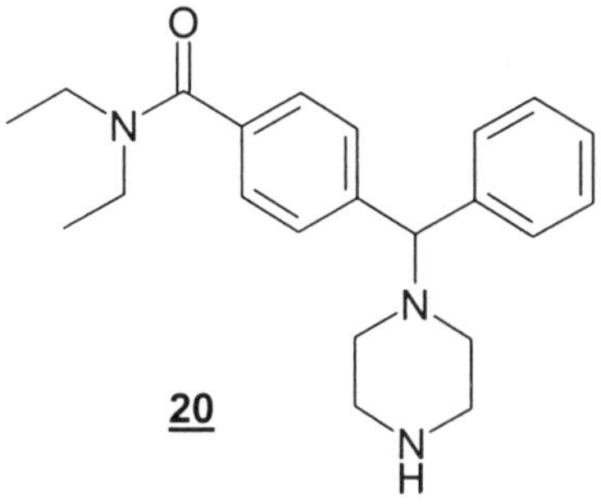

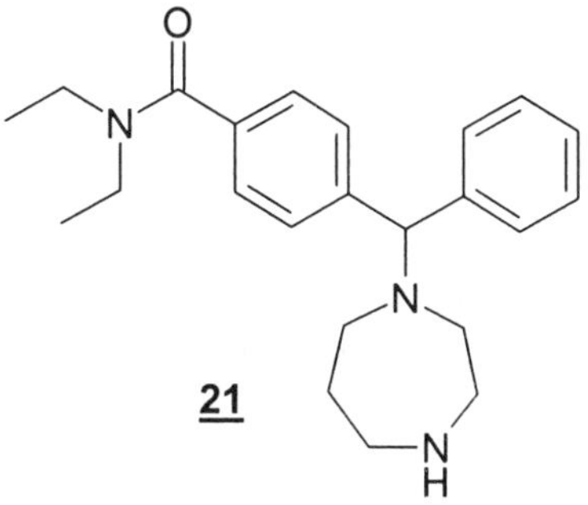

This SAR was further extended to investigate replacements for the phenyl in the head-group by bicyclic heterocycles such as 8-quinolynyl (**22**, δ IC_{50} = 0.5 nM) and 2,2-dimethyl-2,3-dihydro-1-benzofuran-7-yl (**23**, δ IC_{50} = 1 nM). These modifications succeeded in identifying compounds with ~10-fold increased potency relative to SNC-80 **11** together with significantly improved *in vitro* metabolic stability. This translated into an improved PK profile for **23** which had 33% oral bioavailability in rats. In contrast to the N-deallylation metabolic pathway observed with **11**, this series of compounds are N-dealkylated in the amide portion of the molecule.

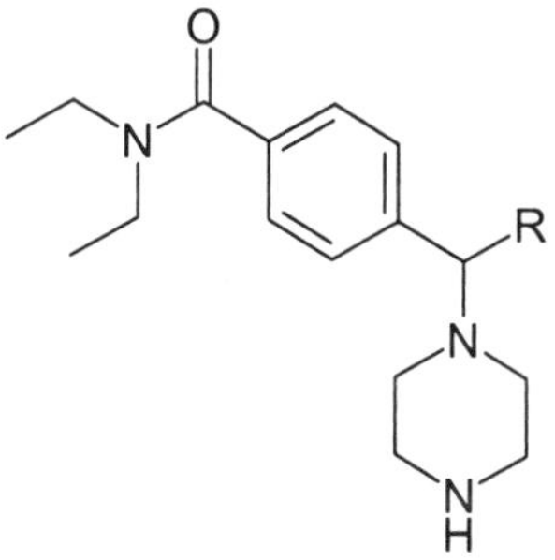

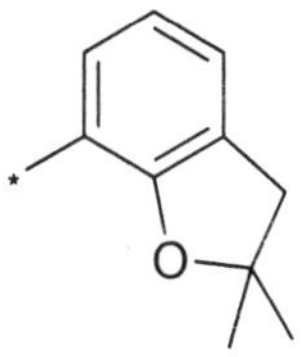

A novel series of δ-opioid agonists derived from the piperazine template has been reported but which now are devoid of stereogenic centres (34). In this case, good δ-opioid binding affinity and selectivity over the μ-subtype are observed. A representative example of the series, **24** (δ IC_{50} = 0.87 nM, μ/δ = 4370) shows subnanomolar activity combined with >4000-fold selectivity over μ- and κ-subtypes. Compound **24** was also significantly more stable in rat liver microsomes which translated into high levels of oral bioavailability in rats.

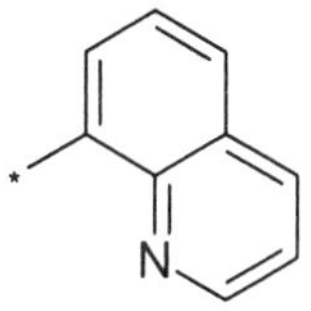

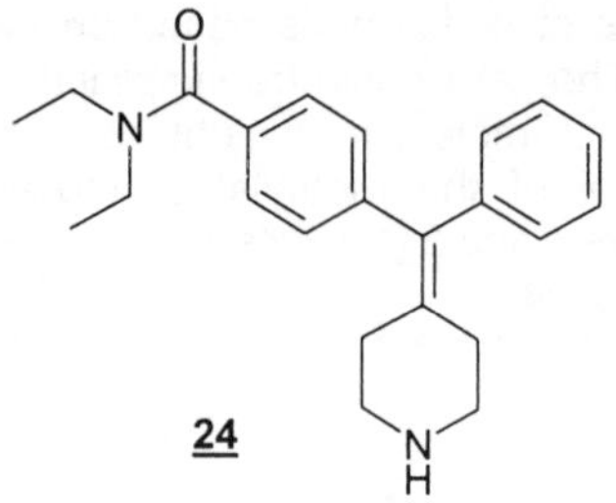

24

A series of potent and selective diarylaminopiperidines (**25-26**) have been disclosed where an aminopiperidine group has replaced the commonly found piperazine moiety (35).   This modification results in a series of compounds that retain good binding affinity for the δ-opioid receptor as well as good μ- selectivity. In particular, **25** shows modest affinity (δ K_i = 25 nM) and only 6-fold selectivity over μ-receptors. This compound also demonstrates efficacy in the mouse acetylcholine bromide-induced abdominal constriction model of pain (ED_{50} = 4.2 μmol/kg p.o.). The limited SAR disclosed so far appears to track with others in this class of ligands as the 3-hydroxy derivative **26** (δ K_i = 0.83 nM; μ K_i = 2763 nM)   has >25-fold greater affinity for the δ-receptor with a concomitant loss of μ-receptor affinity driving improved selectivity.

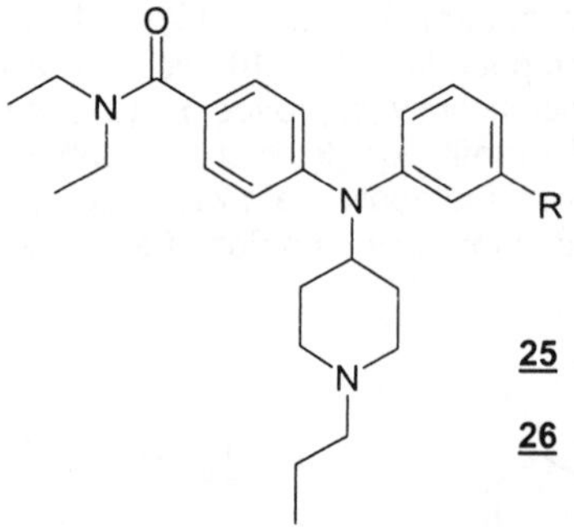

25 R = H

26 R= OH

Another example of the lipophilic class of piperazine analogues, SL-3111 (**27**) exhibits good affinity for the δ-opioid receptor (δ K_i = 8nM; μ/δ >2000) in binding assays (36). However, the selectivity and potency of **27** is significantly reduced in functional tissue strip experiments (EC_{50} = 85 nM; 460-fold selectivity).   Recently disclosed structural modifications describe modifications to the central piperazine toinvestigate the importance of each of the piperazine nitrogens (37).   These modifications confirmed that the distal nitrogen to the benzhydryl portion was

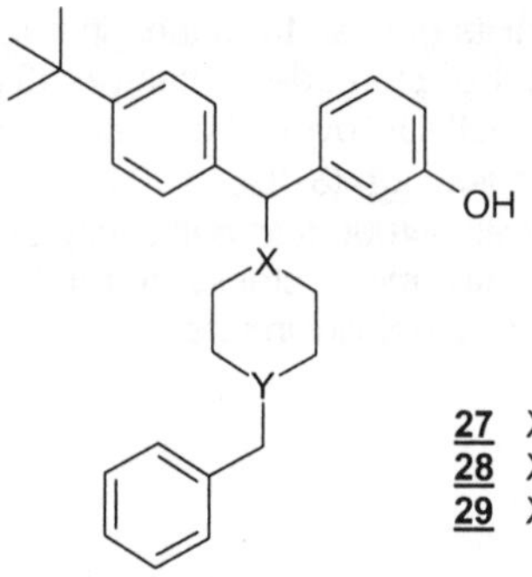

27 X, Y = N
28 X = N, Y = CH
29 X = CH, Y = N

key to δ-opioid activity and that the piperazine conferred selectivity. For example, replacement of the distal nitrogen with CH **28** (δ IC_{50} = 12,640 nM) resulted in a >1,000-fold reduction in affinity, while a similar substitution at the proximal nitrogen in **29** (δ IC_{50} = 33 nM) resulted in a 4-fold reduction in potency relative to **27**.

In addition, incorporation of a range of simple substituents into the central piperazine core was beneficial for potency and selectivity. For example, **30** shows good potency at the δ-receptor (δ IC_{50} = 38 nM) although selectivity data was not reported.

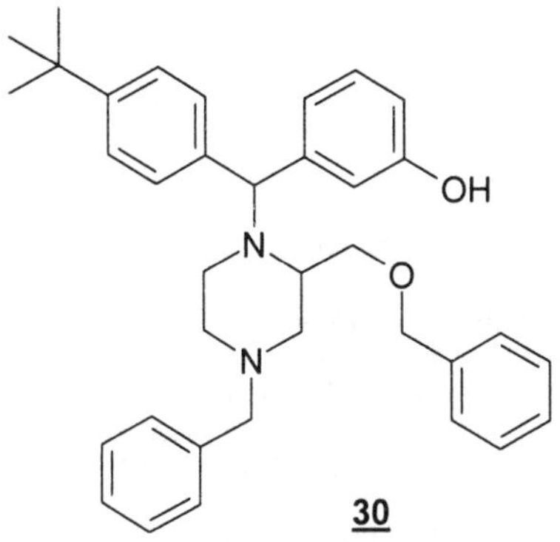

30

Summary – A number of ligands for the δ-opioid receptor have been disclosed over the past few years. The majority of research has focused on compounds derived from the piperazine derivative, BW373U86 (**9**). To date, several compounds combining subnanomolar potency and excellent selectivity have been identified. Much effort has been directed towards the optimisation of physicochemical and pharmacokinetic properties, which has led to the identification of compounds with improved properties. However, there is as yet no reported clinical data to support pre-clinical experiments implicating the activation of the δ-opioid receptor for the treatment of pain or motility impairment. It therefore remains to be seen whether selective activation of δ-opioid receptors will indeed have clinical benefit.

References

1. C.J. Evans, D.E. Keith, H. Morrison, K. Magendzo and R.H. Edwards, Science, 258, 1952 (1992).
2. E. Mannson, L. Bare and D. Yang, Biochem.Biophys.Res.Commun., 202, 1431 (1994)
3. J.B. Wang, P.S. Johnson, A.M. Perisco, A.L. Hawkins, C.A. Griffin and G.R. Uhl, FEBS Letts., 338, 217 (1994).
4. S. Takeda, L.I. Eriksson, Y. Yamamoto, H. Joensen, H. Onimaru and S.G.E. Lindahl, Anesthiology, 95, 740 (2001).
5. J. M. Risdahl, K.V. Peterson and T.W. Molitor, J.Neuroimmunol., 83, 4 (1998).
6. T.K. Einstein and M.E. Hilburger, J.Neuroimmunol., 83, 36 (1998).
7. R. Kanjhan, Clin.Exp.Pharmacol.Physiol., 22, 397 1995.
8. P.A. North, Opioids. A. Herz (ed.), Springer-Verlag, New-York, pp. 773 (1993).
9. J. Mika, R. Przewlocki and B. Przewlocka, Eur.J.Pharmacol., 415, 31 (2001).
10. H.H. Szeto, Y. Soong, D. Wu, N. Olariu, A. Kett, H. Kim and J.F. Clapp, Peptides, 20, 101 (1999)
11. A. Cowan, X.Z. Zhu, H.I. Mosberg and F. Porreca, J.Pharmacol.Exp.Ther., 246, 950 (1988).
12. H.N. Bhargava, R.V. House and P.T. Thomas, Analgesia, 302 (1995).
13. J.L. Schultz and G. Gross, US Patent 6,103,722 (2000)

14. S.P. Bell, M.N. Sac, A. Patel, L.H. Opie and D.M. Yellon, J.Amer.Coll.Cardiol., 36, 2296 (2000).
15. M.P. Hill, C.J. Hille and J.M. Brotchie, Drug News Perspect., 13, 261 (2000).
16. F.I. Carroll, L.L. Howell and M.J. Kuhar., J.Med.Chem., 42, 271 (1999).
17. G. Dondio; S. Ronzoni and P. Petrillo, Expert Opin.Ther.Pat. , 9(4), 353 (1999).
18. R. Schwyzer, Ann.N.Y.Acad.Sci., 297, 3 (1977).
19. P.S. Portoghese, M. Sultana, H. Nagese and A.E. Takemori, J.Med.Chem., 31, 281 (1988).
20. P.S. Portoghese, M. Sultana and A.E. Takemori, J.Med.Chem., 33, 1714 (1990).
21. H. Nagase, H. Wakita, K. Kawai, T. Endoh, H. Matsuura, C. Tanaka and Y. Takezawa, Jpn.J.Pharmacol., 64 (suppl 1), 35 (1997).
22. G. Dondio, S. Ronzoni, C. Farina, D. Graziani, C. Parini, P. Petrillo and G.A.M. Giardina, Il Farmaco, 56, 117 (2001).
23. O. Zimmer, W.A. Wolfgang, W.G. Engelberger and B.-Y. Kogel, Ger. Patent DE 198 57 475-A1 (2000).
24. E.J. Hong, K.C. Rice, S.N. Calderon, J.H. Woods and J.R. Traynor, Analgesia, 3, 269 (1998).
25. J.E. Schetz, S.N. Calderon, C.M. Bertha, K. Rice and F. Porreca, J.Pharmacol.Exp.Ther., 279, 1069 (1996).
26. M.J. Bishop and R.W. McNutt, Bioorg.Med.Chem.Lett., 5, 1311 (1995).
27. S.N. Calderon, R.B. Rothman, F. Porreca, J.L. Flippen-Anderson, R.W. McNutt, H. Xu, L.E. Smith, E.J. Bilsky, P. Davis and K.C. Rice, J.Med.Chem. , 37, 2125 (1994).
28. S.N. Calderon, R.B. Rothman, F. Porreca, J.L. Flippen-Anderson, R.W. McNutt, H. Xu, L.E. Smith, E.J. Bilsky, P. Davis, K.C. Rice, R. Horvath and K. Becketts, J.Med.Chem., 39, 695 (1996).
29. Y. Katsura, X. Zhang, K. Homma, K.C. Rice, S.N. Calderon, R.B. Rothman, H.I. Yamamura, P. Davis, J.L. Flippen-Anderson, H. Xu,, K. Becketts, E.J. Foltz and F. Porreca, J.Med.Chem., 40, 2936 (1997).
30. G.N Maw, Abstr. Pap. MEDI-168, 221st ACS Meeting (2001).
31. P. Norman in Drug News Perspect., 14(9), 551 (2001).
32. N. Plobeck, D. Delorme, Z.-Y. Wei, H. Yang, F. Zhou, P. Schwarz, L. Gawell, H. Gagnon, B. Pelcman, R. Schmidt, S.Y. Yue, C. Walpole, W. Brown, E. Zhou, M. Labarre, K. Payza, S. St-Onge, A Kamassah, P.-E. Morin, D. Projean, J. Ducharme and E. Roberts, J.Med.Chem., 43, 3878 (2000).
33. Z.-Y. Wei, W. Brown, B. Takasaki, N. Plobeck, D. Delorme, F. Zhou, H. Yang, P. Jones, L. Gawell, H. Gagnon, R. Schmidt, S.-Y. Yue, C. Walpole, K. Payza, S. St-Onge, M. Labarre, C. Godbout, A. Jakob, J. Butterworth, A. Kamassah, P.-E. Morin, D. Projean, J. Ducharme and E. Roberts, J.Med.Chem., 43, 3895 (2000).
34. J.R. Carson, R.J. Carmosin, L.J. Fitzpatrick, A.B. Reitz and M.C.Jetter, PCT Int. Appl. WO9933806 (1999).
35. S. Liao, J. Alfaro-Lopez, M.D. Shenderovich, K. Hosohata, H.I. Yamamura and V. Hruby, J.Med.Chem., 41, 4767 (1998).
36. J. Alfaro-Lopez, T. Okuyama, K. Hosohata, P. Davis, F. Porreca, H.I. Yamamura and V. Hruby, Pept.New.Millin. Proc.Am.Pept.Symp. 16th, 38 (2000).

Chapter 17. Selective Glucocorticoid Receptor Modulators

Michael J. Coghlan*, Steven W. Elmore[†], Philip R. Kym[†] and Michael E. Kort[†]

*Eli Lilly and Company, Discovery Chemistry Research & Technologies, DC 0528,
Lilly Corporate Center, Indianapolis, IN 46285,
[†]Abbott Laboratories, Global Pharmaceutical Research and Development
100 Abbott Park Road, Abbott Park, IL 60064

Introduction - Glucocorticoids (GCs) have a pervasive role in human health and physiology. The endogenous members of this family, cortisol and corticosterone, are involved in a breadth of endocrine functions including metabolism of lipids, carbohydrates and proteins, stress response, fluid and electrolyte balance as well as maintenance of immunological, renal and skeletal homeostasis (1-4). Synthetic GCs have long been recognized as effective treatments for inflammatory conditions and immunomodulation. The overriding mode of action of GCs involves regulation of gene expression through the glucocorticoid receptor (GR). Unfortunately, the widespread changes in gene regulation also potentiate GC-mediated homeostatic endocrine functions, leading to the side effects associated with prolonged treatment. Clinical use is further complicated by the cross-reactivity of most commonly used GCs with other steroid hormone receptors resulting in ancillary pharmacology. Consequently, the identification of more selective functional ligands remains a goal of clinical and pharmaceutical research. In this chapter, we detail the methods for the evaluation of selective GR modulators and describe the genesis of new compounds where varying degrees of selectivity have been reported.

Background - The role of GCs in gene transcription has been extensively studied. The discovery and cloning of intracellular receptors (IRs) demonstrated that steroids act in association with specific members of this family (5-7). The GR is a member of the growing genus of IRs (8-10). Endogenous ligands bind to GR and the resulting GR/ligand complex (GRC) regulates gene expression. Other steroid hormone receptors such as estrogen (ER), progesterone (PR), mineralocorticoid (MR), and androgen (AR) also regulate transcription as complexes with endogenous ligands (9). Many natural and synthetic steroids bind to more than one member of this family. In the case of GCs, cross reactivity with MR, AR, and PR can be problematic (11,12).

Since GR-regulated gene transcription is believed to follow two distinct functional pathways, ligands could be further differentiated by selectively following one mode of action. When a steroid or other ligand associates with GR in the cytosol, the resulting GRC is transported into the nucleus where it regulates gene expression via direct action as an endogenous transcription factor or by indirectly affecting the function of existing transcription factors associated with proinflammatory gene expression. The antiinflammatory effects of the GRC are believed to occur by the indirect path, wherein the GRC adopts a conformation possessing an affinity for existing transcription factors such as AP-1 (13) or NFκB (14-17), thereby repressing the transcription of proinflammatory cytokines and other inflammatory mediators (14,18-20). The direct function of GRCs as conventional transcription factors regulates endogenous endocrine function. The dimeric form of the GRC forms a transcription factor that directly binds to specific DNA sequences known as glucocorticoid response elements (GREs). This mode of action is responsible for regulation of the endogenous functions related to routine endocrine and metabolic processes such as gluconeogenesis, bone metabolism, and electrolyte balance. Differentiation of function has been substantially validated *in vitro* and *in vivo* where deletion mutants of the GR dimerization domain (GR^{dim} mice) prevent GRE-mediated gene expression (21,22).

Classic medicinal chemistry optimization of glucocorticoids provided more potent steroids, with dexamethasone **1** and prednisolone **2** emerging as benchmark drugs (23-25). Among the most potent of the functional steroidal GR ligands, dexamethasone (dex) is widely used for *in vitro* validation of putative GR-mediated processes. Although **1** is an enviable standard for evaluation of processes mediated by GR, its *in vivo* side effects make it less attractive for comparisons in efficacy models, especially chronic studies, and it is rarely prescribed. Consequently, other systemic GCs, such as prednisolone (pred), are routinely employed for functional *in vivo* comparisons. These are arguably more relevant since **2** has considerable clinical use, its *in vivo* profile is well characterized, and it has become an approachable benchmark for *in vitro* comparison of receptor binding as well as functional activity in transactivation and transrepression assays. GR antagonists such as RU-486 **3** also provide tools for evaluation in the newer indications resulting from inhibition of GR function. While receptor cross-reactivity remains an issue, RU-486 is a well-known functional antagonist of GR-mediated action in both *in vitro* and *in vivo* assays.

1   **2**   **3**

Therapeutic Indications for Selective Glucocorticoid Modulators - Although GCs are associated with a variety of physiological processes, they are principally used for a broad spectrum of antiinflammatory therapies despite an array of side effects. Following the dual paradigm of GC function, selective transcriptional repressors would be optimal for these therapies. Such agents would fit directly into existing treatment regimens, offering the advantage of reduction of one or more well-characterized side effects. Treatment of inflammatory and autoimmune diseases, allergy, organ transplantation, and immune regulation all represent large opportunities for improved ligands. In some therapies, amelioration of specific side effects would be advantageous. This is particularly true for effects on metabolism in bone, since chronic GC treatment in pediatric asthma and adult rheumatoid arthritis patients demonstrated reduction in growth and bone mass, respectively (26).

While the role of GR-mediated repression is well defined, the potential roles for antagonists of GR function represent an emerging area of investigation. Although there is little corroborative clinical data, GR antagonists have been proposed for the treatment of addiction, depression, dementia, and cognitive function (27-32). Inhibitors of GR function also have potential as modulators of endocrine processes. Tissue-selective GR antagonists have been proposed as potential treatments for type 2 diabetes (33-38). Recent data suggests that a selective GR-antagonist could be an effective treatment for obesity (39). In these cases, tissue selectivity has been a primary goal since systemic inhibition of GR function would result in toxic adrenal insufficiency.

Receptor Selectivity - Since cross-reactivity among steroid hormone receptors or other members of the IR family results in undesired pharmacology, an improved receptor selectivity profile is a primary goal. Steroidal GCs such as **1** and **2** typically bind to MR with comparable affinity, resulting in ambiguity about the effects of GR vs. MR function for a given ligand. Other steroidal GR ligands, such as **3**, compete with reproductive hormones, such as PR, resulting in extremely limited use *in vivo* and only the occasional clinical report of their putative GR-mediated function (40,41). Consequently GR-selective ligands would enable discrimination of GR-related function without the ancillary pharmacology and restrictions of their cross-reactive predecessors. These evaluations are typically done by comparison of the binding affinity measured as Ki or IC_{50} with endogenous ligands for the five major steroid receptors. Instances where cross reactivity is observed in binding assays warrant functional evaluation *in vitro*. This is typically evaluated in cell lines transfected with the IRs in question. Compounds with multiple IR affinities are then evaluated as antagonist / agonists versus reference steroids. Identification of functional cross-reactivity usually dictates the need for a more selective molecule.

<u>Functional Evaluation of Transcriptional Activation and Antagonism</u> - The endogenous function of GRE-mediated gene expression has been demonstrated in transfected cell lines as well as native cells. Reporter gene assays have typically used well known GRE-regulated sequences such as the mouse mammary tumor virus promoter (MMTV) or promoter regions from well-characterized genes known to be activated by GRE sequences such as the promoter region of phosphoenolpyruvate carboxykinase (PEPCK), an enzyme involved in gluconeogenesis. MMTV, PEPCK, or other promoter regions are transfected upstream of reporter gene constructs, such as luciferase (luc), chloramphenicol acetyl transferase (CAT), or alkaline phosphatase (ALP) and candidate GR ligands are used to observe the induction of protein expression mediated by the direct association of the GRC with GREs from the transfectant. Induction of protein expression can also be observed in unaltered cells where upregulation of a number of GRE-dependent genes are used as measures of transcriptional activation. PEPCK, tyrosine amino transferase (TAT), aromatase, and other proteins can be evaluated. The resulting agonist activity would then be compared to assays of transcriptional repression to differentiate function. GRE antagonists are detected using the same cell types and proteins. However, candidate antagonists are evaluated for their ability to inhibit the activation effects of the endogenous GC in the test species (cortisol in non-rodents, corticosterone for murine cell lines). Specific cell types are also susceptible to GR-associated activation. Osteocalcin expression in osteoblasts is downregulated by GCs, while proteins expressed in cell lines from other organs such as liver, breast, and ovary are expressed after exposure to GR ligands. Treatment of primary hepatocytes with GR antagonists allows evaluation of the test compound's ability to block GC mediated upregulation of key gluconeogenic enzymes such as PEPCK and glucose-6-phosphatase (G-6-Pase) or enzymes that convert amino acids to glucose precursors such as TAT or aspartate aminotransferase. Primary hepatocytes have also been used to evaluate the role that GCs play in the induction of key gluconeogenic transcriptional coactivators such as PGC-1 (38,42).

<u>*In Vivo* Models of GR Function</u> - Maintenance of normal endocrine function, metabolism, and gluconeogenesis are the primary *in vivo* functions of GCs. Many of the proteins associated with these processes can be monitored to evaluate the effects of novel ligands on homeostatic function including aromatase and PEPCK enzymes which are clinically assayed as indicators of altered endocrine function. Specific assays associated with glucose synthesis and metabolism have also emerged. In addition to PEPCK, G-6-Pase and TAT are sensitive to GR modulation. Expression of these proteins can be evaluated in serum as well as in specific tissues such as the liver. Serum glucose levels can be used as a convenient measure of altered glucose metabolism. Pred and other classic GCs potentiate glucose levels in rodent glucose tolerance tests. Novel GR ligands can be evaluated in this assay to determine whether the same increases in glucose levels are observed. GR antagonists have been evaluated in animal models of diabetes and obesity, including *ob/ob* mouse (43-45), *db/db* mouse (46,47), *fa/fa* Zucker rat (39), and GK-rat (48). While interpretation of the results of these experiments is complicated by increased levels of endogenous GCs in the transgenic animals (cf. the *ob/ob* and *db/db*), potent GR antagonists such as RU-486 reduced plasma glucose levels, and result in decreased expression of PEPCK, GLUT2, G-6-Pase, and TAT in constitutively obese *db/db* transgenic mice (47). This is

accompanied by normalization of blood glucose levels and a 50% reduction of serum insulin in *ob/ob* mice (44,49). RU-486 has also demonstrated effectiveness in reversing obesity in young *fa/fa* Zucker rats, characterized by a reduction of body weight with decreased fat deposition, the prevention of hyperphagia, and reduction in insulin levels (39).

Despite the plethora of GR-regulated endocrine functions, no correlation has been reported between *in vitro* GRE-mediated protein expression and *in vivo* side effects. Newer models are constantly under development, and improvements in these assays as well as evaluation of GR ligands in the GR^{dim} transgenic animals may provide a clearer association between GRE-mediated *in vitro* and *in vivo* pharmacology.

Functional Evaluation of Transcriptional Repression - Assessment of repression *in vitro* relies on several elements for relevant comparison. These assays are typically done in transfected cell lines such as HepG2 cells, lymphocytes, or those with endogenous genes regulated by transcription factors such as the NFκB or AP-1 (50). *In vitro* studies employ well characterized DNA promoter regions devoid of known GREs upstream of reporter gene constructs. Upon initiation of gene transcription with a number of stimuli such as IL-1β or TNFα, candidate GR ligands are evaluated for their ability to inhibit the expression of these stimulated proteins. In these assays, the choice of the DNA sequence, its expressed protein, the cell line, and the stimulus have been carefully controlled, presumably to limit variability. In all cases to date, reference compounds such as dex and pred are used as standards of efficacy. Repression of AP-1 and NFκB promoted reporter gene constructs in stable or transient transfectants supports the indirect functional effects of GCs. The data from transfected cell lines are typically corroborated with similar reduction of gene expression in unaltered cells. Repression can be evaluated in well-known cell lines such as human skin fibroblasts or in more immunologically relevant cells such as monocytes. Among the models used, IL-1β stimulated IL-6 expression in human skin fibroblasts has been employed to characterize functional GR ligands (51). This effect has been shown to occur by way of a GR-mediated mechanism at the transcriptional level in native human cells, and reference glucocorticoids are effective in this system (52,53). Production of this cytokine is associated with proliferation of B and T cell lines, and circulating IL-6 levels increase in common inflammatory diseases such as rheumatoid arthritis (54). Repression of proinflammatory gene expression has also been demonstrated across a wide spectrum of cell types and gene products. Downregulation of tissue destructive enzymes such as collagenase, cyclooxygenase-2 and other inducible enzymes, adhesion molecules like ICAM, and chemokines exemplified by RANTES and MCP-1 have all been reported for various GCs (55-61). The evolution of these assays and their correlation with *in vivo* effects will rely on the relevance of the targeted clinical endpoint.

In Vivo Models of Inflammation - Once identified *in vitro*, evaluation of antiinflammatory effects of GR ligands *in vivo* falls into an established battery of assays relevant to a particular therapy. Models of rheumatoid arthritis involving acute assays in rodents such as carrageenan-induced paw edema (CPE), chronic evaluation in established adjuvant arthritis (EAA) in the rat, or collagen-induced arthritis (CIA) in the mouse have long been used with correlations to clinical efficacy (62-65). Historically, antiinflammatory activity in these models was determined by comparison of paw weight or volume versus control groups. Today, these assays have evolved to the stage where imaging technology enables detailed evaluation of joint and cartilage damage, and diagnostic study of the diseased joint permits detailed evaluation of tissue destructive enzymes, cytokines, and infiltrating immune cells such as neutrophils and macrophages (63,64).

Asthma models have also been developed for the GC-mediated control of the inflammatory component of this disease. Classic assays such as sephadex-induced eosinophilia or lung edema in rodent models have been standards of GC evaluation. *Ascarus suum* stimulated brochoconstriction and inflammatory cell influx models have also been well documented for evaluation of GCs for the treatment of asthma in a variety of larger animals such as dogs and monkeys.

Models for the control of dermal inflammation would identify treatments for common inflammatory skin conditions such as psoriasis, atopic dermatitis, or contact sensitivity. Acute skin inflammation induced by croton oil or its active component, TPA, in rodents has historically been used to evaluate compounds (66). These assays provide a variety of data, from qualitative measures of edema and skin color to more quantifiable evaluation of cytokine levels and cellular influx. Throughout the development of *in vivo* antiinflammatory assays used to profile classic GCs, the evolution of diagnostic tools for evaluation of the GR-mediated proinflammatory mediators such as TNFα, and IL-6 have enabled more reliable correlations between the repression activities observed *in vitro* with those observed in the more complex setting of *in vivo* evaluation (67). The emerging paradigm suggests that strong repression from novel GR ligands has lead to *in vivo* antiinflammatory activity, although a well-defined correlation has yet to be published.

MEDICINAL CHEMISTRY

It is widely recognized that even the most subtle perturbations to the steroidal nucleus can result in dramatic effects on binding affinity, receptor selectivity, and functional activity within the IR superfamily. This exquisite structural sensitivity has driven the effort to design GR-selective initially steroidal and ultimately non-steroidal analogs. Prototypical of the modified steroid approach is deflazacort **4**, an oxazoline derivative of pred, which has been proposed as an effective substitute for conventional corticosteroid therapies (68). Improved selectivity for GR vs. MR suggests the potential of deflazacort in ameliorating glucose intolerance and loss of cancellous bone mass. Clinical evidence suggests this synthetic glucocorticoid retains the antiinflammatory efficacy of prednisolone with fewer adverse effects on bone and carbohydrate metabolism (69).

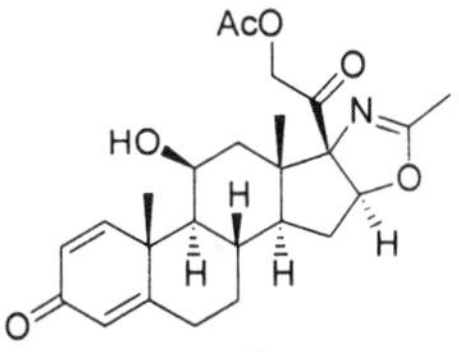

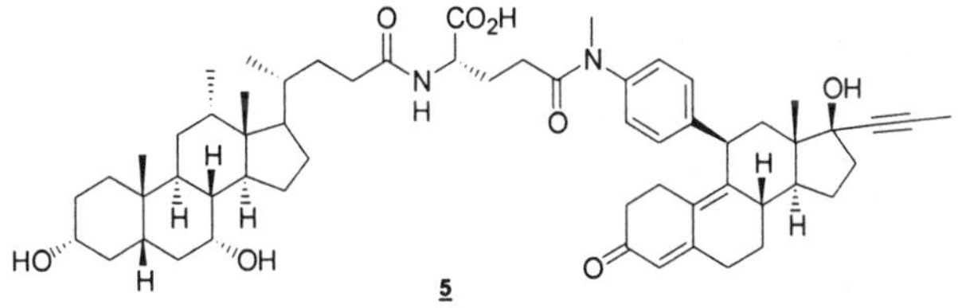

GR Antagonists - The search for GR antagonists has frequently focused on RU-486 **3**, a potent (hGR Ki = 0.5-2.2 nM), *in vivo* active but non-selective (vs. PR and AR) GR ligand (70-73). To deliver therapeutically useful concentrations of a GR antagonist in tissue-specific fashion to the liver for the treatment of type II diabetes, RU-486 has been covalently attached to cholic acid via a glutamic acid linkage (34). It has been posited that liver-specific bile acid conjugates such as **5** would offer the advantage of decreased liver glucose production, enhanced insulin sensitivity, increased half-life of the active entity, with minimal peripheral / systemic exposure. RU-43044, **6**, is a potent *in vitro* antagonist of GR (hGR Ki = 2.4 nM) bearing considerable structural analogy to its predecessor, RU-486 (72,74). Although modestly selective (10-fold), **6** appears to show tissue-selective partial agonism *in vitro*. A limited induction of agonism was seen in osteosarcoma cells with no activation in breast cancer cell lines (75). RU-43044 is inactive *in vivo*. Identification of the required spatial arrangement of key structural elements within the RU motifs enabled the development of tricyclic analogs such as **7** as anti-obesity agents

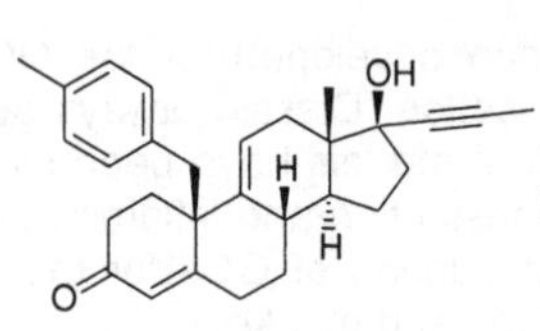

(35). Tricyclic picolinyl derivative **7** (hGR Ki = 2.0 nM) is receptor selective (MR, PR Ki >1 µM), exhibits full functional antagonism (EC$_{50}$ = 25 nM) in human cell lines and has been shown to reduce food intake in *ob/ob* mice (43).

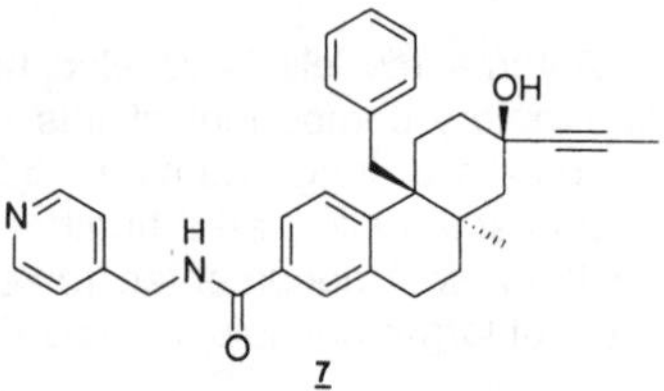

A series of diphenyl ethers, represented by KB285 **8,** has been reported as liver-selective GC antagonists that shows potential for the treatment of diabetes (37,76). KB285 potently binds hGR (IC$_{50}$ = 19 nM) while having little affinity for PR, MR or AR (IC$_{50}$ = 3.8, 2.4 and 5.5 µM, respectively). It antagonizes the effect of **1** (IC$_{50}$ = 0.4 µM) in an MMTV-alkaline phosphatase (ALP) reporter gene assay, and it also reversed the dex-induced increase in TAT activity in hepatoma cells (IC$_{50}$ = 2.5 µM). After an oral dose of 25 mpk, KB285 was found to reduce the fasting serum glucose levels in *db/db* mice by 62% compared to 33% for 75 mpk of RU-486. Corticosterone levels in the KB285-treated mice were only slightly elevated (364 ng/mL vs. 301 ng/mL for control) while those of the RU-486 treated animals were nearly doubled (584 ng/mL). This lack of compound-induced increase in systemic corticosterone levels coupled with the robust glucose lowering effect provides evidence of the functional selectivity of this compound.

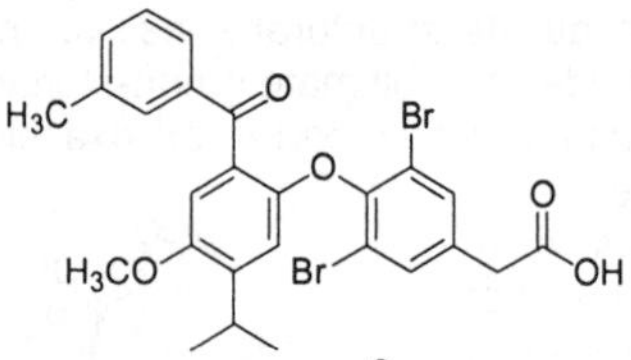

A family of triarylmethanes has emerged from the patent literature as selective nonsteroidal GR ligands; no GR functional data is reported for these compounds (77). Substitution of the aromatic rings has a pronounced effect on GR affinity but not on related IRs. Achiral example **9** exhibits high affinity for GR (K$_i$ = 38 nM) with greater than 100-fold selectivity over PR, MR, AR, and ER (Ki >2500 nM). The discovery of a chromene-based series of non-steroidal GR antagonists has also been recently disclosed (78). Regiochemistry of the methoxy substitution in methylsulfonamide **10** was the critical determinant for establishing GR selectivity with the *m*-tolyl functionality on the pendant aromatic ring providing the optimal balance of potency for GR (IC$_{50}$ = 1.1 nM) and PR cross-reactivity (IC$_{50}$ = 77 nM) (73). Chromene derivative **10** was also reported to antagonize the effect of dex (IC$_{50}$ = 21 nM) in an ALP reporter gene assay (73).

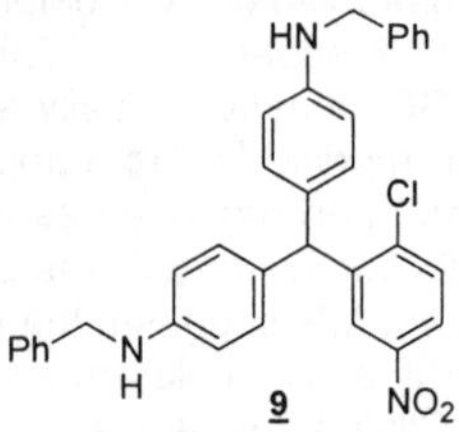

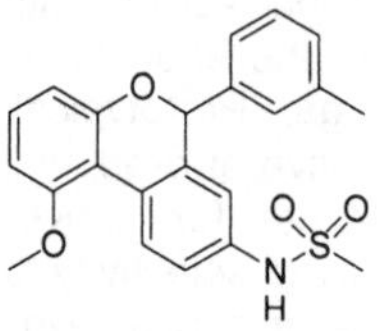

<u>Functional GR Ligands</u> - Triphenylpropanamides such as **11**, **12** and **13** were recently disclosed in a patent application as antiinflammatory compounds that do not demonstrate the side effects normally associated with other GCs (79). The most potent GR

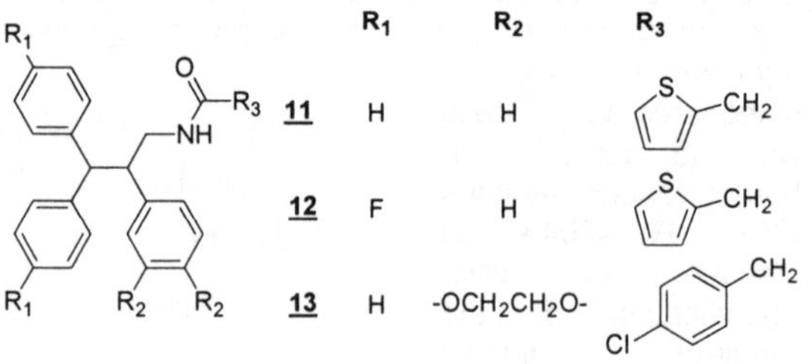

	R$_1$	R$_2$	R$_3$
11	H	H	(2-thienyl)–CH$_2$
12	F	H	(2-thienyl)–CH$_2$
13	H	-OCH$_2$CH$_2$O-	(4-Cl-phenyl)–CH$_2$

binders, **11** and **12** have hGR affinities of 8 and 7 nM, respectively, while having little or no affinity for hPR. Although *in vitro* functional data was not provided, all three compounds exhibited topical antiinflammatory activity that was equipotent to cortisol in an oxazolone ear swelling model. Topical activity of **11** and **12** in the TPA-induced ear edema model was slightly less than that of cortisol with 64.9% and 60.0% inhibition, respectively, compared to 80.0% for cortisol. Although this series was claimed to be devoid of typical GC side effects, no supporting data was provided.

Steroids bearing novel C-17 side chains, RU-24858 **14** and RU-24782 **15**, were identified in a screening effort to discover functionally dissociated GR modulators by comparing potencies for GR-mediated activation and AP-1 repression using reporter gene constructs (80). While these analogs have GR binding potencies equivalent to that of pred, their affinity for other steroid receptors has not been reported. These compounds showed a reduced ability to induce GR-mediated transactivation from the GRE5-tk-CAT reporter in co-transfected HeLa cells. While **15** was able to partially antagonize dex-induced activation, **14** was inactive even at 100 fold higher concentrations. GRE activation in native cells was evaluated by induction of TAT activity in rat HTC cells where RU-24858 elicited only 11% of the dex response. Transrepression potencies and efficacies of these compounds were identical to pred in cotransfection assays using an AP-1 containing collagenase promoter-CAT reporter system, and both **14** and **15** repressed LPS-induced IL-1β secretion in THP1 cells at 95% and 75% the maximal efficacy of dex, respectively. It was later shown that their ability to inhibit TNFα induced IL-6 secretion in murine fibroblasts and HeLa cells was mediated by interfering with NF-κB dependent gene activation (81,82). Compounds **14** and **15** displayed similar *in vivo* antiinflammatory activity to pred in a cotton pellet granuloma test in rats (ED_{50} = 7, 5 and 2.5 mpk, respectively) and the croton oil induced ear edema model (ED_{50} = 10, 2 and 4 μg/ear, respectively). Although there is clear evidence of dissociative behavior *in vitro*, **14** demonstrated no separation between antiinflammatory efficacy and glucocorticoid-induced side effects in the sephadex-induced lung edema model (81). Compared to reference steroids, this compound showed similar antiinflammatory activity as well as a similar decrease in body weight, thymic involution, quantitative osteopenia of the femur growth plate and reduction in systemic osteocalcin levels.

14, R = CN
15, R = SCH₃

The 2,5-dihydro-2,2,4-trimethyl-1H-[1]benzopyrano[3,4-f]quinoline core has emerged as a versatile scaffold for steroid hormone receptor ligands. First reported as PR receptor modulators (83-87), it was found that methoxy substitution at C-10 of the quinoline core imparts complete GR selectivity over other IRs (88,89). Substitution at C-5 was ultimately crucial in determining the functional activity of these GR modulators. The levorotatory enantiomers of this series were almost exclusively responsible for the observed biological effects. Optimal C-5 aryl substitution provided compounds such as A-222977 **16** exhibiting hGR affinity equivalent to pred (Ki = 4.0 nM) and 250- to 1000-fold selectivity over PR, MR, AR and ER (89). Compound **16** possesses pred-equivalent activation activity in both cotransfection assays (MMTV-LUC reporter, IC_{50} = 9.0 nM) as well as in its ability to potentiate aromatase function in native cells (IC_{50} = 140 nM). The repression activity of **16** was also pred-like when evaluated in both cotransfection (E-selectin reporter, IC_{50} = 16 nM) and native cells (IL-6 repression, IC_{50} = 4.0 nM), and **16** also exhibited potent antiinflammatory activity in the sepahadex-induced lung edema model (ED_{50} = 2.8

16, (±) R =

17, (±) R =

18, (-) R =

mpk) compared to pred (ED_{50} = 1.2 mpk). Measures of *in vivo* GC-induced side effects have not been reported for these C-5 aryl analogs. Further SAR studies at C-5 of the quinoline core led to non-aromatic analogs such as **17** and **18** that reportedly show differentiated activation / repression profiles both *in vitro* and *in vivo* (90). Cyclopentyl analog **17** exhibited slightly less potent repression activity compared to pred in a cotransfection assay (E-selectin promoter, IC_{50} = 18 nM and 2.6 nM, respectively) while it lacked the ability to significantly induce GR-mediated activation (MMTV-LUC promoter, 34% of dex). The *in vivo* efficacy and side effect profile of allyl analog **18** has also been recently reported (90, 91). Oral efficacy for this compound was similar to pred in both an acute CPE model (ED_{50} = 6 mpk vs. 2 mpk for pred) and a chronic rat established adjuvant arthritis model (ED_{50} = 5 mpk vs. 2 mpk for pred). Compound **18** did not induce hyperglycemia at efficacious doses and was shown to antagonize the pred-induced hyperglycemic effect. MRI analysis of animals treated with **18** had slightly improved scores of bone and soft tissue injury vs. pred. The animals treated with **18** in the EAA assay also showed less joint injury compared to pred based on a histopathological evaluation of EAA joints.

Conclusion - Despite a half century of glucocorticoid research, a basic understanding of the duality of GR regulated gene transcription has only developed within the last decade. As these insights mature, so has a novel paradigm for generating novel, selective GR modulators. This theory has enabled an accelerated search, and the list of structurally diverse modulators of GR has continued to grow. Reminiscent of traditional steroids, subtle changes in the newer, non-steroidal entities results in wide variation in receptor affinity, functional activity, and tissue selectivity, providing compelling incentive for continued exploration. It can be anticipated that GR ligand development will continue to emphasize complementary phenotypes varying in degrees of dissociation of transcriptional activation and transcriptional repression. At both ends of the functional spectrum, organ targeting to decrease peripheral exposure and address particular therapeutic endpoints is an increasingly important objective. As has been demonstrated in the context of estrogen receptor ligands, more refined therapies wherein tissue-specific agonist and antagonist activities reside in the same molecule may ultimately be achieved for GR. With the published data, there is still a lack of substantial correlation between *in vitro* repression / activation and *in vivo* efficacy / side effects which accentuates the need for a more in depth understanding of the molecular mechanism of transcriptional regulation by the GRC. The unprecedented levels of selectivity demonstrated for the GR-modulators described here may represent the first step toward the realization of selective GR-mediated therapeutic agents.

References

1. L.V. Avioli, C. Gennari, B. Imbimbo and Editors., "Advances in Experimental Medicine and Biology, Vol. 171: Glucocorticoid Effects and Their Biological Consequences," Ed., Plenum Press, New York, N. Y., 1984, 419.
2. J.D. Baxter, Pharmacol. Ther., [B], 2, 605 (1976).
3. J.E. Parrillo and A.S. Fauci, Ann. Rev. Pharmacol. Toxicol., 19, 179 (1979).
4. B.P. Schimmer and K.L. Parker, in "Goodman & Gilman's The Pharmacological Basis of Therapeutics," J. G. Hardman, L. E. Limbird, P. B. Molinoff and R. W. Ruddon, Ed., McGraw-Hill, New York, 1996, p. 1459.
5. R.M. Evans, Science, 240, 889 (1988).
6. M.A. Carson-Jurica, W.T. Schrader and B.W. O'Malley, Endocr. Rev., 11, 201 (1990).
7. B.W. O'Malley and M.-J. Tsai, in "Steroid Horm. Action," M. G. Parker, Ed., IRL, Oxford, UK, 1993, p. 45.
8. C. Weinberger, S.M. Hollenberg, E.S. Ong, J.M. Harmon, S.T. Brower, J. Cidlowski, E.B. Thompson, M.G. Rosenfeld and R.M. Evans, Science, 228, 740 (1985).
9. T.P. Burris, in "Nuclear Receptors and Genetic Disease," T. P. Burris and E. R. B. McCabe, Ed., Academic Press, San Diego, CA, 2001, p. 1.
10. V. Laudet and H. Gronemeyer, "The Nuclear Receptor Facts Book," Academic Press, London, UK, 2002, p. 345.

11. P. Mulatero, F. Veglio, C. Pilon, F. Rabbia, C. Zocchi, P. Limone, M. Boscaro, N. Sonino and F. Fallo, J. Clin. Endocrinol. Metab., <u>83</u>, 2573 (1998).
12. G. Neef, S. Beier, W. Elger, D. Henderson and R. Wiechert, Steroids, <u>44</u>, 349 (1984).
13. P. Herrlich, Oncogene, <u>20</u>, 2465 (2001).
14. R. Brattsand and M. Linden, Aliment Pharmacol. Ther., <u>10</u>, 81 (1996).
15. L.I. McKay and J.A. Cidlowski, Endocr. Rev., <u>20</u>, 435 (1999).
16. B. van der Burg and P.T. van der Saag, Mol. Hum. Reprod., <u>2</u>, 433 (1996).
17. B. van der Burg, J. Liden, S. Okret, F. Delaunay, S. Wissink, P.T. van der Saag and J.-A. Gustafsson, Trends Endocrinol. Metab., <u>8</u>, 152 (1997).
18. S. Heck, Ph.D. thesis, Wiss. Ber. Forschungszent., Karlsruhe, 1998.
19. C. Jonat, H.J. Rahmsdorf, K.K. Park, A.C. Cato, S. Gebel, H. Ponta and P. Herrlich, Cell, <u>62</u>, 1189 (1990).
20. S. Heck, M. Kullmann, A. Gast, H. Ponta, H.J. Rahmsdorf, P. Herrlich and A.C. Cato, EMBO J., <u>13</u>, 4087 (1994).
21. H.M. Reichardt, J.P. Tuckermann, A. Bauer and G. Schutz, Z. Rheumatol., <u>59</u>, 1 (2000).
22. H.M. Reichardt, F. Tronche, S. Berger, C. Kellendonk and G. Schutz, Adv. Pharmacol., <u>47</u>, 1 (2000).
23. R.F. Witzmann, "Steroids, Keys to Life," Van Nostrand Reinhold, New York, 1981.
24. S.L. Ali, Anal. Profiles Drug Subst. Excipients, <u>21</u>, 415 (1992).
25. E.M. Cohen, Anal. Profiles Drug Subst., <u>2</u>, 163 (1973).
26. S.S. Yeap and D.J. Hosking, Curr. Opin. Endocrinol. Diabetes, <u>2</u>, 248 (1995).
27. C. Oberlander and P.V. Piazza, PCT Patent WO/9826783 (1998).
28. B.E. Murphy, D. Filipini and A.M. Ghadirian, J. Psych. Neurosci., <u>18</u>, 209 (1993).
29. A.F. Schatzberg and J.K. Belanoff, PCT Patent WO/9959596 (1999).
30. J.K. Belanoff, B.H. Flores, M. Kalezhan, B. Sund and A.F. Schatzberg, J. Clin. Psychopharm., <u>21</u>, 516 (2001).
31. A.F. Schatzberg and J.K. Belanoff, PCT Patent WO/0137840 (2001).
32. K.D. Jameison, Timothy G., Hum. Psychopharmacol.,<u>16</u>, 293 (2001).
33. T. Deisher, PCT Patent WO/9827986 (1998).
34. T. Apelqvist, J. Wu and K.F. Koehler, PCT Patent WO/0058337 (2000).
35. R.L. Dow, K.K.-C. Liu, B.P. Morgan and A.G. Swick, PCT Patent WO/0066522 (2000).
36. P.R. Kym, B.C. Lane, J.K. Pratt and T. VonGeldern, PCT Patent WO/0116128 (2001).
37. T. Apelqvist, M. Gillner, A. Gustavsson, L. Hagberg, E. Koch, M. Lindberg, B. Pelcman, J. Wu and P.R. Kym, PCT Patent WO/0147859 (2001).
38. S. Herzig, F. Long, U.S. Jhala, S. Hedrick, R. Quinn, A. Bauer, D. Rudolph, G. Schutz, P. Puigerver, B. Spiegelman, M. Montmimy, Nature, <u>413</u>, 179 (2001).
39. S.C.Y. Langley, David A., Am. J. Physiol., <u>259</u>, R539 (1990).
40. D.R. Garrel, R. Moussali, A. De Oliveira, D. Lesiege and F. Lariviere, J. of Clin. Endocrin. Metab., <u>80</u>, 379 (1995).
41. X. Bertagna, H. Escourolle, J.L. Pinquier, J. Coste, M.C. Raux-Demay, P. Perles, L. Silvestre, J.P. Luton and G. Strauch, J. Clin. Endocrin. Metab., <u>78</u>, 375 (1994).
42. J.C. Yoon, P. Peigerver, G. Chen, J. Donovan, Z. Wu, J. Rhee, G. Adelmant, J. Stafford, C. R. Kahn, D.K. Granner, C.B. Newgard, B.M. Spiegelman, Nature, <u>413</u>, 131 (2001).
43. K. Liu, et al., Proceedings of the 222nd ACS National Meeting, Chicago, IL (U.S.), August 26-30, 2001, MEDI-169.
44. T.W. Gettys, P.M. Watson, I. L. Taylor, S.Collins, Int. J. Obes., <u>21</u>, 865 (1997).
45. H.-L. Chen and D.R. Romsos, Am. J. Physiol., <u>271</u>, E151 (1996).
46. B.M. Spiegelman, B. Lowell, A. Napolitano, P. Dubuc, D. Barton, U. Francke, D.L. Groves, K.S. Cook and J.S. Flier, J. Biol. Chem., <u>264</u>, 1811 (1989).
47. J. Friedman, T. Ishizuka, C. J. Farrell, S. E. McCormack, L. M. Herron, P. Hakimi, P. Lechner, J. S. Yun, J. Biol. Chem.,<u>272</u>, 31475 (1997).
48. D. Alvarez, F. Picarel-Blanchot, E. Bertin, A. Pascual-Leone, B. Portha, Am. J. Physiol.,<u>278</u>, E1097 (2000).
49. G.R. Bebernitz, G. Argentieri, B. Battle, C. Brennan, B. Balken, B. F. Burkey, M. Eckhardt, J. Gao, P. Kapa, R.J. Strohshein, H.F. Schuster, M. Wilson, D.D. Xu, J.Med.Chem., <u>44</u>, 2601 (2001).
50. C.M. Bamberger and H.M. Schulte, Eur. J. Clin. Invest., <u>30</u>, 6 (2000).
51. A. Waage, G. Slupphaug and R. Shalaby, Eur. J. Immunol., <u>20</u>, 2439 (1990).
52. A. Ray, D.-H. Zhang, M.D. Siegel and P. Ray, Ann. N. Y. Acad. Sci., <u>762</u>, 79 (1995).
53. A. Ray and P.B. Sehgal, J. Am. Soc. Nephrol., <u>2</u>, S214 (1992).
54. C.B. Cohick, D.E. Furst, S. Quagliata, K.A. Corcoran, K.J. Steere, J.G. Yager and H.B. Lindsley, J. Lab. Clin. Med., <u>123</u>, 721 (1994).
55. M. Pfahl, Endocrine Reviews, <u>14</u>, 651 (1993).

56. R. Newton, J. Seybold, L.M.E. Kuitert, M. Bergmann and P.J. Barnes, J. Biol. Chem., 273, 32312 (1998).
57. H. Inoue and T. Tanabe, Biochem. Biophys. Res. Commun., 244, 143 (1998).
58. R. Newton, L.M. Kuitert, D.M. Slater, I.M. Adcock and P.J. Barnes, Life Sciences, 60, 67 (1997).
59. H. Yoshikawa, Y. Nakajima and K. Tasaka, J. Immunol., 162, 6162 (1999).
60. O.J. Kwon, P.J. Jose, R.A. Robbins, T.J. Schall, T.J. Williams and P.J. Barnes, Am. J. Resp. Cell Mol. Bio., 12, 488 (1995).
61. K.G. Steube, C. Meyer and H.G. Drexler, Leuk. Res., 23, 843 (1999).
62. C.A. Winter, E.A. Risley and G.W. Nuss, Proc. Soc. Exptl. Biol. Med., 111, 544 (1962).
63. P.B. Jacobson, S.J. Morgan, D.M. Wilcox, P. Nguyen, C.A. Ratajczak, R.P. Carlson, R.R. Harris and M. Nuss, Arthr. Rheum., 42, 2060 (1999).
64. R.P. Carlson and P.B. Jacobson, in "In Vivo Models of Inflammation," D. W. Morgan and L. A. Marshall, Eds., Birkhauser Verlag, Boston, 1999, p. 1.
65. W.B. Van Den Berg and L.A.B. Joosten, In Vivo Models of Inflammation, 51 (1999).
66. G. Tonelli, L. Thibault and I. Ringler, Endocrinology, 77, 625 (1965).
67. T. Takii, Yakugaku Zasshi, 121, 9 (2001).
68. B.R. Walker, Clin. Endocrinol. (Oxford), 52, 13 (2000).
69. K. Lippuner, J.P. Casez, F. Horber, E. Dijkhuis and P. Jaeger, J. Am. Soc. Nephr., 9, 692A (1997).
70. R. Rupprecht, J.M.H.M. Reul, B. van Steensel, D. Spengler, M. Soeder, B. Berning, F. Holsboer and K. Damm, Eur. J. Pharmacol., Mol. Pharmacol. Sect., 247, 145 (1993).
71. C.M.P. Pariante, B. D.; Pisell, T. L.; Su, C.; Miller, A. H., J. Endocrinol., 169, 309 (2001).
72. D.C. Philibert, G.; Gaillard-Moguilewsky, M.; Nedelec, L.; Nique, F.; Tournemine, C.; Teutsch, G., Front. Horm. Res., 19, 1 (1991).
73. J. Hartandi, I. Akritopoulou-Zanze, K. Ashworth, M. Winn, D. Arendsen, J. Patel, J. K. Pratt, J. Wang, S. Fung, M. Kalmanovich, M. Grynfarb, A. Goos-Nilsson, B. Lane, T. VonGeldern, P. R. Kym, Proceedings of the 223rd ACS National Meeting, Orlando, FL, MEDI 58.
74. J-M. Robin-Jagerschmidt, B. Guillot, D. Gofflo, B. Benhamou, A. Vergezac, C. Ossart, D. Moras, D. Philibert, Mol. Endocrinol., 14, 1028 (2000).
75. H. K. Fryer, I. Rogatsky, M. J. Garabedian, T. K. Archer, J. Biol. Chem., 275, 17771 (2000).
76. T. Apelqvist and S. Efendic, PCT Patent WO/9963976 (1999).
77. M.J. Coghlan, J.R. Luly, J.M. Schkeryantz and A.X. Wang, US Patent 6166013 (2000).
78. P. Kym, B.C. Lane, J.K. Pratt, T. VonGeldern., PCT Patent WO/0116128 A1 (2001).
79. M. Scott, P.J. Sanfilippo, L. Fitzpatrick, R.F. Cardova, K. Pan, J. Meschino and M. Jetter, PCT Patent WO/9933786 (1999).
80. B.M. Vayssiere, S. Dupont, A. Choquart, F. Petit, T. Garcia, C. Marchandeau, H. Gronemeyer and M. Resche-Rigon, Mol. Endocrin., 11, 1245 (1997).
81. M.G. Belvisi, S.L. Wicks, C.H. Battram, S.E. Bottoms, J.E. Redford, P. Woodman, T.J. Brown, S.E. Webber and M.L. Foster, J. Immunol., 166, 1975 (2001).
82. W.V. Berghe, E. Francesconi, K. De Bosscher, M. Resche-Rigon and G. Haegeman, Mol. Pharmacol., 56, 797 (1999).
83. L. Zhi, C.M. Tegley, E.A. Kallel, K.B. Marschke, D.E. Mais, M. Gottardis and T.K. Jones, J. Med. Chem., 41, 291 (1998).
84. C.L. Pooley, J.P. Edwards, M.E. Goldman, M.W. Wang, K.B. Marschke, D.L. Crombie and T.K. Jones, J. Med. Chem., 41, 3461 (1998).
85. L.G. Hamann, R.I. Higuchi, L. Zhi, J.P. Edwards, X.N. Wang, K.B. Marschke, J.W. Kong, L.J. Farmer and T.K. Jones, J. Med. Chem., 41, 623 (1998).
86. J.P. Edwards, L. Zhi, C.L.F. Pooley, C.M. Tegely, S.J. West, M.-W. Wang, M.M. Gottardis, C. Pathirana, W.T. Schrader, T.K. Jones, J. Med. Chem., 41, 2779 (1998).
87. J.P. Edwards, S.J. West, K.B. Marschke, D.E. Mais, M. Gottardis and T.K. Jones, J. Med. Chem., 41, 303 (1998).
88. M.J. Coghlan, S.W. Elmore, P.R. Kym, M.E. Kort, J.P. Edwards and T.K. Jones, PCT Patent WO/9941256 (1999).
89. M.J. Coghlan, P.R. Kym, S.W. Elmore, A.X. Eang, J.R. Luly, D. Wilcox, M. Stashko, C.-W. Lin, J. Miner, C. Tyree, M. Nakane, P. Jacobson, B. Lane., J. Med. Chem., 44, 2879 (2001).
90. S.W. Elmore, M.J. Coghlan, D.D. Anderson, J.K. Pratt, B.E. Green, A.X. Wang, M.A. Stashko, C.-W. Lin, C.M. Tyree, J.N. Miner, P.B. Jacobson, D.M. Wilcox, B.C. Lane, J. Med. Chem., 44, 4481 (2001).
91. P.B. Jacobson, Proceedings of the Endocrine Society, 83rd Annual Meeting, Denver, CO, June 20-23, 2001, 2-43.

Chapter 18. Inhibitors of p38α MAP Kinase

Sarvajit Chakravarty and Sundeep Dugar
Scios Inc.
820 West Maude Avenue, Sunnyvale, CA 94086.

Introduction – p38 MAP kinases (Mitogen Activated Protein kinases) are intracellular soluble serine threonine kinases. There are four isoforms in the p38 MAP kinase family: p38α, p38β, p38γ and p38δ (1-3). p38α and p38β are expressed in all tissues, p38γ is expressed primarily in skeletal tissue and p38δ is expressed mainly in the lung, kidney, testis, pancreas, and small intestine. These enzymes show high amino acid sequence conservation, particularly in the ATP binding pocket. All p38 MAP kinases have a Thr-Gly-Tyr dual phosphorylation motif and dual phosphorylation is essential for activation of these enzymes. p38α is phosphorylated on Thr-180 and Tyr-182 was thought to be the primary and only mode of p38α activation. However, a recent report has detailed a new mechanism for the activation of p38α that involves autophosphorylation due to its binding to TAB1 (transforming growth factor-β-activated protein kinase-1-binding protein 1) (4). All MAP kinases exhibit exquisite substrate specificity. For example, p38β is twenty times more specific for the transcription factor ATF-2 than p38α, p38δ phosphorylates ATF-2 and Elk-1 but not *c-myc* or *c-jun*. p38α has garnered the most attention for drug discovery efforts as it has been shown to be the primary MAP kinase associated with the production and action of pro-inflammatory cytokines such as interleukin-1 (IL-1) and tumor necrosis factor-α (TNF-α). Disease states in which these cytokines are thought to play a pathophysiological role include rheumatoid arthritis (RA), inflammatory bowel disease (IBD), congestive heart failure (CHF) and psoriasis.

Biochemistry - The sequence of p38α was first reported in 1994 (1). Shortly thereafter, a protein was identified that affected the production and activity of TNF-α and IL-1 and was called Cytokine Suppression Binding Protein (CSBP); it was subsequently renamed p38α (5, 6). p38α kinase is activated as part of a signal transduction cascade in response to cellular stresses growth factors and inflammatory cytokines (Fig. 1) (7). Once activated by its upstream kinase, p38α becomes a part of the signal cascade by subsequently phosphorylating its downstream substrates such as transcription factors ATF-2, Elk-1, and MSK-1. This leads to transcriptional control of the production of primary key pro-inflammatory cytokines. Subsequent to its initial activation p38α can be further activated in an autocrine fashion by inflammatory cytokines. It is due to this ability to modulate pro-inflammatory cytokines that p38α has been implicated in a number of pathophysiological states where inflammatory cytokines are thought to play a causal role. These are the reasons that have made it an attractive target for drug discovery efforts for the treatment of a variety of diseases associated with inflammatory cytokines. A validation of the concept of modulating cytokine activity comes from the success of anti-cytokine therapies, such as etanercept (Enbrel®), infliximab (Remicade®), and anakinara (Kineret®) for the treatment of diseases such as RA. While chronic inflammation, such as in RA, has been the early focus for the development of p38α kinase inhibitors, it should not be surprising that there are several reports that address the potential utility of inhibiting this enzyme in variety of other diseases both acute and chronic including diabetes, cancer, myocardial infarctions (MI), and chronic obstructive pulmonary disease (COPD) (8-20).

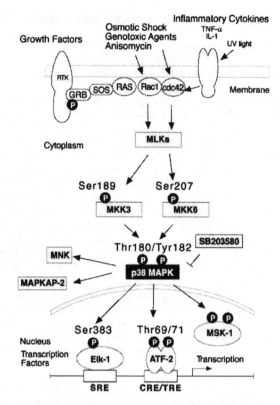

Figure 1. p38 MAP Kinase Signaling Pathway

STRUCTURAL AND MECHANISTIC BASIS FOR INHIBITION OF p38α KINASE

A combination of potent small molecule inhibitors and X-ray crystal structures have contributed to the understanding of the critical structural and chemical requirements for high affinity binding of small molecules to p38α. X-ray crystal structures have shown that p38α has a typical kinase structure, characterized by two distinct domains, the N-terminal domain and the C-terminal domain, connected by a hinge region. The N-terminal domain contains the pocket that binds ATP and the C-terminal domain contains the presumed catalytic site, the metal binding site, and the phosphorylation lip. The substrate binds in the hinge region. Until recently p38α was the only member of the p38 MAP kinase family for which X-ray structure had been solved, p38γ has since been crystallized and its structure in complex with an ATP analog determined (21-28).

The mechanism by which p38α catalyzes the phosphorylation of its substrates has been studied using both synthetic and native substrates (29, 30). These studies show that there is an ordered sequential mechanism of catalysis which is different from most other protein kinases where the mechanism of catalysis is random. Thus, for p38α, the substrate binds first, followed by the binding of ATP, phosphorylation of the substrate, release of ADP, and finally the release of the phosphorylated substrate. The structural implications of this observation are interesting as this suggests that the ATP site may be inaccessible when substrate is not bound to the enzyme and upon the binding of the substrate a conformational change occurs which allows for ATP binding. In addition to studies focussed on understanding the mechanism of catalysis of the enzyme, recent studies have also looked at the kinetics of binding of small molecule inhibitors to p38α. These studies suggest a direct correlation between the *in vitro* cellular activity of these inhibitors and the "tightness" of their binding to p38α (31, 32). This review describes inhibitors of p38α which are known to also inhibit p38β with comparable potency. They do not have any activity against p38γ and p38δ.

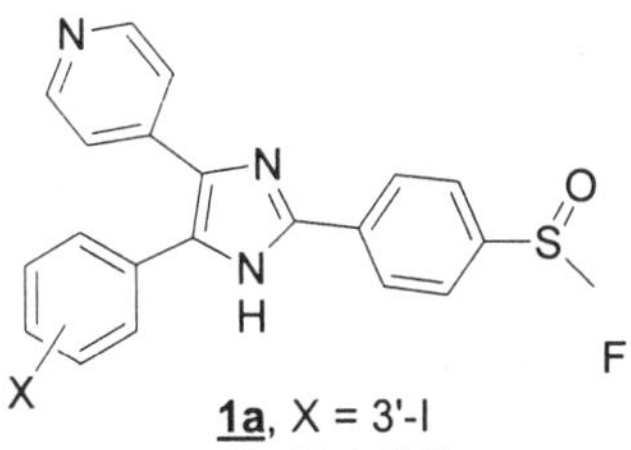

1a, X = 3'-I   **2** X = N, R = CH$_3$ **4**
1b, X = 4'-F   **3** C = CH, R = H   CF$_3$

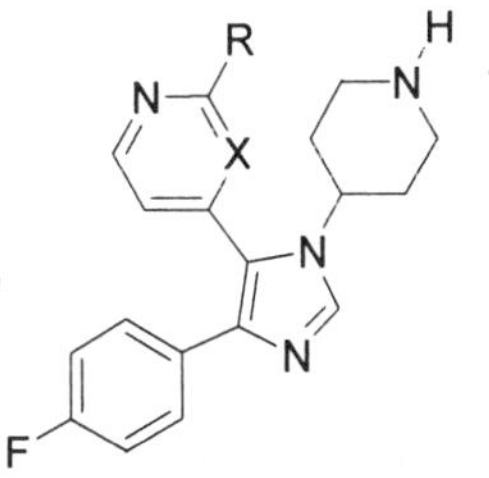

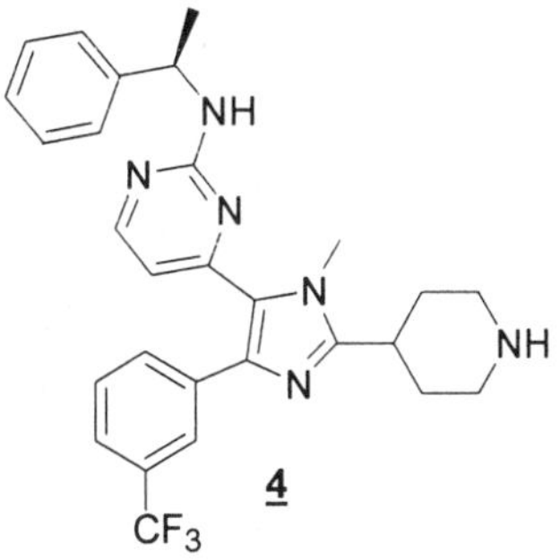

Pyridinyl-imidazoles represented by **1** were the first reported class of p38α inhibitors and proved to be excellent tools in helping elucidate the structural, functional and mechanistic basis for the functioning of this enzyme. The pyridinyl-imidazoles have been shown to be ATP competitive inhibitors. These pyridinyl-imidazoles are typically 2,4,5 tri-substituted (**1b**, SB 203580) or 1,4,5 tri-substituted imidazoles (**2**, SB 220025 and **3**, VK 19,911). A common feature in all of these molecules is the presence of a phenyl ring at the 4 position which fits into a hydrophobic pocket in the ATP binding site of the enzyme formed by Leu76, Leu86, Leu104, and the Lys53...Glu71 ion pair. The second feature common to all of the inhibitors in this general class, is the presence of a pyridine or pyrimidine ring at the 5 position, with the nitrogen at the 4' position of the pyridine ring essential for activity. It has been shown by several X-ray studies that this nitrogen is involved in a critical hydrogen bonding interaction with Met-109 in the p38α hinge region. The interaction with Met-109 is the most common interaction present in all the reported p38α inhibitors for which an X-ray structure is available. The crystal structure also provided confirmation of the observation made earlier from SAR studies that had shown that the presence and the position of the nitrogen atom on the pyridine ring was critical for activity (33). The imidazole ring is in the same region of space as the 5-membered ring of adenine. Water molecules are associated with each of the imidazole nitrogens. Finally, the sulfoxide appears to occupy a similar region of space as the phosphate groups in ATP. Weaker electron density in the X-ray structure suggests significant flexibility in this region which is partially exposed to solvent. Removal of this group results in a 3-fold loss in potency.

In addition to the above mentioned critical hydrogen bonding and hydrophobic interactions, compounds represented by **2** and **3** make an additional salt bridge interaction between the piperidine ring and Asp-168 in p38α (34). In terms of the correlation of structure-activity with actual crystallographic evidence, these inhibitors provide the clearest definition of the requirements for high affinity binding of small molecules to p38α MAP kinase. Another variant of the imidazoles is represented by compound **4** and has been reported to be 350-fold more potent than **1** (36).

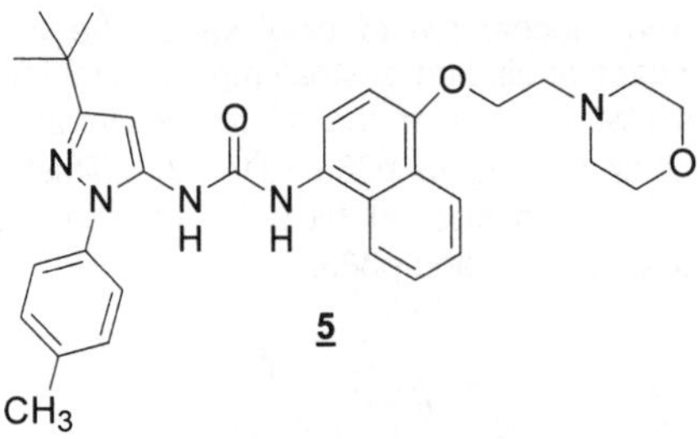

5 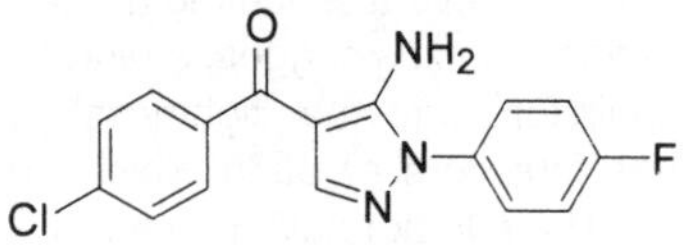   **6**

Two unique p38α inhibitors for which X-ray structures are available are represented by compounds **5** and **6**. Urea **5** is reportedly the first allosteric inhibitor of p38 α in that it uses a binding pocket that is distinct from the ATP binding pocket (34). Another interesting consequence of the binding of this di-aryl urea is a large conformational change in the conserved DFG (Asp-168-Phe-169-Gly-170) region of the kinase. In all currently known structures of p38α, the Phe-169 residue is buried in a hydrophobic pocket within the groove between the N- and the C-terminal domains of the enzyme and defined as the "DFG-in" conformation. The X-ray structure of the complex of compound **5** with p38α shows that the binding of **5** results in the translocation of Phe-169 by 10Å to a new position, and is termed as the "DFG-out" conformation. In this binding mode, none of the nitrogen atoms of the pyrrazole interact with the enzyme. A hydrogen bond is seen between the urea moiety of compound **5** and the side chain of Glu-71. The oxygen of the morpholino group makes the crucial interaction with Met-109. The X-ray structure of compound **6** shows a hydrogen bonding interaction between the carbonyl oxygen of the benzophenone and Met-109 in p38α. In this case the carbonyl oxygen makes the crucial hydrogen bonding interaction with Met-109. The amino group at the 3 position of the pyrazole ring in compound **6** is involved in forming a hydrogen bond with Thr-106.

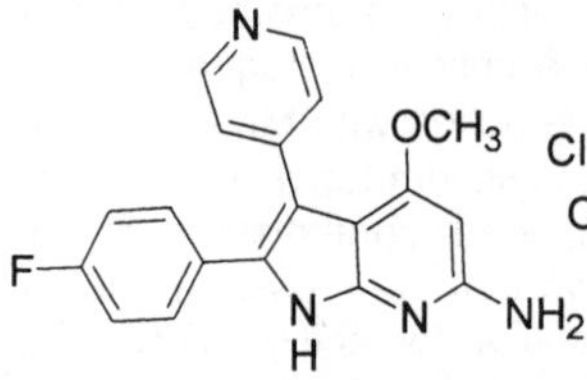

7 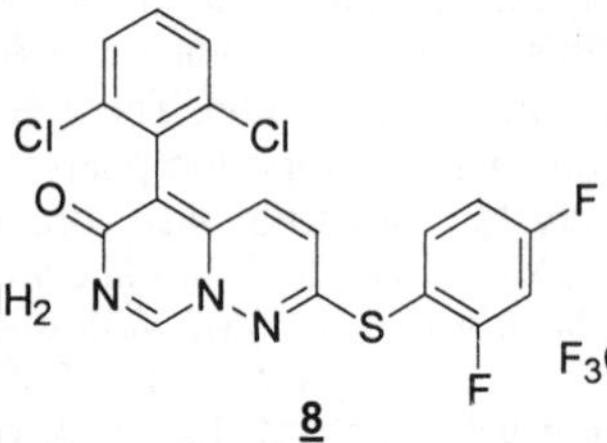   **8** 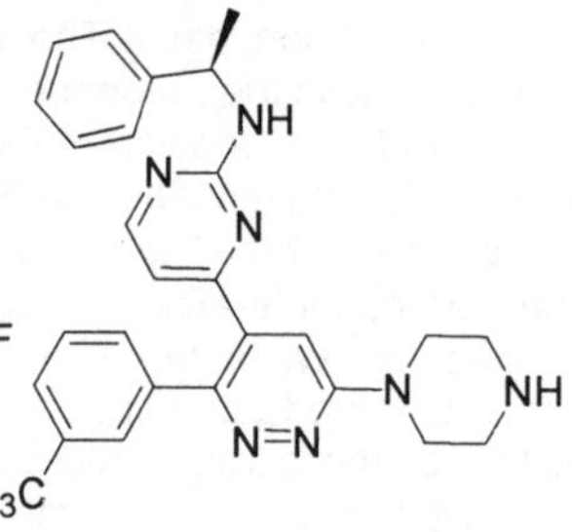   **9**

Recent reports of structure-activity relationship studies have highlighted novel classes of compounds represented by **7-9**. These compounds are highly potent and selective (34-36). X-ray structures and SAR studies have provided ample information that has led to the design of selective and potent inhibitors of p38α. This information continues to provide for the design of more novel classes of p38α inhibitors.

p38α KINASE INHIBITORS IN DEVELOPMENT

Pyrimidinyl imidazoles **10** have been investigated for the treatment of inflammatory disorders, including rheumatoid arthritis. Several compounds in this series have been investigated. As early as 1998, the focus has been on decreasing the affinity of this series of compounds for cytochrome P450 enzymes and has lead to compounds with improved P450 profiles (37-55). Pyrroles **11** have been investigated for their potential

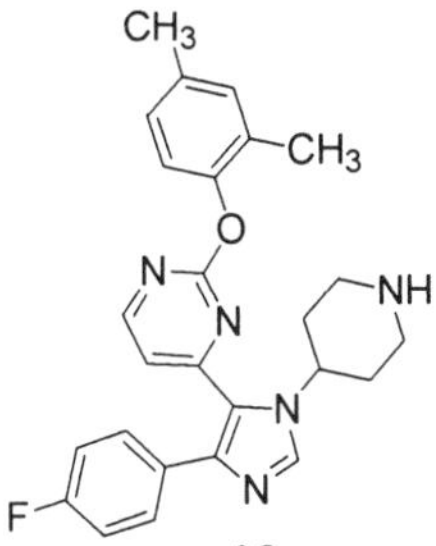

10

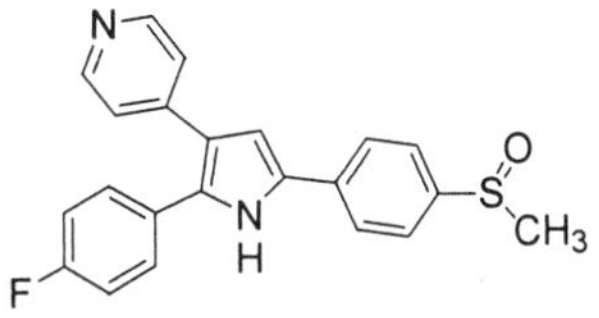

11

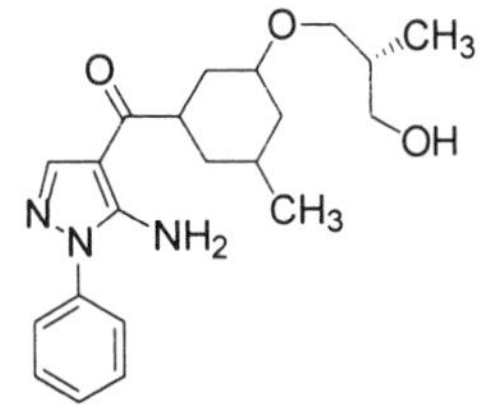

12

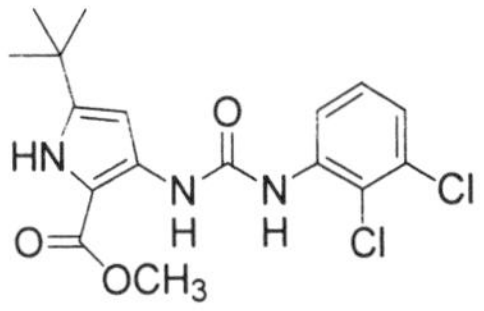

13

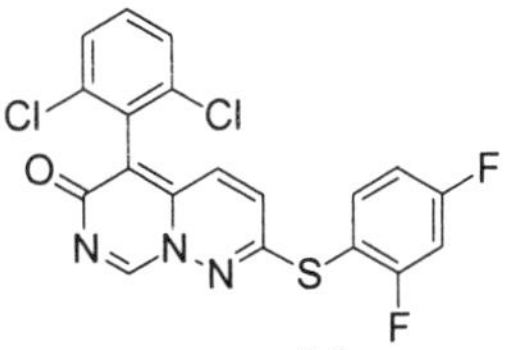

14

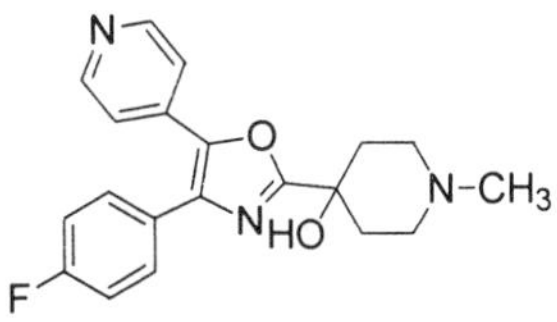

15

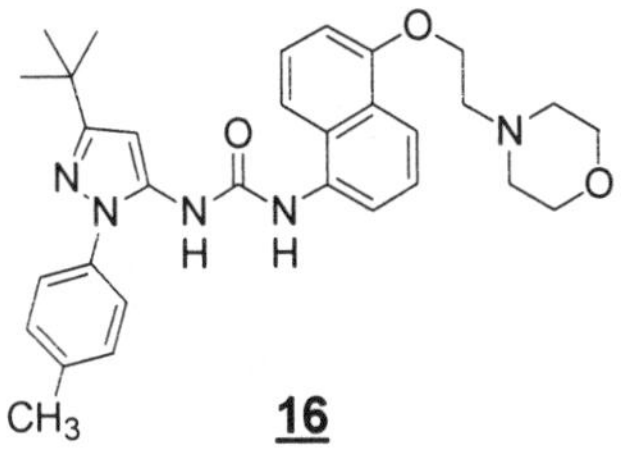

16

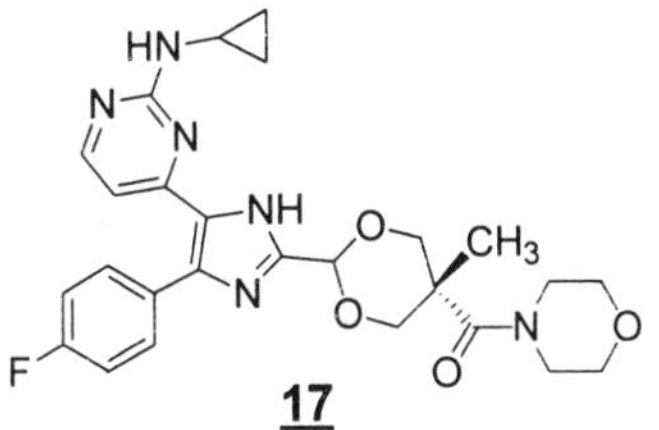

17

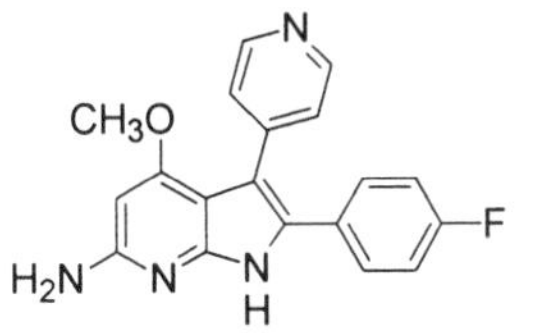

18

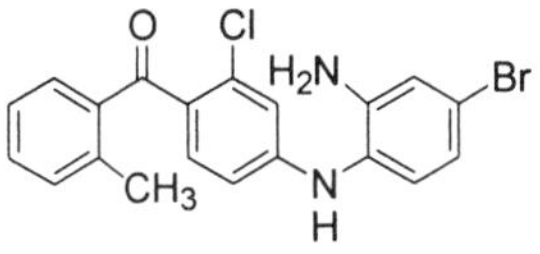

19

in the treatment of arthritis.   Compounds in this class have shown good oral bioavailability in rat and rhesus monkey. These compounds have been reported to have caused significant improvement in disease progression in models of adjuvant-induced arthritis in rats (56).   Amino pyrazoles have been investigated for the treatment of arthritis and other inflammatory diseases. RO-3201195 (**12**) is a lead compound in this series (57).   A series of urea-based inhibitors **13** as potential therapies for autoimmune diseases and associated inflammatory conditions has also been investigated (58-60). VX-745 (**14**) is the lead compound in this novel class of inhibitors and has been evaluated in clinical trials and shown to be efficacious in retarding the progression of disease in humans.   Several analogs of VX-745 have been identified as potential back-up candidates for clinical development (61). A series of 4-hydroxypiperdines (**15**) are being developed for the treatment of inflammatory disorders including rheumatoid arthritis (62). Bis-aryl ureas **16** bind to an allosteric site but are still competitive for ATP.   They are being developed as anti-inflammatory agents (63). The aza-indoles **17** are being developed as novel therapies for chronic inflammatory diseases such as rheumatoid arthritis, inflammatory bowel disease, and psoriasis (64, 65). Imidazole **18** (RPR-203494) is being developed for the treatment of chronic inflammatory diseases is reportedly in Phase I/II clinical trials (66, 67). Finally, amino benzophenone **19** is being developed for topical applications (68).

CLINICAL APPLICATIONS OF p38α MAP KINASE INHIBITORS

There are several p38α inhibitors that are presently being investigated in clinical trials. While details of these studies have yet to be formally reported, two recently published studies evaluated the effect of p38α inhibitors on suppression of pro-inflammatory cytokines and other markers of inflammation, in healthy human volunteers, exposed to endotoxemia, caused by exposure to LPS (64, 69). Summarized below are clinical settings in which p38α kinase inhibition is being evaluated as potential therapy, and others in which the evidence is strong that p38α inhibition may provide significant benefit in the management of disease(s).

Rheumatoid Arthritis - The evidence for the application of p38α inhibition in developing novel anti-inflammatory drugs is most directly related to the role of p38α in inflammatory processes mediated by cytokines such as IL-1 and TNF-α. Both of these cytokines are central regulators of immune and inflammatory responses. They play an important role in the modulation of other inducible, multifunctional cytokines such as IL-6. The benefits of treating patients with therapies that modulate the activity of TNF-α (entanercept and infliximab) and IL-1 (anakinra) have left little doubt in the minds of most researchers as to the importance of developing oral treatments for this disease. Furthermore, the success of COX-2 inhibitors in providing symptomatic pain relief in this disease has added to the potential therapeutic value of p38α inhibitors, as it has been shown that p38α activation leads to the stabilization of COX-2 message. Signaling through the SAPK/MAPK pathways is a typical feature of the chronic synovitis in RA, but not in degenerative joint disease. This points to the fact that MAP kinase signaling is found at distinct sites in the synovial tissue, and is induced by pro-inflammatory cytokines. This could lead to the design of highly targeted therapies (70-76). While there are several therapies that have recently been introduced for the treatment of RA, the need remains for the development of novel therapies that in the least arrest the progression of the disease rather than provide symptom management. It is hoped that p38α kinase inhibitors would be just such a therapy for this highly debilitating disease (77).

Cardiovascular diseases – The heart is a TNF-α producing organ. Both myocardial macrophages and cardiac myocytes synthesize TNF-α. There is accumulating evidence that myocardial TNF-α is an autocrine contributor to myocardial dysfunction and cardiomyocyte death in several settings of heart disease (78-87). The emerging role of the potential benefit of p38α inhibition in cardiovascular diseases comes from several sources, aided in part by a growing understanding of the role of TNF-α and IL-1 in cardiac hypertrophy. An important pathophysiological outcome of stress to cardiac myocytes is a hypertrophic response. This is associated with alterations in gene expression and cell morphology suggesting a significant role of p38α in cardiac myocyte hypertrophy. Selective p38α inhibition could therefore provide a method for the treatment of congestive heart failure (79).

There is evidence of the increasingly important role of IL-1 and TNF-α in the pathogenesis of acute mycocardial infarctions (80). p38α inhibitors have been shown to have beneficial effects in animal models (81, 82). Chronic heart failure, atherosclerosis, viral myocarditis, cardiac allograft rejection, and cardiac dysfunction associated with sepsis are all disorders in which both TNF-α and IL-1 are thought to play a causal role (14-16, 75, 86, 87). The central role of p38α in the regulation of these cytokines supports the hypothesis that inhibitors of p38α could serve as viable therapeutic interventions.

Neurodegenerative diseases -There are several studies that point to the potential of developing p38α inhibitors for the treatment of CNS disorders (88). TNF-α and IL-1 are produced by activated astrocytes and microglia under a variety of neuropathological conditions, including Alzheimer's disease (AD) and multiple sclerosis (MS) (89-91). p38α mediates signal transduction in the transcriptional and post-transcriptional gene expression of TNF-α and inducible nitric oxide synthase (iNOS) in immunoregulatory cells of the CNS. The cytotoxicity associated with glutamate receptor activation is proposed to be p38α mediated. The role of p38α activation in the hyperphosphorylation of tau in a variety of abnormal tau inclusions, suggests that this kinase may play a role in the development of degenerative diseases with tau pathology (92). The p38α kinase inhibitor, SD-282, abolishes the increased neuronal vulnerability in APP751 transgenic mice. SD-282 also suppresses the expression of inducible nitric-oxide synthase and the binding activity of activator protein 1 (AP-1) (93). These findings suggest that compounds preventing activation of p38α in microglia may reduce neuronal vulnerability in AD.

Another clinical application of p38α inhibitors could be their use in the setting of ischemic stroke injury. p38α inhibitors have been shown to reduce infarct size and neurological deficits in animal models of focal ischemia (94). While the role of p38α inhibition in the treatment of CNS disorders is getting increased attention, there clearly is the need for more work in this area, before a clear clinical path for p38α inhibitors can be charted.

Diabetes – Elevated glucose levels activate the p38α MAP kinase pathway in different ways. While moderate hyperglycemia can activate the p38α MAP kinase pathway by a PKC-δ dependent pathway, glucose at extremely high concentrations can activate the p38α MAP kinase pathway in a PKC-independent pathway (95). In cultures of adult rat sensory neurons, high glucose activated p38α kinase expression and resulted in cellular damage. In the dorsal ganglia of rats from a model of type I diabetes, p38α activation was observed. This data implicates the MAP kinase pathway in the etiology of diabetic neuropathy both via direct effects of glucose and glucose-induced oxidative stress (96). The p38α pathway is also activated by events that occur in the setting of diabetes, such as protein kinase C (PKC) up-regulation and cellular stresses (osmotic stress and redox changes). Substrates of activated p38 MAPK include transcription factors that are involved in the microvascular complications of diabetes. The activity of AP-1, a transcription factor complex that regulates several genes involved in diabetic nephropathy, is reversed when the p38 MAPK pathway is inhibited (97).

Cancer – The underlying inflammation associated with certain forms of cancer have been related to activation of the MAPK pathway. Cachexia, a chronic state of negative energy balance and muscle wasting that is a severe complication of cancer and chronic infection is mediated by cytokines, though the mechanism of how these molecules affect energy expenditure is unknown. Recent studies have also shown that cytokines activate PPARγ coactivator-1 (PGC-1) through phosphorylation by p38α MAP kinase, resulting in stabilization and activation of PGC-1 protein (9).

Tamoxifen (TAM) has been used in the treatment of breast cancer for over a decade. The observed clinical efficacy of TAM has been attributed to both arrest of cell growth and induction of apoptosis within breast cancer cells. Although the primary mechanism of action of TAM is believed to be through antagonism of the estrogen receptor, research indicates that TAM may be effective *via* additional, non-ER mediated mechanisms. These include the modulation of signaling proteins such as protein kinase C, calmodulin, transforming growth factor-beta, the protooncogene *c-myc*, caspases and mitogen-activated protein kinases (MAPK), including p38α (98).

The invasive phenotype of cancers critically depends on the expression of proteases such as the type IV collagenase (MMP-9). Several growth factors and oncogenes have been found to increase protease expression. This frequently requires the activation of the transcription factor AP-1. The tumor promoter TPA has been shown to induce MMP-9 expression via the p38α pathway. Dominant negative p38α reduces MMP-9 promoter activity in CAT assays, while a construct encoding an activating mutation in the MKK-6 (an activator of p38α activity) potently stimulates it. Compound **1**, reduces MMP-9 expression/secretion and in vitro invasion of cancer cells. This evidence suggests that the SAPK/MAPK signaling cascade, stimulates MMP-9 expression in an AP-1-dependent fashion (8). Inhibitors of p38α could reduce the invasive phenotype of several cancers.

<u>Conclusions</u> – p38α kinase inhibitors provide a tremendous opportunity for the development of agents targeting a variety of human diseases through its central role in acute and chronic inflammation. This central role may also raise concerns of its key role in immunocompetence. Only as compounds progress through clinical development will the true nature of p38 inhibition be fully realized.

References

1. J. Han, J. D. Lee, Y. Jiang, Z. Li, L. Feng and R. J. Ulevitch, Sci., <u>265</u>, 808 (1994).
2. J. Han, J. D. Lee, Y. Jiang, Z. Li, L. Feng and R.J. Ulevitch, J.Biol.Chem. <u>271</u>, 17920 (1996).
3. Z. Li, Y. Jiang, R. J. Ulevitch and J. Han, Biochem.Biophys.Res.Commun. <u>228</u>, 334 (1996).
4. B. Ge, H. Gram, F. Di Padova, B. Huang, L. New, R. J. Ulevitch, Y. Luo and J. Han, Sci., <u>295</u>, 1291 (2002).
5. J.C. Lee, J. T. Laydon, P. C. McDonnell, T. F. Gallagher, S. Kumar, D. Green, D. McNulty, M. J. Blumenthal, J. R. Heys, and S. W. Landvatter, Nature, <u>372</u>, 739 (1994).
6. T. F. Gallagher, G. L. Seibel, S. Kassis, J. T. Laydon, M. J. Blumenthal, J. C. Lee, D. Lee, J. C. Boehm, S. M. Fier-Thompson, J. W. Abt, M. E. Soreson, J. M. Smietana, R. F. Hall, R. S. Garigipati, P. E. Bender, K. F. Erhard, A. J. Krog, G. A. Hofmann, P. L. Sheldrake, P. C. McDonnell, S. Kumar, P. R. Young and J. L. Adams, Bioorg.Med.Chem., <u>5</u>(1), 49 (1997).
7. K. Ono and J. Han, Cell Signal, <u>12</u>(1), 1 (2000).
8. T.A. Purves, T., A. Middlemas, S. Agthong, E. B. Jude, A. J. Boulton, P. Fernyhough and D. R. Tomlinson, FASEB J. <u>15</u>(13), 2508 (2001).
9. M. Igarashi, H. Wakasaki, N. Takahara, H. Ishii, Z. Y. Jiang, T. Yamauchi, K. Kuboki, M. Meier, C. J. Rhodes and G. L. King, J.Clin.Invest., <u>103</u>(2), 185(1999).
10. M. M. Manson, K. A. Holloway, L. M. Howells, E. A. Hudson, S. M. Plummer, M. S. Squires and S. A. Prigent, Biochem.Soc.Trans., <u>28</u>(2), 7 (2000).
11. Q. Ding, Q. Wang and B.M. Evers, Biochem.Biophys.Res.Commun., <u>284</u>(2), 282 (2001).
12. U. K. Misra, G. Gawdi, G. Akabani and S. V. Pizzo, Cell Signal, <u>14</u>(4), 327 (2002).
13. Y. S. Guo, M. R. Hellmich, X. D. Wen and C. M. Townsend, Jr., J.Biol.Chem., <u>276</u>(25), 22941 (2001).
14. F. T. Gao, T. L. Yue, D. W. Shi, T. A. Christopher, B. L. Lopez, E. H. Ohlstein, F. C. Barone and X. L. Ma, Cardiovasc.Res., <u>53</u>(2), 414 (2002).
15. M. M. Mocanu, G. F. Baxter, Y.Yue, S. Critz, D. Yellon, Basic Res.Cardiol., <u>95</u>, 472 (2000).
16. A. Nakano, M. V. Cohen, S. Critz and J. M. Downey, Basic Res.Cardiol., <u>95</u>(6), 466 (2000).
17. M. Joyeux, A. Boumendjel, R. Carroll, C. Ribuot, D. Godin-Ribuot and D. M. Yellon, Cardiovasc.Drugs Ther., <u>14</u>(3), 337 (2000).
18. A. Nakano, C. P. Baines, S. O. Kim, S. L. Pelech, J. M. Downey, M. V. Cohen and S. D. Critz, Circ.Res., <u>86</u>(2), 144 (2000).
19. D. C. Underwood, R. R. Osborn, S. Bochnowicz, E. F. Webb, D. J. Rieman, J. C. Lee, A. M. Romanic, J. L. Adams, D. W. Hay and D. E. Griswold, Am.J.Phsiol., <u>279</u>(5), L895 (2000).
20. P. J. Barnes, Respiration, <u>68</u>(5), 441 (2001).
21. S. Bellon, M.J. Fitzgibbon, T. Fox, H. Hsiao, K. Wilson, Struct.Fold Des.,1999. <u>7</u>, 1057 (1999).
22. J. C. Lee, S. Kassis, S. Kumar, A. Badger and J.L. Adams, Pharmacol.Ther., <u>82</u>, 389 (1999).
23. T. Fox, J. T. Coll, X. Xie, P. J. Ford, U. A. Germann, M. D. Porter, S. Pazhanisamy, M. A. Fleming, V. Galullo, M. S. Su and K. P. Wilson, Protein Sci., <u>7</u>(11), 2249 (1998).
24. Z. Wang, B. J. Canagarajah, J. C. Boehm, S. Kassisa, M. H. Cobb, P. R. Young, S. Abdel-Meguid, J. L. Adams and E. J. Goldsmith, Structure, <u>6</u>(9), 1117 (1998).
25. R. J. Gum, M. M. McLaughlin, S. Kumar, Z. Wang, M. J. Bower, J. C. Lee, J. L. Adams, G. P. Livi, E. J. Goldsmith and P. R. Young, J.Biol.Chem., <u>273</u>(25), 15605 (1998).

26. L. Tong, S. Pav, D. M. White, S. Rogers, K. M. Crane, C. L. Cywin, M. L. Brown and C. A. Pargellis, Nat.Struct.Biol., 4(4), 311 (1997).
27. Z. Wang, P. C. Harkins, R. J. Ulevitch, J. Han, M. H. Cobb and E. J. Goldsmith, Proc.Natl.Acad.Sci. U.S.A., 94(6), 2327 (1997).
28. K. P. Wilson, M. J. Fitzgibbon, P. R. Caron, J. P. Griffith, W. Chen, P. G. McCaffrey, S. P. Chambers and M. S. Su, J.Biol.Chem., 271(44), 27696 (1996).
29. L. A. Shewchuk, A. Hassell, B. Wisely, W. Rocque, W. Holmes, J. Veal and L. F. Kuyper, J. Med.Chem., 43(1),133 (2000).
30. S. Pav, D. M. White, S. Rogers, K. M. Crane, C. L. Cywin, W. Davidson, J. Hopkins, M. L. Brown, C. A. Pargellis and L. Tong, Protein Sci., 6(1), 242 (1997).
31. J. Lisnock, A. Tebben, B. Frantz, E. O'Neill, G. Croft, S. O'Keefe, B. Li, C. Hacker, S. deLaszlo, A. Smith, B. Libby, N.Liverton, J.Hermes, P.LoGrasso, Biochem., 37, 16573 (1998).
32. P. V. LoGrasso, B. Frantz, A. M. Rolando, S. J. O'Keefe, J. D. Hermes and E. A. O'Neill, Biochem., 36(34), 10422 (1997).
33. R. L. Thurmond, S. A. Wadsworth, P. H. Schafer, R. A. Zivin and J. J. Siekierka, Eur.J.Biochem., 268(22), 5747 (2001).
34. C. Pargellis, L. Tong, L. Churchill, P. F. Cirillo, T. Gilmore, A. G. Graham, P. M. Grob, E. R. Hickey, N. Moss, S. Pav and J. Regan, Nat.Struct.Biol., 9(4), 268 (2002).
35. F. G. Salituro, U. A. Germann, K. P. Wilson, G. W. Bemis, T. Fox and M. S. Su, Current Medicinal Chemistry, 6, 807 (1999).
36. N. J. Liverton, J. W. Butcher, C. F. Claiborne, D. A. Claremon, B. E. Libby, K. T. Nguyen, S. M. Pitzenberger, H. G. Selnick, G. R. Smith, A. Tebben, J. P. Vacca, S. L. Varga, L. Agarwal, K. Dancheck, A. J. Forsyth, D. Fletcher, B. Frantz, W. A. Hanlon, C. Harper, S. J. Hofsess, M. Kostura, J. Lin, S. Luell, E. O'Neill, S. O'Keefe, J.Med.Chem., 42, 2180 (1999).
37. J. J. Haddad, Curr.Opin.Invest.Drugs, 2(8), 1070 (2001).
38. J. R. Henry, K. C. Rupert, J. H. Dodd, I. J. Turchi, S. A. Wadsworth, D. E. Cavender, B. Fahmy, G. C. Olini, J. E. Davis, J. L. Pellegrino-Gensey, P. H. Schafer and J. J. Siekierka, J.Med.Chem., 41, 4196 (1998).
39. C. J. McIntyre, G. S. Ponticello, N. J. Liverton, S. J. O'Keefe, E. A. O'Neill, M. Pang, C. D. Schwartz and D. A. Claremon, Bioorg.Med.Chem.Lett., 12, 689 (2002).
40. J. G. Steadman and A. K. Takle, PCT Patent Applic. WO00/166539 (2000).
41. A. Gaiba, A. K. Takle and D. M. Wilson, PCT Patent Applic. WO00/166540 (2000).
42. J. L. Adams, J. C. Boehm, R.F. Hall, J.J. Taggart, PCT Patent Applic. WO/00164679 (2000).
43. D. K. Dean, P. J. Lovell and A. K. Takle, PCT Patent Applic. WO/00138324 (2000).
44. E. A. Irving and A. A. Parsons, PCT Patent Applic. WO00/064422 (2000).
45. J. L. Adams, PCT Patent Applic. WO00/019824 (2000).
46. J. L. Adams and D. Lee, PCT Patent Applic. WO00/010563 (2000).
47. G. Z. Feuerstein, PCT Patent Applic. WO96/01440 (1996).
48. T. F. Gallagher, J.C. Boehm and J.L. Adams, PCT Patent Applic. WO99/32121 (1999).
49. J. L. Adams, D. Lee and S.A. Long, PCT Patent Applic. WO98/57966 (1998).
50. J. L. Adams, T. Gallagher and K. Osifo, PCT Patent Applic. WO98/56377 (1998).
51. J. L. Adams, T. Gallagher, K. Osifo and J. Boehm, PCT Patent Applic. WO98/25619 (1998).
52. J. L. Adams and J.C. Boehm, PCT Patent Applic. WO97/25046 (1997).
53. J. L. Adams, T. F. Gallagher, J. Sisko, Z.Q. Peng, K. Osifo and J.C. Boehm, PCT Patent Applic. WO96/40143 (1996).
54. J. L. Adams, T.F. Gallagher, R.S. Garigpati, K. Osifo and J.C. Boehm, J. Sisko, Z.Q. Peng, and J.C. Lee, PCT Patent Applic. WO96/21452 (1996).
55. J. L. Adams, J. C. Boehm, S. Kassis, P. D. Gorycki, E. F. Webb, R. Hall, M. Sorenson, J. C. Lee, A. Ayrton, D. E. Griswold and T. F. Gallagher, Bioorg.Med.Chem.Lett., 8, 3111 (1998).
56. S. E. de Laszlo, D. Visco, L. Agarwal, L. Chang, J. Chin, G. Croft, A. Forsyth, D. Fletcher, B. Frantz, C. Hacker, W. Hanlon, C. Harper, M. Kostura, B. Li, S. Luell, M. MacCoss, N. Mantlo, E. A. O'Neill, C. Orevillo, M. Pang, J. Parsons, A. Rolando, Y. Sahly, K. Sidler, and S. J. O'Keefe, Bioorg.Med.Chem.Lett., 8(19), 2698 (1998).
57. Pharmaprojects, PJB Publications, Ltd., London (2002).
58. J. Dumas, U. Khire, T. Lowinger, M. Monahan, R. Natero, J. Renick, B. Ridel, W. Scott, R. Sibley, R. Smith and J. Wood, PCT Patent Applic. WO00/41698 (2000).
59. J. Dumas, U. Khire, T.B. Lowinger, B. Riedl, W.J. Scott, J.E. Wood, H. Hatoum-Mokdad, A. Redman and R. Sibley, PCT Patent Applic. WO99/32110 (1999).
60. S. Miller, J. Dumas, U. Khire, T.B. Lowinger, B. Riedl, W.J. Scott, R.A. Smith, J.E. Wood, D. Gunn, H. Hatoum-Mokdad, M.Rodriguez, R.Sibley, PCT Patent Applic. WO99/32463 (1999).
61. F. Salituro, G. Bemis and J. Cochran, PCT Patent Applic. WO99/64400 (1996).
62. L. Revesz, F. E. Di Padova, T. Buhl, R. Feifel, H. Gram, P. Hiestand, U. Manning and A. G. Zimmerlin, Bioorg.Med.Chem.Lett., 10(11), 1261 (2000).

63. M. L. Cappola,G.W. Gereg and S. Way, PCT Patent Applic. WO02/07772 (2002).
64. J. W. Fijen, J. G. Zijlstra, P. De Boer, R. Spanjersberg, J. W. Cohen Tervaert, T. S. Van Der Werf, J. J. Ligtenberg and J. E. Tulleken, Clin.Exp.Immunol., 124(1), 16 (2001).
65. S. A. Wadsworth, D. E. Cavender, S. A. Beers, P. Lalan, P. H. Schafer, E. A. Malloy, W. Wu, B. Fahmy, G. C. Olini, J. E. Davis, J. L. Pellegrino-Gensey, M. P. Wachter and J. J. Siekierka, J.Pharm.Expt.Ther., 291(2), 680 (1999).
66. A. J. Collis, M. L. Foster, F. Halley, C. Maslen, I. M. McLay, K. M. Page, E. J. Redford, J. E. Souness and N. E. Wilsher, Bioorg.Med.Chem.Lett., 11(5), 693 (2001).
67. L. M. McLay, F. Halley, J. E. Souness, J. McKenna, V. Benning, M. Birrell, B. Burton, M. Belvisi, A. Collis, A. Constan, M. Foster, D. Hele, Z. Jayyosi, M. Kelley, C. Maslen, G. Miller, M. C. Ouldelhkim, K. Page, S. Phipps, K. Pollock, B. Porter, A. J. Ratcliffe, E. J. Redford, S. Webber, B. Slater, V. Thybaud and N. Wilsher, Bioorg.Med.Chem., 9(2), 537 (2001).
68. E. R. Ottosen and H.Z. Dannacher, PCT Patent Applic. WO01/05749 (2001).
69. S. R. Bartlett, R. Sawdy and G. E. Mann, J.Physiol., 520(2), 399 (1999).
70. J. Branger, B. van Den Blink, S. Weijer, J. Madwed, C. L. Bos, A. Gupta, C. L. Yong, S. H. Polmar, D. P. Olszyna, C. E. Hack, S. J. van Deventer, M. P. Peppelenbosch and T. van Der Poll, J.Immunol., 168(8), 4070 (2002).
71. J. C. Lee, S. Kumar, D. E. Griswold, D. C. Underwood, B. J. Votta and J. L. Adams, Immunopharmacology, 47(2-3), 185 (2000).
72. A. M. Badger, D. E. Griswold, R. Kapadia, S. Blake, B. A. Swift, S. J. Hoffman, G. B. Stroup, E. Webb, D. J. Rieman, M. Gowen, J. C. Boehm, J. L. Adams and J. C. Lee, Arthritis Rheum., 43(1), 175 (2000).
73. M. Suzuki, T. Tetsuka, S. Yoshida, N. Watanabe, M. Kobayashi, N. Matsui and T. Okamoto, FEBS Lett., 465(1), 23 (2000).
74. J. R. Jackson, B. Bolognese, L. Hillegass, S. Kassis, J. Adams, D. E. Griswold and J. D. Winkler, J.Pharmacol.Exp.Ther., 284(2), 687 (1998).
75. A. M. Badger, J. N. Bradbeer, B. Votta, J. C. Lee, J. L. Adams and D. E. Griswold, J.Pharmacol.Exp.Ther., 279(3), 1453 (1996).
76. G. M. Schett, M. Tohidast-Akrad, J. S. Smolen, B. J. Schmid, C. W. Steiner, P. Bitzan, P. Zenz, K. Redlich, Q. Xu and G. Steiner, Arthritis Rheum., 43(11), 2501 (2000).
77. H. M. Lorenz and J. R. Kalden, Z.Rheumatol., 60(5) 326 (2001).
78. N. H. Bishoprict, P. Andreka, T. Slepak, K. A. Webster, Curr.Opin.Pharmacol., 1, 141 (2001).
79. D. R. Meldrum, Am.J.Physiol, 274(3), R577 (1998).
80. P. Andreka, Z. Nadhazi, G. Muzes and N. H. Bishoprict, Orv.Hetil, 142(32), 1717 (2001).
81. C. A. Dinarello and B. J. Pomerantz, Blood Purif., 19(3), 314 (2001).
82. T. A. Fischer, S. Ludwig, E. Flory, S. Gambaryan, K. Singh, P. Finn, M. A. Pfeffer, R. A. Kelly and J. M. Pfeffer, Hypertension, 37(5) 1222 (2001).
83. P. H. Sudgen, Ann.Med., 33(9), 611 (2001).
84. Z. B. Lei, Z. Zhang, Q. Jing, Y. Qin, G. Pei, B. Cao, X.Y. Li, Cardiovasc.Res., 53, 524 (2002).
85. N. Ohashi, A. Matsumori, Y. Furukawa, K. Ono, M. Okada, A. Iwasaki, T. Miyamoto, A. Nakano and S. Sasayama, Arterioscler.Thromb.Vasc.Biol., 20(12), 2521 (2000).
86. I. Sano, T. Takahashi, T. Koji, H. Udono, K. Yui and H. Ayabe, J.Heart Lung Transplant., 20(5), 583 (2001).
87. D. Jarrar, P. Wang, G.Y. Song, W.G. Cioffi, K.I. Bland, I.Chaudry, Ann.Surg., 231, 399 (2000).
88. M. Hull, K. Lieb and B. L. Fiebich, Curr.Med.Chem., 9(1), 83 (2002).
89. S. L. Dunn, E. A. Young, M. D. Hall and S. McNulty, Glia, 37(1), 31 (2002).
90. J. Zhu, C. A. Rottkamp, A. Hartzler, Z. Sun, A. Takeda, H. Boux, S. Shimohama, G. Perry and M. A. Smith, J. Neurochem., 79(2), 311 (2001).
91. T. G. D'Aversa, K. M. Weidenheim and J. W. Berman, Am.J.Pathol, 160(2), 559 (2002).
92. C. Atzori, B. Ghetti, R. Piva, A. N. Srinivasan, P. Zolo, M. B. Delisle, S. S. Mirra and A. Migheli, J.Neuropathol.Exp.Neurol., 60(12), 1190 (2001).
93. M. Koistinaho, M. I. Kettunen, G. Goldsteins, R. Keinanen, A. Salminen, M. Ort, J. Bures, D. Liu, R. A. Kauppinen, L.S. Higgins, J. Koistinaho, Proc.Natl.Acad.Sci.U.S.A., 99, 1610 (2002).
94. F. C. Barone, E. A. Irving, A. M. Ray, J. C. Lee, S. Kassis, S. Kumar, A. M. Badger, J. J. Legos, J. A. Erhardt, E. H. Ohlstein, A. J. Hunter, D. C. Harrison, K. Philpott, B. R. Smith, J. L. Adams and A. A. Parsons, Med.Res.Rev., 21(2), 129 (2001).
95. W. A. Wilmer, C. L. Dixon and C. Hebert, Kidney Int., 60(3), 858 (2001).
96. P. Puigserver, J. Rhee, J. Lin, Z. Wu, J. C. Yoon, C. Y. Zhang, S. Krauss, V. K. Mootha, B. B. Lowell and B. M. Spiegelman, Mol.Cell, 8(5), 971 (2001).
97. S. Mandelkar and A. N.Kong, Apoptosis, 6(6), 469 (2001).
98. C. Simon, M. Simon, G. Vucelic, M. J. Hicks, P. K. Plinkert, A. Koitschev and H. P. Zenner, Exp.Cell Res., 271(2), 344 (2001).

Editor: Janet M. Allen,
Inpharmatica, London, United Kingdom

Chapter 19. Expanding and Exploring Cellular Pathways for Novel Drug Targets

Malcolm P. Weir, John P. Overington and Marlon Schwarz
Inpharmatica Ltd.
60 Charlotte Street, London, W1T 2NU

Introduction – Over the past decade, there has been an enormous growth in understanding of biological processes at a cellular and molecular level, driven by advances in a wide range of techniques and by the ready availability of the complete genome sequence of yeast, fly and worm and of the majority of the human genome sequence (1,2). These advances have led to the burgeoning field of "functional genomics", which is defined here as the use of scalable technologies in combination with genome sequence to solve biological problems. These techniques have generated large datasets of mRNA expression levels or protein-protein interactions, which have often proved to have little real impact. This situation has dramatically improved in the past 2-3 years as experimental design and the methods themselves have matured, often reinforced by parallel advances in data analysis and genome annotation as necessary aids to interpretation (3). Systematic, rapid description of the metabolic and signalling pathways that constitute the machinery of the cell has become a realistic prospect, as has the discovery of drug targets within them. This review will outline the emerging concepts that are transforming the practice of molecular cell biology, some of the technologies that underpin this transformation, and the likely future impact on drug discovery.

PATHWAYS, NETWORKS AND DISEASES

Classical biochemistry gave rise to the concept of metabolic pathways, and many drug targets are at control points in such pathways, for example the Cox-1 and Cox-2 enzymes that are at the apex of prostanoid biosynthesis (4). Molecular cell and developmental biology have led to a massive growth in understanding of signalling pathways through which cell surface and intracellular processes are connected, and which are typically governed by protein-protein interactions. Such pathways, given their pivotal role in biological mechanisms and, therefore in pathophysiology, are seen as a rich source of drug targets and, in the case of the cancer target bcr-abl tyrosine kinase, has led to a launched small-molecule inhibitor (5). Although there are a number of signalling targets in pharma or biotech pipelines (typically kinases or caspases, that have arisen from the study of intracellular pathways), these have generally been biologically validated through academic research and so are common to many companies. It is far from clear whether many of these novel targets will prove druggable or efficacious, and a more systematic exploration of pathways for the *best* targets is needed; to do so they will need to be evaluated in as comprehensive a biological context as possible, both for the likely effects of antagonism/agonism on disease (i.e. their biological validity) and their potential for binding a selective, orally available small molecule (i.e. their druggability). Before some applications of large-scale pathway expansion and analysis are outlined in the next section, it is worth considering the nature of biological pathways and networks in general terms, so that the target discovery challenge can be appreciated.

SYSTEMS BIOLOGY

Whilst genome sequencing and related activities are giving us the basic list of genes and proteins that are fundamental to the cellular machinery, they do not tell us how this machinery is assembled nor how it operates on a macroscopic level. Systems biology is an emerging field which seeks to explain the behaviour of protein pathways through consideration of large sets of components, not just small clusters or strands abstracted from the whole. The field draws its methods by analogy to engineering and computer science and the modelling of complex systems. A full description of systems biology and the mathematical study of complexity in general are beyond the scope of this review but two cardinal features, structure and robustness, are briefly summarised here (6,7).

Systems structure - At a fundamental level, any system comprises a "parts list" of components, and for the cell these are primarily proteins, whose sequences and 3D structures provide the basic list, and interacting nucleic acid and small-molecule metabolites. Signalling pathways are generally built from multi-protein complexes that vary in space and time, and which are further combined to create networks of interacting pathways. Such physical networks comprise a cell map which is the next highest level of complexity from the sequence/structure description of the proteome (8). Structural connectivities between proteins are probably the most powerful single predictor of cellular function, since they define discrete protein complexes or "modules", for example those involved in cell cycle control or apoptosis, that to a first approximation are interchangeable amongst many cell types or organisms (9). A further level of control and complexity is afforded by the cytoskeleton which acts as a dynamic scaffold for channeling reactions, as do scaffold or anchor proteins that are key components in the assembly of signalling pathways; compartmentalization in organelles provides another level of organisation (10).

The layers of structural organisation that build up biological systems are illustrated in Figure 1 alongside their approximate relationships to function.

Robustness - Biological processes are typically robust to environmental or genetic change, exhibiting adaptability and only rarely failing catastrophically (7). An analysis of biological networks suggests that they follow a power law distribution, that relatively few nodes are highly interconnected, and that these are associated with lethal mutations (11); in general, networks display redundancy, partial overlap, modularity and feedback regulation which provide mechanisms for "buffering" change (12). Much of the complexity seen in higher organisms may be attributed to the need for elaborate regulatory systems that maintain fine control and homeostasis, which will scale geometrically as the components list (genes) increases; by analogy, much of the engineered complexity of a Boeing 777 is required for robustness rather than basic flight capabilities (13).

Biological robustness provides a qualitative explanation for the frequently observed lack of efficacy seen for unprecedented targets; it may also be a saviour with respect to toxicity of what might otherwise be expected to be impossibly "dirty" drugs. The challenge from a systems angle is to select targets that will occupy the middle ground between essentiality and redundancy, and that will rebalance dysregulated pathways in disease.

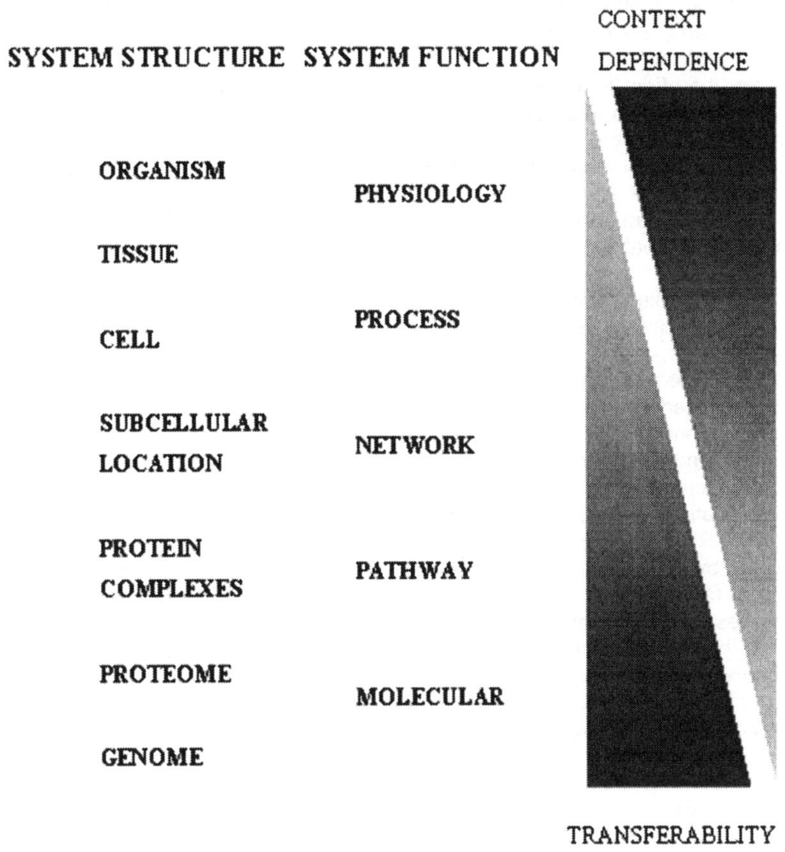

SYSTEM STRUCTURE   SYSTEM FUNCTION

CONTEXT
DEPENDENCE

ORGANISM

PHYSIOLOGY

TISSUE

CELL

PROCESS

SUBCELLULAR
LOCATION

NETWORK

PROTEIN
COMPLEXES

PATHWAY

PROTEOME

MOLECULAR

GENOME

TRANSFERABILITY

Fig.1. Schematic view of biological system structure and its broad relationship to function. Increased levels of structural complexity approximate to a corresponding increase in functional complexity and dependence on biological context; conversely the transferability of functional "components" between system structures (for example between cell types or organisms) is much greater at reduced levels of complexity.

Modelling pathways and cellular systems is an active field, and in spite of the complexity and lack of completeness of the structural components list, simulations are showing promise and delivering meaningful results under certain circumstances. For example, the effect of a mutant cardiac sodium channel was simulated and the basis for its elongation of the QT interval seen in the electrocardiogram; this modelling exercise incorporated physiological, anatomical, cellular, gene expression and electrophysiological information (14). The use of symbolic programming languages to represent modules and functional descriptions is an approach that will enable large and incomplete sets of information to be handled usefully as in the case of the EcoCyc E.coli metabolic pathway database (15).

The principles outlined above are nicely illustrated by the eukaryotic gene expression pathway (16). Large-scale mapping of protein-protein interactions, genetic pathway expansion, biochemical reconstitution experiments and structural studies have pieced together many aspects of this process, which is now seen to comprise highly regulated subdivisions of a continuous process. For example capping of the 5' end of nascent mRNA occurs at an early checkpoint such that uncapped transcripts are not extended; a series of specific protein-protein interactions and phosphorylation events coordinate this checkpoint. Pre-mRNA splicing, elongation and 3' polyadenylation are similarly coordinated through assembly along the C-terminal domain of RNA polymerase II, which itself comprises 52 heptapeptide repeats that are regulated by phosphorylation. Co-repressors and co-activators of transcription factors regulate transcription initiation through chromatin remodelling, invoking histone acetylases, deacetylases, methyltransferases and kinases; co-regulators play critical roles in the tissue-specific responses of nuclear hormone receptors, and so do their antagonists and agonists (17). A multi-protein complex, the Mediator, governs transduction of co-regulator information to the RNA polymerase, probably acting as a physical bridge; the human mediator complex comprises about 25 subunits, which vary in composition around a common core and act as a "control panel" for integration and transduction of diverse cell-specific signals, so displaying modularity and finely tuned regulation (18).

FUNCTIONAL GENOMICS APPLICATIONS

In principle, any cell biology technique can be automated and resultant datastreams integrated and interpreted in order to focus precise manual experimentation. However, pathway mapping and target validation ultimately requires synthesis of information from many sources (biochemistry, structural biology, mRNA and protein expression, tissue distribution, immunohistochemistry, gene knockout, regulated transgenics, model organism genetics etc) not all of which are readily scaled in a meaningful way. This review will concentrate on examples from gene expression analysis, the first functional genomic technology to be robustly scaled, and from recent advances in the proteome-level identification of protein complexes by mass spectrometry.

DNA Microarrays - One of the first convincing uses of genome-wide DNA microarrays was for monitoring downstream signalling during the yeast pheromone response (19). This pathway involves the archetypal G-protein coupled receptor/MAP kinase cascade, which in turn was found by expression analysis to be linked to three other MAPK pathways activated by cell surface stress, high osmolarity and filamentous growth. A series of 46 gene deletion/overexpression experiments were correlated with significant expression changes in 383 transcripts, and the results clustered to reveal functional relationships. As well as pathway-specific expression of sets of genes, higher-order relationships between processes were revealed, for example sequential activation of the pheromone and protein kinase C regulated pathways. Extension of this approach to human cells has clear implications for pathway expansion, cross-talk and mechanistic studies of drug action (20).

An example of pathway cross-talk elucidated by the use of a 9984-element cDNA microarray is provided by a study of TRAIL-mediated gene expression in breast carcinoma cells (21). TRAIL is an apoptosis-inducing member of the tumour necrosis factor family, which was found to induce three sets of genes (early, middle and late) spanning a 24h period. The early set includes a wide range of proteins not previously known to be associated with TRAIL; the middle set includes known members associated with the TNF pathway as well as novel observations, and the late set strongly correlates unexpectedly with induction of the interferon pathway. Subsequent combination of TRAIL and interferon-beta *in vitro* synergistically induced apoptosis and caspase activation in breast cancer cells. The authors concluded from this study that multiple levels of cross-talk exist between these two diverse cytokine pathways, which has implications for target discovery and combination therapy.

Microarrays have also been used to describe the effects of overexpression of the transcription factor EGR1 on prostate carcinoma cells (22). EGR1 is naturally overexpressed in prostate cancer, and its target genes were found in the cell line to include several growth factors (including insulin-like growth factor and platelet-derived growth factor) as well as neuroendocrine genes, proteases and signalling proteins.

Examples of specific identification of potential drug targets using DNA microarrays that have reached the literature are sparse, but one clear case is that of superoxide dismutase (SOD) in cancer cell killing (23). SOD eliminates superoxide and so protects cells from free-radical mediated damage; it also has abnormally low levels of activity in cancer cells, rendering them particularly sensitive to free-radical induced damage. Oestrogen derivatives were found serendipitously to kill human leukaemia cells, but not normal lymphocytes; 2-methoxyoestradiol caused this effect, and in a search for candidate targets CuZnSOD mRNA was found to be 2-fold increased. Biochemical investigation showed this to be due to a decrease in cellular SOD activity and consequent feedback upregulation of SOD expression. Limited SAR and further functional studies confirmed the view that methoxyoestradiol selectively kills cancer cells through SOD inhibition.

<u>Proteome analysis</u> - Protein mass spectrometry is now able to detect very low levels of protein and, when combined with proteolytic digestion and genome database searching, can unambiguously identify proteins at very high throughput (24). When applied to immunoprecipitates or other affinity purified protein complexes, this technique has led to the assignment of cellular function to hundreds of human proteins over the past few years, including novel drug targets such as caspase-8 and I-kappa kinase 2 in the tumour necrosis factor (25, 26). It has proved possible to scale up this approach by coupling high-throughput cellular expression and affinity purification to gel electrophoresis, liquid chromatography and mass spectrometry.

Two large-scale analyses in yeast exemplify this technology (27,28). Gavin et al, using affinity tagged genes (1,739 in total) under control of their natural promoters, isolated 232 distinct multi-protein complexes. Ninety-eight of these were already known and present in the Yeast Protein Database ; the remaining 134 complexes were novel. Complexes ranged in size between 2 and 82 different protein components, with a typical size of around 5. Because the novel complexes generally contain some proteins of known function, it was possible to propose functional roles for associated members within the complex based on circumstantial evidence; 231 proteins with no previous annotation were assigned a function by this process. For example, the complex that polyadenylates mRNA comprised 21 proteins, 4 of which had no previous annotation of any kind. The protein complexes observed frequently contained common components that point to interconnections between them, indicating that a complex network of functional relationships exists through which a variety of cellular processes are effected. This network is dynamic, in that complexes can assemble and disassemble and any given complex may show variable composition, enabling coupling through multiple pathways. For example the protein phosphatase PP2A was found to be bound to cell-cycle regulators, and in a separate complex, to proteins involved in cellular morphogenesis. Higher-level (complex-to-complex) mapping suggested grouping of complexes that belong to similar biological processes, for example intermediary metabolism or cell cycling. The study suggested the presence of orthologous complexes (not just orthologous proteins) between yeast and man, consistent with conservation of key functional units defining a "core proteome". As discussed above for gene expression pathways, there is already evidence for variations in the composition of complexes (modules) in metazoans depending on cellular context, effectively "paralogous" complexes, trends that are likely to be reinforced as studies such as this are extended. Not all proteins tested could be assigned to complexes – around 20% could not, probably in part due to interference by the affinity tags used as apart of the purification method.

In a similar study, Ho et al., starting with 725 "bait" proteins, detected 3,617 associated proteins corresponding to 25% of the yeast proteome (28). The average success-rate in identifying known complexes was 3-fold higher than for large-scale yeast two-hybrid experiments (Gavin et al. found around a 5-fold improvement), probably due to greater physiological relevance and the cooperative stability of multi-protein complexes (yeast two-hybrid approaches detect only pairwise interactions, although this technique has the virtues of very high scalability at relatively low cost and detection of weak interactions). Their general findings were similar to Gavin et al. They additionally showed that 275 of the detected complexes contained two or more interaction partners within the same biological process as defined by Gene Ontology, reinforcing the concordance of physical and functional networks (29). They also found that the network conformed to the expected power-law distribution referred to above (11). The authors focussed on signalling proteins (kinases, phosphatases and regulatory subunits) which enabled them to identify many novel connections of possible regulatory significance. For example an extensive network was assembled around the cyclin-dependant kinase Cdc28 with negative and positive regulators and links to other pathways. The DNA-damage response network was described in depth, revealing many known and new pathways that dictate cell cycle progression, transcription, protein degradation and DNA repair; members of a yeast E3 ubiquitin ligase complex not evident from simple bioinformatics were assigned based on comparison with mammalian orthologues and more detailed bioinformatics analysis. Finally, the probable upstream regulators, substrates and downstream effects of the protein kinase Dun-1, a known member of the DNA damage response process but of previously unclear role, were identified as part of this single global analysis. Extension of this approach to human cells, either by orthology or direct application, is bound to reveal many potential targets for drug intervention; in many cases they will be once or twice removed from the "disease pathway" and so will not be detected by more linear approaches.

Even without the use of parallelised expression and purification, substantial progress can be made by focussed use of biochemical isolation combined with mass spectrometry to build up a comprehensive picture of signalling complexes. This approach is readily applicable to mammalian cells. Grant and Husi took this approach to describe the multiprotein complexes that process neural information and encode memory (30). It was postulated that signaling complexes influence learning, and that the simple model of the N-methyl-D-Aspartate receptor (NR), a membrane protein sitting at the post-synaptic side of the synapse, injecting Ca^{2+} into the dendrite to activate a number of cytosolic enzymes, is too simplistic. The first evidence came from isolating a variant form of a protein PSD-95 that binds directly to the NR, and produces marked changes in synaptic plasticity and learning in a mouse model. A large-scale isolation of NR-PSD-95 complexes was carried out by biochemical methods, and the proteins identified by a combination of immunoblotting and mass spectrometry. Over 75 proteins were identified that could be joined into a meaningful picture consistent with genetic observations. This assemblage of proteins provides a firm framework for understanding the molecular basis of learning, and for how it might be regulated naturally or through intervention.

COMBINING FUNCTIONAL GENOMICS INFORMATION

It stands to reason that the information content from orthogonal functional genomics technologies will be greater in combination than in isolation, and there have been several recent examples that underscore this point, either from observational or experimental studies. Marcotte et al. used a combination of phylogenetic profiles, domain fusion analysis, mRNA expression analysis, metabolic function and protein-protein interaction data to assign a general function to around 1200 previousy uncharacterised yeast proteins (31). Gerstein et al. showed a progressive improvement in prediction of known yeast protein-protein associations through sequential addition of information on mRNA expression, knockout, essentiality and subcellular localization (32). Expression data alone was a relatively weak indicator of complex formation. Ge et al. statistically correlated mRNA expression and protein-protein interaction data, and showed that a) there was a significant association between the two orthogonal datasets and b) the combined analysis gave a clearer picture of pathways than either dataset alone (33). The same group extended this approach to combine protein-protein interaction and large-scale phenotypic analysis of the nematode worm Caenorhabditis elegans by RNA interference assays as applied to the DNA damage response pathway, identifying 12 worm orthologs and 11 novel genes in this pathway, including the hBCL3 otholog which is often altered in chronic lymphocytic leukaemia (34). Taking a different experimental approach, Ideker et al. systematically perturbed gene expression in the yeast galactose utilization pathway and quantified cellular changes by large-scale measurement of mRNA and protein expression. They combined this information with known protein-protein interactions to produce an integrated network describing the regulation of galactose utilization, and its connections to other metabolic pathways (35). These approaches, both experimental and computational, are extensible to human cell systems and therefore to target identification and prioritisation.

SEARCHING FOR DRUGGABLE TARGETS

From the medicinal chemistry perspective, it is essential to consider at an early stage how probable it is that a biologically valid target will be "druggable" – ie amenable to selective inhibition by a small molecule, preferably one that is orally active. If precedent is any kind of indicator, this is a rare event; it has been estimated that there are only 200 to 400 protein targets accountable for all launched pharmaceuticals in the US and UK pharmacopoeias, many of which are not even conventional small molecules (36, 37). Combinatorial chemistry and high-throughput screening have not as yet made much impact on this pattern; the vast majority of novel NCEs in Pharma pipelines still fall within precedented classes (38).

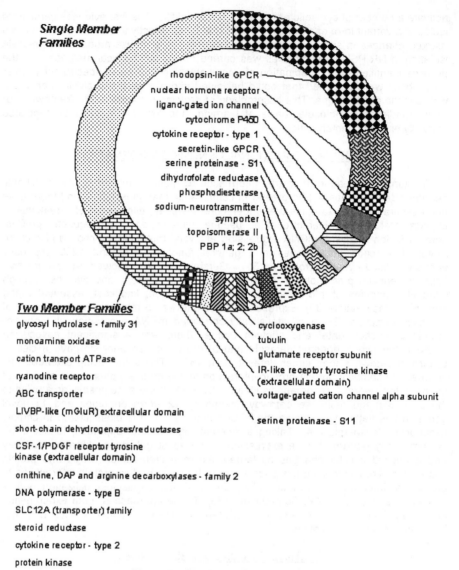

Figure 2. The Drugged Proteome

To date, and excluding non-specific or nucleic-acid binding drugs, the authors find that 238 protein targets in total (194 human, 44 microbial) account for all known launched US and UK pharmaceuticals for which there is a known molecular mechanism. 50 of these are rhodopsin-like GPCRs (17 peptidergic and the rest binding small endogenous molecules i.e. the metabolome) but in contrast, 77 are singletons such as HMG-CoA reductase. The drugged proteome is distributed across 109 gene families, 76 of which have known 3D structure, minimally defined by having a BLAST homologue in the Protein Databank. Historically therefore, drugged families are rare (<1% of total gene families), but when they do recur in biologically valid pathways could be exploited from a sound structural and functional base.

Signalling pathways have a preponderance of kinases against which it is relatively easy to find attractive leads but much harder to find drugs (there are only two launched kinase inhibitors to date, although there are many more attractive prospects in pipelines), probably due to network robustness or lack of selectivity stemming from the large size of this gene family and the molecular similarity of its drug binding sites. Screening or design of small-molecule tools for biological validation and as a handle on lead discovery promises to streamline target prioritisation, a field that has been termed chemical genetics or chemogenomics. This coupled with protein engineering can be used to more precisely delineate function, an approach that has been developed for kinases (39).

In general there are two approaches to the *in silico* analysis of pathway data for potential druggable targets a) assignment to precedented domain families which contain binding sites of a type that have been previously drugged or b) detection of sites with structural properties that will predispose them to drug binding. The former approach covers to date 114 structural families (Fig2), members of any one of which may occur in genetic and functional genomic datasets collected without a family bias. Given that biological validation is the slow step in most novel discovery programmes, and that there is no *a priori* reason to believe that valid targets will happen to fall in the few families that are major operational foci for most companies (typically GPCRs, ion channels, proteases, kinases and nuclear hormone receptors), it will be vital to select and prosecute targets across the whole range of precedented families. This argument can be extended to unprecedented families when the presence of a site inferred by structural homology and binding site analysis to have general properties consistent with binding a drug-like molecule may give cause to promote it for screening (40-42). It is reasonable to expect there will be many such targets, since the historic drugged set is strongly biased by classical medicinal chemistry and pharmacology methods, focussed as they were on natural ligand related leads. The combination of functional genomics, *in silico* pathway analysis and selection of druggable targets coupled to the screening of appropriate chemical diversity promises to revolutionise drug discovery in a way that no single technology can ever achieve alone.

References

1.   International Human Genome Sequencing Consortium, Nature, 409, 860 (2001).
2.   J.C.Venter, M.D.Adams, E.W.Myers, P.W.Li, R.J.Mural, G.G.Sutton, H.O.Smith, M.Yandell, C.A.Evans, R.A.Holt et al., Science, 291, 1304 (2001)
3.   M.J.Cunningham, S.Liang, S.Fuhrman, J.J.Seilhamer & R.Somyogyi, Ann. NY Acad. Sci., 919, 52 (2000)
4.   M.Abramovitz and K.M.Metters, Ann.Rev.Med.Chem.. 33, 223 (1998).
5.   B.J. Druker, Trends Mol. Med., 8(4):S14 (2002).
6.   S.Eker, M.Knapp, K.Laderoute, P.Lincoln, J.Meseguer and K.Sonmez in "Pacific Symposium on Biocomputing 2002", R.B.Altman, A.K.Dunker, L.Hunter, K.Lauderdale and T.E.Klein, Eds. World Scientific Publishing, 2001.
7.   H.Kitano, Science, 295, 1662 (2002).
8.   W.P Blackstock and M.P.Weir, Trends. Biotechnol., 17, 121 (1999).
9.   L.H. Hartwell, J.J. Hopfield, S.Leibler and A.W. Murray, Nature, 402, C47 (1999).
10.   G.Weng, U.S.Bhalla and R.Iyengar, Science, 284, 92 (1999).
11.   H.Jeong, S.P. Mason, A.-L. Barabasi and Z.N.Oltvai, Nature, 411, 41 (2001).
12.   J.L.Hartman, B.Garvik and L.Hartwell, Science, 291, 1001 (2001).
13.   M.E. Cseste and J.C.Doyle, Science, 295, 1664 (2002).
14.   D.Noble, Science, 295, 1678 (2002).
15.   P.D. Karp, Science, 293, 2040 (2001).
16.   G. Orphanides and D. Reinberg, Cell, 108, 439 (2002).
17.   N.J.McKenna and B.W.O'Malley, Cell, 108, 465 (2002).
18.   N.Woychik and M.Hampsey, Cell, 108, 453 (2002).
19.   C.J.Roberts, B.Nelson, M.J. Marton, R.Stoughton, M.R.Meyer, H.A.Bennett, Y.D.He, H.Dai, W.L.Walker, T.R.Hughes, M.Tyers, C.Boone and S.H.Friend, Science, 287, 873 (2000) .
20.   R. Ulrich and S. Friend, Nature Drug Discovery, 1, 84, (2002).

21.   C.Kumar-Sinha, S.Varambally, A.Sreekumar and A.M.Chinnaiyan, J.Biol.Chem., <u>277</u>, 575 (2002).
22.   J.Svaren, T.Ehrig, S.A.Abdulkadir, M.U.Ehrengruber, M.A.Watson and J.Milbrandt, J.Biol.Chem., <u>275</u>, 38524 (2000).
23.   P.Huang, L.Feng, E.A.Oldham, M.J.Keating and W.Plunkett, Nature, <u>407</u>, 390 (2000).
24.   M.Mann, R.C.Hendrickson and A.Pandey, Ann.Rev.Biochem., <u>70</u>, 437 (2001)
25.   M.Muzio, A.M.Chinnaiyan, F.C.Kischkel, K.O'Rourke, A.Shevchenko, J.Ni, C.Scaffidi, J.D.Bretz, M.Zhange, R.Gentz, M.Mann, P.H.Krammer, M.E.Peter and V.M.Dixit, Cell, <u>85 (6)</u>, 817 (1996).
26.   F.Mercurio, H.Zhu, B.W.Murray, A.Shevchenko, B.L.Bennett, J.Li, D.B.Young, M.Barbosa, M.Mann, A.Manning and A.Rao, Science, <u>278 (5339)</u>, 860 (1997).
27.   A.C.Gavin, M.Bösche, R.Krause, P.Grandi, M.Marzioch, A.Bauer, J.Schultz, J.M.Rick, A.-M.Michon, C.-M.Cruciat et al., Nature, <u>415</u>, 141 (2002).
28.   Y.Ho, A.Gruhler, A.Heilbut, G.D.Bader, L.Moore, S.-L.Adams, A.Millar, P.Taylor, K.Bennett, K.Boutilier et al., Nature, <u>415</u>, 180 (2002).
29.   The Gene Ontology Consortium, Genome Res., <u>11</u>, 1425 (2001).
30.   H.Husi and S.G.Grant, J.Neurochem., <u>77</u>, 281 (2001).
31.   E.M.Marcotte, M.Pellegrini, M.J.Thompson, T.O.Yeates and D.Eisenberg, Nature, <u>402</u>, 83 (1999).
32.   M.Gerstein, N.Lan and R.Jansen, Science, <u>295</u>, 284 (2002).
33.   H.Ge, Z.Liu, G.Church and M.Vidal, Nature Genetics, <u>29</u>, 482, (2001)
34.   S.Boulton, A.Gartner, J.Reboull, P.Vaglio, N.Dyson,D.Hill and M.Vidal, Science, <u>295</u>, 127, (2002)
35.   T.Ideker, V.Thorsson, J.A.Ranish, R.Christmas, J.Buhler, J.K.Eng, R.Bumgarner, D.R.Goodlett, R.Aebersold and L.Hood, Science, <u>292</u>, 929 (2001).
36.   M.Swindells, J.P.Overington and M.P.Weir, Drug Discovery Today, <u>7</u>, (2002).
37.   J.Drews and St.Ryser, Nature Biotechnol., <u>15</u>, (1997)
38.   Pharma Business, <u>48</u>, (2002)
39.   A.C.Bishop, O.Buzko and K.M.Shokat, Trends.Cell.Biol., <u>11</u>, 167 (2001).
40.   R.Laskowski, N.M.Luscombe, M.B.Swindells and J.M.Thornton, Protein Sci., <u>5</u>, 2438 (1996).
41.   G.P. Brady and P. Stouten, J. Comput.-Aided Mol. Design, <u>14</u>, 383 (2000)
42.   A.W.E.Chan and M.P.Weir, Chemical Innovation, <u>31</u>, 12 (2001)

Chapter 20. Searching for Alzheimer's Disease Therapies In Your Medicine Cabinet: The Epidemiological and Mechanistic Case For NSAIDs and Statins

Robert B. Nelson
Pfizer Global Research and Development
Eastern Point Road, Groton, CT 06340

Introduction – Alzheimer's disease (AD) is a chronic neurodegenerative disorder that progressively robs its victims of their intellect, memory, and personality. With the demographic aging of the population, AD is becoming an ever-larger public health burden. There are, unfortunately, no approved therapies for AD able to block progression of the disease. The current pharmacological arsenal against AD includes only the cholinesterase inhibitors, which seek to ameliorate the symptoms of AD by augmenting activity of compromised cholinergic neurons.

Despite this bleak clinical picture, an intriguing story has emerged in recent years from epidemiological studies asking whether the use of any currently marketed drugs is associated with a decreased risk of AD. Two classes of drugs have consistently emerged in the affirmative for this question. The first are the non-steroidal anti-inflammatory drugs (NSAIDs), commonly used either acutely in the treatment of mild to moderate pain, swelling, and fever, or more chronically in the treatment of rheumatoid arthritis. The second is a class of popular cholesterol-lowering drugs known collectively as statins, used as a primary preventative to block the development of atherosclerosis. This epidemiological record has spawned a growing literature that seeks to explore the mechanisms by which both of these classes of drugs might have their purported protective effects in AD.

Since observational epidemiological studies do not prove causality, large placebo-controlled clinical trials in AD patients are clearly needed for both the NSAIDs and the statins. However, epidemiological studies *have* often described causal links and therefore provide appropriate starting points for exploring potential mechanisms that underlie theses connections. Three examples of epidemiological findings that have led to the establishment of a causal association through clinical trials include the negative association made between aspirin use and myocardial infarction, the negative association between folic acid intake in the first trimester and incidence of neural tube defects, and the positive association between estrogen use and increased risk for venous thromboembolism (1, 2).

Due to space limitations, several epidemiological connections between decreased risk of AD and marketed pharmaceuticals or supplements are omitted here, including estrogen, antioxidants, and H2 blockers. This review focuses on NSAIDs and statins as there is evidence of a class effect, a defined molecular target to which the class effect could likely be ascribed, and significant new developments over the past year.

THE EPIDEMIOLOGICAL RECORD: DRUGS ASSOCIATED WITH DECREASED RISK OF AD

One of the histopathological hallmarks of AD is the neuritic plaque, a dense proteinaceous deposit comprised largely of the amyloid β-peptide (Aβ), a proteolytic product of the amyloid precursor protein (APP). Features of the neuritic plaque include accumulation of microglia around a central amyloid core, a local cytokine-mediated acute phase response, and activation of the complement cascade (3). All of these features are indicative of a largely local inflammatory response that is associated with damaged neurons and potentially exacerbates the pathological processes underlying AD. This presence of a local inflammatory response in AD brain led to the first retrospective study linking anti-inflammatory drug use with lowered risk of AD back in 1990 (4).

NSAID studies – To date more than 20 epidemiological studies have explored the connection between use of anti-inflammatory therapies and decreased risk of AD. Most of these studies are retrospective in nature and therefore subject to recall bias, i.e. whether subjects interviewed are able to accurately recall their usage of prescription and non-prescription drugs years after the fact, and whether the types of questions asked during interviews may bias their recall. Nonetheless, the overall consensus of these studies, which include identical twin and sibling comparisons, is consistent evidence for decreased risk of AD with extended NSAID use (5). This was evidenced in particular through a meta-analysis of 17 retrospective studies that had been completed as of 1996 (6). The general conclusions from these reports are as follows: First, NSAID use over an extended period is associated with a lower incidence of developing AD. Second, acetylsalicylic acid (aspirin) and other salicylates, which are considered in a separate category from NSAIDs, show a weaker but discernible association with decreased risk of AD. It is of note that most aspirin therapy is for low-dose cardiovascular treatment, while NSAID use is more often associated with much higher pain-reducing or anti-arthritic doses. Third, acetaminophen is not associated with reduced AD risk in any of the studies. This is consistent with acetaminophen being considered in a separate class, i.e. a fever-reducer and painkiller that lacks local anti-inflammatory properties associated with NSAIDs.

Prospective studies avoid many of the reporting bias or selection bias concerns engendered by retrospective studies. Two prospective studies examining the link between NSAID usage and incidence of AD confirmed the link between use of NSAIDs and decreased risk of AD. The earlier study, from the Baltimore Longitudinal Study of Aging, found that among those with 2 or more years of reported NSAID use, the relative risk of AD was reduced 60% vs. 35% for those with less than 2 years of NSAID use (7). The most recent study, drawn from the Rotterdam Study, found that relative risk was reduced 80% among those with 1 or more years of NSAID use vs. 17% for those with 1 to 23 months of NSAID use (8). In both prospective studies, an insignificant trend toward lower incidence in AD was found among users of oral salicylates.

Placebo-controlled clinical trials using NSAIDs have been run, but are largely inconclusive based on the small numbers of patients studied. A six-month study of patients with a preliminary diagnosis of AD found a significant decrease in cognitive decline in those treated with indomethacin (1) vs. placebo (9). However, only 14 patients were enrolled in each arm of the study. A 6-month diclofenac (2) trial failed to detect differences between diclofenac- and placebo-treated patients on measures of cognitive decline (10). This study also suffered from low patient numbers and a high withdrawal rate in the diclofenac treated group (12 completers).

Statin studies – Statins are a class of drugs that inhibit 3-hydroxy-3-methylglutaryl coenzyme A (HMG-CoA) reductase, a key enzyme in the synthesis of cholesterol. The epidemiological story surrounding statin use and risk of AD is much more recent than the NSAID story and based upon two retrospective studies published in late 2000. The first article examined over 57,000 patient records from three sites and concluded that prevalence of probable AD in the cohort taking statins was 60 to 73% lower than that in the total patient population, or compared with patients taking other medications typically used in the treatment of hypertension or cardiovascular disease (11). Surprisingly, simvastatin (**3**), in contrast to pravastatin (**4**) and lovastatin (**5**), did not contribute to this effect though it was speculated that the more recent introduction of simvastatin to the market contributed to this lack of an association. The second study was substantially smaller but also reported a large decrease in incidence of AD (71%) among statin users (12). There were too few statin users to break out the analysis in this study, but it is of note that 75% of statin use was accounted for by simvastatin, the drug that failed to show an association in the first retrospective study. In the latter study, no association between use of other lipid-lowering agents and incidence of AD was detected. Atorvastatin was represented in too few patients to analyze, due to the brief period of time this agent has been on the market.

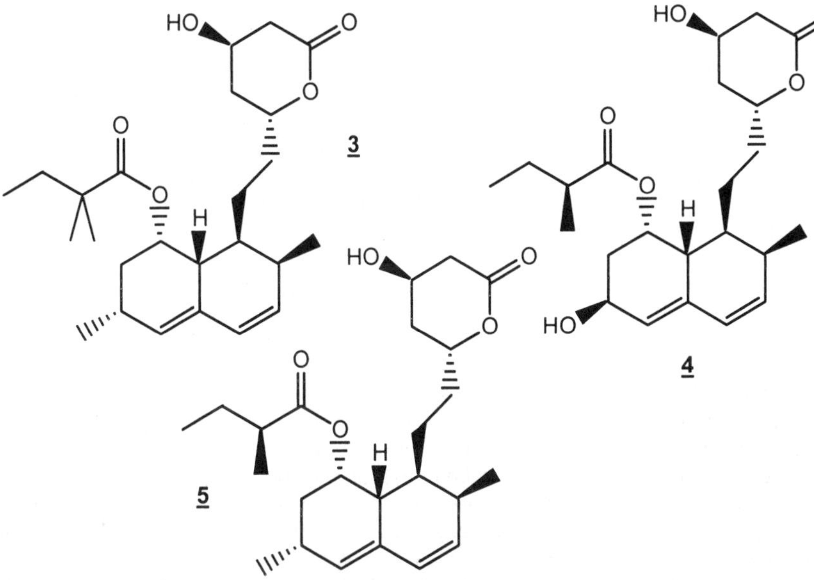

These retrospective data are consistent with other epidemiological data pointing to a relationship between cholesterol and risk of AD (13-15). Hypercholesterinemia is implicated as an important risk factor for AD in cross-sectional analyses. Longitudinal studies have suggested a relationship between elevated midlife cholesterol and late-life cognitive impairment or AD.

HOW MIGHT NSAIDS AND STATINS AFFECT THE PATHOGENESIS OF AD?

Convergent lines of evidence suggest that altered production, aggregation, and deposition of the Aβ peptide, a 40 to 42 residue proteolytic fragment of APP, is a key event in the pathogenesis of AD. Generation of the longer form of this peptide, Aβ42, is selectively increased in all presenilin mutations analyzed to date and in most of the APP mutations known to cause early-onset familial AD. Since Aβ42 is thought to initiate extracellular deposits in brain, and since this peptide aggregates very readily *in vitro*, it is considered the most likely culprit in disease pathogenesis.

In the last year, one of the more provocative hypotheses to emerge attempting to explain how both NSAIDs and statins might be related to AD pathogenesis is through their common ability to lower extracellular Aβ levels. This has led to several reports testing the hypothesis in both *in vitro* and *in vivo* models.

The Aβ-lowering hypothesis: Evidence implicating NSAIDs – Surprisingly little literature exists exploring mechanistic links between NSAIDs and Aβ production, considering how long the epidemiological link between AD and NSAIDs has been appreciated. This past year, an article appeared reporting that a subset of NSAIDs specifically lowers levels of Aβ42 (vs. Aβ40) in a variety of cells by as much as 80% (16). This effect was not seen for all NSAIDs and did not appear to be mediated by inhibition of cyclooxygenase (COX) activity, the principal target of NSAIDs. To address pharmacological relevance, the NSAID ibuprofen (**7**) was acutely administered to mice that overexpress mutant APP, produce Aβ42, and form amyloid deposits. Ibuprofen lowered brain levels of Aβ42 in these animals ~40% after a single dose.

While the hypothesis that NSAIDs may be linked to AD through lowering Aβ levels is an attractive one, it has a number of important caveats. First, the potency of those NSAIDs lowering Aβ42 is 2-3 orders of magnitude weaker against whole cells than their corresponding potency inhibiting COX activity. While the concentrations used against whole cells are in rough agreement with plasma levels found in humans, <1% of most NSAIDs are in the active, protein-unbound fraction in human plasma. CSF and brain levels of NSAIDs are extremely low and estimated to reach equilibrium with the unbound fraction in plasma (17-20). There has been reluctance to accept physiological relevance of other putative mechanisms ascribed to NSAIDS (e.g. PPARγ agonism, NFkB inhibition; see below) due to this same disparity between *in vitro* potency and free fraction of drug *in vivo*. Second, only ibuprofen among the NSAIDs has demonstrated the ability to lower Aβ levels in a transgenic mouse (16). These data complement previous observations that chronic administration of ibuprofen decreases amyloid plaque load and improves behavioral indices in APP transgenic animals (21, 22). However, ibuprofen has alternative activities ascribed to it, including its ability to decrease intracellular accumulation of H_2O_2 and to inhibit generation of lipid peroxides and reactive oxygen species (23, 24). It will be important to establish a relationship between Aβ-lowering *in vivo* and those NSAIDs implicated in lowering risk of AD. Lastly, the recent Rotterdam prospective study demonstrating a decreased risk of AD with prolonged NSAID use monitored which specific NSAIDs were taken through pharmacy records, since until 1995 all NSAIDs in the Netherlands had to be obtained by prescription (8). In that study, which showed an 80% decrease in risk of AD with NSAID use >1 year, 72% of NSAIDs used were reported *not* to be Aβ42-lowering agents *in vitro* by Koo and colleagues while only 23% were (1% were not categorized) (16, 25). Such a dramatic decrease in risk of AD in a landmark study where the preponderance of NSAIDs used were reported not to affect Aβ levels would indicate caution at this point in concluding that NSAIDs are linked to AD solely through their ability to lower Aβ levels in brain.

Alternative Mechanisms For NSAIDs – It could be regarded as serendipitous that NSAID usage was ever linked to a decreased risk of AD. While a sizeable histopathological record strongly indicates that AD brain is the site of a local, chronic inflammatory response, NSAIDs would not a priori be predicted to have efficacy against this inflammatory response. This is because NSAIDs are largely used to treat symptoms of peripheral inflammatory disorders, such as pain, swelling, and fever, that are not symptoms of AD. NSAIDs lack the disease-modifying properties of a large heterogeneous assemblage of agents (the disease-modifying anti-rheumatic drugs or DMARDs) that slow the progression of inflammatory disorders such as rheumatoid arthritis, but that are also associated with substantial toxicity and limited toleration.

Recognizing this disconnection between symptomatic effects in peripheral inflammatory settings and potential disease-modifying effects in AD, several studies have sought to explore alternative mechanisms of action for the NSAIDs. The salicylates, for example, have been reported to inhibit NFkB, a signal transduction protein implicated in inflammatory response signalling pathways (26). In a similar vein, NSAIDs have recently been reported to act as agonists for the nuclear receptor PPARγ (27, 28). The concentration of agents required for these in vitro effects are orders of magnitude higher than those required to inhibit COX under the same conditions (similar to the in vitro data linking NSAIDs to Aβ-lowering). It is unlikely then that these activities would be seen in vivo, since the free levels of NSAIDs achieved in CSF are much lower than those used in vitro to see these activities.

Perhaps the potential role of COX in AD is being considered in an inappropriate context. The orthodox view of COX as a leukocyte-associated inflammatory mediator has expanded into non-inflammatory realms, as a wider role for COX activity in such functions as angiogenesis and cancer comes to light. In the CNS, high constitutive COX2 expression has unexpectedly been found in neurons. The potential function of neuronal COX2 has accordingly become an area of growing interest (29, 30). Recent studies indicate that COX2 expression levels in neurons of the hippocampal formation, a brain region at high risk for neuronal loss in AD, may be a predictor of progression of early AD before neurodegeneration occurs (31). The increase in neuronal COX2 expression is concomitant with an increase in total neuronal Aβ content. Interestingly, COX2 expression in AD does not behave in the same manner as other inflammatory markers in AD. For example, expression of the major histocompatibility antigen HLA-DR appears only in later stages of the disease and is restricted in distribution to activated microglia (32).

In contrast to the hypothesis that NSAIDs have COX-independent Aβ-lowering effects, there is a sizeable literature suggesting that COX2 activity and Aβ levels have a reciprocal relationship to each other. A recent provocative study reports that crosses made between a mouse overexpressing COX2 via a neuron-specific promoter and an APP Tg Aβ-depositing mouse results in mice that have an increased rate and extent of Aβ deposition (33). This report is consistent with in vitro studies reporting that overexpression of COX2 in the NG108-15 cell line increases mRNAs for APP and release of Aβ by greater than two-fold (34). In this latter study, both indomethacin and the COX-2 selective inhibitor, JTE-522, were able to block the COX2-mediated increase in Aβ release. There is also evidence that PGE2, one of the stable prostanoids generated from COX2, can stimulate APP expression, although this latter study is in astrocytes rather than neurons (35).

So how might COX2 be functionally important in the neuronal loss that occurs during AD? The picture that emerges from studies on COX2-overexpressing or COX2-lacking neurons or mice is that COX2 plays a role in Aβ production/clearance, Aβ toxicity, and toxicity related to excitatory amino acid receptor activation. Primary neurons cultured from COX2-overexpressing transgenic mice are more susceptible than their wild-type counterparts both to Aβ-induced toxicity and to toxicity induced by the excitatory amino acid analog kainate (30, 36). Both forms of toxicity *in vitro* have been linked to a state of oxidative stress in the cell. Corroborating this association, other investigators report that the COX2-selective inhibitor S-2474 inhibits Aβ-induced death of primary cortical neurons (37). Demonstrating the reciprocal nature of this relationship, Aβ and kainate increase COX2 expression in primary neurons from wildtype mice, while kainate administration increases neuronal COX2 expression *in vivo* (38-40). Together these data suggest a multi-faceted interaction between Aβ, COX2, and neuronal excitotoxicity.

The Aβ-lowering hypothesis: Evidence implicating statins – The main hypothesis emerging for the role of statins in AD is that regulation of cholesterol levels rather than inhibition of HMG-CoA reductase *per se* is an important mechanistic link in modulating Aβ levels. Both simvastatin and lovastatin have been shown to reduce intracellular and extracellular levels of Aβ42 and Aβ40 levels in primary cultures of either cortical or hippocampal neurons (41). Moreover, guinea pigs treated with high doses of simvastatin over the course of 3 weeks show a 50% reduction of Aβ levels in cerebrospinal fluid and brain homogenates. This reduction is reversible upon termination of statin treatment (41). BM15.766 (**6**), a drug that inhibits the enzyme catalyzing the last step of cholesterol biosynthesis (7-dehydrocholesterol-delta7-reductase) and also reduces plasma cholesterol levels, was recently shown to

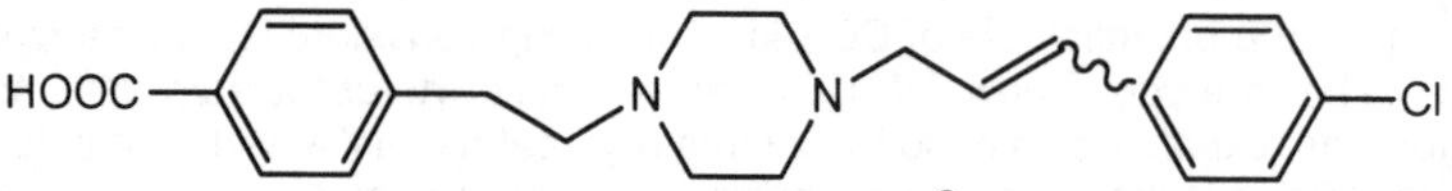

6

decrease brain Aβ peptide levels and β-amyloid load by greater than twofold in transgenic mice exhibiting an Alzheimer's β-amyloid phenotype (42). In this latter study, a strong, positive correlation between the amount of plasma cholesterol and brain Aβ was observed.

A causal role for cholesterol in regulating extracellular Aβ levels is indicated by several recent cell culture studies (43-47). The mechanisms by which cholesterol regulates Aβ levels remains unknown, although cholesterol has been linked to regulation of α- and β-secretase, two of the enzymes important in generating and degrading Aβ from APP (48-50). Cholesterol regulates Aβ levels *in vitro*. The net result of these putative activities of cholesterol would be to increase Aβ production.

Recent evidence suggests that the ratio of free cholesterol to cholesterol esters may be a more important factor in regulating Aβ production than total cholesterol levels (51). Aβ production is decreased in cells lacking acyl CoA:cholesterol acyltransferase (ACAT), the enzyme that converts free cholesterol to cholesterol esters. Moreover, inhibitors of ACAT such as CP-113,818 (**7**) decrease Aβ production, an effect that can be abrogated by addition of cholesterol esters. Other mechanisms that have been

7

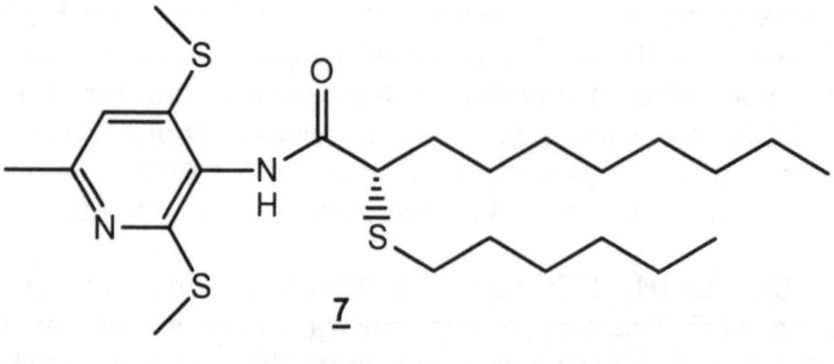

proposed for how cholesterol might affect Aβ levels are cholesterol's acceleration of Aβ aggregation in aqueous solution, or when Aβ is added to membranes of varying cholesterol content (52, 53). Reduction in cholesterol has been shown to protect cultured PC12 cells (cell line having neuron-like properties) against Aβ-induced toxicity (54). Finally, cholesterol and Aβ tend to co-accumulate in late endosomes of cells having defects in cholesterol transport (55). Abnormalities in this organelle are one of the earliest morphological signs of neuronal pathology in AD. The points in the cholesterol biosynthesis pathway at which different inhibitors have been shown to decrease Aβ release from cells are illustrated in the figure below.

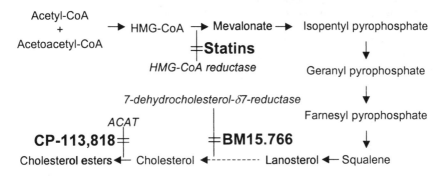

<u>Alternative Mechanisms For Statins</u> – As outlined above, the prevailing thought for how statins may interfere mechanistically with AD pathogenesis is through HMG-CoA reductase-mediated decreases in cholesterol production, resulting in decreased Aβ generation. There are alternative mechanisms ascribed to statins that might also be germane to the pathogenesis of AD. *In vivo,* statins show immunosuppressive properties, and *in vitro,* statins suppress T-cell responses and chemokine synthesis in peripheral blood mononuclear cells (56-60). One potential mechanism for this effect that appears to be independent of HMG-CoA reductase inhibition is the ability of statins to block activation of leukocytes by selectively binding to a regulatory site on the integrin leukocyte function antigen-1 (61). While it is unclear whether this anti-inflammatory effect is physiologically important, statins have also been shown to decrease plasma levels of the non-specific inflammatory marker, C-reactive protein in humans (62).

CLINICAL CONSIDERATIONS FROM THE EPIDEMIOLOGICAL DATA

<u>Primary Prevention or Therapeutic Intervention?</u> – One of the difficulties in using epidemiological data as a basis for running clinical trials is that epidemiological associations between use of a drug and decreased incidence of a disease often represent primary preventive effects, whereas clinical trials are most often designed to detect therapeutic intervention effects. The populations followed in epidemiological studies are generally asymptomatic for the disease in question, so that decreased incidence of a disease in such populations may represent the ability of the agent to block a pathogenic process in its early pre-symptomatic stages. Clinical AD trials generally select patients with some degree of symptomology and ask whether a given treatment causes symptomatic improvement or slows the disease process relative to its normal progression rate. If an agent that is effective against early pre-symptomatic pathogenic events is not able to alter late stage post-symptomatic pathogenic events, then a protective effect detected in an epidemiological study might not be detected in a clinical trial. For economic and practical reasons, it is seldom feasible to run clinical trials recapitulating the "design" of an epidemiological study.

The distinction between primary prevention and therapeutic intervention is of particular concern with a disease such as AD, since the etiology of AD occurs over several decades, and only in later stages do noticeable clinical signs appear. Using amyloid pathology as an example, the earliest detectable events appear to be release of Aβ peptide from neurons. This process, which often starts in the sixth or seventh decade of life, can remain largely asymptomatic until the eighth or ninth decade of life. In the early stages, "diffuse" extracellular Aβ deposits form that have little obvious deleterious effect on the surrounding brain tissue (63, 64). The appearance of the dense-core neuritic plaques, usually many years later, temporally corresponds more closely to the onset of clinical symptoms. So, for example, a treatment that blocks the generation and release of Aβ from neurons may show good efficacy if chronically administered in the pre-symptomatic stages of AD when Aβ is first being released from neurons, but have little effect once symptoms begin and massive extracellular deposits of Aβ already exist. The use of statins in treating atherosclerosis draws a useful parallel. Both AD and atherosclerosis are age-related diseases in which abnormal accumulation of a normal metabolite (Aβ in AD, cholesterol in atherosclerosis) ultimately leads to appearance of clinical symptoms. Like Aβ, cholesterol deposition begins years, even decades before clinical symptoms appear, and statins are much more effective at blocking the formation of atherosclerotic plaques thought to be the proximal cause of ischemic damage in atherosclerosis than they are in resolving the plaques once they are formed. Similarly, the epidemiological record for NSAIDs and AD suggests that greater protection is seen when individuals are cognitively asymptomatic, and multiple years separate NSAID usage from a diagnosis of AD. Thus, the design of clinical trials able to measure efficacy through primary preventive effects will be a major challenge in the development of therapeutics for AD.

Critical Need for Biomarkers – It is self-evident that the successful development of pre-morbid markers of disease pathogenesis will be critical in bringing AD therapeutics to market that act through primary prevention. This issue is underscored for therapeutics inspired by epidemiological studies, since such studies are more likely to identify agents working through primary prevention than therapeutic intervention. As with atherosclerosis, the greater challenge could come in making the leap from biomarker to causal factor. Plasma cholesterol levels are now widely accepted as a causal risk factor in the development of atherosclerosis, but this was a hotly debated scientific question until statins became available to establish this causal link in the clinic. In a similar vein, monitoring plasma and/or cerebrospinal fluid levels of Aβ holds some promise of providing a pre-morbid marker for the development of AD, but it would be difficult to argue that a potential therapeutic for AD should be approved solely on its ability to lower Aβ levels without better evidence of a causal link between Aβ levels and development of AD. Again, this will be a major challenge facing the pharmaceutical industry in the development of therapeutics for AD.

WILL AD DRUGS AGAINST NOVEL TARGETS EMERGE FROM THE NSAID/STATIN EPIDEMIOLOGICAL DATA?

If clinical efficacy were ultimately demonstrated for existing statins and/or NSAIDs in large placebo-controlled AD clinical trials, it would still seem unlikely that the penultimate efficacy against AD rests in these particular agents. More likely is a scenario in which the mechanisms targeted by these agents have a serendipitous impact on the disease process in AD, but are not the optimal molecular targets either in terms of efficacy or safety. Rather, these mechanistic targets, if validated in the clinic, should form the starting point in seeking to better understand the relationship between molecular mechanism and disease process, and from there to identify new molecular targets that directly regulate the pathogenic process in AD.

NSAIDs – The first and most fundamental question to be addressed for the NSAIDs is whether or not their purported protection against AD is mediated through inhibition of COX activity. As reviewed above, separate bodies of literature argue both for and against this hypothesis. For any argument of mechanism to be convincing, it must progress from cell culture to animal and demonstrate that pharmacokinetic properties associated with a given drug, including its brain and CSF exposure at pharmacologically relevant doses, are consistent with observed *in vitro* potencies for the purported mechanism. It is also important to show an appropriate relationship across multiple agents from the class between *in vivo* CNS exposure and *in vitro* potency against the putative mechanism of action. It is always helpful to include an *in vivo* marker for the ostensive mechanism (in this case COX inhibition) as a comparator.

One possibility for the NSAID hypothesis that deserves consideration is that COX activity *per se* does not mediate the NSAID protective effect, but that the true molecular target is closely related. COX is one of many enzymes involved in arachidonic acid metabolism. NSAIDs bind COX1/COX2 in a competitive manner largely by mimicking arachidonic acid metabolites (i.e. all NSAIDs are both lipophilic and acidic). Since these features describe the bulk of known arachidonic acid metabolites, it is possible that NSAIDs also alter the activity of other enzymes involved in arachidonic acid metabolism and that these enzymes are more central to protection against AD. Such enzymes could either serve functions closely linked to that of COX (e.g. PGE2 synthases) or be found in different pathways from COX (e.g. the lipoxygenases).

To test these hypotheses, it is necessary to establish model systems that recapitulate elements of the disease and test the effects of NSAIDs and other modulators of arachidonic acid metabolism in those models. Such models have been established mainly around APP processing and production of the Aβ peptide, but could also be applied to models of Aβ aggregation/fibrillization, tau hyperphosphorylation and cytoskeletal disruption, endosomal/lysosomal abnormalities, or any other aberrant events occurring in AD.

Statins – Despite the brief history of studies seeking to understand the mechanism by which statins have their putative protective effects in AD, the data already points strongly toward cholesterol and its metabolites as key players. Using regulation of Aβ as the model outcome measure, these studies suggest that not only the statins, but also other pharmaceuticals capable of inhibiting cholesterol biosynthesis may have efficacy in AD. Moreover, if emerging stories are correct, cholesterol esters rather than cholesterol itself may be closer to the final common pathway of pathology. In this context, ACAT inhibitors present another therapeutic opportunity for optimizing efficacy while minimizing side effect liability.

From the scientific standpoint, establishing that cholesterol metabolism is an important component of AD is only a first step. The mechanism(s) by which cholesterol and its metabolites might increase a predilection toward developing AD is not known. Understanding this connection may open completely new avenues in the treatment of AD that do not involve cholesterol modulation *per se*.

Conclusion – Epidemiological links between use of marketed pharmaceuticals and decreased incidence of AD provide an attractive alternative starting point for the design of novel pharmaceuticals to treat AD. While in and of themselves such studies do not prove causality, they provide important clues in attempting to understand the molecular pathogenesis for a disease whose etiology is still poorly understood. If proven to be clinically efficacious, these pharmaceuticals provide potentially large advantages over novel therapies currently in development. Besides being immediately available for therapeutic use in AD, their historical use in the market offers much information about safety and tolerability relative to untried agents targeting novel mechanisms. These established agents also provide "positive controls" to use in attempting to validate relevant models of disease process. While there are potential pitfalls in attempting to connect mechanism to disease process when studying epidemiological links, a rigorous attention to pharmacology, pharmacokinetics, pharmacodynamics, and structure/activity relationships can help avoid the risk of prematurely ascribing mechanisms that are based on pharmacologically irrelevant *in vitro* phenomena.

Serendipity points the way with epidemiological studies, but need not have the final word. Finding unanticipated therapeutic effects for existing pharmaceuticals should provide a launching point for a better understanding of the mechanistic underpinnings of the therapeutic effect and lead to the design of newer, safer, and more efficacious pharmaceuticals against the same or related molecular targets.

References

1. Program BCDS, BMJ, 1, 440 (1974).
2. A. Milunsky, H. Jick, S. S. Jick, C. L. Bruell, D. S. MacLaughlin, K. J. Rothman, W. Willett, JAMA, 262, 2847 (1989).
3. P. L. McGeer, E. G. McGeer, Neurobiology of Aging, 22, 799 (2001).
4. P. L. McGeer, E. McGeer, J. Rogers, J. Sibley, Lancet, 335 (1990).
5. J. C. Breitner, B. A. Gau, K. A. Welsh, B. L. Plassman, W. M. McDonald, M. J. Helms, J. C. Anthony, Neurology, 44, 227 (1994).
6. P. L. McGeer, M. Schulzer, E. G. McGeer, Neurology, 47, 425 (1996).
7. W. F. Stewart, C. Kawas, M. Corrada, E. J. Metter, Neurology, 48, 626 (1997).
8. A. in t' Veld, A. Ruitenberg, A. Hofman, L. J. Launer, C. M. van Duijn, T. Stijnen, M. M. Breteler, B. H. Stricker, New Engl.J.Med., 345, 1515 (2001).
9. J. Rogers, L. C. Kirby, S. R. Hempelman, D. L. Berry, P. L. McGeer, A. W. Kaszniak, J. Zalinski, M. Cofield, L. Mansukhani, P. Willson, F. Kogan, Neurology, 43, 1609 (1993).
10. S. Scharf, A. Mander, A. Ugoni, F. Vajda, N. Christophidis, Neurology, 53, 197 (1999).
11. B. Wolozin, W. Kellman, P. Ruosseau, G. G. Celesia, G. Siegel, Arch. Neurol., 57, 1439 (2000).
12. H. Jick, G. L. Zornberg, S. S. Jick, S. Seshadri, D. A. Drachman, Lancet, 356, 1627 (2000).
13. A. Hofman, A. Ott, M. M. Breteler, M. L. Bots, A. J. Slooter, F. van Harskamp, C. N. van Duijn, C. Van Broeckhoven, D. E. Grobbee, Lancet, 349, 151 (1997).
14. M. Kivipelto, E. L. Helkala, T. Hanninen, M. P. Laakso, M. Hallikainen, K. Alhainen, H. Soininen, J. Tuomilehto, A. Nissinen, Neurology, 56, 1683 (2001).
15. I. L. Notkola, R. Sulkava, J. Pekkanen, T. Erkinjuntti, C. Ehnholm, P. Kivinen, A. Tuomilehto, A. Nissinen, Neuroepidemiology, 17, 14 (1998).
16. S. Weggen, J. L. Eriksen, P. Das, S. A. Sagi, R. Wang, C. U. Pietrzik, K. A. Findlay, T. W. Smith, M. P. Murphy, T. Butler, D. E. Kang, N. Marquez-Sterling, T. E. Golde, E. H. Koo, Nature, 414, 212 (2001).
17. B. Bannwarth, P. Netter, J. Pourel, R. J. Royer, A. Gaucher, Biomed. Pharmacother., 43, 121 (1989).
18. B. Bannwarth, P. Netter, F. Lapicque, P. Pere, P. Thomas, A. Gaucher, Eur.J.Clin.Pharmacol., 38, 343 (1990).
19. R. A. Ferrari, S. J. Ward, C. M. Zobre, D. K. Van Liew, M. H. Perrone, M. J. Connell, D. R. Haubrich, Eur.J.Pharmacol., 179, 25 (1990).
20. W. M. O'Brien, Am.J.Med., 75, 32 (1983).
21. G. P. Lim, F. Yang, T. Chu, P. Chen, W. Beech, B. Teter, T. Tran, O. Ubeda, K. H. Ashe, S. A. Frautschy, G. M. Cole, J.Neurosci., 20, 5709 (2000).
22. G. P. Lim, F. Yang, T. Chu, E. Gahtan, O. Ubeda, W. Beech, J. B. Overmier, K. Hsiao-Ashe, S. A. Frautschy, G. M. Cole, Neurobiology of Aging, 22, 983 (2001).
23. D. Zapolska-Downar, A. Zapolska-Downar, H. Bukowska, H. Galka, M. Naruszewicz, Life Sciences, 65, 2289 (1999).
24. Z. Lambat, N. Conrad, S. Anoopkumar-Dukie, R. B. Walker, S. Daya, Metabolic Brain Disease, 15, 249 (2000).
25. E. Koo, M. Hao, T. E. Golde, D. R. Galasko, WO Patent 01/78721 A1 (2001).

26. M. Grilli, M. Pizzi, M. Memo, P. Spano, Science, 274, 1383 (1996).
27. C. K. Combs, D. E. Johnson, J. C. Karlo, S. B. Cannady, G. E. Landreth, J.Neurosci., 20, 558 (2000).
28. J. M. Lehmann, J. M. Lenhard, B. B. Oliver, G. M. Ringold, S. A. Kliewer, J.Biol.Chem., 272, 3406 (1997).
29. G. Tocco, J. Freiremoar, S. S. Schreiber, S. H. Sakhi, P. S. Aisen, G. M. Pasinetti, Exp. Neurol., 144, 339 (1997).
30. L. Ho, C. Pieroni, D. Winger, D. P. Purohit, P. S. Aisen, G. M. Pasinetti, J. Neurosci. Res., 57, 295 (1999).
31. L. Ho, D. Purohit, V. Haroutunian, J. D. Luterman, F. Willis, J. Naslund, J. D. Buxbaum, R. C. Mohs, P. S. Aisen, G. M. Pasinetti, Arch. Neurol., 58, 487 (2001).
32. J. D. Luterman, V. Haroutunian, S. Yemul, D. Purohit, P. S. Aisen, R. Mohs, G. M. Pasinetti, Arch. Neurol., 57, 1153 (2000).
33. Z. Xiang, G. Pasinetti, Soc. Neurosci. Abstr., 27, 652.10 (2001).
34. K. Kadoyama, Y. Takahashi, H. Higashida, T. Tanabe, T. Yoshimoto, Biochem. Biophys. Res. Comm., 281, 483 (2001).
35. R. K. K. Lee, S. Knapp, R. J. Wurtman, J. Neurosci., 19, 940 (1999).
36. K. A. Kelley, L. Ho, D. Winger, J. Freire-Moar, C. B. Borelli, P. S. Aisen, G. M. Pasinetti, Am.J.Pathol., 155, 995 (1999).
37. T. Yagami, K. Ueda, K. Asakura, T. Sakaeda, T. Kuroda, S. Hata, Y. Kambayashi, M. Fujimoto, Brit.J.Pharmacol., 134, 673 (2001).
38. G. M. Pasinetti, P. S. Aisen, Neuroscience, 87, 319 (1998).
39. E. J. Kim, J. E. Lee, K. J. Kwon, S. H. Lee, C. H. Moon, E. J. Baik, Brain Res., 908, 1 (2001).
40. T. L. Sandhya, W. Y. Ong, L. A. Horrocks, A. A. Farooqui, Brain Res., 788, 223 (1998).
41. K. Fassbender, M. Simons, C. Bergmann, M. Stroick, D. Lutjohann, P. Keller, H. Runz, S. Kuhl, T. Bertsch, K. von Bergmannn, M. Hennerici, K. Beyreuther, T. Hartmann, Proc.Natl.Acad.Sci.USA., 98, 5856 (2001).
42. L. M. Refolo, M. A. Pappolla, B. Malester, J. LaFrancois, T. BryantThomas, R. Wang, G. S. Tint, K. Sambamurti, K. Duff, Neurobiology Of Disease, 7, 321 (2000).
43. M. Simons, P. Keller, B. De Strooper, K. Beyreuther, C. G. Dotti, K. Simons, Proc.Natl. Acad. Sci. USA., 95, 6460 (1998).
44. T. Mizuno, C. Haass, M. Michikawa, K. Yanagisawa, Biochim. Biophys. Acta, 14, 119 (1998).
45. T. Mizuno, M. Nakata, H. Naiki, M. Michikawa, R. Wang, C. Haass, K. Yanagisawa, J.Biol.Chem., 274, 15110 (1999).
46. D. S. Howland, S. P. Trusko, M. J. Savage, A. G. Reaume, D. M. Lang, J. D. Hirsch, N. Maeda, R. Siman, B. D. Greenberg, R. W. Scott, D. G. Flood, J.Biol.Chem., 273, 16576 (1998).
47. E. R. Frears, D. J. Stephens, C. E. Walters, H. Davies, B. M. Austen, NeuroReport, 10, 1699 (1999).
48. E. Kojro, G. Gimpl, S. Lammich, W. Marz, F. Fahrenholz, Proc.Natl.Acad.Sci.USA., 98, 5815 (2001).
49. S. Bodovitz, W. L. Klein, J.Biol.Chem., 271, 4436 (1996).
50. M. Racchi, R. Baetta, N. Salvietti, P. Ianna, G. Franceschini, R. Paoletti, R. Fumagalli, S. Govoni, M. Trabucchi, M. Soma, Biochem.J., 322, 893 (1997).
51. L. Puglielli, G. Konopka, E. Pack-Chung, L. A. Ingano, O. Berezovska, B. T. Hyman, T. Y. Chang, R. E. Tanzi, D. M. Kovacs, Nature Cell Biology, 3, 905 (2001).
52. N. A. Avdulov, S. V. Chochina, U. Igbavboa, C. S. Warden, A. V. Vassiliev, W. G. Wood, J.Neurochem., 69, 1746 (1997).
53. A. Kakio, S. I. Nishimoto, K. Yanagisawa, Y. Kozutsumi, K. Matsuzaki, J.Biol.Chem., 276, 24985 (2001).
54. S. S. Wang, D. L. Rymer, T. A. Good, J.Biol.Chem., 276, 42027 (2001).
55. T. Yamazaki, T. Y. Chang, C. Haass, Y. Ihara, J.Biol.Chem., 276, 4454 (2001).
56. J. A. Kobashigawa, S. Katznelson, H. Laks, J. A. Johnson, L. Yeatman, X. M. Wang, D. Chia, P. I. Terasaki, A. Sabad, G. A. Cogert, New Engl.J.Med., 333, 621 (1995).
57. K. Wenke, B. Meiser, J. Thiery, D. Nagel, W. von Scheidt, G. Steinbeck, D. Seidel, B. Reichart, Circulation, 96, 1398 (1997).
58. S. Kurakata, M. Kada, Y. Shimada, T. Komai, K. Nomoto, Immunopharmacology, 34, 51 (1996).
59. M. Romano, L. Diomede, M. Sironi, L. Massimiliano, M. Sottocorno, N. Polentarutti, A. Guglielmotti, D. Albani, A. Bruno, P. Fruscella, M. Salmona, A. Vecchi, M. Pinza, A. Mantovani, Laboratory Investigation, 80, 1095 (2000).

60. I. Inoue, S. Goto, K. Mizotani, T. Awata, T. Mastunaga, S. Kawai, T. Nakajima, S. Hokari, T. Komoda, S. Katayama, Life Sci., 67, 863 (2000).
61. G. Weitz-Schmidt, K. Welzenbach, V. Brinkmann, T. Kamata, J. Kallen, C. Bruns, S. Cottens, Y. Takada, U. Hommel, Nature Medicine, 7, 687 (2001).
62. I. Jialal, D. Stein, D. Balis, S. M. Grundy, B. Adams-Huet, S. Devaraj, Circulation, 103, 1933 (2001).
63. C. Duyckaerts, M. A. Colle, F. Dessi, Y. Grignon, F. Piette, J. J. Hauw, J.Neural. Transmission. Supplementum, 53, 119 (1998).
64. H. M. Wisniewski, C. Bancher, M. Barcikowska, G. Y. Wen, J. Currie, Acta Neuropathologica, 78, 337 (1989).

Chapter 21. Matrix Metalloproteinases and the Potential Therapeutic Role for Matrix Metalloproteinase Inhibitors in Chronic Obstructive Pulmonary Disease (COPD)

Kevin M. Bottomley and Maria G. Belvisi
Inpharmatica Ltd, 60 Charlotte Street, London W1T 2NU, UK and Respiratory
Pharmacology, Cardiothoracic Surgery, Imperial College School of Medicine,
National Heart and Lung Institute, Dovehouse Street, London SW3 6LY, UK

Introduction – For more than 20 years, the search for drugs based on direct, competitive inhibition of Matrix Metalloproteinases (MMPs) has continued in various pharmaceutical research centers. Although very potent inhibitors have been identified, some of which have been the subject of clinical studies, there has been a notable lack of success in bringing these to the market (1-3). The primary therapeutic targets have been cancer, periodontal disease and arthritis but recent developments in the understanding of the molecular and cellular actions of members of MMPs, both in matrix metabolism but also as modulators of both enzyme and cytokine function suggest that inhibitors of particular members of the MMP family could be beneficial in the treatment of other pathologies (4,5).

Metzincins – The MMPs belong to a super-family of metalloproteins known as the Metzincins. The name derives from consensus sequence and structural features, specifically a "HExxH" zinc binding motif (zincin) and C-terminal conserved methionine residue which forms a conserved structure known as a "met-turn" (6). The metzincin superfamily contains a number of families but three of these include metalloproteinases of current pharmaceutical interest. These are the MMPs, ADAMs (a disintegrin and metalloprotoeinase-like) and ADAMTSs (ADAMs which include one or more thrombospondin (TS) domains).

MMP structure and function - The MMPs, are a family of zinc containing endoproteases which have been traditionally characterised by their collective ability to degrade all components of extracellular matrix, at neutral pHs. The zinc ion is necessary for the catalytic integrity of the protease. The catalytic mechanism of MMPs has been described but the essential components are 3 histidines, a glutamic acid and a water molecule (7,8). Key amino acid residues are arranged in a highly characteristic sequence: HExxHxxGxxH. The triad of histidine residues co-ordinate the zinc ion, which in combination with a glutamic acid forms the critical protein sequence components of the catalytic mechanism. The glutamic acid residue (adjacent to the first, most N-terminal histidine residue) acts as a neutrophile, which with an associated water molecule promotes cleavage of the substrate peptide scissile bond. The conserved glycine residue C-terminal to the 2nd co-ordinating histidine allows a sharp turn, permitting the most C-terminal histidine in the triad to associate with the zinc ion. C-terminal to the zinc ion binding motif is the conserved methionine residue which is responsible for the "met-turn" in metzincin structures and provides a hydrophobic base for the histidine triad (6). The enzymes also require calcium for stabilisation of the protein tertiary structure.

In addition to the conserved catalytic motif, the metzincins generally contain a conserved cysteine residue which lies N-terminal to the catalytic motif in the "pro-peptide" domain. This conserved residue co-ordinates with the zinc ion, displacing the water molecule essential for a functional catalytic mechanism. These enzymes (in common with most proteinases) are produced as inactive zymogens requiring either the displacement of, or more generally, removal of the pro-peptide domain for full expression of proteolytic activity. This processing of the enzyme for the majority of metzincins occurs at or beyond the cell surface. However, there is a subset of enzymes that are potentially processed inside the cell and then exported to the membrane as active enzymes. For these, activation occurs in the Golgi by the action of furin, or furin-like enzymes which cleave at a furin recognition site (RxKR) located at the junction between the pro-peptide and the catalytic domains (10). MMP23 while containing a shortened pro-domain and a cysteine which would correspond to the residue responsible for conferring latency in other MMPs shows some sequence divergence in the adjacent residues (11). MMP23 also contains a furin sensitive sequence between the pro-peptide and catalytic domains suggesting that this enzyme is activated within the cell.

The C-terminal domain (where present) shows the most sequence diversity between MMPs and participates variously in enzyme localisation to extracellular matrix, tissue inhibitor of metalloproteinases (TIMP) binding and/or substrate recognition (12). The C-terminal domain is generally a hemopexin. The exception is MMP-23 where the C-terminal domain contains a cysteine-rich, proline-rich, IL-1 type II-like sequence (13). The 3D structures of the C terminal domains of MMP-1, MMP-2 and MMP-13 have been solved and are essentially the same: a 4 bladed β-propeller (14-16). This type of structure is normally associated with proteins involved in protein-protein interactions. The hemopexin domain is attached to the catalytic domain by a disordered hinge region, which has been taken to suggest a flexible interaction between the catalytic and the hemopexin domain. However, in two full length structures, the relative positions of the two domains appear almost superimposed implying there may be a preferred relative relationship between the two domains (14,17).

For metalloelastase (MMP12), activation of this MMP is followed by immediate loss of the hemopexin domain. The active enzyme appears to consist of the catalytic domain alone. Metalloelastase is important for macrophage migration through extracellular matrix. (18).

There are 6 reported human membrane associated MMPs (MT1-MMP to MT6-MMP). These contain a membrane anchoring region C-terminal to the hemopexin domain. MT1-MMP, MT2-MMP, MT3-MMP and MT5-MMP contain transmembrane regions, consisting of a stretch of hydrophobic amino acids followed by a short intra-cellular sequence, while MT4-MMP and MT6-MMP are anchored on the cell surface by a glycophosphatidyl inositol-anchoring sequence. All the MT-MMPs containing furin sensitive sequences at the junction of the pro-peptide and catalytic domains have the potential to arrive at the cell surface in a catalytically active form. However, MT1-MMP requires the pro-domain (either bound or unbound to MT1-MMP) to activate MMP2. The pro-domain is required to form part of the cell surface complex (MT1-MMP)-TIMP2-MMP-2 necessary for the successful activation of MMP-2. It has been proposed that the pro-domain of MT1-MMP may be removed in the Golgi but remains non-covalently associated with the enzyme during its journey to the cell surface where it is then able to participate in the activation of MMP-2 (19).

For gelatinases A and B (MMP-2 and MMP-9) there is an insertion of 3 fibronectin type II modules at the N-terminal side of the catalytic domain. This inserted domain will associate with gelatin (a feature which allows easy purification of these enzymes) and it is speculated that its function is to augment the enzyme-substrate interaction (20).

ADAMs and ADAMTSs - Both protein families are members of the matrixin superfamily and members of these families are of interest both for their known biological functions and for their sensitivity to competitive inhibitors of the MMPs. The ADAMs comprise a pro-peptide, catalytic (functional and non-functional) and C-terminal disintegrin domain, cysteine rich region, a membrane anchoring transmembrane region and a cytoplasmic tail.   Of the 30 reported ADAMs, 17 have functional metalloproteinase domains. ADAM-17 (also known as TNFα converting enzyme) is active at the cell surface where it is active in producing the soluble version of the cytokine.   ADAM-17 is currently the subject of concerted pharmaceutical research to identify inhibitors, the precedented success of anti-TNFα therapies in the treatment of rheumatoid arthritis indicates a key role for TNFα in the pathology of this disease and suggests that inhibitors of ADAM-17 capable of preventing the production of the soluble form of TNFα may be of therapeutic value (21, 22).

ADAMTSs are secreted.   Like ADAMs, they comprise pro-peptide, catalytic and disintegrin domains, with an additional series of thrombospondin-like type I repeats, positioned variously in the catalytic and/or the C-terminal domain.   Like the MMPs, they appear to be important in extra-cellular matrix homeostasis including pro-collagen processing, metabolism of cartilage aggrecan and recently reported normal metabolism of von Willebrand factor (9).   Again substrate competitive inhibitors of MMPs have been demonstrated to inhibit those ADAMTSs (ADAMTS-4 and ADAMTS-5/11) involved in IL-1α induced metabolism of cartilage aggrecan (23).

Control of MMP activity - As with most proteinases, MMP activities are controlled at many levels including transcription, translation, secretion, activation, inhibition and metabolism.   In general, expression of MMPs is tightly regulated at the transcriptional level. Induction of MMP transcription is, in many cases, co-ordinated with expression of cofactors and endogenous inhibitors, for example MT1-MMP and TIMP2 (24).   A number of polymorphisms have been demonstrated in genes for MMPs and these may affect expression levels. Attempts have been made to link these polymorphisms using either human pedigree or association studies to disease states or prevalence (25). The reported pedigree studies implicating ADAMTS-13 in familial thrombocytopenic purpura are a good example of how polymorphisms that lead to a change in the activity of the gene product can be demonstrated to have a disease association (9).   Human genetic studies have the potential to link human gene to a disease, an important level of validation for a putative disease target.

Activation of metalloproteinases is an additional important mechanism for regulating activity of the enzyme.   Enzyme activation generally requires removal of the pro-peptide domain.   However, the means of removing the pro-peptides varies from sequence to sequence and usually requires the presence of a number of accessory proteins (12).   The variety and complexity of some of the activation events suggests a fine level of control is exerted on metalloproteinase proteolytic activity. Most MMPs are exported to the cell surface where they either remain membrane anchored or are secreted.   Exceptions are MMP-8 (neutrophil collagenase), MMP-9 and MMP-12 (metalloelastase).   MMP-8 and MMP-9 are stored in specific granules of neutrophils and MMP-12 is stored by macrophages and secreted by a protein kinase C-mediated plasmin, and/or thrombin regulated mechanism (26). MMP activity is closely regulated by the presence of specific endogenous inhibitors: Tissue Inhibitors of Matrix metalloproteinases (TIMPs).   Four TIMPs are known; TIMP1,2,3 & 4.   These are approximately 21-29kDa and these bind and inhibit MMPs with 1:1 stoichiometry (26). They are often co-expressed with MMPs which enforces the conviction that MMP proteolytic activity is tightly regulated.   TIMPs can also participate in the activation of specific MMPs, such as MMP2.   α2-macroglobulin a general serum protease inhibitor is also capable of inhibiting active MMPs.

DOMAIN STRUCTURE OF MMPS AND OTHER MATRIXINS REFERRED TO IN THIS REVIEW

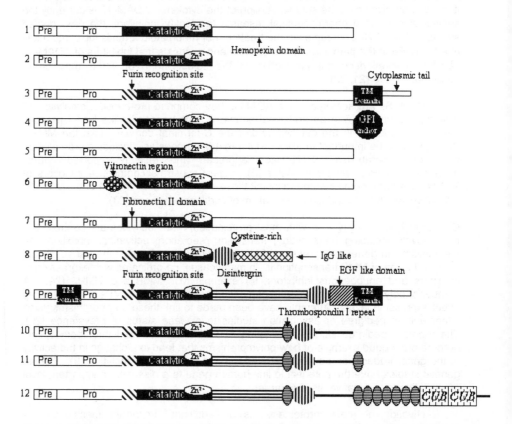

1:      MMP-1 (collagenase 1), MMP-2 (stromelysin 1), MMP-8 (collagenase 2), MMP-10 (stromelysin 2), MMP-12 (metalloelastase), MMP-13 (collagenase 3), MMP-18,19,20 (enamelysin), MMP-27
2:      MMP-7 (matrilysin), MMP-26 (epilysin)
3:      MMP-14 (MT1-MMP), MMP-15 (MT2-MMP), MMP-16 (MMT3-MMP), MMP-24 (MT5-MMP)
4:      MMP-17 (MT4-MMP), MMP-25 (MT6-MMP)
5:      MMP-11 (stromelysin 3), MMP-28
6   MMP-21
7:      MMP-2 (gelatinase A), MMP-9 (gelatinase B)
8:      MMP-23
9:      ADAM-17 (TACE)
10:     ADAMTS-4 (aggrecanase)
11:     ADAMTS-5 (aggrecanase)
12:     ADAMTS-13

MMPs and Matrix Metabolism - The matrixins are able to metabolise many components of extra-cellular matrix and this has suggested both physiological and pathological roles (12).   They are important in development, including bone development and vascular remodelling.   There have been many studies involving either gene knockout or MMP over expression studies, however these have generally failed to produce pronounced phenotypes (possibly indicating a high degree of redundancy in matrixin mediated metabolism of extra-cellular matrix). The exceptions are ablation of collagenolytic activity associated with inactivation of the MT1-MMP gene in mice, this produced mice with arthritis, osteopenia, dwarfism and fibrosis of soft tissues (28). Also MMP-12 (metalloelastase) null mice demonstrated protection from cigarette smoke induced emphysema (29). Uncontrolled MMPs activity has been implicated in a number of other diseases including periodontal disease, inflammatory bowel disease and COPD (30, 31)

Synthetic Inhibitors - MMP inhibitors generally consist of two basic features, a zinc chelating ligand and a chemical moiety which binds the substrate recognition cleft of the enzyme.   Five ligands have been traditionally used; these are in order of decreasing potency - hydroxamic acid, reverse hydroxamic acid, thiol, carboxylic acid and phosphinic acid (7). The phosphinic acid based inhibitors incorporate features both to the N- and C- terminal to the sissile bond of the substrate. For the remaining ligands, binding corresponds to the C-terminal of the substrate, with the zinc binding ligand positioned to correspond to the sissile amide. For these, there are broadly two subclasses, those based on the substrate peptide sequence and a series based on (D)-amino acids (sulphonamide-hydroximates) (32). Substrate peptide based hydroximates have produced many potent inhibitors of varying selectivity's against different members of the MMP family (30). Examples of these have been the subject of extensive clinical trails for cancer and rheumatoid arthritis (Marimastat and Trocade respectively (1,7,30). In both cases, the clinical results failed to provide evidence of therapeutic benefit. For Marimastat there has been the also a marked musculoskeletal toxicity reported which limited both the level and duration of dosing. The same toxicity has not been reported for Trocade. The first series of new non-peptide MMP inhibitors was reported by Ciba-Geigy in 1994 (33). This new template has provided patent freedom and basis for combinatorial chemistry approaches to the identification of new inhibitors (32). A number of inhibitors based on the sulphonamidie-hydroximate scaffold have been tested in the clinic, most notably AG3340 (prinomastat), and RS-130,830 for cancer and arthritis respectively (7).   Development of AG3340 was stopped, for treatment of both advanced non-small cell lung cancer and prostate cancer, due to the lack of effectiveness in patients with late-stage disease. Patients on high doses also reported musculoskeletal toxicity. Clinical development has also stopped for RS-130,830. The last two years has witnessed a number of high profile clinical failures in the MMP inhibitors field, primarily though lack of efficacy in cancer and arthritis trials with in some cases examples of unacceptable toxicity. The apparent coincidence in the toxicities observed with Marimastat and AG3340, two potent non-specific, structurally divergent hydroximate inhibitors would suggest that this particular musculoskeletal toxicity is mechanism rather than chemical structure based. Since no similar toxicities have been reported for Trocade, a potent but relatively selective hydroximate based substrate analogue, it could be inferred that the musculoskeletal toxicity is not a function of the zinc ligand but a consequence of inhibition of those metalloproteinase(s) not inhibited by Trocade (33). Some MMP inhibitors continue in clinical trials, such as the thiol peptide mimetic, BMS-275291, and Periostat, a tetracycline with MMP inhibitory activity, which is on the market for the treatment of periodontitis.   Recent reported studies in phase II AIDS-related Kaposi's sarcoma indicate that there is promise with Col-3, a non-antimicrobial tetracycline analogue. This has been reported to be "reasonably well tolerated" and that "anti-tumor activity noted" although dose related photo-toxicity was observed (35).

Disease General - Aberrant MMP activity has been implicated in a number of pathologies, but rheumatoid arthritis and cancer are the two areas that have traditionally received most attention. In rheumatoid arthritis, progress of the disease is associated with progressive destruction of both the aggrecan and collagen II components of articulating cartilage. The cleavage of aggrecan is characteristic of the action of aggrecanase (ADAMTS-4 and ADAMTS-5/11) and type II collagen is cleaved at a specific locus which is distinctive for the actions of collagenases 1, 2 and 3 (MMP-1, 8, 13). In cancer, angiogenesis which is associated with tumour growth and degradation of extra-cellular matrix may be mediated by MMPs. Additionally, there are many studies which have shown various MMPs mRNAs to be induced in either cancer cells or adjacent stoma cells at least showing an association with enhanced levels of these proteins and the pathologies. Arthritis and cancer have to date been the principal focus for clinical trials involving small molecular weight inhibitors of MMPs (1, 6).

MMPs and COPD - Chronic obstructive pulmonary disease (COPD) is a major cause of morbidity and mortality in adults, and is increasing in many parts of the world. It represents a huge burden on health care resources and there is increasing interest in developing new therapies to combat disease progression. COPD is defined as a chronic, slowly progressive disorder characterised by airways obstruction (FEV_1 <80% predicted and FEV_1/FVC ratio <70%) which does not change markedly over several months. The term COPD embraces chronic obstructive bronchitis, with obstruction of small airways, and emphysema, with enlargement of air spaces, destruction of lung parenchyma, loss of lung elasticity, and closure of small airways (36).

The diagnosis of COPD is usually made relatively late in the progression of the disorder when there is already an appreciable fall in FEV_1 and symptomatic deterioration, as the early stages of the disease are relatively asymptomatic. It is most often associated with cigarette smoking or the genetic defect α-anti-trypsin deficiency. The disease is characterised by cough, sputum production, dyspnea, impaired gas exchange and airflow limitation. The latter results from the narrowing of the airway lumen due to inflammatory scarring and loss of alveolar support to an airway wall. Therefore, the ultimate aim is to find drugs which can reverse this process and re-grow the damaged lung.

The pathogenesis of airway obstruction/emphysema is multi-factorial, involving macrophages, neutrophilic airway inflammation, protease-antiprotease imbalance, oxidative stress, and recurrent infection. The protease-antiprotease imbalance hypothesis is thought to play a key role in cigarette smoke-induced chronic lung disease, and has received considerable attention in recent years (36). It proposes that an antiprotease 'shield' protects the normal lung from locally elaborated proteases and that emphysema is the result of an abnormal increase in proteases and/or reduction in pulmonary antiproteases, which leads to parenchymal destruction. As only between 10 and 20% of smokers manifest respiratory symptoms due to impaired lung function, it has been suggested that genetic polymorphisms may account for this disease variability. Interestingly, it has been shown that MMP-1 and MMP-12 haplotypes were associated with rate of decline of lung function and so it has been suggested that polymorphisms in the MMP-1 and MMP-12 genes, but not MMP-9, are either causative factors in smoking related lung injury or are in linkage disequilibrium with causative polymorphisms (37). In contrast, other workers have suggested that a polymorphism of MMP-9 could act as an intrinsic factor determining sensitivity to smoking in the development of pulmonary emphysema (38). A polymorphism of the TIMP-2 gene has been documented which may lead to downregulation of TIMP-2 activity thereby increasing the activity of MMPs and resulting in the degradation of tissue matrix (39).

With regard to the expression of MMPs in the lung, several researchers have shown increased concentrations of MMP-1 and MMP-9 in bronchoalveolar lavage fluid and an increase in the expression and activity of MMP-1, MMP-2 and MMP-9, but not

MMP-12 in the lung parenchyma of patients with emphysema (36, 40, 41). The main cellular sources of enzymatic activity in the lower respiratory tract are neutrophils and alveolar macrophages, both of which are increased in the smoker's lung. Many studies have focused on the proteolytic potential of neutrophils, which includes neutrophil elastase and MMPs with collagenase and gelatinase activity (MMP-9, MMP-8). However, some studies have implicated alveolar macrophages as the major inflammatory effector cells in the lungs of patients with emphysema. In fact, researchers have found that there is a direct relationship between the number of alveolar macrophages and lung destruction, and that MMPs are produced during macrophage-dependent lung injury. Thus, recently there has been considerable interest in the potential role of macrophage-derived MMPs, in matrix degradation in emphysema. Human macrophages are known to produce several MMPs, including MMP-1, MMP-2, MMP-3, MMP-9, and MMP-12 (36).

Additional studies using transgenic and gene-targeted mice have supported expression data generated in human normal and diseased tissues and confirmed a role for MMPs in the pathogenesis of emphysema. In particular, studies in MMP-12 knock out mice have demonstrated a role for MMP-12 in the development of smoke-induced emphysema (29). Although, as mentioned earlier, the role of macrophage-derived MMP12 in the genesis of human emphysema is still far from clear. Other studies, in MMP-1 gene targeted mice, have proposed that MMP-1 contributes to the airway enlargement that characterizes emphysema but the accumulation of collagen (rather than degradation) at sites of emphysematous damage does not lend support to this claim (42). In complete contrast, studies of gene knock-out mice have implicated certain MMPs in the lung repair process. MMP-7 deficient mice demonstrate a severe wound-repair defect and MMP-9 gene deletion studies have suggested that MMP-9 is required for alveolar bronchiolization (31).

Conclusions: - The data discussed suggests that elevated levels of many MMPs are present in chronically inflamed tissues in diseases such as COPD. These data together with those generated from gene targeting and deletion studies in mice strongly implicate MMPs in the pathophysiology of such inflammatory diseases. A role for MMP inhibitors in the treatment of inflammatory diseases has been suggested but many of the recently clinically tested drugs appear hampered by a lack of efficacy, selectivity and suggested mechanism based toxicities. Given the 'protective' role demonstrated for certain MMPs in wound repair processes it seems feasible that effective drugs with an acceptable therapeutic ratio need to be developed which target certain 'disease specific' MMPs. Therefore, it appears that more research needs to be performed in order to identify which MMPs in particular may be causal factors in the pathophysiology of diseases such as COPD.

References

1.   R. Hoekstra, F.A.L.M. Eskens and J. Verwij, Oncologist, 6, 415 (2001).
2.   A. Dove, Nat. Med., 8 95 (2002).
3.   L.M. Coussens, B. Fingleton and L.M. Matrisian, Science, 295, 2389 (2002).
4.   H. Nagase and F.J. Woessner. Jr, J. Biol. Chem., 274, 21494 (1999).
5.   L.J. McCawley and L.M. Matrisian., Curr. Opin. Cell Biol., 13, 534 (2001).
6.   W. Bode, F.X. Gomis-Ruth and W. Stockler, FEBS Lett., 331, 134 (1993).
7.   J.W. Skiles, N.C. Gonnella and A.Y. Jeng, Curr. Med. Chem., 8, 425 (2001).
8.   B. Lovejoy, A.M. Hassell, M.A. Luther, D. Weigl and S.R Jordan, Biochemistry, 33, 8207, (1994).
9.   G.G. Levy, W.C. Nichols, E.C. Lian, T. Foroud, J.N. McClintick, B.M. McGee, A Y A.Y. Yang, D.R. Siemieniak, K.R. Stark, R. Gruppo, R. Sarode, S.B. Shurin, V. Chandrasekaran, S.P. Stabler,. H. Sabio,. E.E. Bouhassira, J.D. Upshaw, D. Ginsburg and H.M. Tsai, Nature, 413, 488 (2001).
10.   D. Pie and S.J. Weiss, Nature, 375 244 (1995).
11.   G. Velasco, A.M. Pendás,.A. Fueyo, V. Knäuper, G. Murphy and C. López-Otín, J.Biol.Chem., 274, 4570. (1999)
12.   M.D. Sternlicht and Z. Werb, Annu Rev. Cell.Biol., 17 463 (2001).

13. R. Gururajan, J. Grenet, J.M. Lahti, and V.J. Kidd, Genomics, 52 101 (1998).
14. J. Li, P. Brick, M.C.O'Hare, T. Skarzynski, L.F. Lloyd, V.A. Curry, I.M. Clark, H.F. Bigg, B.L. Hazleman and T.E. Cawston., Structure, 3, 541 (1995)
15. A.M. Libson, A.G. Gittis, I.E. Collier, B.L. Marmer, G.I. Goldberg and E.E. Lattman., Nat. Struct. Biol., 2, 938 (1995)
16. F.X. Gomis-R, U. Gohlke, M. Betz, V. Knäuper, G. Murphy, C. López-Otín, W. Bode, J.Mol.Biol., 264 556.
17. E. Morgunova, A. Tuuttila, U. Bergmann, M. Isupov, Y. Lindquist, G. Schneider and K. Tryggvason., Science, 284, 1667 (1999).
18. J.M. Shipley, R.L. Wesselschmidt, D.K. Kobayashi, T.J. Ley, and S.D. Shapiro,. Proc. Natl.Acad.Sci.USA, 93, 3942 (1996)
19. J. Cao, M. Hymowitz, C. Conner, W.F. Bahou and S Zucker., J.Biol.Chem., 275, 29648. (2000)
20. K. Briknarová, M. Gehrmann, L. Bányai, H. Tordai, L. Patthy, and M. Llinás., J.Biol. Chem., 276, 27613 (2001).
21. T. Shaw, J.S. Nixon and K.M. Bottomley., Expert Opin.Investig. Drugs, 9, 1469, (2000).
22. J.G. Conway, R.C. Andrews, B. Beaudet, D.M. Bickett, V. Boncek, T.A. Brodie, R.L. Clark, R.C. Crumrine, M.A. Leenitzer, D.L. McDougald, B. Han, K. Hedeen, P. Lin, M. Milla, M. Moss, H. Pink, M.H. Rabinowitz, T. Tippin, P.W. Scates, J. Selph, S.A. Stimpson, J. Warner, and J.D. Becherer, J.Pharmacol.Exp.Ther., 298, 900 (2001).
23. K.M. Bottomley, N. Borkakoti, D. Bradshaw, P.A. Brown, M.J. Broadhurst, J.M. Budd, L. Elliott, P. Eyers, T.J. Hallam, B.K. Handa, C.H. Hill, M. James, H.W. Lahm, G. Lawton, J.E. Merritt, J.S. Nixon, U. Röthlisberger, A. Whittle, and W.H. Johnson., Biochem. J., 323, 483 (1997).
24. J. Lohi, K. Lehti, H, Valtanen, W.C. Parks and J. Keski-Oja., Gene, 242, 75 (2000).
25. Z. Yong, M.R. Spitz, L. Lei, G.B. Mills and X. Wu. Cancer Res., 61, 7825 (2001).
26. S.L. Raza, L.C. Nehring, S.D. Shapiro and L.A. Cornelius., J.Biol.Chem., 275. 41243 (2000).
27. F.X. Gomis-Ruth, K. Maskos, M. Betz, A. Bergner, R. Huber, K. Suzuki, N. Yoshida, H. Nagase, K. Brew, G.P. Bourenkov, H. Bartunik, W. Bode., Nature, 389. 77 (1997).
28. K. Holmbeck, P. Bianco, J. Caterina, S. Yamada, M. Kromer, S.A. Kuznetsov, M. Mankani, P.G. Robey, A.R. Poole, I. Pidoux, J.M. Ward, and H. Birkedal-Hansen,. Cell, 99, 81 (1999).
29. R.D. Hautamaki, D.K. Kobayasahi, R.M. Senior and S.D. Senior, Science, 277, 2002 (1997).
30. K.M. Bottomley. W.H. Johnson and D.S. Walter, J.Enzy.Inhib, 13, 79 (1998).
31. W.C. Parks and S.D. Shapiro., Resp.Res., 2, 10, (2000).
32. J.S. Skotnicki, J.I. Levin, A. Zask and L.O. Killar in "Metalloproteinases as Targets for Anti-Inflammatory Drugs" K.M.K. Bottomley, D. Bradshaw and J.S. Nixon, Ed., Birkhauser Verlag, Basel, 1999, p 17.
33. E.M. O'Byrne, D.T.Parker, E.D. Roberts, R.L. Goldberg, L.J. MacPherson, V. Blancuzzi, D. Wilson, H.N. Singh, R. Ludewig, V.S. Ganu, Inflamm. Res. 44 Suppl 2, S117-8 (1995).
34. E.J. Lewis, J. Bishop, K.M. Bottomley, D. Bradshaw, M. Brewster, M.J. Broadhurst, P.A. Brown, J.M. Budd, L. Elliott, A.K. Greenham, W.H. Johnson, J.S. Nixon, F. Rose, B. Sutton, and K. Wilson, Br.J.Pharmacol., 121 540 (1997).
35. M. Cianfrocca T.P. Cooley, J.Y. Lee, M.A. Rudek, D.T. Scadden, L. Ratner, J.M. Pluda, W.D. Figg, S.E. Krown, B.J. Dezube, J.Clin.Onc.,20, 153, (2002).
36. P.J. Barnes. New Engl.J.Med.,343, 269, (2000).
37. L Joos, J.-Q. He, M.B. Sheperdson, J.E. Connett, N.R. Anthonisen, P.D. Pare and A.J. Sandford. Human Molecular Genetics, 11, 569 (2002).
38. N Minematsu, H. Nakamura, H. Tateno, T. Nakajima and K Yamaguchi. Biochem. Biophys. Res. Comm., 289, 116, (2001).
39. K. Hirano, T. Sakamoto, Y. Uchida, Y. Morishima, K. Masuyama, Y. Ishii, A. Nomura, M. Ohtsuka and K. Sekizawa, Eur.Respir.J., 18, 748 (2001).
40. K. Imai, S.S. Dalal, E.S. Chen, R. Downey, L.L. Schulman, M. Ginsburg and J. D'Armiento, Am.J.Respir.Dis., 163, 786 (2001).
41. K. Ohnishi, M. Takagi, Y. Kurokawa, S. Satomi and Y.T. Konttinen. Lab.Invest., 78, 1077 (1998).
42. J. D'Armiento, S.S. Dalal, Y. Okada, R.A. Berg and K. Chada. Cell, 71, 955 (1992).

Chapter 22. Fc Receptor Structure and the Design of Anti-inflammatories: New Therapeutics for Autoimmune Disease

Geoffrey A. Pietersz, Maree S. Powell, Paul A. Ramsland and P. Mark Hogarth
Austin Research Institute
Austin and Repatriation Medical Center, Studley Road, Heidelberg, Victoria 3084,
Australia

Introduction - Inflammation caused by immune complexes in the blood vessels (vasculitis), the kidney (glomerulonephritis) and in the joints (arthritis) is a major cause of morbidity in autoimmune diseases. It is also one of the major mechanisms of tissue destruction in diseases including rheumatoid arthritis especially in the extra articular disease, immune thrombocytopenia (ITP), systemic lupus erythematosus (SLE) and Wegner's granulomatosis. Whilst clear that in autoimmune diseases the likely initiation of the autoimmune process is a corruption of T cell regulation, the tissue damage and the perpetuation of disease are due to immune complexes. In the last decade, the role of Fc receptors as major initiators of inflammation caused by immune complexes has become more precisely defined with the use of recombinant Fc receptor (FcR) and gene knock-out or transgenic mice (1-8).

Fc receptors are cell surface glycoproteins of inflammatory leukocytes and lymphocytes that specifically bind the Fc portion of immunoglobulins (1-8). The receptors have been identified for each of the immunoglobulin classes and the best characterised of these are the IgE receptor, FcεRI, the IgA receptor, FcαRI and the IgG receptors, FcγRI, FcγRII, FcγRIII. For the most part, the binding of immunoglobulins to these receptors either as immune complexes or as monomeric Ig and its subsequent cross linking with antigen activates inflammatory cells leading in normal circumstances to the elimination of pathogens. However, this process in autoimmunity leads to powerful pathological inflammatory responses and severe tissue destruction.

VALIDATION OF FcγRIIa AS A TARGET FOR ANTI-INFLAMMATORY COMPOUNDS

Studies have shown that recombinant soluble Fc receptors can inhibit the antibody dependent Type II and Type III hypersensitive reactions *in vivo*, providing the first direct evidence of the role of Fc receptors in this type of inflammation (1). Subsequent experiments using recombinant receptors and genetically manipulated mice have shown that immune complex induced inflammation is Fc receptor dependent in, for example, vasculitis, glomerulonephritis and, also, in models of rheumatoid arthritis (1-3). These results are consistent with the clinical evidence showing the presence of antibody secreting cells or immune complexes in the joints or tissues of patients, as well as, the classical studies of Dixon examining hypersensitivity reactions which implied a role for cell surface Fc receptors (1-5).

FcγRIIa is unique - The IgG receptor FcγRIIa is unique among Fc receptors (5-9). It is the most widely distributed of all receptors, being present on all leukocytes with the exception of T and B lymphocytes. It is also the only Fc receptor present on platelets. It is found only in humans and the higher primates. Biochemically, it is also unique in that all of the signalling motifs are contained in the cytoplasmic chain of this receptor. In the other leukocyte receptors, signal transduction requires the association with other subunits, such as the common FcRγ chain and also the β chain shared among several IgG receptors and the high affinity IgE receptor (7-12).

In addition, FcγRIIa is central to the signalling of other receptors especially on neutrophils where these depend on FcγRIIa for signal transduction and, thereby, cell activation (13). This is of major importance in Type III hypersensitivity reactions where the initial inflammatory cells are neutrophils. Whilst immune complex binding occurs via FcγRIII and FcγRIIa, neutrophil activation via Fc receptors occurs only through FcγRIIa.

STRUCTURE OF THE FcγRIIa RECEPTOR

The crystal structure of FcγRIIa is composed of two extracellular C2 subtype Ig-like domains (Figure 1). The association of the domains is acute with a 52° angle

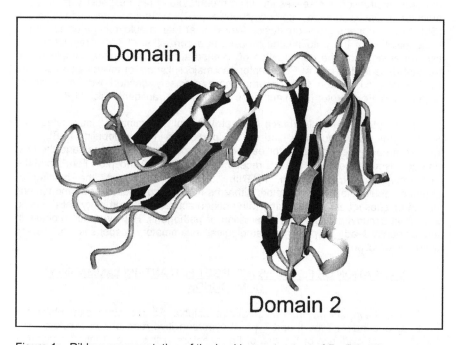

Figure 1: Ribbon representation of the backbone structure of FcγRIIa (6).

between the domains. This domain arrangement is achieved by several key interactions spanning the A' strand, AB loop and G strand of domain one (D1), the A and B strand and the BC loop of domain two (D2). Importantly, hydrophilic side chains and buried water molecules co-operate to form an extensive hydrogen bonded network within this interface (6). This tight association is common to the other members of the FcR family. Crystal structures of FcγRIIb, FcγRIII and FcεRI all display acute angles between the two Ig-like domains (14-16). Indeed, the superimposition of alpha carbon skeletons for FcγRIIa, FcγRIII and FcεRI are all very similar [see Figure 3; upper panel]. This domain arrangement forces (i) both the NH_2 and COOH termini toward the membrane and (ii) the BC, C'E and FG loops of domain 2, all known to be critical to ligand recognition, away from the membrane at the 'top' of the receptor. It is, noteworthy, however that, in contrast to the FcγR's and FcεRI, FcαRI binds immunoglobulin through D1 and uses the same loops in this domain as used in D2 of the other leucocyte Fc receptors (17).

The description of the crystal structures of FcγRIII complexed with IgG1 Fc fragment and FcεRI complexed with IgE Fc fragment (see below) supports the biochemical and genetic data that originally identified the ligand binding surface (18-27). Interestingly, these structures of FcR complexed with ligand found that the receptor ectodomains remain rigid with little movement upon binding of ligand. Also, in both structures, the stoichiometry between receptor and ligand was 1:1; neither FcγRIII nor FcεRI have been seen to dimerize under a variety of conditions. This is in contrast to FcγRIIa that has been found to form a crystallographic dimer (Figure 2).

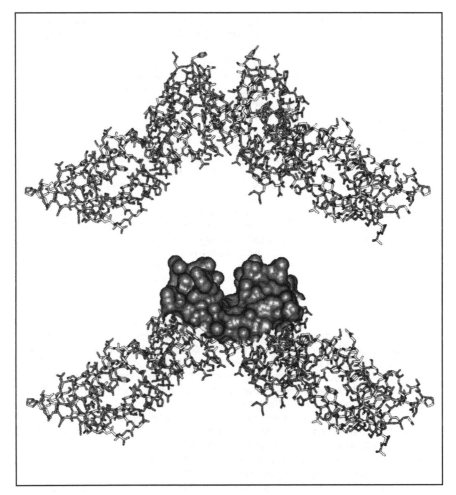

<u>Figure 2</u>: The FcγRIIa dimer and the partial molecular surface of the dimer depicting the dimer groove (6).

The two FcγRIIa molecules in the dimer associate via the C'CFG sheets and the CC' loops of each D2. Although there is limited biochemical data that supports FcγRIIa dimerization, stoichiometric analysis of FcγRIIa and ligand using equilibrium gel filtration analysis as a physiological process found that the receptor can bind ligand with a 1:1 or 2:1 stoichiometry (28). Dimerization of FcγRIIa juxtaposes the ligand binding surfaces of each monomer to create one large binding surface of

630Å^2; similar to that of an antigen-Fab complex (6). Also, dimerization of FcγRIIa forces the binding sites to be parallel to the cell membrane and, thereby, more accessible to ligand than a monomeric version of the receptor where the ligand binding site would be presented at an oblique angle to the membrane. Furthermore, the FcγRIIa dimer interface has shape (S_c = 0.80) and electrostatic complementarity (EC_{FS} = 0.52) within the limits seen for functional interactions (6). Thus, FcγRIIa dimerization may be a feature of interaction with ligand. However, it remains to be elucidated how FcγRIIa dimerization affects receptor function, and how dimerization may affect the localisation of the receptor within the plasma membrane.

Structure of FcγRIII and FcεRI Fc Receptors -- Analysis of the ligand bound receptor complexes has indicated that both FcεRI and FcγRIII interact with their respective ligands in a similar way (18-20). The total solvent accessible areas for interaction are 1453 and 1850 Å^2 for FcγRIII and FcεRI respectively, with the slightly larger area providing a more extensive contact surface that allows for a higher affinity interaction. The interaction between FcγRIII and IgG1 Fc involves identical residues from the Fc portion that interact with different, unrelated surfaces of the receptor. The hinge loop between D1 and D2, the BC, C'E, FG and C' β strand all contribute to the binding of ligand. The interaction with ligand is a combination of salt bridges, hydrogen bonds and hydrophobic interactions. The interface between FcγRIII and one half of the Fc region is dominated by hydrogen bonding contributed by residues Leu[235] to Ser[239] of the Fc region of IgG, identified from earlier mutagenesis studies as the main contact site. Importantly, it is now firmly established that two areas of interaction occur between FcγRIII and ligand in the complex structure. The Pro[329] of Cγ2-A of IgG interacts with Trp[87] and Trp[110] of FcγRIII to create a 'proline sandwich'. The second site includes residues Leu[234] to Ser[239] of the lower hinge of both IgG1 chains. In previous structures of IgG this region of the immunoglobulin has been disordered but is easily defined in the complex of receptor and ligand (18-19). It is also noteworthy that the two areas of interaction may also explain the biphasic pattern of interaction when examining real-time interactions with ligand using surface plasmon resonance.

A comparison of the FcγRIII:IgG1 complex with the FcεRI:IgE, the only other solved Fc receptor: ligand complex, is interesting. Like the FcγRIII: IgG1 complex, the solvent accessible area of the FcεRI:IgE interaction is divided into two regions with 740Å^2 coming from the IgE "equivalent" of the IgG lower hinge and 1110Å^2 from elsewhere in the IgE Fc. The area outside the lower hinge equivalent region contributes approximately twice the area seen for the interaction between FcγRIII and IgG1 and adopts a very different conformation than that of IgG Fc, which may enable a more extensive interaction with receptor. Salt bridges, hydrogen bonds and hydrophobic interactions also dominate the interaction between FcεRI and IgE throughout both sites. The first site is the interaction with one Cε3 IgE domain with the C-C' region of FcεRI D2. The second site refers to the interaction between the second Cε3 domain of IgE binding with the D1-D2 interface of FcεRI and involves the cluster of four surface exposed tryptophan residues (20). The structural complex identified that similar areas of the receptor were involved in the interaction with ligand and confirmed earlier biochemical studies that identified the crucial areas of the IgE Fc (29-31).

To date, a solved complex of IgG and FcγRIIa does not exist. However, despite the differences in sequence between FcγRIIa and other Fc receptor and their respective ligands, mutagenesis studies suggest that the similar regions are likely to be involved in the interaction with IgG. In addition, the apparent conservation of a "tryptophan sandwich" in both FcγRIII and FcεRI involving residues found within the

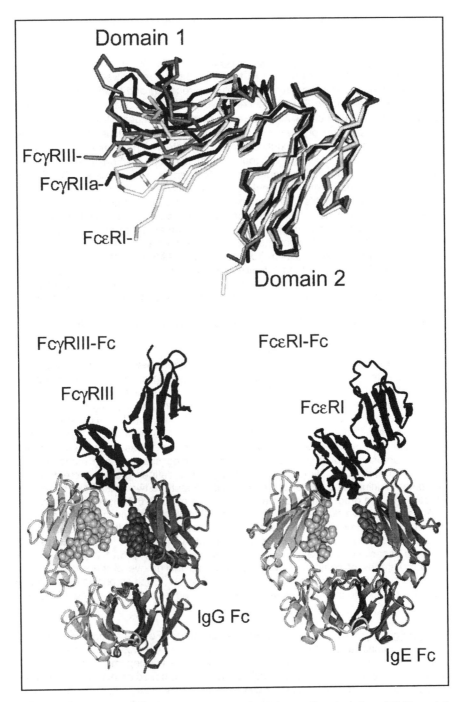

Figure 3: Superpositions (Cα) of the immunoglobulin binding domains of FcR crystal structures. Lower panels: Ribbon representations of the crystal structures of the FcγRIII and FcεRI bound to their Fc ligands (18-20). Carbohydrates are shown as cpk structures.

receptor D2 may imply that this is a motif for interaction with ligand which is conserved in FcγRIIa.

Dimerization of FcγRIIa but not other Fc Receptors – As indicated above, the dimer configuration of FcγRIIa is unique. At this time, there is no solved complex of IgG and FcγRIIa however if the interaction between receptor and ligand were to occur precisely as seen for FcγRIII, the dimer could not form because of a clash of the receptor with Cγ2 of the IgG Fc. However, modelling suggests that relatively little change in the position of the Cγ2:FcγRIIa would allow the dimer to form. This obvious difference in receptor configuration is consistent with the major difference in the biology of these receptors. Signal transduction through FcγRIII and FcεRI is dependent on their association with FcR-γ chain homodimers (7). These homodimers are themselves covalently associated and contain the Immunoreceptor Tyrosine Activating Motifs (ITAMs) that are essential to the induction of cell activation. There is no requirement for these homodimers in FcγRIIa-dependent activation as the signalling motifs have been incorporated into its cytoplasmic tail, i.e. the ligand binding ecto-domains and signalling motifs are contained in the one polypeptide. FcγRIIa dimerization juxtaposes the ITAMs in each monomer that in turn ensures dimerization of the signalling motifs, which is an obligatory requirement for signalling in immunoreceptors systems of this type. It will therefore be of great interest to determine how ligand interactions affect or are affected by dimerization.

NOVEL APPROACHES TO THE DEVELOPMENT OF ANTI-INFLAMMATORY COMPOUNDS

Rational Drug Design – A number of strategies can be utilised to design antagonists to the FcγRIIa/IgG interaction. These involve screening of peptide or non-peptide combinatorial libraries, phage display or use of anti-receptor antibodies. An alternative strategy is to use a structure based drug design approach to the development of small chemical antagonists specific for FcγRIIa. The structure of FcγRIIa demonstrates that this receptor is a dimer in contrast to the highly related FcγRIIb or the FcγRI or FcγRIII (6, 32). Thus, the receptor has several features that make it amenable to drug design. The IgG binding sites of each monomer are juxtaposed and, in the dimerization process, a well defined groove is created that abuts the binding sites and also creates a hole about the two-fold axis of symmetry (Figure 2). The groove - approximate dimensions of 23x7x6 Angstrom tapering to a 4-Angstrom groove floor – together with the binding site provides several suitable target areas for drug design. A variety of side chains, including those involved in the interaction with IgG, are juxtaposed and were the initial targets for the drug design process especially phenylalanine (Phe[124]) and lysine (Lys[120]) (5,21-27). As an initial screen, libraries of compounds were docked into the groove of the receptor such that interactions involved the phenylalanine and lysine. Such screening generated a number of compounds that were further modified and tested in the screening assays.

In vitro Screening – The choice of screening assays for FcγRIIa receptor antagonists is of prime importance. Since the physiological role of dimerization or their point of association in cells is not known, conventional assays utilising immobilised receptor assays is a poor choice. We have utilised two assays based on human platelets, as FcγRIIa is the only Fc receptor expressed on these cells. Such a choice can overcome the interference caused by the other Fc receptors. Binding of heat aggregated human immunoglobulin (HAGG) or immune complexes to FcγRIIa receptors on platelets activates them and induces the expression of P-selectin, which can be measured by flow cytometry and is utilised as one of the assay systems. The platelet aggregation assay can also be utilised, as aggregation is the final endpoint of activation of platelets by immune complexes. Specificity controls include a number of other agonists such as ADP, arachidonic acid or collagen that aggregate platelets via a non-FcγRIIa receptor.

In Vivo Efficacy – FcγRIIa is absent in mice and receptor antagonists cannot be tested in inbred strains of mice. A transgenic mouse expressing FcγRIIa could be used for *in vivo* evaluation of receptor antagonists. The collagen-induced arthritis model (CIA) is used as an *in vivo* model for evaluating anti-arthritic drugs (33). Even though the strain of transgenic mice is on a resistant H-2^b haplotype, these mice develop CIA with similar features of the disease to the DBA/1 strain. As the transgenic mice also express FcγRIIa on platelets (as in humans), the same mice can be used to test the efficacy of the inhibitors against immune thrombocytopenia.

DESIGN OF INHIBITORS TO OTHER Fc RECEPTORS

The solved 3D structures of complexes of human FcγRIII and IgG1, human FcεR1 and IgE also offer opportunities in the design of antagonists of receptor:Ig interactions. In both cases, no receptor dimer is indicated and, indeed, the complexes show interactions over a broad surface which is largely featureless. Such arrangements make design of effective small chemical entities (SCE) difficult which is a general problem in protein receptor: protein ligand interactions. None-the-less with the advent of modern chemistries, the possibilities of targeting multiple sites with multispecific SCEs maybe possible.

The case for targeting the IgE receptor is clear – inhibition of the IgE:FcεR1 interaction will inhibit allergic reactions but the extraordinary high affinity ($>10^{-10}$ M^{-1}) makes the practical use of any SCE in receptor: ligand antagonism problematic. In the case of FcγRIII, the case for it as a target in autoimmunity remains to be clarified. All mouse studies to date have been performed in the absence of FcγRIIa. The features – structurally, biochemically and biologically that make FcγRIIa the "ideal" target (described above) are sufficiently compelling to warrant targeted effort against this receptor in the development of novel anti-inflammatories.

Conclusion – The increasing body of evidence that Fc receptors are key receptors in the development of inflammation caused by immune complexes now makes these valid targets for new therapeutics in the treatment of autoimmune diseases and in the case of the IgE receptor, allergy. The powerful revolution in drug discovery and development technologies especially in relation to structure based approaches now enables an entirely or partly rational approach to the generation of receptor antagonists. It is clear in the immediate future that the design of compounds directed towards the uniquely human FcγRIIa with its dominant role in activation of platelets and leucocytes by immune complexes will be fertile ground for novel antiinflammatory approaches.

References

1. F. Ierino, M. Powell, I.F.C. McKenzie and P.M. Hogarth, J.Exp.Med., 178, 1617 (1993).
2. J.V. Ravetch and S. Bolland, Ann.Rev.Immunol., 19, 275 (2000).
3. A. Gavin and P.M. Hogarth in "Human IgG Fc Receptors," J. van de Winkel and P. Capel, Eds., 1995, p. 271.
4. I. Roitt, J. Brostoff and D. Male in "Immunology" Mosby International Ltd (1993).
5. M. Hulett and P.M. Hogarth, Adv.Immunol., 57, 1 (1994).

6.   K. Maxwell, M. Powell, M. Hulett, P.A. Barton, I.F.C. McKenzie, T.P.J. Garrett and P.M. Hogarth, Nat.Struct.Biol., 6, 437 (1999).
7.   M. Daëron, Ann.Rev.Immunol., 15, 203 (1997).
8.   J.C. Cambier, Proc.Natl.Acad.Sci., U S A. 94, 5993 (1997).
9.   M.L. Hibbs, L. Bonadonna, B.M. Scott, I.F.C. McKenzie and P.M. Hogarth, Proc.Natl. Acad.Sci., USA, 85, 2240 (1988).
10. P.M. Hogarth, E. Witort, M. Hulett, C. Bonnerot, J. Even, W. Fridman and I.F.C. McKenzie, J.Immunol., 146, 369 (1991).
11. D.W. Sears, N. Osman, B. Tate, I.F.C. McKenzie and P.M. Hogarth, J.Immunol., 144, 371 (1990).
12. H. Metzger, J.Immunol., 149, 1477 (1992).
13. F.Y. Chuang, M. Sassaroli and J.C. Unkeless, J.Immunol., 164, 350 (2000).
14. P. Sondermann, R. Huber and U. Jacob, EMBO., 18,1095 (1999).
15. Y. Zhang, C.C. Boesen, S. Radev, A.G. Brooks, W-H. Fridman, C. Sautes-Fridman and P.D. Sun, Immunity. 13, 387 (2000).
16. S.C. Garman, J-P. Kinet and T.S. Jardetzky, Cell. 95, 951 (1998).
17. B.D. Wines, C.T Sardjono, H.M. Trist, C.S. Lay, P.M. Hogarth, J. Immunol., 166, 1781 (2001).
18. P. Sondermann, R. Huber, V. Oosthuizen and U. Jacob, Nature. 406, 267 (2000).
19. S. Radev, S. Motyka, W-H. Fridman, C. Sautes-Fridman and P.D. Sun, J. Biol. Chem., 276,16469 (2001).
20. S.C. Garman, B.A. Wurzburg, S.S. Tarchevskaya, J.P. Kinet and T.S. Jardetzky, Nature, 406, 259 (2000).
21. M.D. Hulett, I.F.C. McKenzie and P.M. Hogarth, Eur.J.Immunol., 23, 640 (1993).
22. M.D. Hulett, E. Witort, R.I. Brinkworth, I.F.C. McKenzie and P.M. Hogarth. J.Biol. Chem., 269, 15287 (1994).
23. M.D. Hulett, E. Witort, R.I. Brinkworth, I.F.C. McKenzie and P.M. Hogarth, J.Biol. Chem., 270, 21188 (1995).
24. B.D. Wines, M.S. Powell, P.W.H.I. Parren and P.M. Hogarth, J.Immunol., 164, 5313 (2000).
25. M.D. Hulett, R. Brinkworth, I.F.C. McKenzie and P.M. Hogarth, J.Biol.Chem., 274,13345 (1999).
26. M. Hulett, N. Osman, I.F.C. McKenzie and P.M. Hogarth, J.Immunol., 147, 1863 (1991).
27. M.D. Hulett and P.M. Hogarth, Mol.Immunol., 35, 989 (1998).
28. P. Sondermann, U. Jacob, C. Kutscher and J. Frey, Biochem., 38, 8469 (1999).
29. J.M. McDonnell, R. Calvert, R.L. Beavil, A.J. Beavil, A.J. Henry, B.J. Sutton, H.J. Gould, D. Cowburn. Nat.Struct.Biol., 8, 437 (2001).
30. A.J. Henry, J.M. McDonnell, R. Ghirlando, B.J. Sutton, H.J. Gould, Biochemistry, 39, 7406 (2000).
31. B.J. Sutton, R.L. Beavil, A.J Beavil, Br. Med.Bull, 56, 1004 (2000).
32. M.S. Powell, P.A. Barton, D. Emmanouilidis, B.D. Wines, G.M. Neumann,G.A. Pietersz, K.F. Maxwell, T.P.J. Garrett, and P.M. Hogarth, Immunology Letters, 68, 17 (1999).
33. S.E. McKenzie, S.M. Taylor, P. Mallandi, H. Yuhan, D.L. Cassel, P. Chien, E. Schwartz, A.D. Schreiber, S. Surrey and M.P. Reilly, J.Immunol., 162, 4311 (1999).

Chapter 23. Tumor classification for tailored cancer therapy

Fiona McLaughlin and Nick LaThangue
Prolifix Ltd
91 Milton Park, Abingdon, Oxon OX14 4RY

Introduction – Cancer is an unmet clinical disease that remains the second largest killer in the Western world, with a steady rise in both the occurrence and death rates due to cancer throughout the last century. Cancer is a genetic disease which, together with its multi-stage nature and the realization that many environmental and dietary factors can influence the disease, has wide ranging implications for its treatment and prevention in the 21st century. Although chemotherapy is currently the most promising treatment in clinical use, its success is greatly limited by a severe lack of understanding of tumor biology. This leads to an inability to predict the response of a given tumor type to a particular treatment regime, and consequently a considerable failure rate in clinical oncology.

The inherent heterogeneity of tumors makes classification frequently difficult and often impossible. Current methods for classifying tumors rely on pathological and clinical diagnosis to determine the stage of progression of the tumor, prediction of the course of disease, interval to metastasis and response to treatment. In this respect, the concept of a tailored therapeutic approach based upon a particular sub-type of a disease has received much attention with the advent of the wealth of genetic data resulting from the Human Genome Project. The identification of new genes involved in processes such as cell proliferation, differentiation, apoptosis and tumor suppression has, and continues to increase our understanding of the basic molecular and biological mechanisms of tumor growth. Such advances provide a significant aid to classification and treatment and therefore the overall prognosis of cancer as a disease.

This review focuses on the technologies which offer the most promise in enhancing the ability to classify tumors, based on gene and protein expression profiling. The application of pharmacogenomics to profile both tumors and patients with the aim of defining the most appropriate treatment regime, with minimum toxicology and maximum efficacy, will also be discussed.

CURRENT METHODS TO CLASSIFY CANCER – SUCCESS STORIES AND BOTTLE NECKS

Historically, both solid tumors and leukemia have been classified according to criteria identified through clinical experience. Through histological approaches, tumors are broadly classified into carcinoma (epithelial), sarcoma (connective tissue) or lymphoma (lymphoid tissue), with sub-classifications dependent on the tumor type. A histological grade is then assigned to a tumor depending on the degree of differentiation and mitotic activity. In addition, designation of a particular tumor stage is established depending on information gleaned from its size, nodal status and metastases. Classification still relies very heavily on an assessment of cell morphology which, although often accurate, frequently fails to offer predictions regarding either the final clinical outcome or the patient's ability to respond to a particular treatment regime. Acute leukemia demonstrates this point. The widely used 1940's classification is based on the ability to follow variations in outcome of disease and cell morphology based on the resemblance of blasts to normal haematopoiesis (1,2). Subsequently, the classification became more sophisticated with the introduction of immunocytochemistry in the 1960's which allowed classification based on expression of enzymes such as myeloperoxidase and more recently mast cell tryptases (3,4). Acute leukemias have been further classified into acute myoblastic leukemia (AML) and acute lymphoblastic leukemia (ALL), depending on the haematopoeitic precursor from which the leukemia arose. The majority of pathological classification systems of acute myeloid leukemia (AML) are influenced by the criteria proposed by the French-American-British (FAB) Cooperative Group. More recently the World Health Organization (WHO) proposed a further classification of leukemias and lymphomas which includes the FAB recommendations and additional diagnosis which involved correlation of disease with specific cytogenetic data (5,6). Cytogenetic analysis in particular has aided the further subclassification of acute leukemias dependent on a particular chromosomal translocation. For example, in the case of T cell acute lymphoblastic leukemia (T-ALL) almost 25% of patients have a chromosomal rearrangement in the stem cell leukemia (SCL or TAL-1) gene (7). This may be a deletion (12-26% of T-ALL) or t(1;14)(p32;q11) chromosomal translocation, in which SCL is transposed from its normal position on chromosome 1 into the T-cell receptor alpha/delta chain locus on chromosome 14. Similarly, in the case of B cell leukemias, many chromosomal translocations occur such as t(9:22)[BCR-ABL], t(1;19)[E2A-PBX1], t(12;21)[TEL-AML1] (8,9). Identification of these translocations has led to possibilities not only for increased classification but also for design of novel oncology drugs based around the newly discovered target genes at the translocation breakpoints. In the case of BCR-ABL (the Philadelphia chromosome), the fusion results in a constitutively active tyrosine kinase present in virtually all patients with chronic myeloid leukemia (CML) and in approximately 20% of patients with ALL. This has led to the development of a small molecule inhibitor of the BCR-ABL tyrosine kinase (ST1571, Gleevec, Novartis, Basel, Switzerland) which is a competitive inhibitor of the ATP-binding site of the enzyme. ST1571 is highly selective for cells expressing BCR-ABL, causing growth arrest or apoptosis, whilst leaving normal haematopoetic cells unaffected (10,11). In a recent Phase I trial, 11% of patients with CML were in remission following treatment with ST1571 continued out to 349 days (12,13).

Whilst Gleevec exemplifies the advances that have been made in cancer classification over the years, several crucial issues remain with the current guidelines. Early diagnosis of prostate cancer, currently using expression of prostate specific antigen (PSA) allows curative surgery as a treatment option. Despite this surgical intervention, approximately 30% of patients will relapse due to metastases present at the time of surgery (14,15). The current classification tools for prostate cancer namely PSA expression, Gleason score and tumor stage do not allow adequate prognosis of low to intermediate risk patient (16-18). The difficulties in diagnosing of patients with early stage disease are also apparent in patients with epithelial ovarian cancer, where a potential biomarker, cancer antigen-125 (CA-125), is found in 85% of patients with ovarian cancer (19). Again, only 50% of patients with early stage cancer exhibit high levels of CA-125 and, thus, as a single marker, CA-125 fails to fulfill the criteria for an efficient diagnostic tool.

Overall, whilst methods of cancer classification using information from histological, clinical and epidemiological studies have played a major role in enhancing the probability of a correct prognosis of disease, much remains to be achieved in developing new, robust, reliable and predictive prognostic tools.

MOLECULAR CLASSIFICATION – A WAY FORWARD?

Despite the advances in cancer research throughout the last century, the annual mortality rate rose to 5.2 million people worldwide in the years around 1990 (20). Clearly, the development of more robust classification systems for diagnosis, prognosis and tailored therapies is pivotal to advance cancer treatment and clinical efficacy. Molecular analysis, measuring changes in the DNA, RNA and protein levels provide powerful approaches to meeting these needs. A significant amount of research over the past 5-10 years has focused on changes in mRNA expression patterns in human tumors. Initial studies aimed to identify genes with changes in expression pattern in cancer relative to normal tissue, potentially identify novel biomarkers for disease classification and progression studies, as well as new targets for drug discovery. There are many difficulties with this simplistic approach, in particular, measurement of one gene is rarely sufficient for an accurate diagnosis and prognosis of tumor type due to the cellular complexity of each tumor, variation between patient populations and variation in expression profile at different stages of the disease.

The advances in DNA array technology over recent years allow robust quantitative analysis of several thousand mRNAs from a single patient sample. Ideally, these arrays should contain every single gene in the human genome for generating unbiased data.   At a technical level, tumor samples isolated from patients are snap frozen in liquid nitrogen to preserve the RNA which is sensitive to degradation (Figure 1). Methodologies do not yet exist to allow good quality RNA to be extracted from formalin fixed sections and so the availability of tissue samples can be a limiting factor for some studies. Once isolated, RNA is transcribed into complementary RNA or reverse transcribed into complementary cDNA and fluorescently labeled. Corresponding RNA from a reference sample, such as normal tissue where available, is also labeled with a different fluorophore. Labeled probes are hybridized to arrays containing several thousand cDNAs or oligonucleotides. In cDNA micro-array, polymerase chain reaction (PCR) is used to generate fragments of full length cDNAs of interest. Many cDNA micro-arrays are available from commercial suppliers and are widely used by both the academic and commercial researchers. Oligonucleotide arrays have several advantages over cDNA arrays. Flexibility in design of oligos for each gene allows greater specificity with less chance of cross-hybridization with non-related genes, standardization with regards to length of oligo and discrimination of splice variants. Commercially available oligonucleotide arrays are also available (Affymetrix) and are becoming more widely used as the costs of chip technology and software decrease.

Figure 1. DNA micro-array procedure

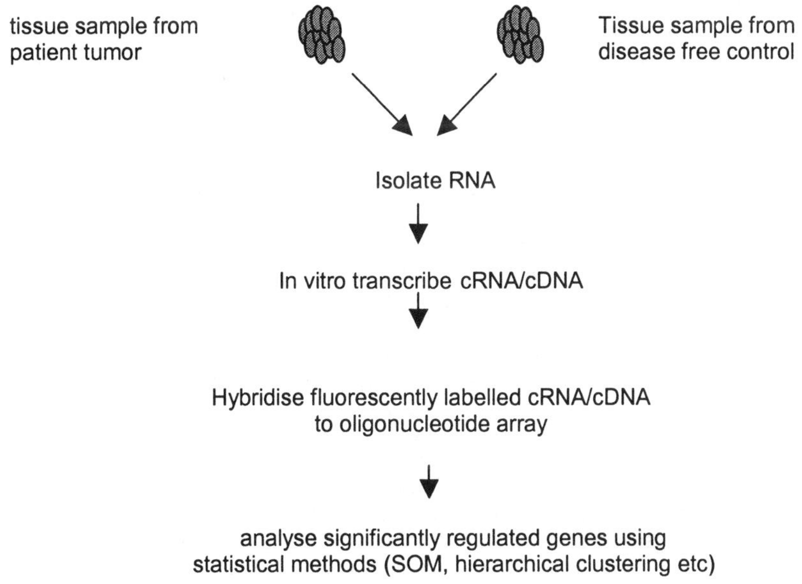

tissue sample from   Tissue sample from
patient tumor   disease free control

Isolate RNA

In vitro transcribe cRNA/cDNA

Hybridise fluorescently labelled cRNA/cDNA
to oligonucleotide array

analyse significantly regulated genes using
statistical methods (SOM, hierarchical clustering etc)

Having proceeded through the technicalities of sample preparation, probe preparation and hybridization, an enormous volume of data is generated which has to be stored and organized ready for statistical analysis. Fundamental to any analysis is provision of high quality data from studies which achieve low signal to noise ratios and with relevant negative and positive controls. Several algorithms exist which allow interpretation of these results. Two common statistical methods to analyze micro-array data are known as unsupervised and supervised hierarchical learning (Figure 2) (21-24). In the case of unsupervised learning, categorization of data is based on clustering genes (or data points) together which have similar features or expression profiles (if comparing tumor samples from several patients). Using this system, no assumptions are made to bias the cluster formation. In the parallel approach of supervised learning, prior information is utilized and a training data set aids the identification of particular classes or features which allow distinctions to be made (25). The biological and/or clinical relevance of these types of analysis of data is paramount. Current approaches are limited by the necessity to select a subset of the human genome for analysis and the heterogeneity of tumor samples. As a result, the most convincing studies are those which use the most patient samples per study, with the best quality input data generated for analysis.

A large number of studies have been undertaken to identify gene expression profiles of a range of tumor types with the aim of furthering classification of tumors for both prognosis and treatment protocol. A selection of these studies will be discussed here, which demonstrate the potential and limitations of this technology.

Gene expression profiling for tumor diagnosis – In a recent study by Ramaswany et al., the authors set out to demonstrate the feasibility of correctly diagnosing tumors from a heterogeneous set representing the most common tumors, using only a single reference database (26). The study attempted to diagnose 218 tumor samples which contained 14 different tumor types using gene expression signatures, where tumor samples were carefully chosen from patients with information on standard classification (histological and clinical) and biopsies taken from primary sites in the majority of cases. RNA was hybridized to Affymetrix GeneChips containing over 16,000 probe sets. Both unsupervised and supervised learning methods were used to analyse the data set. These primarily epithelial solid tumors are molecularly complex and the unsupervised learning method was unable to cluster the data accurately in a manner which allowed the tissue of origin to be distinguished. The most likely explanation for the failure of this technique is the biological variation in the input gene expression data. This is compounded by the inability to make any selection criteria based on knowledge of non-neoplastic cell contribution to the tumor genotype. Having concluded that a totally random approach was not sufficiently robust for tumor classification, the authors compared several algorithms for supervised learning; weighted voting, k-nearest neighbor and Support Vector Machine (SVM) (27-29). All three algorithms generated significant classification predictions, with SVM being the most consistently accurate. Importantly, using supervised learning methods, tumors containing neoplastic and non-neoplastic elements can now be classified with respect to tissue of origin.

Despite the successes of this approach, several limitations are also evident; most notably the inability to classify poorly differentiated tumors. The poorly differentiated tumors referred to were 20 adenocarcinomas from breast, lung, colon, ovary and uterus. Only 30% of these tumors were correctly classified regarding tissue of origin, which is statistically the same as would be expected by chance alone. The key implication of this finding is that the gene expression profiles of poorly differentiated tumors are statistically different than their differentiated counterparts, possibly because they originate from a distinct precursor cell, and as such represent a new subset of tumor for classification. Thus, all tumors originating

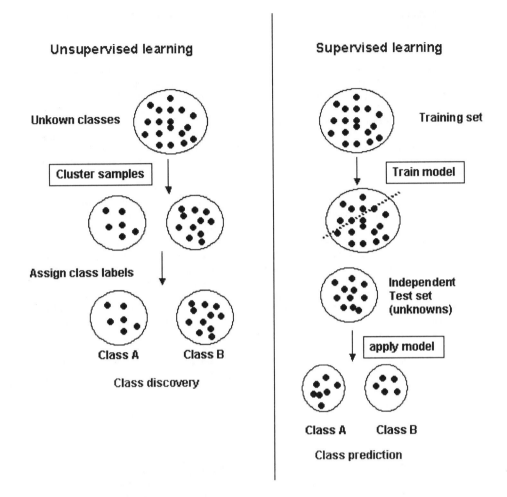

Figure 2: Unsupervised and supervised learning. In the unsupervised learning model, tumor samples are clustered together based on similarities in their gene expression data. Distinctions can then be made which allow assignment of classes to the tumor samples. In the supervised learning model, existing data from known class types is used to develop a training model which can then be used to classify unknown data (30).

from breast should not be classified together and should be further distinguished as poorly or well differentiated breast tumors.

A more recent study by Yeoh et al. focused on classification of acute lymphoblastic leukemias and the ability of DNA micro-arrays to identify previously undiscovered subtypes and to aid diagnosis and understanding of the biology of existing leukemias (8). Leukemias are in many ways more straight-forward to study than solid tumors, primarily due to the fact that the target leukemic cells are readily available for study in the bloodstream. As a result, much is known about classification of leukemia using existing technologies such as cytogenetics, morphology and immunohistochemistry. Despite these advantages, only recently have distinctions been made between AML and ALL using a single platform technology of DNA micro-array rather than a combination of historical methods (31). In the Yeoh study, 327 diagnostic bone marrow (leukemic blast) samples from patients with pediatric ALL were analyzed using DNA micro-array technology (Affymetrix 12,600 probe chips). Here using leukemic samples, unsupervised learning methods allowed identification of 6 major leukemia subtypes, based on similarities in gene expression between members of each group. Even within the cases which were sufficiently heterogeneous and could not be assigned to a known subgroup, one distinct class of 14 cases was identified which had a significant similarity of gene expression. This new class did not exhibit any cytogenetic abnormality and had either normal, pseuodiploid or hyperdiploid karyotypes. Looking in detail at the top 40 discriminating genes for each leukemia sub-type, lineage distinctions can be made (for example T-ALL from B-ALL) but different stages of differentiation within B cells for example, cannot be distinguished. Importantly, in this study, a small subset of the vast amount of data generated from analysis of thousands of genes on a chip was verified using a combination of real time PCR (for 5 genes) and flow cytometry (for 9 cell surface antigens). Although these are small numbers compared with the total number of genes studied, the correlation of expression of all genes identified by micro-array compared with real time PCR or FACS suggests that micro-array technologies and analysis can provide data which is clinically relevant for diagnostic purposes.

Using supervised learning (SVM) in this leukemia study with 215 cases taken as a training set, accurate classification of the remaining 112 samples was 96%. In addition, a surprisingly low number of genes was required for each classification ranging from one gene for T-ALL to up to 20 genes for other classes.

Gene expression profiling for tumor prognosis - The large amount of information that can be gleaned from DNA micro-array studies has already increased our ability to classify tumors for the purpose of diagnosis, and many ongoing studies will no doubt add to the statistical robustness of this technology. Taking the classification process a step further, scientists and clinicians are beginning to unravel the first layers of the more complex area of tumor prognosis, with the final goal being tailored therapy reflecting an increasingly accurate classification system.

Diffuse large B-cell lymphoma (DLBCL), which makes up 30-40% of adult non-Hodgkin lymphomas, is curable in less than 50% of patients (32). Current prognosis is centered around the International Prognostic Index (IPI), using clinical parameters, with the aim of ranking patients according to their likely survival on standard chemotherapy and identifying patients in need of more intensive treatment (33). DLBCL provides an ideal opportunity to compare the merits of classification of the standard IPI with the emerging DNA micro-array platform technology. Two groups to date have published gene expression profiles of DLBCL (31,33). In the first study by Alizadeh et al., unsupervised learning allowed classification of DLBCL into two subgroups depending on the cells of origin – B cells from within germinal centers or activated B cells from peripheral blood (34). In addition, the germinal center-like DLBCLs had a more favorable outcome than the peripheral blood-like DLBCLs. In the more recent study by Shipp et al., a supervised learning algorithm was used based on previous success of this technique in distinguishing between AML and ALL (30). After initially demonstrating that supervised learning could allow distinctions to be made between DLBCL from a related germinal center B cell lymphoma known as follicular lymphoma, the authors progressed to differentiating between subsets of DLBCL patients with different outcomes. This study used a relatively small number of patients - 58 total, of whom 32 were cured and 26 were refractory and subsequently fatal. Using sets of gene predictors (between 8 and 16 genes per set), statistically significant outcome predictions were generated. In contrast to this finding, the IPI categorization correctly predicted high risk patients but incorrectly predicted outcome of many patients who were at intermediate or low risk. This is not to suggest that the use of gene expression profiles should replace existing classification strategies such as IPI. Rather the results imply that the technologies should be used in a parallel approach to define the outcome for DLBCL patients with as much accuracy as possible and to aid in the design of realistic treatment regimes for individuals.

Classification of breast tumors provides considerable hurdles due to the differences in tumor response to therapy and outcome between patients with apparently the same stage of disease. More alarmingly, up to 80% of patients treated with chemotherapy or hormone therapy for prevention of metastases would have survived without this treatment regime (35, 36). In a focused effort to classify breast tumors for tailored therapy, a recent report outlines the potential advantages of utilizing molecular expression profiling (37). Using 98 primary breast samples from patients under the age of 55, unsupervised learning could separate these patients to some extent into those with good and those with poor prognosis (38, 39). Subsequently, using a complex three step supervised learning approach, a panel of 70 genes was selected (from an initial 25,000 genes on the micro-array) to achieve an accurate prediction of outcome. This was the case for 83% of the 78 lymph node negative patients selected for the study. Using this panel of 70 genes to predict outcome in a separate group of tumor samples from 19 young patients, only 2 out of 19 were misclassified. The implication from this study is that by assigning a poor prognosis signature at an early stage in diagnosis, the number of patients given adjuvant therapy would be reduced as more accurate predictions of the likelihood of metastases would be possible. This would prevent both potential side effects of adjuvant therapy as well as reduce the healthcare cost burden.

In a slightly different approach to classify prostate cancer, Dhanasekaran et al. made use of the availability of reference samples from normal adjacent prostate (NAP) tissue from patients with prostate cancer and prostate tissue from men with no apparent pathology (40). These reference samples were used to compare the gene expression profile of normal or benign prostatic hyperplasia (BPH) samples, with neoplastic samples from patients with localized prostate tumor or metastatic prostate tumor (40). The aim of this study was to identify prognostic biomarkers rather than identify a subset of genes to be used clinically to define outcome, although the benign samples BPH and NAP did cluster separately to the neoplastic samples and distinctions could be found between metastatic and non-metastatic samples.

The issues to be addressed by further classification of prostate cancer are separate from those for breast cancer, since the success rate of surgery in prostate patients is approximately 70%. The emphasis here is to identify biomarkers indicative of relapse following surgery. Although the Dhanasekaran study identified several potential biomarkers, these were not found to be correlated with outcome in an independent, albeit smaller, study by Singh et al. (41). As with breast tumors, future studies must focus on combining larger patient numbers for molecular profiling of samples, together with existing methods for histological and morphological classification.

Proteomic approaches for tumor screening and classification – Compared with DNA micro-array technology, proteomics is a relative newcomer to the classification arena. Arguably, changes in protein expression and post-translational modification are the quintessential indication of tumor origin, metastatic potential and response to therapy. It is known, for example, that many of the proteins involved in regulation of cell cycle, frequently deregulated in all cancer types, are subject to modifications including phosphorylation, methylation, prenylation, acetylation, ubiquitination and sumolation (42-49). As with DNA micro-array technology several years ago, proteomics is a labor intensive procedure involving two-dimensional polyacrylamide electrophoresis (2D PAGE), often linked to MALDI-TOF (matrix assisted laser desorption time-of-flight) or SELDI-TOF (surface-enhanced laser desorption and ionization time-of-flight) mass spectrometry. It is now possible to use this as a relatively high throughput approach to identify low molecular weight proteins, with the advantage of allowing the investigator to choose between biological fluids such as serum, urine, ascites fluid and/or tumor tissue as the sample of interest.

Using a commercially available SELDI-TOF system (Ciphergen), proteomic patterns have been identified which allow distinctions to be made between benign and malignant ovaries (50). Serum samples were obtained from 50 women with ovarian cancer (representing early and late stage cancer) and 100 control samples (including patients with ovarian cysts and endometriosis). Proteomic analysis was carried out in two phases. In phase I, samples from patients with ovarian cancer and control samples are analyzed by SELDI-TOF, compared, clustered and a discriminatory pattern of 5-20 proteins identified. In phase II, this pattern was applied to a new set of blind serum samples and each sample is assigned to one of three predictive categories: unaffected, cancer or no match/new cluster. All of the 50 ovarian cancer samples were correctly identified, including the early stage tumors, and 63 of 65 controls were correctly identified. From this study, a proteomic approach to screen for ovarian cancer in the general population would have 95% specificity and would therefore not be acceptable, due to the low prevalence of the disease in the population as a whole (too high a rate of false positives). It is expected that as the number of studies increases, the quality of the data will be improved by new methods to reduce background noise and variation in patient samples and, as with DNA technology, the training data set becomes larger and more robust. The current proteomic approach is almost, but not sufficiently, accurate to be used as a stand alone screening tool for ovarian cancer. However, if used in conjunction with existing ultrasound technology, the predictive value may reach the desired 100% level.

Proteomics has applications in characterization of all human tumors and has been used by researchers to identify new biomarkers for lung cancer (51). For some cancer, a single biomarker may be sufficient for early diagnosis and subsequent treatment regime, whilst for the majority, no single marker is significantly accurate for such diagnosis. Hanash et al. analyzed tumor and serum samples from patients with lung cancer and, using 2D PAGE, identified proteins which induce auto-antibodies, often detectable in serum from cancer patients. One such protein identified was termed PGP9.5 and has been detected in approximately 80% of lung tumors. It is not envisaged that PGP9.5 would be used in isolation, but as part of a cohort of proteins, or the corresponding auto-antibodies, for diagnosis of lung cancer.

Proteomic technologies offer much promise for both diagnosis and prognosis of all cancers, in addition to the already acknowledged applications in toxicology (52). The advantages in terms of access to sample material and relevance of studying protein rather than RNA must be weighed up against the disadvantages of throughput, accuracy and variability. Whether a technology which achieves a high success rate in a laboratory can be translated to a useful clinical tool will depend on the advances in both the bioinformatics and accessibility, together with ease of application in the clinical arena.

The contribution of pharmacogenetics to tailored therapy – So far the contribution of profiling changes in expression of RNA and protein from tumor samples have been discussed, with the aim of segregating tumors into categories which will either not respond to current therapy, or will respond to one particular chemotherapy regime. Measuring such changes in gene expression can be collectively called pharmacogenomics. Analyses of the differences in inherited genetic profile which affect response to drugs and susceptibility to disease is termed pharmacogenetics. As with transcriptional profiling, pharmacogenetics approaches frequently aim to identify fingerprints of DNA variants which are correlated with disease or drug response. The analyses of single nucleotide polymorphisms (SNPs) allows identification of disease associated polymorphisms within genes conferring susceptibility, increased therapeutic efficacy and adverse events. As would be expected, many SNPs have recently been identified which confer susceptibility to a range of cancers (53-55).

Mass spectroscopy can, as with other profiling systems, aid the higher throughput identification of SNPs which is necessary for the progression of this technology (56). SNP mapping on a large scale is not a trivial task. DNA has to be obtained from patient samples, SNP analysis must be feasible on a practical scale and with realistic costs per patient. Once the data has been collected, systems have to be developed for optimal storage, organization and interpretation of SNP data. Pharmacogenetics is already in use to identify patients who will respond to a particular drug therapy, or for whom adverse events are more probable (57). Many more applications of this technology are expected in the near future; for example, commercially available systems allow profiling tools such as antibody micro-arrays (Clontech),

Conclusion – This review has described the recent advances in technologies to tailor therapy for cancer patients. Although the majority of progress comes from DNA micro-array technology, proteomics is close behind with technological improvements making large scale analysis a reality for the future. A combination of profiling gene expression and SNPs from cancer patients generates vast amounts of information for both scientists and clinicians. Generating and collating good quality information is the first hurdle to be overcome. Developing accurate algorithms for statistical analysis which give relevant correlations of data clusters with patient information is the second hurdle. A few examples of transcriptional profiling leading to highly significant predictions of tumor type or prognosis are available, but improvements can clearly be made on the accuracy of predictions. These advances can be made in the laboratory, however the real hurdles will be encountered when these tools are transferred to the clinic.

When this profiling data is used to screen patients for the presence of a particular tumor, the ethical considerations are no different than using any surrogate biomarker as a screening tool. Decisions will be made relating to use as a single agent or in conjunction with an existing screening method such as histology or cytogenetics. However, if the information generated from profiling technologies is to be used to stream patients for particular treatment regimes or for entry into clinical trials, then there are ethical implications to be addressed. Targeted therapy for cancer already occurs, most notably in the Gleevec and Herceptin clinical trials (13,58). Using panels of genes on micro-arrays to select patients for treatment is a step on from this strategy, as no single gene is significantly correlated with the disease to be used for prognosis. Larger patient populations will be necessary for the statistical programmes to learn from existing trial data sets and the predictive value of the data should continue to improve.

The use of pharmacogenetic data to increase the efficacy of a drug in clinical trials is much debated. Such trials would in theory involve smaller patient numbers and be more cost effective. However, a balance must be struck between increasing efficacy (and reducing adverse effects) versus reducing the target population for marketing and sales purposes.

The advances which have been made in recent years have made individualized cancer treatment a realistic option for the future. Whilst this will not be the case for all tumor types, several examples already exist, such as ALL and breast tumors, where this is a real opportunity to tailor tumor therapy. It is important to acknowledge that transcriptional profiling and proteomics will not replace existing methods to classify tumors, rather they should be used in conjunction to provide significant improvements in current classification potential for patients.

References

1. S.K. Farber, R.D. Mercer, R.F. Sylvester and J.A. Wolff, N.Engl.J.Med., 238, 787, 1948
2. C.E. Forkner, Leukemia and Allied Disorders 1938, Macmillan, New York
3. J.M. Bennett and T.-F. Dutcher, Blood, 33, 341-347, 1969.
4. W.R. Sperr, J.H. Jordan, M. Baghestanian, H.P. Kiener, P. Samorapoompichit, H. Semper, A. Hauswirth, G.H. Schernthaner, A. Chott, S. Natter, D. Kraft, R. Valenta, L.B. Schwartz, K. Geissler, K. Lechner and P. Valent, Blood, 98, 2200-2209, 2001.
5. D.R. Head, Curr. Opin. Oncol., 14, 19-23, 2002
6. D.A. Arber, M.L. Slovak, L. Popplewell, V. Bedell, D. Ikle, and J.D. Rowley, Am.J. Clin,Pathol., 117, 306-313, 2002
7. R.O. Bash, S. Hall, C.F. Timmons, W.M. Crist, M. Amylon, R.G. Smith, and R. Baer, Blood, 86, 666-676, 1995.
8. E.-J.R. Yeoh, S.A. Shurlteff,S.A. W.K. Williams, D. Patel, R. Mahfouz, F.G. Behm, S.C. Raimondi, M.V. Reling, A. Patel, C. Cheng, D. Campana, D. Wilkins, X. Zhou, J. Li, H. Liu, C.H. Pui, W.E. Evans, C. Naeve, L. Wong, L. Downing, Cancer Cell, 1, 133-143, 2002.
9. T.R. Golub, G.F. Barker, S.K. Bohlander, S.W. Hiebert, D.C. Ward, P. Bray-Ward, E. Morgan, S.C. Raimondi, J.D. Rowley and D.G. Gilliland. Proc.Natl.Acad.Sci.U S A, 92, 4917-4921, 1995.
10. B.J. Druker, S. Tamura, E. Buchdunger, S. Ohno, G.M. Segal, S. Fanning, J. Zimmerman, and N.B. Lydon, Nat. Med., 2, 561-566, 1996.
11. M.W. Deininger, J.M. Goldman, N. Lydon, and J.V. Melo, Blood, 90, 3691-3698,1997.
12. B.J. Druker, M. Talpaz, D.J. Resta, B. Peng, E. Buchdunger, J.M. Ford.,N.B. Lydon, H. Kantarjian, R. Capdeville, S. Ohno-Jones, and C.L. Sawyers, N.Engl.J.Med., 344, 1031-1037, 2001.
13. B.J. Druker, C.L. Sawyers, H. Kantarjian, D.J. Resta, S.F. Reese, J.M. Ford, R. Capdeville, and M. Talpaz, N.Engl.J.Med., 344, 1038-1042, 2001.
14. W.W. Roberts, E.J. Bergstralh, M.L. Blute, J.M. Slezak, M. Carducci, M. Han, J.I. Epstein, M.A. Eisenberger, P.C. Walsh, and A.W. Partin, Urology, 57, 1033-1037, 2001.
15. S.G. Roberts, M.L. Blute, E.J. Bergstralh, J.M. Slezak, and H. Zincke, Mayo Clin. Proc., 76, 576-581, 2001.
16. H.J. Jewett, Urol. Clin. North Am., 2, 105-124, 1975.
17. D.F. Gleason, Cancer Chemother. Rep., 50, 125-128, 1966.

18. T.A. Stamey, N. Yang, A.R. Hay, J.E. McNeal, F.S. Freiha, and E. Redwine, N.Engl.J.Med., 317, 909-916, 1987.
19. I. Jacobs and R.C. Bast, Jr. Hum. Reprod., 4, 1-12,1989.
20. P. Pisani, D.M. Parkin, F. Bray, and J. Ferlay, Int. J. Cancer, 83, 18-29, 1999.
21. Avanzolini, G., Barbini, P., and Gnudi, G. (1990) Int.J.Biomed.Comput., 25, 207-221
22. A. Babic, G. Bodemar, U. Mathiesen, H. Ahlfeldt, L. Franzen, and O. Wigertz, Medinfo., 8 Pt 1, 809-813, 1993.
23. M.P. Brown, W.N. Grundy, D. Lin, N. Cristianini, C.W. Sugnet, T.S. Furey, M. Ares, and D. Haussler, Proc.Natl.Acad.Sci.USA., 97, 262-267, 2000.
24. T. Hastie, R. Tibshirani, D. Botstein, and P. Brown, Genome Biol., 2, RESEARCH0003, 2001.
25. E.J. Moler, M.L. Chow, and I.S. Mian, Physiol. Genomics, 4, 109-126, 2000.
26. S. Ramaswamy, P. Tamayo, R. Rifkin, S. Mukherjee, C-H. Yeang, M. Angelo, C. Ladd, M.Reich, E. Latulippe, J.P. Mesirov, T. Poggio, W. Gerald, M. Loda and E.S. Lander, Proc. Natl. Acad. Sci. USA., 98, 15149-15154, 2001.
27. C.H. Yeang, S. Ramaswamy, P. Tamayo, S. Mukherjee, R.M. Rifkin, M. Angelo, M. Reich, E. Lander, J. Mesirov, and T. Golub, Bioinformatics, 17 Suppl 1, S316-322, 2001.
28. A. Ben-Dor, L. Bruhn, N. Friedman, I. Nachman, M. Schummer, and Z. Yakhini, J.Comput.Biol., 7, 559-583, 2000.
29. T.S. Furey, N. Cristianini, N. Duffy, D.W. Bednarski, M. Schummer, and D. Haussler, Bioinformatics, 16, 906-914, 2000.
30. S. Ramaswamy and T. Golub, J.Clin.Oncol., 20, 1932-1941, 2001.
31. T.R. Golub, D.K. Slonim, P. Tamayo, C. Huard, M. Gaasenbeek, J.P. Mesirov, H. Coller, M.L. Loh, J.R. Downing, M.A. Caligiuri, C.D. Bloomfield and E.S. Lander, Science, 286, 531-537, 1999.
32. M.A. Shipp, K.N. Ross, P. Tamayo, A.P. Weng, J.L. Kutok, R.C. Aguiar, M. Gaasenbeek, M. Angelo, M. Reich, G.S. Pinkus, T.S. Ray, M.A. Koval, K.W. Last, A. Norton, T.A. Lister, J. Mesirov, D.S. Neuberg, E.S. Lander, J.C. Aster, and T.R. Golub, Nat. Med., 8, 68-74, 2002.
33. M.A. Shipp, Blood, 83, 1165-1173, 1994.
34. A.A. Alizadeh, M.B. Eisen, R.E. Davis, C. Ma, I.S. Lossos, A. Rosenwald, J.C. Boldrick, H. Sabet, T. Tran, X. Yu, J.I. Powell, L. Yang, G.E. Marti, T. Moore, J. Hudson, L. Lu, B.B. Lewis, R. Tibshirani, G. Sherlock, W.C. Chan, T.C. Greiner, D.D. Weisenburger, J.O. Armitage, R. Warnke, L.M. Staudt, Nature, 403, 503-511, 2000.
35. Early Breast Cancer Trialists Collaborative Group, Lancet, 352, 930-942, 1998.
36. Early Breast Cancer Trialists Collaborative Group, Lancet, 351, 1451-1467, 1998.
37. L.J. van 't Veer, H. Dai, M.J. van de Vijver, Y.D. He, A.A. Hart, M. Mao, H.L. Peterse, K. van der Kooy, M.J. Marton, A.T. Witteveen, G.J. Schreiber, R.M. Kerkhoven, C. Roberts, P.S. Linsley, R. Bernards, and S.H. Friend, Nature, 415, 530-536, 2002.
38. J.D. Brenton, S.A. Aparicio, and C. Caldas, Breast Cancer Res., 3, 77-80, 2001.
39. C.M. Perou, T. Sorlie, M.B. Eisen, M. van de Rijn, S.S. Jeffrey, C.A. Rees, J.R. Pollack, D.T. Ross, H. Johnsen, L.A. Akslen, O. Fluge, A. Pergamenschikov, C. Williams, S.X. Zhu, P.E. Lonning, A.L. Borresen-Dale, P.O. Brown, and D. Botstein, Nature, 406, 747-752, 2000.
40. S.M. Dhanasekaran, T.R. Barrette, D. Ghosh, R. Shah, S. Varambally, K. Kurachi, K.J. Pienta, M.A. Rubin and A.M. Chinnaiyan, Nature, 412, 822-826, 2001.
41. D.F. Singh, K. Ross, D.G. Jackson, J. Manola, C. Ladd, P. Tamayo, A.A. Renshaw, A.V. D'Amico, J.P. Richie, E.S. Lander, M. Loda, P.W. Kantoff, T.R. Golub and W.R.Sellers, Cancer Cell, 1, 203-209, 2002.
42. A.J. Obaya and J.M. Sedivy, Cell. Mol. Life Sci., 59, 126-142,2002.
43. S.B. Baylin, M. Esteller, M.R. Rountree, K.E. Bachman, K. Schuebel and J.G. Herman, Hum. Mol. Genet., 10, 687-692, 2001.
44. F. Tamanoi, J. Kato-Stankiewicz, C. Jiang, I. Machado and N. Thapar, J.Cell Biochem. Suppl., 37, 64-70, 2001.
45. H.M. Chan, M. Krstic-Demonacos, L. Smith, C. Demonacos and N.B. La Thangue, (2001) Nat. Cell Biol., 3, 667-674, 2001.
46. C. Wang, M. Fu, S. Mani, S. Wadler, A.M. Senderowicz and R.G. Pestell, Front. Biosci., 6, D610-629, 2001.
47. H.M. Chan, N. Shikama and N.B. La Thangue, Essays Biochem., 37, 87-96, 2001.
48. C.M. Pickart, Mol. Cell, 8, 499-504, 2001.
49. S. Muller, C. Hoege, G. Pyrowolakis and S. Jentsch, Nat.Rev.Mol.Cell Biol., 2, 202-210, 2001.

50.   E.F. Petricoin, A.M. Ardekani, B.A. Hitt, P.J. Levine, V.A. Fusaro, S.M. Steinberg, G.B. Mills, C. Simone, D.A. Fishman, E.C. Kohn and L.A. Liotta, Lancet, 359, 572-577, 2002.
51.   S. Hanash, F. Brichoryand D. Beer, Dis. Markers, 17, 295-300, 2001.
52.   L.R. Bandara and S. Kennedy, Drug Discov. Today, 7, 411-418, 2002.
53.   B. Bharaj, A. Scorilas, E.P. Diamandis, M. Giai, M.A. Levesque, D.J. Sutherland and B.R. Hoffman, Breast Cancer Res. Treat., 61, 111-119, 2000.
54.   W.W. Wang, A.B. Spurdle, P. Kolachana, B. Bove, B. Modan, S.M. Ebbers, G. Suthers, M.A. Tucker, D.J. Kaufman, M.M. Doody, R.E. Tarone, M. Daly, H. Levavi, H. Pierce, A. Chetrit, G.H. Yechezkel, G. Chenevix-Trench, K. Offit, A.K. Godwin and J.P. Struewing, Cancer Epidemiol. Biomarkers Prev., 10, 955-960, 2001.
55.   H. Primdahl, F.P. Wikman, H. von der Maase, X.G. Zhou, H. Wolf and T.F. Orntoft, J. Natl. Cancer Inst., 94, 216-223, 2002
56.   E.A. Panisko, T.P. Conrads,, M.B. Goshe and T.D. Veenstra, Exp. Hematol., 30, 97-107, 2002.
57.   F. Innocenti and M.J. Ratain, Eur. J. Cancer, 38, 639-644, 2002
58.   P. Carter, Nature Rev. Cancer, 1, 118-129, 2001.

SECTION VI. TOPICS IN DRUG DESIGN AND DISCOVERY

Editor: George L. Trainor, DuPont Pharmaceuticals Company, Wilmington, Delaware

Chapter 24. Advances in Technologies for the Discovery and Characterization of Ion Channel Modulators: Focus on Potassium Channels

Valentin K. Gribkoff[1] and John E. Starrett, Jr.[2]
[1]Neuroscience Drug Discovery and [2]Neuroscience Chemistry
Bristol-Myers Squibb Pharmaceutical Research Institute
5 Research Parkway, Wallingford, Connecticut 06492

Introduction- The existence of ion channels as discrete regulatory entities has been known indirectly for many years. Over the last two decades advances in molecular biological, electrophysiological and imaging techniques have allowed for the cloning and expression of many ion channels and the visualization of their roles in cell regulation. Currently, very few drugs have been developed that target specific ion channels, although *post hoc* analysis of drug actions have revealed therapeutic effects of many compounds mediated through modulation of ion channels. This is particularly true in the CNS, where ion channels are important regulatory components but their value as drug targets has, until recently, been under-appreciated (1).

Ion channels are pore-forming proteins that permit the passage of ionic species across cell membranes. They are particularly important in the regulation of cellular electrical properties of excitable cells such as myocytes and neurons, but also have important functions in many other cell types. The number of known ion channels is constantly increasing. They can be roughly grouped by virtue of the ionic species they regulate or their lack of ion selectivity. Within each ion channel class they can be further grouped by their molecular structure and major mechanism(s) of regulation, such as transmembrane voltage or second messenger activity. Other ion channels are directly regulated by the action of known endogenous ligands (ligand-gated ion channels). This latter category, as it is presently defined, will not be the subject of this review.

Due to their critical roles in cell function, non-ligand gated ion channels have become targets of interest for the development of therapeutic agents (2). Unlike G-protein-coupled receptors and other 'classical' drug targets, however, the discovery of potential ion channel drugs has relied until quite recently on serendipity and relatively slow (but highly accurate) electrophysiological techniques. The regulation of ion channels is complex, the nature of the sites of ligand interaction ill-defined in most cases, and the end point of their modulation is a change in channel open probability. Most ion channel assays therefore must rely on functional measures related to ion flux, including electrophysiology, or indirect measures of ion flux or transmembrane potential, such as fluorescent ion-dependent or voltage-sensitive dyes (3). In the present report we will focus, where applicable, on recent advances one class of ion channel, those selective for the passage of potassium (K^+).

K^+ channels have been a drug target for many years. Modulators of K^+ currents had originally been discovered by *post hoc* examination of the mechanism of action of known drugs. However, the discovery of relatively potent and specific openers and blockers of these currents, before the cloning of K_{ATP} or other potassium channels, gave credence to the notion that drugs targeting particular K^+ channels could be discovered (4,5). The difficulty of this attempt, due to the inefficiency of screening methods, meant that almost a decade would elapse between the discovery of K_{ATP} openers and specific openers of another type of potassium channel, the large-conductance calcium-activated (maxi-K or BK) potassium channel (6,7). Subsequently, it has been found that many K^+ channels are potentially 'druggable' targets. Small molecule modulators (including openers or activators) have now been reported for other K^+ channels, such as KCNQ channels (8). These findings have resulted in considerable effort to discover and improve techniques for high throughput screening (HTS) of K^+ channels. In addition, drug interactions with certain K^+ channels have resulted in undesirable side effects and the recall of approved drugs (9). These K^+ channels are involved in regulation of action potential duration in the heart; key among these is the hERG channel (10). Regulatory agencies now expect quantification of hERG interaction as part of the safety assessment of any clinical candidate, and pharmaceutical companies need HTS techniques to eliminate hERG and other undesirable ion channel interactions early in the discovery process for any drug candidate.

Recent technological advances in at least 4 areas have significantly changed the ability of drug researchers as well as basic scientists to discover effective modulators of K^+ channels. Molecular cloning has identified a large number of unique K^+ channels (α subunits) in a number of classes, along with a large and growing number of accessory (β) subunits. Heterologously expressed in *Xenopus* oocytes, or in clonal cell lines, these channels have expanded our ability to search for channel-specific modulators. Other advances have now allowed for the use of moderate to high throughput fluorescence- and radioisotope-based assays to directly or indirectly measure specific ion fluxes or their effects on membrane potential. Advances in electrophysiological techniques have also allowed for the development of *higher* throughput assays to directly visualize the effects of modulators on specific channel currents. While not currently at the level of high-throughput fluorescence assays, electrophysiological assays allow for the secondary validation of molecular 'hits' in other assays. Finally, the accumulation of published structure-activity relationships (SAR) in relation to specific K^+ channels has allowed for the creation of directed chemical libraries rich in K^+ channel modulators tractable to combinatorial synthesis. We will be briefly examining all of these advances and some of the limitations of their use in this review.

STEPS TOWARD A MOLECULAR PHARMACOLOGY OF K^+ CHANNELS

The greatest initial steps in the quest to discover small molecule modulators of K^+ channels resulted from advances in molecular biological techniques. Improved cloning and expression techniques, as well as the availability of genetic databases containing sequence information on large numbers of whole and partial genes, resulted in the cloning and expression of a large number of K^+ channels in a relatively short period (11). Representative members of most families of known K^+ channels have now been cloned, resulting in identification of the molecular substrate of many physiologically important K^+ currents. In addition, genes coding for associated proteins, including a large number of K^+ channel β-subunits, have also been cloned (12). Expression of these genes in cell systems have allowed the development of clonal cell lines for use in electrophysiological analyses as well as HTS assays.

At a basic research level, much more information is now available concerning the localization and function of many K^+ channels. Information is also becoming available concerning the roles of β-subunits and other non-pore forming channel proteins. Currently, relatively little is publicly known about the effects of these accessory proteins and β-subunits on the pharmacology of particular K^+ channels. An exception is the K_{ATP} channels, K^+ channels which couple the metabolic state of the cell to its excitability via the suppression of channel opening by ATP. It is now known that these important channels are formed by a tetramer of pore-forming inwardly rectifying Kir6.x subunits coupled with 4 sulfonylurea receptor subunits (13). It is the latter accessory subunits that confer much of the known modulator pharmacology on the functional channel. Large-conductance calcium-activated (maxi-K or BK) K^+ channels are also opened by a variety of small molecules, and blocked by a series of scorpion toxins and alkaloids (14); recent evidence suggests that blocker pharmacology is significantly affected by β-subunit interaction, but less is known about the effects on opener pharmacology (15). These examples serve to illustrate the critical role of channel constitution in determining the pharmacology of particular K^+ channels. Native channels expressed in cell lines can obviate this problem (relative to clonal cell lines), but this is a difficult goal for drugs targeting cells such as central neurons that are diverse in their patterns of channel expression and are difficult to culture. Defining the constitution of native K^+ channels in different tissues and their differential pharmacologies is an area of future refinement that is just now being appreciated in high throughput screening for these targets.

HIGH THROUGHPUT ASSAYS FOR ION CHANNELS

There have been a number of recent reviews of techniques for ion channel HTS (16, 17). In the current report we will restrict discussion to those areas which are currently employed to screen K^+ channel drug candidates, or which appear to offer the most promise in the near future. In general, HTS of K^+ channel targets requires a reporter system capable of accurately responding to changes in K^+ conductance produced by changes in channel open probability. Advantages of these techniques over current electrophysiological assays (see below) include their throughput and their sampling of large numbers of cells. While in theory radioligands could be used to perform classical binding assays for K^+ channel ligands, few small molecule K^+ channel modulators have an affinity high enough to be useful in this type of assay. Perhaps most important, until more is understood about the nature of modulator interactions with ion channels, a high affinity ligand for a particular channel is no guarantee that other compounds from novel chemotypes will bind at the same site and displace the radioligand. Data from such experiments are hard to interpret. Therefore K^+ channel screening assays usually rely on indirect but functional assessments of channel opening or block. Most assays depend on one of the following: redistribution of a radioisotope that can pass through the channel ($^{86}Rb^+$ for K^+ channel assays) or fluorescence-based assays detecting changes in transmembrane voltage using voltage-sensitive dyes. While a number of ion-sensitive dyes exist for Ca^{2+}, this technology is not yet available for K^+ channels.

Radioactive flux assays using cells pre-loaded with $^{86}Rb^+$ are amenable to HTS formats, but the temporal resolution is relatively slow (18). Disposal of radioactive waste resulting from these assays is also environmentally problematic and expensive. Recently, however, a spectroscopic method was developed for the detection of non-radioactive Rb^+, an advance that could lead to more widespread use of this technique (19). A recent study comparing assays for detecting inhibitors of the hERG channel found that relative to fluorescent potential-sensitive dyes, the non-radioactive Rb^+ flux assay was sensitive and resulted in a low false-positive rate (20).

Fluorescence-based assay systems, using efficient fluorescence imagers amenable to HTS, generally employ voltage-sensitive dyes to detect changes in K^+ channel conductance. One technique uses voltage-dependent distribution of a dye such as the oxonol bis-(1,3-dibutylbarbituric acid) trimethine oxonol ($DiBAC_4(3)$). These dyes are anionic and pass readily through the membrane, responding to changes in net intracellular charge. Their utility as voltage reporters results from their increased fluorescence in the hydrophobic intracellular compartment, allowing for quantitation of their intracellular concentration and calculation of the change in potential of a population of cells (21). Drawbacks of this type of assay are the slow rate of response of these dyes, and the high potential for false-positive HTS results due to fluorescent compounds or test compound interactions with the dye. Another technique uses fluorescence resonance energy transfer (FRET) between membrane-bound donors and different lipophilic oxonol voltage-dependent acceptor dyes to generate a ratio reflecting membrane voltage (22). The advantage over other voltage-sensing dye assays is the greater temporal resolution, although new advances in the form of novel proprietary voltage-sensitive dye kits may eliminate this advantage (23).

Detection of K^+ channel functional block, or voltage-dependent opener action, requires activation of the channels during HTS. A shortcoming of these techniques for K^+ channel HTS is their usual reliance on changing ionic distribution to induce membrane depolarization. In most cases this means increasing extracellular K^+ concentration to induce voltage-dependent channel activation, which depolarizes but also proportionately reduces driving force for intracellular K^+. It is also a relatively slow process most useful for steady-state analyses; K^+-induced depolarization usually cannot be used to activate channels which display voltage-dependent desensitization. Techniques for electrically stimulating these assay systems are coming on line, which will broaden the applicability of K^+ channel HTS. However, at least in the foreseeable future HTS techniques using fluorescent probes will have neither the temporal resolution nor the freedom from test-compound interference possible using direct electrophysiological assays.

INCREASING THE THROUGHPUT OF ELECTROPHYSIOLOGICAL ASSAYS

Electrophysiological techniques are still the most direct means of measuring the biophysical characteristics of ion channels. They allow for the direct control of experimental conditions, such as transmembrane voltage, that are important to the characterization of K^+ channels and their pharmacological modulation. The temporal resolution of electrophysiological recordings is extremely fast, and new techniques for drug delivery allow for very rapid solution exchanges. The perennial problem with electrophysiological assays is their technical difficulty and their low throughput. Advances in voltage- and patch-clamp techniques have improved throughput by increasing the probability of experimental success. However, these improvements have not changed the fundamental fact that these techniques usually involve protracted manipulation of one cell/channel at a time for each experiment. Very recently there have been attempts to change this paradigm, resulting in the first commercially available apparatus for performing higher throughput electrophysiological screening (for review see 24).

Electrophysiological efficiency can be improved by automation in at least two ways. Most electrophysiological assessment of ion channel function either involves cell penetration for single or two-electrode voltage-clamping, the *Xenopus* oocyte expression system is an example of this (25), or some form of high-resistance seal formation of glass electrodes to cell membranes (patching) (26). This is followed by an experimental protocol using voltage- or current-clamp to measure current or voltage changes, respectively, resulting from ion channel activation/modulation. The effect of compounds on the open probability of an ion channel can be assessed in a variety of ways, including the effect at the resting membrane potential or across a number of membrane voltages. One technique for improving throughput involves automation and improvement of the actual experimental manipulation - the cell penetration or seal formation. A significant advance was also made with the advent of computer-controlled amplifiers, now available commercially from a number of vendors and incorporated into higher throughput systems now reaching the market.

In the case of oocytes, automation of the mRNA injection also can improve throughput, since this is a time consuming step that must be accomplished prior to the experiments to allow for channel expression. Each of these improvements can modestly increase throughput, but do not change the fundamental sequential nature of the assay system.

To date, most of the commercially available electrophysiological equipment is limited by this serial nature of the experimental preparations. In other words each has some advantages over totally manual control of these functions, and 'hit rate' can be significantly improved, but they still require the experimenter/controller to perform a single experiment at a time, then move to the next cell. These techniques have a relatively high degree of variability, since each experiment is performed on a single cell or membrane patch. Depending on the nature of the preparation and the questions asked of the screen, multiple test compounds may be applied to the preparation serially. Currently available oocyte systems allow for automated injection of mRNA, computer controlled impalement of oocytes as well as automated experimental protocols and data collection. Cell-based systems for automated patch clamping are also now available. Like the oocyte systems, these have achieved a degree of experimental automation, but still require experienced electrophysiologists for experimental setup and data interpretation. Systems such as these have distinct advantages over the hands-on approach used by most experimenters. They can be set up to work without the presence of the experimenter, and the automated nature of the drug infusions, data collection and analyses allow for many more experiments/day. However, the fact remains that except by arranging a number of these systems in parallel, an expensive prospect with significant space requirements, they only allow a more efficient version of slow serial recording. At best such a system allows for a modest level of throughput. To the degree that one is concerned about the effects of repeated unique drug applications to individual cells, the number of compounds examined prior to shifting to another cell may be significantly reduced.

The advantages of electrophysiological assays over other techniques rest in the ability to ask detailed questions about the nature of the interaction of a test compound with the channel of interest. Such experiments require detailed protocols to determine, for example, the effects of a modulator on the current-voltage relationship of a voltage-sensitive channel. These sorts of experiments do not lend themselves as well to the application of many compounds in succession to the same cell or preparation. To greatly improve the throughput in this type of experiment, significant improvements will be required in the ability to record from multiple preparations in parallel without sacrificing accuracy. There has been some progress in this area recently, but even with the increase in throughput afforded by a parallel system, they do not approach high-throughput status and electrophysiological assays remain a secondary technique with at best modest throughput. All of these systems, serial and parallel, are hampered by the need to record using glass electrodes. These are time-consuming to manufacture, are in many cases difficult to store without significantly impacting performance, and contribute to the relatively low 'hit-rate' of most electrophysiological assays. To truly improve the throughput of electrophysiological assays will require new technologies obviating the need for individual electrodes.

Several companies are currently developing 'chip'-based whole-cell patch assay systems for electrophysiology (24). This emerging technology will rely upon perforated chips covered with a suitable substrate such as glass or silicon, which will be the site of seal formation when cells are pulled into the perforations using pressure. The whole-cell configuration would be attained after seal formation again using pressure. In this protocol chambers corresponding to the extracellular and intracellular compartments could be set up on either side of the chip, and compounds applied to the appropriate chamber. Since the chambers can be quite small, the entire setup can in principle be reduced to a multi-well format. The primary limitations under these conditions are the number of amplifier channels, the fluidics involved in drug delivery, and the question of 'chip' cost.

TARGETING K$^+$ CHANNELS THROUGH DEVELOPMENT OF MODULATOR-ENRICHED SMALL MOLECULE LIBRARIES

A very well written and thorough perspective on a decade of progress in the medicinal chemistry and biology of K$^+$ channels appeared in 2001 (27). The goal of this section is to review data that has appeared in the journal literature in 2001 and 2002, examine the nuances of structural modification on channel selectivity, and to gain insight into how combinatorial chemistry is beginning to impact ion channel research.

In contrast to structure-activity relationships of agonists and antagonists of well characterized receptors such as serotonin receptors, the nature of interactions between ion channel modulators and their targets is not as well understood. As discussed above, even with the advent of cloned K$^+$ channels, routine screening of modulators using electrophysiological techniques has been quite challenging. As a consequence, much of the literature surrounding K$^+$ channel modulators has relied on screening in tissue strips or in whole animal systems. However reports are now appearing wherein compounds are screened using electrophysiological methods. Due to limitations of throughput associated with screening, most of the chemistry approaches to K$^+$ channel modulation have consisted of small, focused libraries.

One recurring theme among K$^+$ channel modulators is that subtle changes in a given chemotype can greatly effect the channel selectivity and the resulting pharmacology. For example, the benzopyran chromokalim (**1**) opens K$_{ATP}$ channels in smooth muscle (28). Replacement of the lactam functionality with a N-methyl sulfonamide affords chromanol 293 B (**2**), which blocks the slow component of the delayed rectifier (I$_{KS}$) found in cardiac myocytes (29). Further modification and optimization of chromanol 293 B has resulted in the discovery of HMR 1556 (**3**), which is in development as an antiarrhythmic drug (30). HMR 1556 was screened using *Xenopus* oocytes injected with c-RNA encoding for the human minK (KCNE1) protein, which is the β-subunit of KvLQT1 (KCNQ1). When expressed in oocytes, minK associates with the endogenously expressed XKCNQ1 and forms a functional I$_{KS}$ channel.

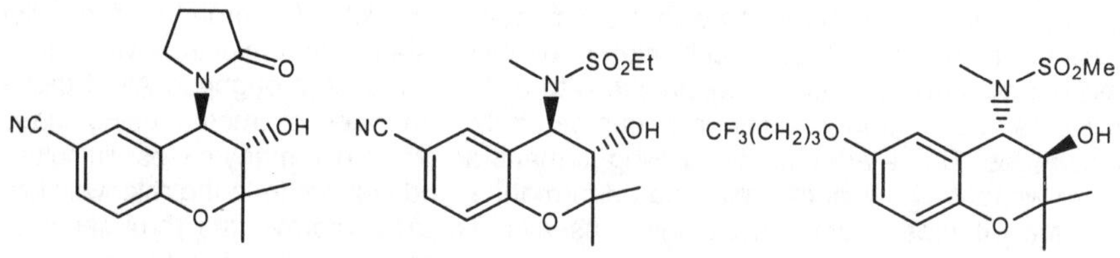

       1   **2**   **3**

Combining some of the structural elements of rilmakalim (**4**) and bimkalim (**5**) resulted in a series of 6-sulfonyl-4-pyridone chromenes which are K$^+$ channel openers as measured by their ability to relax pre-contracted aortic and tracheal tissue (31). Sulfonamides **6** and **7** are representative of this potent class of 4-pyridonechromenes.

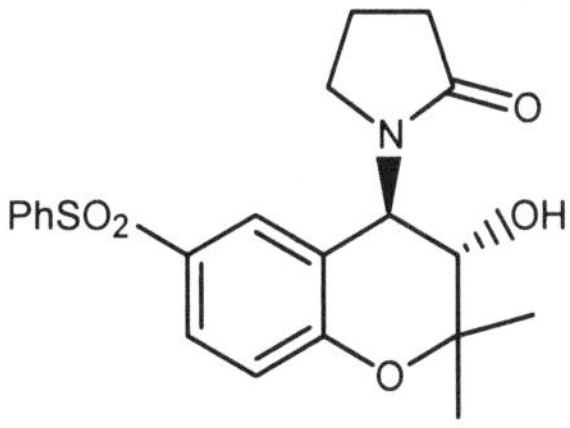

4 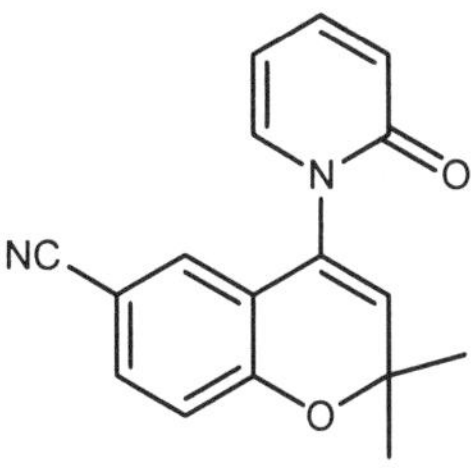   **5** 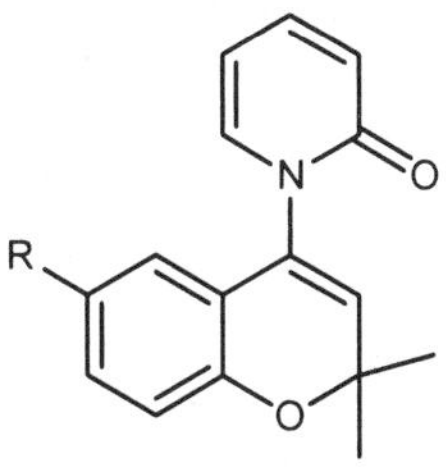   **6** (R = -SO$_2$NHPh)
                              7 (R = -SO$_2$N(CH$_3$)Ph)

A series of naphthopyrans was found to relax precontracted aortic tissue, but the relaxation was not sensitive to a variety of K$^+$ channel blockers, including glibenclamide (K$_{ATP}$), apamin (SK) or charybdotoxin (maxi-K). The relaxation property of naphthopyran **8** was attenuated by 4-aminopyridine, indicating that it was acting through opening of voltage-sensitive K$^+$ channels (32). Cyclization of the oxygen and nitrogen of the chromokalim chemotype resulted in a series of benzopyranooxazines which may be of utility for the treatment of urinary incontinence. Sulfonamide **9** relaxed precontracted bladder tissue selectively as compared to porteal vein (33). The relaxation could be inhibited by the addition of glibenclamide, indicating that the relaxation was mediated by opening of K$_{ATP}$ channels.

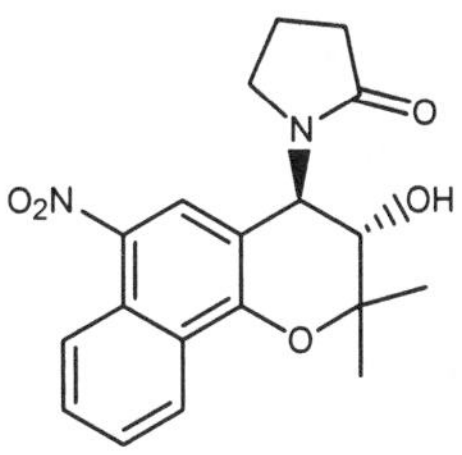

8 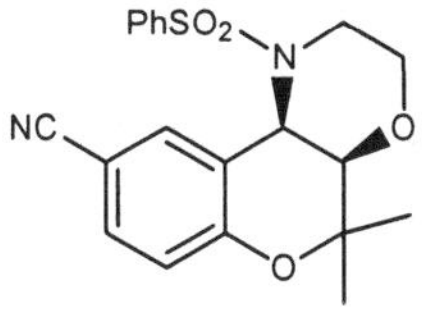   **9**

In attempting to prepare indazolone **10** as a bladder relaxant, reductive cyclization of hydrazone **11** gave instead benzofuranoindole **12** as the major product (34). Evaluation of **12** and related analogs indicated that they relaxed bladder smooth muscle, but that the relaxation was not sensitive to glyburide (a K$_{ATP}$ blocker). The activity could be reversed by the addition of iberiotoxin, indicating that the bladder relaxation activity was mediated through the opening of maxi-K channels. This chemotype represents a radical departure from previously disclosed chemotypes of maxi-K modulators as shown below.

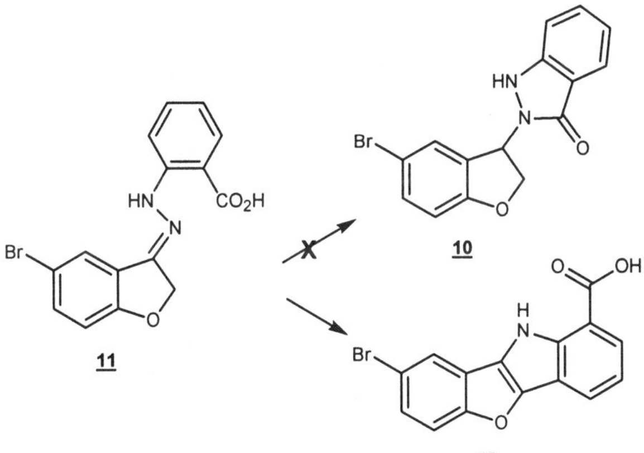

11   **10**   **12**

The benzimidazolone NS-004 (**13**) was identified in 1992 as an opener of maxi-K channels (35). Replacement of the benzimidazolone nitrogen with a fluoro-substituted carbon provided oxindole BMS-204352 (**14**), which has been taken into PhaseII/III trials in stroke (36). Interestingly, BMS-204352 has also been reported to open KCNQ4 and KCNQ5 channels (8,37). The corresponding phenolic demethylated oxindoles in which the fluorine has been replaced by a hydroxy or hydrogen (**15** and **16**, respectively) have also been reported to open cloned maxi-K channels expressed in oocytes (38).

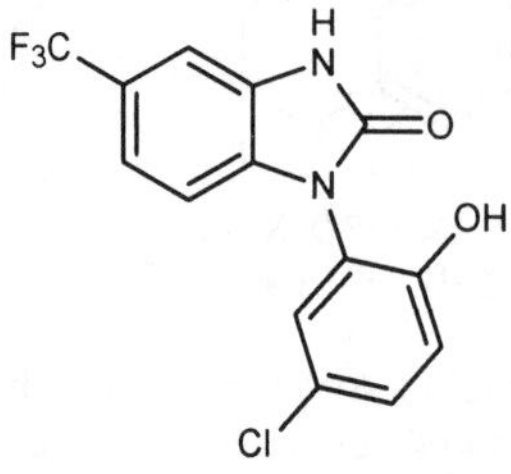

| **13** | **14** | **15** (R = OH)
16 (R = H) |

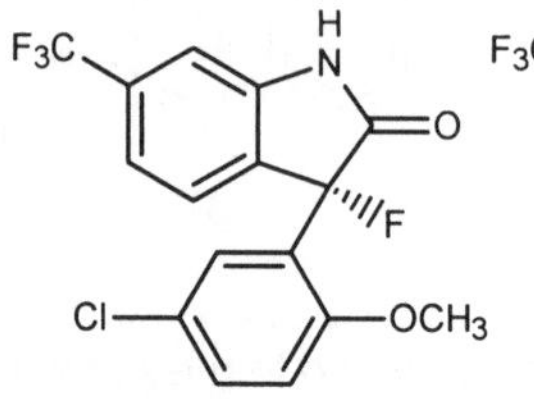

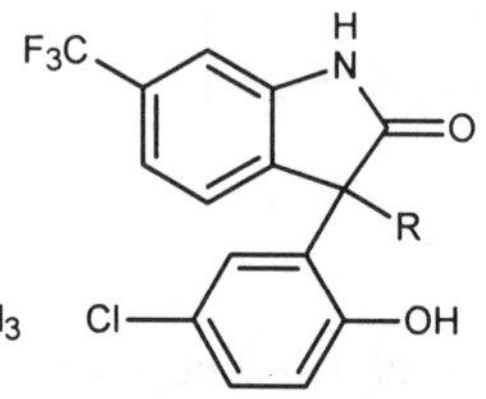

Deannulation of the benzimidazolone/oxindole chemotype provided triazolone **17** and related analogs substituted on the aromatic rings (39). The triazolones opened cloned maxi-K channels expressed in *Xenopus* oocytes and several were found to be potent smooth muscle relaxants, however the relaxant activity did not completely correlate with the maxi-K channel opening.

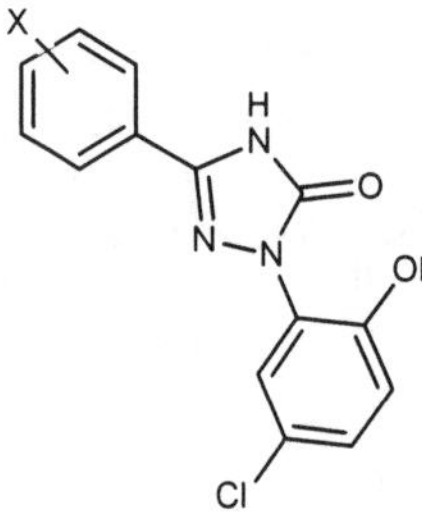

17

A series of haloquinazolinones was prepared and evaluated as putative K_{ATP} openers on isolated pancreatic endocrine tissue (40). Although not all of the compounds in this series exhibited a relaxation profile consistent with selective opening of K_{ATP} channels, 6-chloroquinazoline **18** was determined to relax tissue primarily by activation of K_{ATP} channels. This chemotype is of particular interest with respect to combinatorial libraries, as two independent accounts have recently appeared which employ solid phase techniques to prepared substituted quinazolinones (41,42). Starting with Wang resin **19**, attachment to a dichloroquinone gave **20**, which afforded 6-oxo-2-aminoalkyl quinazolinones **21** upon further elaboration.

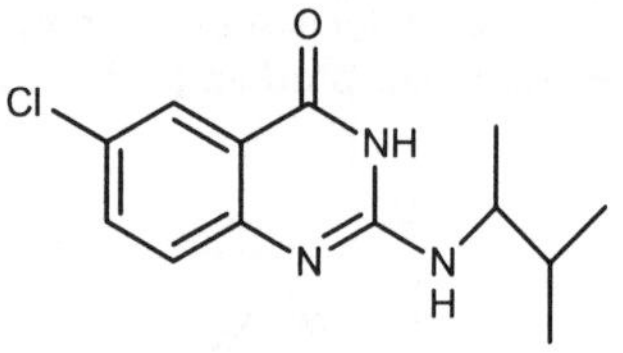

18

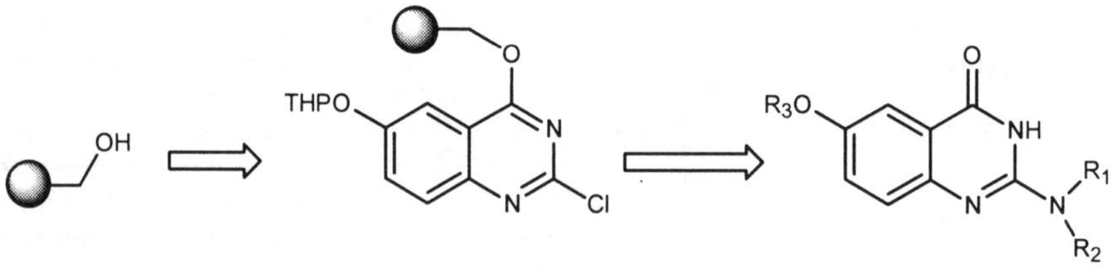

| **19** | **20** | **21** |

In a second resin-based quinazolinone synthesis, formyl resin was loaded with an amine to give **22**, which was acylated to give amide **23** and derivatized to afford quinazolinone **24** with three points of diversity. Although these solid phase libraries were not reportedly prepared for the specific purpose of evaluating ion channel activity, they may find utility as scaffolds for unique ion channel modulators.

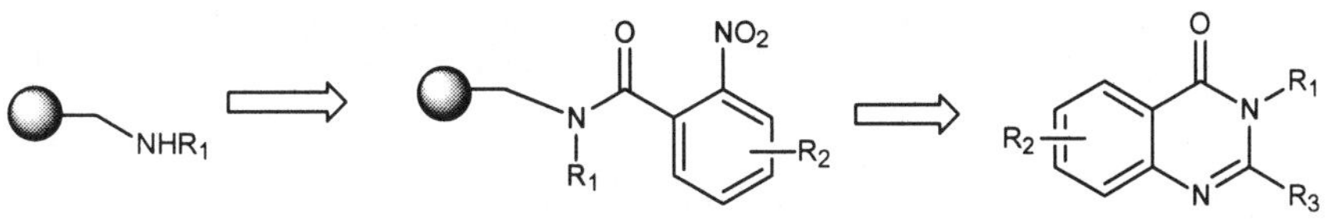

22   23   24

To prepare a library of benzimidazolones, amine resin **25** was acylated, reduced and further acylated to provide nitrofluorobenzamide **26**, which was elaborated to afford dihydrobenzimidazolone **27** with three points of diversity (43). No biological activity was reported in this chemical account. However, it is an interesting method and may prove to be useful in further characterizing this class of K^+ channel openers, especially as some of the high throughput assays discussed in the current review become more commonplace.

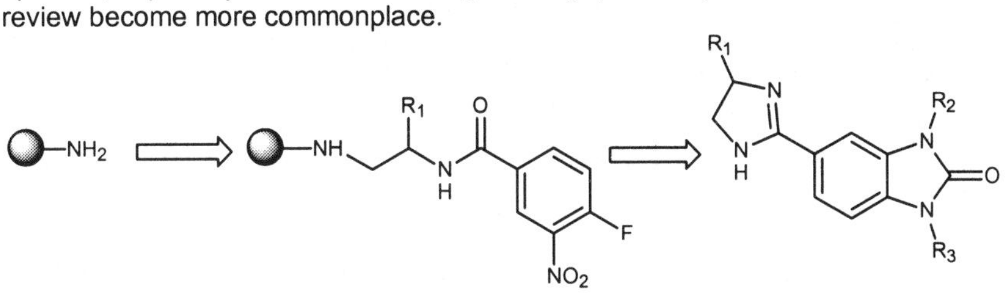

25   26   27

Conclusion – The K^+ channels represent a diverse class of drug target for the development of therapies in a wide range of diseases. Recent developments in the area of HTS for K^+ channels and other ion channel targets have allowed the screening of large libraries for new modulator chemotypes. Fluorescence techniques employing voltage-sensitive dyes remain the most amenable to HTS, but advances in electrophysiological techniques have increased throughput in these direct assays of K^+ channel function. These assays have also increasingly found utility in screening all drug candidates for undesirable K^+ channel interactions such as hERG. Examination of K^+ channel modulator chemotypes has revealed structural bases for modulation of specific K^+ channels, with many chemotypes ideal for combinatorial syntheses. The molecular basis of modulator function in most K^+ channels is currently unknown.

References

1. V.K. Gribkoff and J.E. Starrett, Jr., Exp. Opin. Pharmcother. 1, 1 (1999).
2. J.I. Vandenberg and S.C.R. Lummis, Trends Pharmacol. Sci., 21, 409 (2000).
3. J.E. Gonzalez and P.A. Negulescu, Curr. Opin. Biotechnol., 9, 624 (1998).
4. K.S. Atwal, J. Cardiovasc. Pharmacol., 24 suppl. 4, S12 (1994).
5. R.L. Engler and D.M. Yellon, Circulation, 94, 2297 (1996).
6. M.C. McKay, S.I. Dworetzky, N.A. Meanwell, S.P. Olesen, P.H. Reinhart, I.B. Levitan, J.P. Adelman and V.K. Gribkoff, J. Neurophysiol., 71, 1873, (1994)
7. V.K. Gribkoff, J.T. Lum-Ragan, C.G. Boissard, D.J. Post-Munson, N.A. Meanwell, J.E. Starrett, Jr., E.S. Kozlowski, J.L. Romine, J.T. Trojnacki, M.C. McKay, J. Zhong, S.I. Dworetzky, Mol. Pharmacol., 50, 206 (1996).
8. R.L. Schroder, T. Jespersen, P. Christophersen, D. Strobaek, B.S. Jensen and S.P. Olesen, Neuropharmacol., 40, 888 (2001).
9. R.S. Kass and C. Cabo, Proc. Natl. Acad. Sci. U.S.A., 97, 12329 (2000).
10. M.T. Keating and M.C. Sanguinetti, Cell, 104, 569 (2001).

11. W.A. Coetzee, Y. Amarillo, J. Chiu, A. Chow, D. Lau, T. McCormack, H. Moreno, M.S. Nadal, A. Ozaita, D. Pountney, M. Saganich, E. Vega-Saenz de Miera, B. Rudy, Ann. NY Acad. Sci., 868, 233 (1999).
12. O. Pongs, T. Leicher, M. Berger, J. Roeper, R. Bahring, D. Wray, K.P. Giese, A.J. Silva and J.F. Storm, Ann. NY Acad. Sci., 868, 344 (1999).
13. L. Aguilar-Bryan and J. Bryan, Endocr. Rev., 20, 101 (1999).
14. V.K. Gribkoff, J.E. Starrett, Jr. and S.I. Dworetzky, Adv. Pharmacol., 37, 319 (1997).
15. P. Meera, M. Wallner and L. Toro, Proc. Natl. Acad. Sci. U.S.A., 97, 5562 (2000).
16. L.C. Mattheakis and A. Savchenko, Curr. Opin. Drug Discov. Devel. 4, 124 (2001).
17. P.J. England, Drug Disvovery Today, 4, 391 (1999).
18. W. Hu, J. Torval, P. Cervoni, M.R. Ziai and PT Sokol, J. Pharmacol. Toxicol. Methods, 34, 1 (1995).
19. G. Terstappen, Anal. Biochem., 272, 149 (1999).
20. W. Tang, J. Kang, X. Wu, D. Rampe, L. Wang, H. Shen, Z. Li, D. Dunnington and T. Garyantes, J. Biomol. Screening, 6, 325 (2001).
21. A. Waggoner, J. Membrane Biol., 27, 317 (1976).
22. J. Gonzalez and R. Tsien, Chem. Biol., 4, 269 (1997).
23. D.F. Baxter, M. Kirk, A.F. Garcia, A. Raimondi, M.H. Holmqvist, K.K. Flint, D. Bojanic, P.S. Distefano, R. Curtis and Y. Xie, J. Biomol. Screening, 7, 79 (2002).
24. J. Xu, X. Wang, B. Ensign, M. Li, L. Wu, A. Gui and J. Xu, Drug Disc. Today, 6, 1278 (2001).
25. C.A. Wagner, B. Friedrich, I. Setiawan, F. Lang and S. Broer, Cell Physiol Biochem. 10, 1 (2000).
26. O.P. Hamill, A. Marty, E. Neher, B. Sakmann and F.J. Sigworth, Pfluger's Arch., 391, 85 (1981).
27. M.J. Coghlan, W.A. Carroll and M. Gopalakrishnan, J. Med. Chem., 44, 1627 (2001).
28. N.S. Cook and U. Quast in "Potassium Channels. Structure, Classification, Function and Therapeutic Potential," N.S. Cook, Ed., Ellis Horwood, Chichester, West Sussex, England, 1990, p 214.
29. A.E. Busch, H. Suessbrich, S. Waldegger, E. Sailer, R. Greger, J.-J. Land, F. Lang, K.J. Gibson and J.G. Maylie, Plugers Arch., 432, 1094 (1996).
30. U. Gerlach, J. Brendel, H.-J. Lang, E.F. Paulus, K. Weidmann, A. Bruggemann, A.E. Busch, H. Suessbrich, M. Bleich and R. Greger, J. Med. Chem., 44, 3831 (2001).
31. E. Salamon, R. Mannhold, H. Weber, H. Lemoine and W. Frank, J. Med. Chem. 45, 1086 (2002).
32. W.-F. Chiou, S.-Y. Li, L.-K. Ho, M.-L. Hsien and M-.J. Don, Eur. J. Med. Chem. 37, 69 (2002).
33. H.-I. Chiu, Y.-C. Lin, C.-Y. Cheng, M.-C. Tsai and H.C. Yu, Bioorg. Med. Chem., 9, 383 (2001).
34. J.A. Butera, S.A. Antane, B. Hirth, J.R. Lennox, J.H. Sheldon, N.W. Norton, D. Warga, and T.M Argentieri, Bioorg. Med. Chem. Lett., 11, 2093 (2001).
35. S.-P. Oleson and F. Watjen, Eur. Patent Appl. EP 0477819 (1992).
36. V.K. Gribkoff, J.E. Starrett, Jr., S.I. Dworetzky, P. Hewawasam, C.G. Boissard, D.A. Cook, S.W. Frantz, K. Heman, J.R. Hibbard, K. Huston, G. Johnson, B.S. Krishnan, G.G. Kinney, L.A. Lombardo, N.A. Meanwell, P.B. Molinoff, R.A. Myers, S.L. Moon, A. Ortiz, L. Pajor, R.L. Pieschl, D.J. Post-Munson, L.J. Signor, N. Srinivas, M.T. Tabor, G. Thalody, J.T. Trojnacki, H. Wiener, K. Yeleswaram, S.W. Yeola, Nature Med., 7, 471 (2001).
37. D.S. Dupuis, R.L. Schroder, T. Jespersen, J.K. Christensen, P. Christophersen, B.S. Jensen and S.P. Olesen, Eur. J. Pharmacol., 437, 129 (2002).
38. P. Hewawasam, M. Erway, S.L. Moon, .J. Knipe, H. Weiner, C.G. Boissard, D.J. Post-Munson, Q. Gao, S. Huang, V.K. Gribkoff, and N.A. Meanwell, J. Med. Chem., 45, 1487 (2002).
39. P. Hewawasam, M. Erway, G. Thalody, H. Weiner, C.G. Boissard, V.K. Gribkoff, N.A. Meanwell, N. Lodge and J.E. Starrett, Jr., Bioorg. Med. Chem. Lett., 12, 1117 (2002).
40. F. Somers, R. Ouedraogo, M.-H. Antoine, P. de Tuillo, B. Becker, J. Fontaine, J. Damas, L. Dupont, B. Rigo, J. Delarge, P. Lebrun and B. Pirotte, J. Med. Chem., 44, 2575 (2001).
41. C. Weber, A. Bielik, G.I. Szendrei and I. Greiner, Tet. Lett., 43, 2971 (2002).
42. D.J.R. O'Mahony and V. Krchnak, Tet. Lett., 43, 939 (2002).
43. A.N. Acharya, J.M. Ostresh and R.A. Houghten, Tetrahedron, 58, 2095 (2002).

Chapter 25. Microwave-Assisted Chemistry as a Tool for Drug Discovery

Carolyn D. Dzierba and Andrew P. Combs
Bristol-Myers Squibb Co.
Experimental Station, Wilmington, DE, 19880-0500

Introduction - Accelerating the drug discovery process has been the topic of numerous automation, informatics, molecular design, genomics, proteomics, structural biology, microbiology and medicinal chemistry symposia in the past decade. Each discipline has made significant advances in its field toward the ultimate goal of discovering novel therapeutic agents in shorter timeframes than previously possible. Impressive advances in parallel tasking technologies and automation have driven these achievements.

Medicinal chemistry has benefited tremendously from the technological advances in the field of combinatorial chemistry. This discipline has been the innovative machine for the development of methods and technologies which accelerate the design, synthesis, purification and analysis of compound libraries. These new tools have significantly impacted both lead identification and lead optimization in the pharmaceutical industry. Large compound libraries (100-100,000 members) can now be designed and synthesized to provide valuable leads for new therapeutic targets. Once a chemist derives a suitable high-speed synthesis of a lead it is now possible to synthesize and purify hundreds of molecules in parallel to discover new leads and/or derive structure-activity relationships (SAR) in unprecedented timeframes.

The bottleneck of parallel synthesis is typically optimization of reaction conditions to afford the desired products in suitable yields and purities. Since many reaction sequences require a heating step for extended time periods, these optimizations are often difficult and time-consuming. Microwave-assisted heating has been shown to be an invaluable optimization method since it dramatically reduces reaction times, typically from days or hours to minutes or seconds. Many reaction parameters can be evaluated in a few days to optimize the desired chemistry. Compound libraries can then be rapidly synthesized using the new methodology. Thus, microwave-assisted heating provides an effective tool for the acceleration of the medicinal chemistry process from compound synthesis to library production.

The impact of microwave-assisted organic synthesis on medicinal chemistry is just beginning to be realized, but is clearly evident in the rapid increase in numbers of lectures and published articles in the last few years focusing on the use of this technology to drive the discovery and optimization of new leads. A brief overview of the latest advances in these methods will be discussed herein. Recent reviews of microwave-assisted chemistry are excellent and highly recommended reading for those interested in a more in depth examination of the fundamental aspects of this technology (1-4).

INSTRUMENTATION

The recent advent of microwave heating sources coupled to autosampling instrumentation or fitted with multi-vessel apparatus has enabled medicinal chemists to rapidly and safely optimize synthetic methods and increase sample throughput. Microwave irradiation of compounds containing a dipole affords an extremely rapid increase in temperature compared to conventional heating methods. The microwave super-heating phenomena is primarily due to continuous repolarization of the molecular dipoles of the individual molecules with the applied oscillating electromagnetic radiation (microwaves). Early claims of "non-thermal heating effects" using microwave irradiation have been reinterpreted as the effects of super-heating (5). Several technological advances in microwave instrumentation from the traditional multi-mode domestic microwave have enabled their expanded use and safety. Several new multi-mode and single-mode microwave instruments are commercially available today. These instruments allow for safe and reliable temperature and pressure control of samples during the microwave heating process. Systems are available which allow the chemist to perform up to 50 reactions in parallel in one microwave source to speed up the optimization and production processes. Alternatively, single vessel instruments integrated with a liquid handling robot are available. These instruments allows for unattended sequential microwave experiments. Optimization of reaction conditions is easily performed since reagents, temperatures, solvents and reaction times can all be programmed to run unattended.

MICROWAVE METHODOLOGY

Microwave-assisted organic synthesis has been applied in many formats ranging from traditional solution-phase synthesis to solid-phase and solvent free reactions. These methods have been used extensively to rapidly construct libraries of compounds for lead generation and optimization. Examples utilizing each of these techniques will be discussed herein.

Solution-Phase Synthesis - Microwave reactions are often carried out using volatile solvents. When multiple vessels are heated simultaneously, care must be taken to ensure that there is not significant loss of solvent or reagents during the reaction, which could lead to a lack of reproducibility or even cross contamination. Though sealed vessels are often used to obviate these potential problems, this also creates a possible risk of explosion. Expandable reaction vessels which can accommodate pressure build-up during heating is one way which has been used to address this hazard (6). To demonstrate the usefulness of such an apparatus, a library of 4(5)-sulfanyl-1-H-imidazoles, **1**, was prepared. Thus, ethanol solutions of the appropriate aldehyde, alkyl bromide and 2-oxothioacetamide were combined in vessels containing a movable piston to allow for expansion during heating. The apparatus was heated in a multi-mode laboratory microwave for 4 x 2 min giving compounds with an overall purity of 76% in an average yield of 68%. The same library generated using conventional heating at 70 °C for 12 h and gave products with an average purity and yield comparable to the microwave-generated library (78% and 63% respectively). Although the purity and yields were similar between the two methods, the amount of time required to generate the library was significantly reduced by using microwave heating.

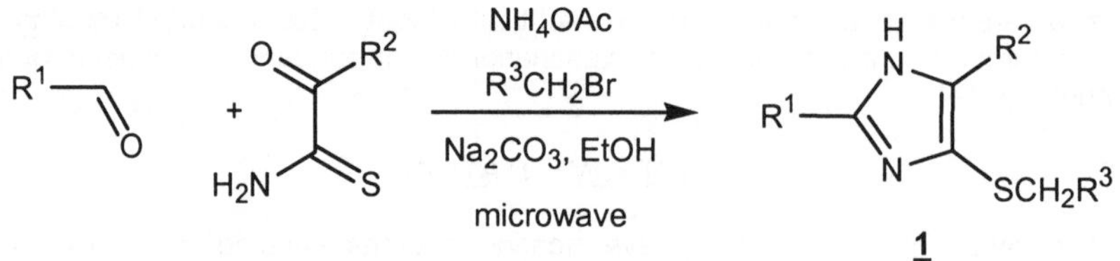

1

Another method to deal with pressure build-up in microwave reactions is to use a microwave instrument with a built in pressure and temperature sensor. These systems shut off the microwave irradiation when a set limit is exceeded. A mono-mode microwave synthesizer, which has such a sensor, has been employed to facilitate the synthesis of a library of dihydropyrimidines, **2**, employing the Biginelli multicomponent condensation (7). The building blocks were delivered to the sealed reaction vials (with Teflon lined aluminum crimp tops) using the liquid handler of the instrument. The vials were then irradiated in series (using the automated transfer arm) for 10-20 min at 100-120 °C. All of the products were obtained with >90% purity with an average yield of 52%. Using the automation available on the microwave synthesizer, the entire library of 48 compounds could be synthesized unattended within 12 h.

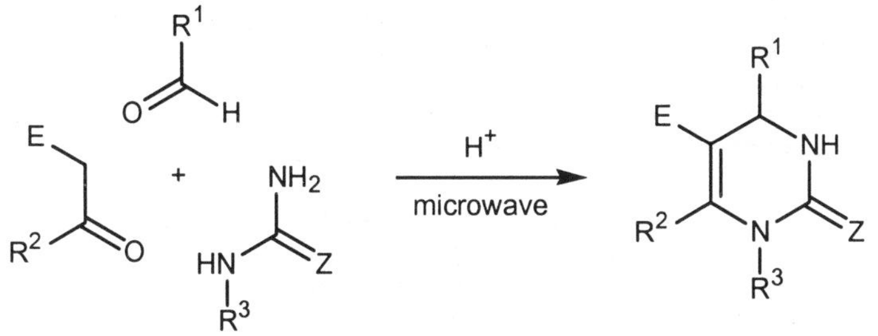

2

Solvent-Free Synthesis - Microwave-assisted organic synthesis can also be carried out under solvent-free conditions. By running the reactions dry, the hazards of using volatile organic solvents in a microwave oven can be eliminated. For example, the synthesis of substituted imidazoles, **3**, has been reported via condensation of a 1,2-dicarbonyl compound with an aldehyde and an amine using acidic alumina impregnated with ammonium acetate to catalyze the reaction (8). Here the acidic alumina acts as both a catalyst and a media onto which the reagents can be adsorbed. The alumina/ammonium acetate was impregnated with the building blocks and the dry mix was irradiated at 130 W for 20 min in open vessels in a domestic microwave oven. The purity of the 8 products ranged from 67-82%.

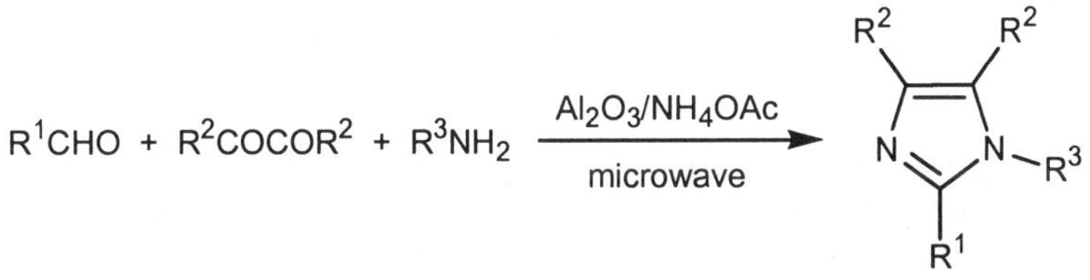

$$R^1CHO + R^2COCOR^2 + R^3NH_2 \xrightarrow[\text{microwave}]{Al_2O_3/NH_4OAc}$$

3

A "dry-media" synthesis of substituted pyridines, **4**, was carried out in a 96-well plate format (9). The building blocks (12 aldehydes x 8 1,3-dicarbonyl compounds) were distributed into each well of a fritted 96-well plate containing bentonite/ammonium nitrate and irradiated in a domestic microwave oven for 5 min at 70% power. The products were extracted with ethyl acetate and HPLC analysis

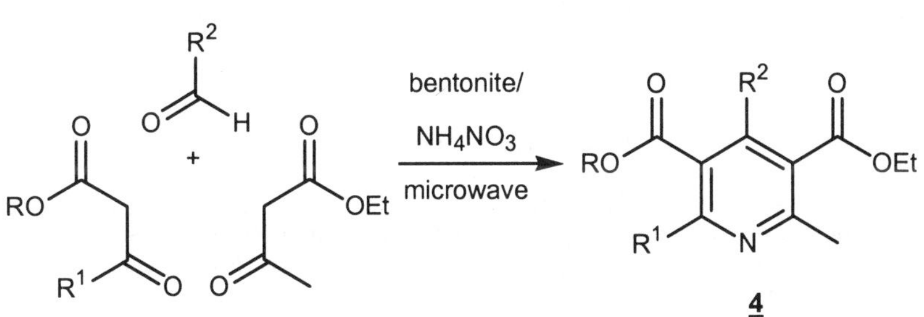

4

of the resulting solutions showed a purity of $\geq$70% for all products. In all cases, no unreacted starting materials were observed even though there is typically a significant temperature gradient across the wells. The use of the 96-well format with automated liquid handling allowed for a rapid and efficient synthesis of these compounds.

Solid-Phase Synthesis - Microwave-assisted solid-phase combinatorial synthesis is also gaining popularity. Solid-phase reactions offer the advantages of being able to drive a reaction to completion by the addition of excess reagents and ease of product isolation. A drawback is that prolonged reaction times over the corresponding solution-phase reactions are often required. However, by employing microwave heating the reaction times can be significantly decreased.

An example of this is illustrated by the esterification of 1,3-dicarbonyl compounds onto Wang resin followed by Knoevenagel condensation to form substituted enones, **5**, (10). For the first step, microwave flash heating of Wang resin with 21 different 1,3-dicarbonyl compounds at 170 °C gave complete conversion in 1-10 min (*vs* hours using conventional heating). Subsequent treatment of the polymer bound enones with aldehydes and piperidinium acetate at 125 °C for 30-60 min (*vs* 1-2 days using conventional heating) gave the Knoevenagel condensation products with >95% conversion in all cases. These reactions were all carried out in open vessels in a microwave reactor which allows for the irradiation of up to 50 vessels at once.

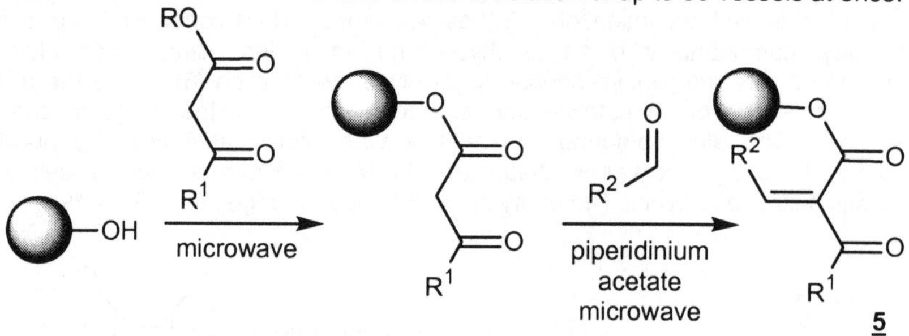

The displacement of amides from safety-catch linkers can be improved by employing microwave irradiation (11). In this case, 10 differentially-substituted sulfonamide-linked resins in fritted 96-well plates and were treated with 88 different amines of varying nucleophilicity. The plates were subjected to microwave irradiation (4 plates, or 352 compounds at a time) in a domestic microwave for 1 min to form the amides, **6**. In spite of the fact that a temperature gradient of almost 20 °C could be measured across the plate, an average purity of >90% was obtained for the entire library.

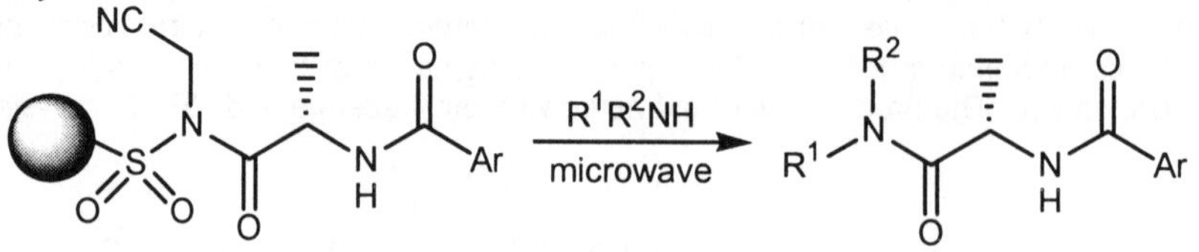

Another method of immobilizing reagents onto a solid-support is to create spatially addressed arrays of compounds on cellulose or polypropylene membranes. Microwave irradiation has been used to facilitate the preparation of an 8,000-member library of 1,3,5-triazines, **7**, on a cellulose membrane using SPOT-synthesis (12). After conventional construction of the triazine core from cynauric chloride, substitution of the final chlorine atom was achieved by spotting the building block

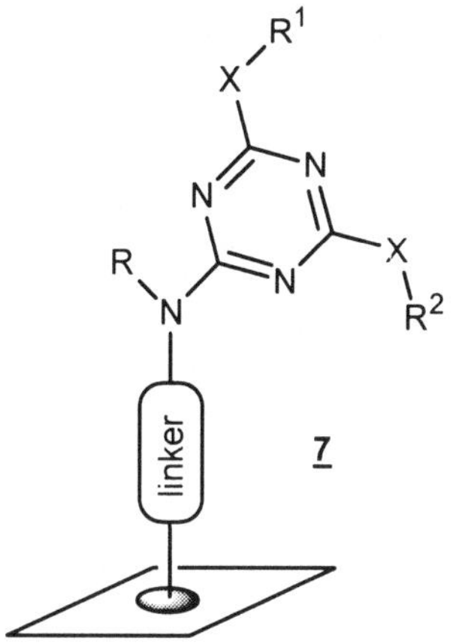

7

onto the membrane and subjecting the membrane to microwave irradiation for 6 min. This period of time was sufficient to drive the reactions to >95% completion, whereas with conventional heating at 80 °C for 1 h, only 50% conversion was observed.

MEDICINAL CHEMISTRY APPLICATIONS

Microwave-assisted organic synthesis is beginning to have an impact on drug discovery. Synthesis of small libraries directed toward specific biological targets can be facilitated using this technique. Thus, development of SAR for a known scaffold and discovery of compounds with improved biological properties can be made more efficient.

Condensation Reactions – Thiadiazepines (13) and benzodiazepines (14–17) are known to interact with a wide range of pharmacological targets. Thus, there has been considerable interest in the efficient synthesis of these compounds. The "dry-media" technique has been used to rapidly synthesize a series of thiadiazepines (18). Basic alumina impregnated with 1-amino-2-mercapto-5-substituted triazoles and substituted chalcones was irradiated in a domestic microwave oven for 1-2 min to afford the corresponding thiadiazepines, **8**. The average yield of the products was 94% (1-2 min reaction time) compared to 67% (10-18 h reaction time) using conventional heating. Additionally, this represents an improvement over the microwave-assisted solution-phase synthesis of the same products. When the reactions were run in the microwave oven in acetone solvent, the average yield was 76% with reaction times of 7-10 min. The thiadiazepines were screened for antibacterial and antimicrobial activity. Four compounds from the library where shown to have significant activity against *A. niger* and *A. flavus*.

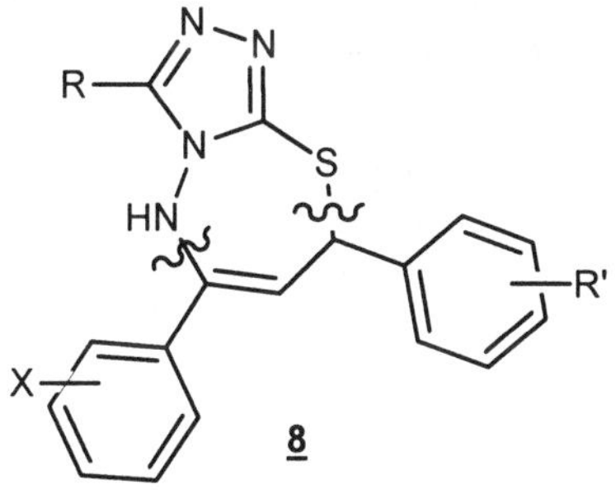

8

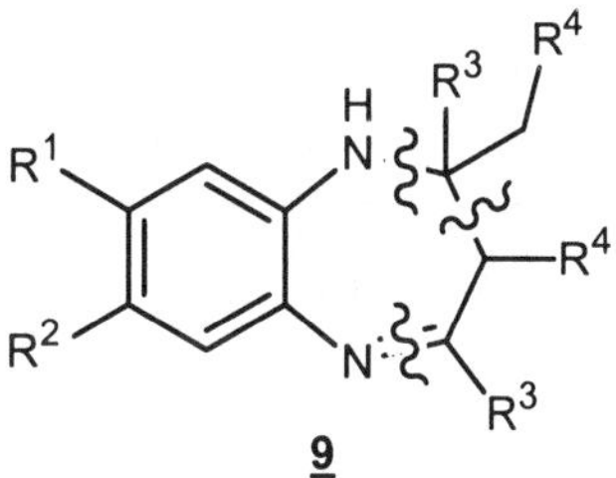

9

Benzodiazepines, **9**, have also been made under solvent-free conditions by condensation of substituted o-phenylenediamines with two equivalents of a ketone in the presence of a catalytic amount of acetic acid (19). Microwave irradiation of the reaction mixtures in a domestic oven for 2-7 min gave the products in 90-99% yields.

Several drugs and natural products have recently been prepared in an efficient manner using microwave-assisted condensation reactions as a key step. The cytotoxic alkaloid luotonin A, **10**, has been synthesized by a microwave-assisted cyclocondensation of 3-oxo-1H-pyrrolo[3,4-b]quinoline with isatoic anhydride to form the central ring of the pentacyclic core (20). Irradiation at 450 W for 6 min gave luotonin A in 85% yield. The antifungal agent, Nortopsentin D, **11**, was prepared by the condensation of a ketoamide with ammonia to form the 2,4-disubstituted imidazole ring (21). Using conventional heating, the reaction proceeds at 130 °C in 14 h to give Nortopsentin D in 25% yield. The reaction was optimized to 75% yield by microwave heating for 40 min. Thalidomide, **12**, could easily be prepared in 85% yield by condensation of thiourea with N-phthaloyl-L-glutamic acid by irradiation for 15 min (22). Conventional heating at 200 °C for 50 min provides Thalidomide in only 54% yield. Sildenafil (Viagra™, **13**) was synthesized utilizing microwave heating to facilitate the final cyclization step to form the pyrimidinone ring (23). Irradiation for 10 min at 120 °C provided the compound in quantitative yield.

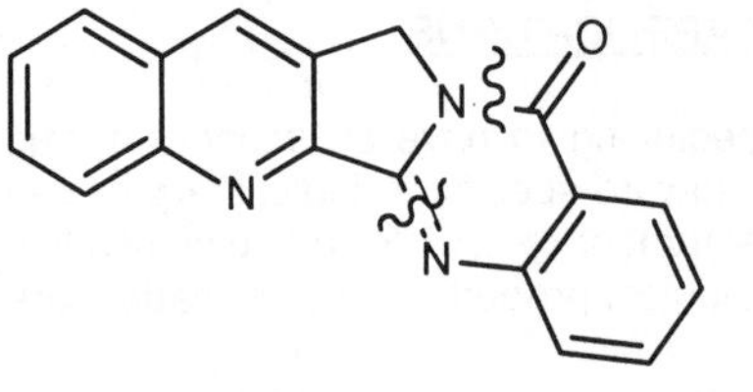

10

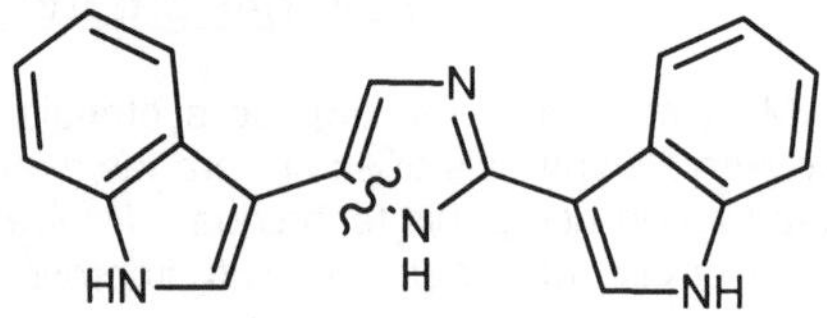

11

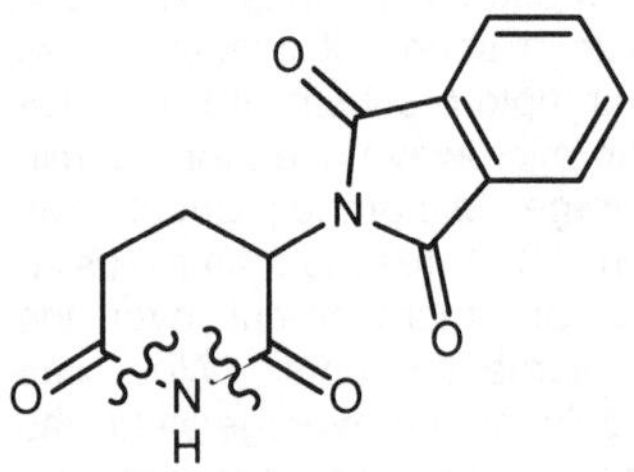

12

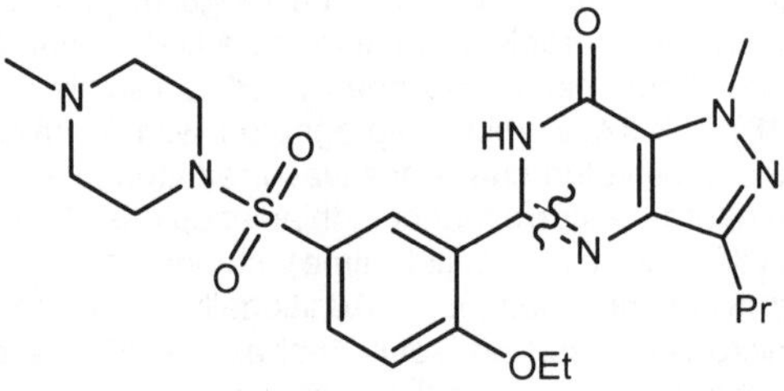

13

Palladium Couplings - There have been numerous publications on the use of microwave heating to improve palladium catalyzed carbon-carbon bond forming reactions. This microwave-assisted technology has been employed to identify potent HIV-1 protease inhibitors by modifying P1/P1' benzyloxy side chains on a diol isostere scaffold via Suzuki, Stille, and Heck couplings (24). The p-bromobenzyloxy derivative **14** was treated with arylboronic acids or aryl stannanes in the presence of a catalytic amount of palladium(0). The mixtures were irradiated in sealed vials for 2-4 min at 45-60 W. The Suzuki coupling products were obtained in 85-96% yield,

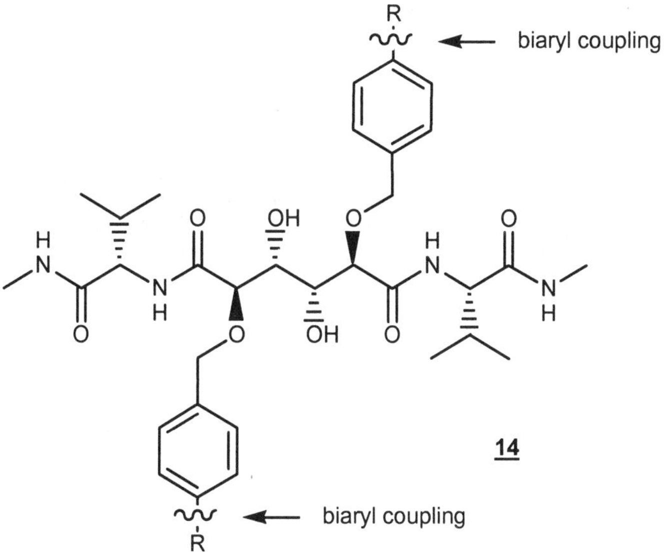

14

while the Stille products were obtained in 50-54% yield. Interestingly, phenylethyl 9-BBN could also be used as a coupling partner under these conditions, albeit in modest yield (38%). The authors were thus able to establish SAR for the series in a rapid and efficient manner.

In a similar fashion, microwave heating was used to accelerate the coupling of cyanide to aryl bromides (25). A variety of aryl bromides were reacted with zinc cyanide in the presence of palladium(0) in sealed tubes under microwave irradiation for 2-2.5 min at 60 W. The reactions proceeded cleanly to give the aryl nitriles in 78-95% yield. Using conventional thermal heating, the same reactions required reaction times of 2-16 h to achieve similar yields. Additionally, the authors were able to convert the aryl nitriles into tetrazoles, a common carboxylic acid bioisostere. The aryl nitriles were irradiated in sealed vessels in the presence of sodium azide for 10-25 min at 20 W. Because the by-products of this reaction can be explosive, the authors warn that this transformation should only be carried out in vessels that have a pressure release cap in an oven with the cavities enclosed in an explosion proof case. The two step procedure can also be carried out in one pot. Conversion of the bromo-precursor **14** to the aryl nitrile for 2.5 min, addition of sodium azide, and heating for an additional 40 min gave an 82% yield of the tetrazole after chromatography.

A microwave-assisted palladium coupling was also used to establish SAR for another class of HIV-1 protease inhibitors (26). To explore the unique binding mode of cyclic sulfamide inhibitors **15**, several symmetrical and nonsymmetrical compounds were synthesized and analyzed for their ability to inhibit HIV-1 protease. Suzuki coupling of 2-thienylboronic acid in the presence of catalytic palladium using microwave irradiation gave the thienyl-products within 4 min and in good yields (48-73%).

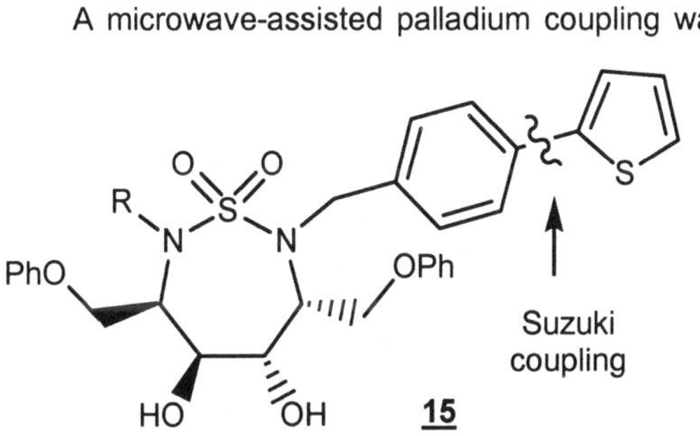

<u>Nucleophilic Substitution Reactions</u> - Another reaction which has been accelerated by microwave heating is the nucleophilic substitution of alkyl halides with amines. Fluoroquinolone antibacterials have been synthesized by microwave-assisted nucleophilic amination of quinolone carboxylic acids (27). For example, ciprofloxacin, **16**, has been conveniently prepared by coupling of a 7-chloro-6-fluoroquinolone carboxylic acid with piperazine using microwave heating for 20 min in 84% yield and >99% purity. Even less reactive pyrrolidines could be coupled in this fashion in high yields (78-94%) with excellent purities (92-99%). Similarly, antibacterial quinolines, **17**, were synthesized via coupling of 7-chloro-quinolines with 1,3,4-thiadiazoles or oxadiazoles (28). The starting materials were adsorbed onto basic alumina and irradiated for 2-3 min to provide six compounds, all of which showed promising antibacterial activity.

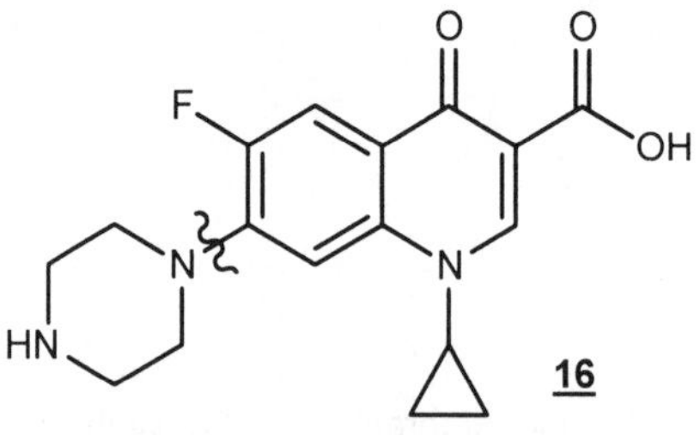

In another case, a library of substituted amines directed toward herpes simplex virus (HSV) has been synthesized based on a lead with modest activity (IC_{50} = 3 µM vs HSV-1 helicase ATPase and 11 µM antiviral activity in an HSV-1-plaque reduction assay) (29). Irradiation of a mixture of an alkyl iodide with a variety of piperidines and piperizines gave a library of 60 compounds. From this collection, three compounds showed improved activity over the initial lead. For example, **18** has the following activities: IC_{50} = 1.1 µM vs HSV-1 helicase ATPase and 4.9 µM antiviral activity in a HSV-1-plaque reduction assay.

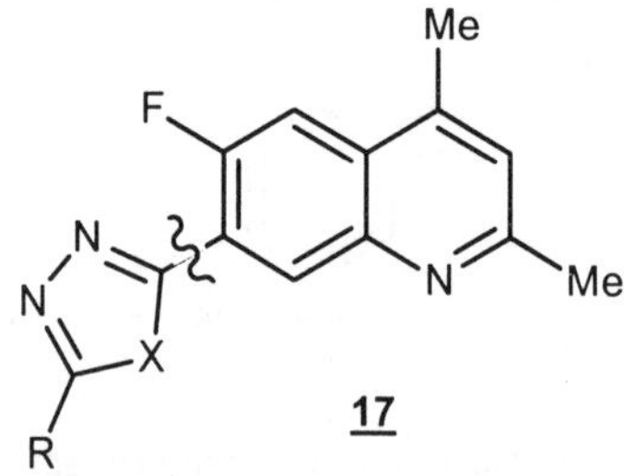

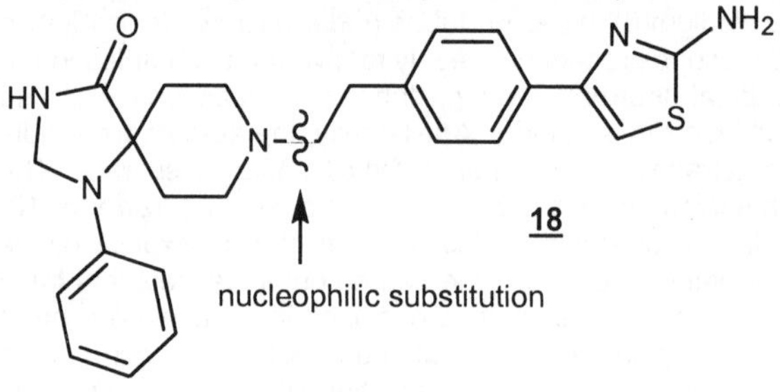

nucleophilic substitution

Nucleophilic substitution accelerated by microwaves has also been used to synthesize a library of compounds to identify a selective ligand for the 5-HT$_{1A}$ receptor based on initial piperazine leads (30). In this case, microwave heating was utilized for each step of the synthesis. In addition to the convenience and speed of optimization, better yields of the final products were obtained by using microwave rather than conventional heating.   Of the 22

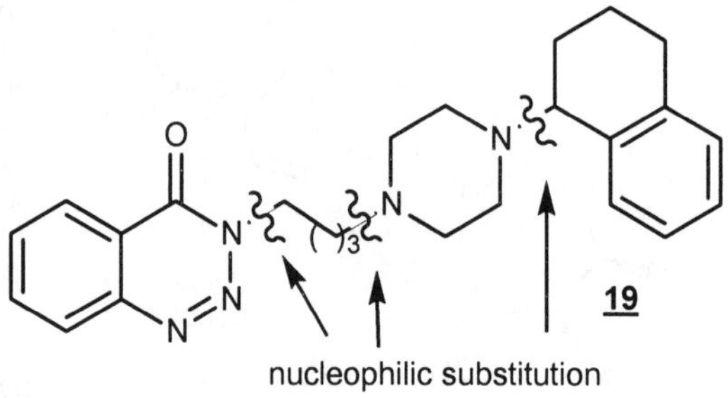

nucleophilic substitution

compounds screened for 5-HT$_{1A}$ binding affinity, $(\pm)$ **19** showed the best activity and selectivity (5-HT$_{1A}$ K_i = 60 nM, no activity vs 5-HT$_{2A}$ or 5-HT$_{2C}$). When the two enantiomers were separated and analyzed, (-) **19** was found to be the more potent isomer with a 5-HT$_{1A}$ K_i = 36 nM.

Conclusion - The competition among pharmaceutical companies to discover novel therapeutic agents for unmet medical needs has demanded the acceleration of the drug discovery process.   The mission of medicinal chemistry is to design and synthesize novel chemical entities with potent biological activities. The optimization of these challenging new chemistries and production of compounds for screening often constitute the bottleneck for the medicinal chemist. Microwave-assisted organic synthesis has emerged as a powerful tool for the rapid optimization of reaction conditions and library production.   The impact of this technology on shortening the timeframe for discovery of new therapeutic agents by medicinal chemists will thus likely be seen in the next few years.

References

1. A. Lew, P. O. Krutzik, M. E. Hart and A. R. Chamberlin, J. Comb. Chem., 4, 95 (2002).
2. M. Larhed and A. Hallberg, Drug Discovery Today, 6, 406 (2001).
3. B. Wathey, J. Tierney, P. Lidstrom and J. Westman, Drug Discovery Today, 7, 373 (2002).
4. P. Lidström, J. Tierney, B. Wathey and J. Westman, Tetrahedron, 57, 9225 (2001).
5. R. N. Gedye and J. B. Wei, Can. J. Chem., 76, 525 (1998).
6. C. M. Coleman, J. M. D. MacElroy, J. F. Gallagher and D. F. O'Shea, J. Comb. Chem., 4, 87 (2002).
7. A. Stadler and C. O. Kappe, J. Comb. Chem., 3, 624 (2001).
8. A. Y. Usyatinsky and Y. L. Khmelnitsky, Tetrahedron Lett., 41, 5031 (2000).
9. I. C. Cotterill, A. Y. Usyatinsky, J. M. Arnold, D. S. Clark, J. S. Dordick, P. C. Michels and Y. L. Khmelnitsky, Tetrahedron Lett., 39, 1117 (1998).
10. G. A. Strohmeier and C. O. Kappe, J. Comb. Chem., 4, 154 (2002).
11. B. M. Glass and A. P. Combs in "High-Throughput Synthesis," I. Sucholeiki, Ed., Marcel Dekker, Inc., New York, N.Y., 2001, p. 123.
12. D. Scharn, H. Wenschuh, U. Reineke, J. Schneider-Mergener and L. Germeroth, J. Comb. Chem., 2, 361 (2000).
13. M. A. Khalil, O. A. El-Sayed and H. A. El-Shamy, Arch. Pharm. (Weinheim, Ger.), 326, 489 (1993).
14. R. K. Smalley in "Comprehensive Organic Chemistry," Vol. 4, D. Barton and W. D. Ollis, Ed., 1979, p. 565.
15. H. Schutz in. "Benzodiazepines". Springer, Heidelberg, 1982.
16. J. K. Landquist in "Comprehensive Heterocyclic Chemistry," Vol. 1, A. R. Katritzky and W. C. Rees, Ed., Pergamon, Oxford, 1984, p. 166.
17. L. O. Randall and B. Kappel in "Benzodiazepines," S. Garattini, E. Mussini and L. O. Randall, Ed., Raven Press, New York, 1973, p. 27.
18. M. Kidwai, P. Sapra, P. Misra, R. K. Saxena and M. Singh, Bioorg. Med. Chem., 9, 217 (2001).
19. M. Pozarentzi, J. Stephanidou-Stephanatou and C. A. Tsoleridis, Tetrahedron Lett., 43, 1755 (2002).
20. J. S. Yadav and B. V. S. Reddy, Tetrahedron Lett., 43, 1905 (2002).
21. P. M. Fresneda, P. Molina and M. A. Sanz, Synlett, 218 (2001).
22. J. A. Seijas, M. P. Vázquez-Tato, C. Gonzaláz-Bande, M. M. Martínez and B. Pacios-López, Synthesis, 999 (2001).
23. I. R. Baxendale and S. V. Ley, Bioorg. Med. Chem. Lett., 10, 1983 (2000).
24. M. Alterman, H. O. Andersson, N. Garg, G. Ahlsén, S. Lövgren, B. Classon, U. H. Danielson, I. Kvarnström, L. Vrang, T. Unge, B. Samuelsson and A. Hallberg, J. Med. Chem., 42, 3835 (1999).
25. M. Alterman and A. Hallberg, J. Org. Chem., 65, 7984 (2000).
26. W. Schaal, A. Karlsson, G. Ahlsén, J. Lindberg, H. O. Andersson, U. H. Danielson, B. Classon, T. Unge, B. Samuelsson, J. Hultén, A. Hallberg and A. Karlén, J. Med. Chem., 44, 155 (2001).

27. P. G. Reddy and S. Baskaran, Tetrahedron Lett., 42, 6775 (2001).
28. M. Kidwai, K. R. Bhushan, P. Sapra, R. K. Saxena and R. Gupta, Bioorg. Med. Chem., 8, 69 (2000).
29. C. N. Selway and N. K. Terrett, Bioorg. Med. Chem., 4, 645 (1996).
30. G. Caliendo, F. Fiorino, E. Perissutti, B. Severino, S. Gessi, E. Cattabriga, P. A. Borea and V. Santagada, Eur. J. Med. Chem., 36, 873 (2001).

SECTION VII. TRENDS AND PERSPECTIVES

Editor : Annette M. Doherty
Pfizer Global Research & Development

Chapter 26. To Market, To Market - 2001

Patrick Bernardelli, Bernard Gaudillière and Fabrice Vergne
Pfizer Global Research & Development
Fresnes, France

In 2001, a total of 25 new therapeutic chemical and biological entities (18 NCEs and 7 NBEs) were introduced for the first time (1-5). Last year was the least productive of the last decade since the number of new therapeutic entities has ranged from 27 (in 1998) to 44 (in 1994).

During this period, most new drugs were marketed for the first time in the US (13 new launches) followed by Europe and Japan with 6 and 4 respectively. These drugs originated for the most part from the US (14 of the 25). The remaining ones were discovered in Europe (with 8), most notably in Switzerland (with 4), and Japan (with 3). Interestingly, the development of 3 of the 25 therapeutic entities originated from academic laboratories. Last year, Novartis and Mitsubishi, the most productive companies, were at the origin of 2 approved substances that they also marketed. Aventis, Roche and Sepracor created 2 new molecular entities that were further developed and first launched last year by themselves or a licensee.

There were 5 new antiinfective agents launched in 2001 making this the most productive therapeutic area. Cancidas® (caspofungin acetate) is the first echinocandin, a new class of antifungal agents, to be launched for the treatment of invasive aspergillosis, a life-threatening fungal infection. Ketek® (telithromycin) was also the first of a new class of antibiotics called ketolides to reach the market. Viread® (tenofovir disoproxil fumarate) and Valcyte® (valganciclovir hydrochloride) are prodrugs that are rapidly hydrolyzed into antiviral agents after absorption. Viread® is the first nucleotide analog reverse transcriptase inhibitor launched for the treatment of HIV. A major advancement in severe sepsis therapy was achieved with the introduction of Xigris® (drotrecogin alfa) a few days after approval by the FDA.

In the field of allergic and respiratory diseases, 3 NCEs were marketed. Clarinex® (desloratadine) is the active metabolite of the second-generation antihistamine Claritin® (loratadine). Another nonsedating antihistamine, Xusal® (levocetirizine), the eutomer of Zyrtec® (cetirizine) was first launched for the treatment of allergic rhinitis and chronic urticaria. Cleanal® (fudosteine) is a new mucoactive cysteine derivative for the treatment of bronchitis and respiratory congestion.

In the cardiovascular area, 2 major new drugs were introduced: Tracleer® (bosentan), the first commercialized endothelin receptor antagonist for patients with pulmonary arterial hypertension, and Natrecor® (nesiritide), a recombinant form of human B-type natriuretic peptide, as the first new treatment for acute congestive heart failure in over a decade.

The CNS field was represented with a new antipsychotic agent Lullan® (perospirone hydrochloride) for the treatment of schizophrenia, and Radicut® (edaravone) a lipophilic antioxidant for the improvement of neurologic recovery after acute brain infarction.

In the anticancer domain, the first-in-class antiproliferative agent Gleevec® (imatinib mesylate) reached the market for chronic myelogenous leukemia after a record FDA review time of 2.5 months. The humanized monoclonal antibody Campath® (alemtuzumab) was introduced shortly after approval for the treatment of B-cell chronic lymphocytic leukemia.

Two ophthalmic drugs of the prostaglandin class, Lumigan® (bimatoprost) and Travatan® (travoprost), were launched in 2001 to treat glaucoma.

The vitamin D analog falecalcitriol was co-marketed as Horner® by Taisho and Fulstan® by Sumimoto and Kissei as a new therapy against secondary hyperthyroidism. Ondeva®, a combination of the progestin trimegestone and estrogen, was introduced as hormone replacement therapy for the treatment of menopausal symptoms and the prevention of osteoporosis.

Unlike the previous year, in 2001 there was only one NCE launched in gastroenterology. Although approval has not yet been obtained in the European Union and in the US, Novartis launched Zelmac® (tegaserod maleate), a new 5-HT$_4$ receptor partial agonist, against irritable bowel disease, in Mexico.

Relpax® (eletriptan), a fast-acting antimigraine drug, is the seventh member of the triptan class.

Kineret® (anakinra) is the most recent antiarthritic drug launched. This recombinant human IL-1 receptor antagonist balances the destruction of bone and cartilage mediated by IL-1. Novos®, a combination of recombinant osteogenic protein-1 involved in the development of bone and cartilage and a collagen carrier, can be surgically implanted to allow healing of long bone nonunion fracture.

Last year, Replagal® and Fabrazyme®, 2 recombinant forms of the enzyme α-galactosidase A expressed in different cell lines, were commercialized for the treatment of the rare life-threatening Fabry's disease.

Finally, 14 additional biological entities were launched in 2001 but they are not considered as NBEs:
- natural products or vaccines containing natural antigenic extracts such as CTC-111 from Teijin (Anact C®, purified human plasma-derived activated protein C as antithrombotic), Ceprotin® from Baxter (purified human plasma-derived protein C against hereditary protein C deficiency), OrCel® from Ortec (a bilayered cellular matrix for dermal tissue repair), Myobloc® from Elan (a purified botulinum toxin type B for the treatment of cervical dystonia), Hep A Vi® from Aventis (purified typhoid Vi polysaccharide and inactivated hepatitis-A virus for the prevention of hepatitis-A and typhoid infections), and Melacine® from Schering (a melanoma vaccine consisting of cell lysate of two types of human melanoma cell);
- new formulation or new modification of already existing products such as darbepoetin alfa from Amgen (Aranesp®, a supersialyated analog of recombinant human EPO with prolonged half-life for the treatment of anemia in chronic renal failure), and peginterferon alfa-2a from Roche (Pegasys®, a pegylated analog of interferon alfa-2a for the treatment of chronic hepatitis C), and agalsidase beta from Genzyme (Fabrazyme®, another recombinant form of α-galactosidase A expressed in CHO cells to treat Fabry's disease, launched shortly after Replagal®);
- recombinant forms of already launched proteins such as choriogonadotropin alfa from Serono (Ovidrel®, recombinant human chorionic gonadotropin for the anovulatory infertility in female patients), lutropin alfa from Serono (Luveris®, recombinant human

luteinizing hormone for the treatment of women with an endocrine form of infertility), trafermin from Scios (Fiblast Spray®, recombinant human basic fibroblast growth factor for the treatment of skin ulcers), DWP-401 from Daewoong (Easyef liq®, recombinant human epidermal growth factor for the treatment of diabetic ulcer), and rasburicase from Sanofi-Synthelabo (Fasturtec®, a recombinant form of urate oxidase for the treatment and prophylaxis of acute hyperuricemia).

Acemannan (wound healing agent)   (6-10)

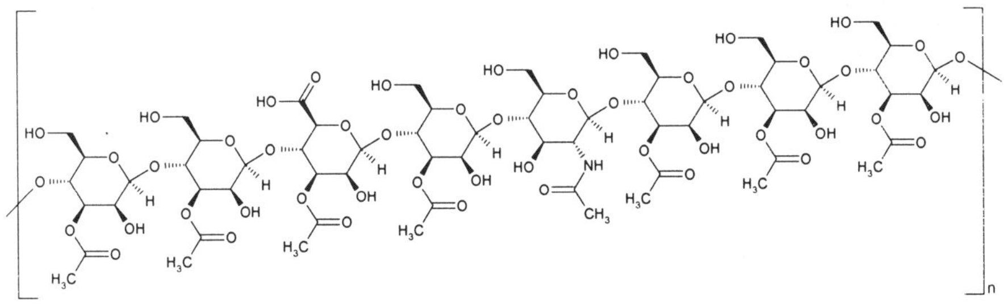

Country of Origin : US   Trade Name :   Acemannan hydrogel
Originator :   Carrington Laboratories   CAS Registry No. : 110042-95-0
First Introduction : US   Molecular Weight : 1-2 million Da
Introduced by :   Carrington Laboratories

Acemannan Hydrogel was marketed in the US as a wound healing agent for the care of ulcers, burns and post-surgical incisions. Acemannan is a complex water soluble poly-manno-galacto acetate isolated from *Aloe vera* by alcohol precipitation and purified by filtration to limit the size of the polysaccharide polymers. Pharmacological studies demonstrated immunostimulatory activity through the activation of monocytes and macrophages triggering enhanced secretion of cytokines such as interferon, interleukins, TNF-alpha, GM-CSF and some prostaglandins. It was proposed that the adjuvant activity of this agent could be partly related to its ability to induce phenotypic and functional maturation of immature dendritic cells. In normal healthy males, acemannan administered orally up to 3200 mg/day for 6 days did not show any acute oral toxicity. Acemannan is currently undergoing clinical trials for the treatment of AIDS.

Agalsidase alfa (Fabry's disease)   (11-15)

Country of Origin : US   Class :   Recombinant protein
Originator :   Transkaryotic Therapies   Type : α-galactosidase A
First Introduction : Sweden   Molecular Weight :   100 kDa
Introduced by :   Transkaryotic Therapies   Expression system : Human fibroblast
Trade Name :   Replagal   cells
CAS Registry No :   104138-64-9   Manufacturer : Transkaryotic Therapies

Fabry's disease is a rare genetic glycolipid-storage disorder characterized by the deficiency of the lysosomal enzyme α-galactosidase A. This enzyme participates in the catabolism of the glycosphingolipid globotriaosylceramide (Gb_3) by specifically cleaving its terminal α-linked galactose residue. Progressive accumulation of Gb_3 and related glycosphingolipids in vascular endothelial lysosomes of the kidneys, heart, skin and brain leads to several chronic symptoms such as debilitating neuropathic pain and characteristic cutaneous lesions called angiokeratomas. As vital organs are affected with age, premature death usually occurs in the fourth or fifth decade of life due to renal, cardiac or cerebral complications. Until now, there has been no treatment available for this disease other than therapies dedicated to symptom control. Last year, two recombinant forms of the glycoprotein α-galactosidase A named agalsidase alfa, Replagal® from Transkaryotic Therapies and agalsidase beta, Fabrazyme® from Genzyme (cf. introduction) were launched in the European Union as long-term enzyme replacement therapy for patients diagnosed with Fabry's disease. Agalsidase alfa, produced by transfected human skin fibroblasts and extensively purified from the culture medium by five chromatographic steps and a viral filtration step, was the first to be marketed. Agalsidase alfa is delivered to its lysosomal site of action via the recognition of its mannose-6-phosphate (M6P) residues by specific M6P receptors on the cell surface followed by endocytosis. In clinical trials, twice weekly intravenous infusions of agalsidase alfa over six months to one year have been found to be safe and effective in reversing the accumulation of microvascular endothelial deposits of globotriaosylceramide in the kidneys, heart and skin.

Alemtuzumab (anticancer)   (16-19)

Country of Origin :	UK	Class :	Humanized monoclonal antibody
Originator :	Cambridge University		
First Introduction :	US	Type :	Immunosuppressant
Introduced by :	Millenium/ILEX Oncology/Schering AG	Molecular Weight :	21-28 kDa
		Expression system :	CHO cells
Trade Name :	Campath	Manufacturer :	Berlex Laboratories
CAS Registry No :	146705-13-7		

Alemtuzumab was first introduced in the US for the treatment of B-cell chronic lymphocytic leukemia (CLL) in patients who have been treated with alkylating agents and have failed fludarabine therapy. Alemtuzumab is a humanized monoclonal antibody of the IgG1 isotype specific for the glycoprotein CD52 expressed on the cell surface of over 95% of normal and malignant B and T lymphocytes and monocytes. It was genetically engineered by adding six hypervariable regions from the heavy-chain and light-chain variable domains of an IgG2a rat monoclonal antibody onto a human IgG1 immunoglobulin molecule. The exact mechanism of action of alemtuzumab is unknown but the antibody probably causes the lysis of lymphocytes via complement fixation and antibody-dependent cytotoxicity. In a phase III clinical trial with 93 fludarabine-resistant CLL patients, treatment with alemtuzumab gave a positive response in 31 patients with 2 complete remissions and 29 partial remissions. This study demonstrated a higher response rate in patients having malignancy confined to the blood and bone marrow. Alemtuzumab does not seem to penetrate well into solid tissues and it was found to be concentrated in the blood which seems to correlate with the fact that it appears to be inefficient against tumor cells from lymph nodes and extranodal masses. The half-life of the antibody ranged between 23 and 30 hours. Premedication with acetaminophen and diphenhydramine is effective in reducing infusion-related reactions such as fevers, rigors, rash, nausea or hypotension. The most significant adverse effect is lymphopenia resulting in increased risk of opportunistic infections. As a consequence, prophylactic treatment with antibiotics is recommended for patients under alemtuzumab therapy.

Anakinra (antiarthritic)   (20-23)

Country of Origin : US
Originator :   Amgen/University of
                     Colorado
First Introduction : US
Introduced by :   Amgen
Trade Name :   Kineret
CAS Registry No : 143090-092-0

Class :   Recombinant protein
Type :   Interleukin-1 receptor
                  antagonist
Molecular Weight :   17.3 kDa
Expression system : *E. coli*
Manufacturer :   Amgen

Anakinra was launched in the US as a new daily subcutaneous injection therapy for the reduction of signs and symptoms of moderately to severely active rheumatoid arthritis (RA) in adults who have failed to respond to one or more disease-modifying antirheumatic drugs. Anakinra, N^2-L-methionylinterleukin-1 receptor antagonist (human isoform x reduced), is the first recombinant non-glycosylated human IL-1 receptor antagonist (IL-1ra). It was isolated from human monocytes, cloned and expressed in *Escherichia coli*. IL-1ra is an endogenous cytokine that blocks the binding of the pro-inflammatory cytokine IL-1 to its receptor thereby balancing the cartilage destruction and bone resorption mediated by IL-1. In RA patients, the amount of endogenous IL-1ra in the synovial joint is not sufficient to neutralize the detrimental effects of an excessive level of IL-1. In collagen-induced arthritic mice, continuous i.p. infusion of IL-1ra suppressed the established arthritis, reduced cartilaginous destruction and restored the synthetic function in articular cartilage chondrocytes. In clinical trials with RA patients, anakinra was found to attenuate disease severity and to reduce joint destruction over 6 to 12 months. Treatment with anakinra was generally well tolerated, although skin reactions at the injection site were reported. Experiments in rat showed that subcutaneous or intradermal injection of high doses of anakinra or its vehicle caused a non-immunologically mediated cutaneous mast cell degranulation.

Bimatoprost (antiglaucoma)   (24-28)

Country of Origin : US
Originator :   Allergan
First Introduction : US, Brazil
Introduced by :   Allergan
Trade Name :   Lumigan
CAS Registry No. : 155206-00-1
Molecular Weight : 415.57

Bimatoprost was first introduced in the US and Brazil as Lumigan®, an ophthalmic solution (0.03%) for the reduction of elevated intraocular pressure (IOP) in patients with open-angle glaucoma or ocular hypertension, a proven risk of glaucomatous visual field loss. Bitamoprost is a new agent belonging to the $PGF_{2\alpha}$ analog class of prostamides launched in this indication after latanoprost, the most efficaceous topical medication currently available. This synthetic fatty acid amide can be prepared by esterification of 17-phenyl-18,19,20-trinorprostaglandin $F_{2\alpha}$ followed by ethylamidation. Bimatoprost, as related prostamides, is devoid of typical activities associated with $PGF_{2\alpha}$ analogs; it exhibits a unique pharmacological profile in contracting the feline iris sphincter with an EC_{50} of 34 nM without interacting with any known prostanoid receptors. Thus it mimics the action of endogenous prostamides on specific receptors that lower IOP by increasing aqueous humor outflow through both pressure-insensitive and pressure-sensitive pathways without reducing humor formation. In a 3-month controlled clinical trial of efficacy and safety in patients with elevated IOP, bimatoprost demonstrated lower mean intraocular pressures at every time point throughout the study, as well as a good tolerance and systemic safety profile compared to latanoprost.

Bosentan (antihypertensive)   (29-34)

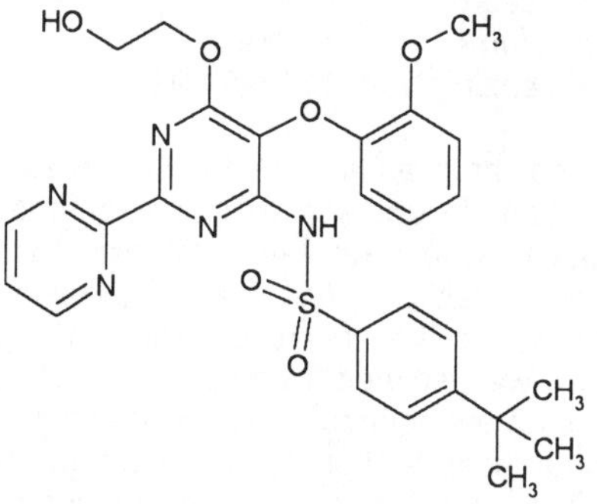

Country of Origin : Switzerland
Originator :   Roche
First Introduction : US
Introduced by :   Actelion/Genentech
Trade Name :   Tracleer
CAS Registry No : 147536-97-8
Molecular Weight : 551.62

Bosentan was introduced in the US as a twice-daily oral treatment for pulmonary arterial hypertension. It can be synthesized in five steps via condensation of diethyl (2-methoxyphenoxy)malonate with pyrimidine-2-carboxamidine to give the precursor of the symmetrical central dichloropyrimidine ring which is then successively treated with the potassium salt of 4-*tert*-butylbenzenesulfonamide and the sodium salt of ethylene gycol. Bosentan is the first endothelin (ET) receptor antagonist to be launched. ET-1, the most potent endogenous vasoconstrictor known, has been demonstrated to play a major role in the functional and structural changes observed in pulmonary hypertension. Bosentan is a mixed ET_A and ET_B receptor antagonist that inhibits the pulmonary arterial vasoconstricting effect of ET-1 predominantly mediated via ET_A receptors on smooth muscle cells. In a hypoxia-induced model of pulmonary hypertension in rat, it reduced the development of pulmonary hypertension as well as right ventricular hypertrophy and prevented pulmonary arterial remodeling. In clinical trials, patients treated with bosentan showed a 20% increase in exercise capacity compared to placebo as measured by the six minute walk test. Bosentan not only improved the distance walked by patients but also significantly decreased mean pulmonary artery pressure, mean pulmonary vascular resistance, mean capillary wedge pressure and mean right atrial pressure. It demonstrated a beneficial selectivity for the pulmonary vasculature since it had no significant effect on mean aortic blood pressure and systolic vascular resistance. The compound is hepatically metabolized into three major metabolites by CYP3A4 and 2C9 and almost exclusively eliminated in the bile. Although large interspecies differences in systemic plasma clearance was observed (1.5 mL/min/kg in dogs to 72 mL/min/kg in rabbits), a satisfactory systemic clearance (2 mL/min/kg) was measured in human. The most frequent adverse effect was reversible elevation of liver transaminases. This adverse reaction appears to be due to intracellular accumulation of cytotoxic bile salts resulting from inhibition of the hepatocanalicular bile salt export pump by bosentan.

Caspofungin acetate (antifungal)   (35-40)

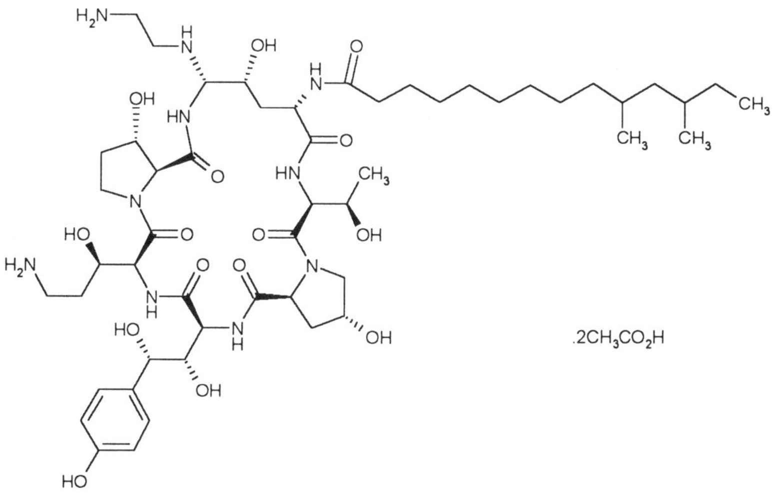

.2CH$_3$CO$_2$H

Country of Origin :	US	Trade Name :	Cancidas
Originator :	Merck & Co	CAS Registry No. :	179463-17-3
First Introduction :	US	Molecular Weight :	1213.43
Introduced by :	Merck & Co		

Caspofungin is the first in a new class of antifungal agents named echinocandin to be launched in the US for the parenteral treatment of invasive aspergillosis in patients refractory to, or intolerant of other antifungal therapies such as amphotericin-B, lipid formulation of amphotericin-B and/or itraconazole. Caspofungin is a water soluble, stable, aza-substituted semisynthetic derivative of pneumocandin B, a fermentation product of the fungus *Zalerion arboricola*. Since fungi and humans are both eukaryotes, new antifungal agents must focus on fungus-specific targets in order to minimize potential toxicity in the human host. In this respect, the lipopeptide caspofungin, inhibits the synthesis of 1,3-beta-D-glucans present only in the fungal cell wall, which leads to the specific lysis of the pathogenic cells. The compound is fungicidal rather than fungistatic, an unusual feature among antifungal agents. Caspofungin has activity *in vitro* and *in vivo* against a range of *Candida* species (including azole and amphotericin B resistant *Candida*) and also against *Aspergillus*, an important pathogen which is not susceptible to fluconazole. Caspofungin was generally more active than amphotericin B, flucytosine, fluconazole and itraconazole against *Candida* species. *Cryptococcus* is not susceptible to caspofungin. Caspofungin has a half-life in human beings around 10 hours. Unlike the azoles, caspofungin does not inhibit cytochrome P450. The drug has been given intravenously with loading dose of 70 mg, followed by 50 mg daily to be effective in invasive aspergillosis refractory to prior antifungal therapy. Most of the side effects were mild and comprised fever, thrombophlebitis, headache, nausea, rash and flushing.

Crotalidae Polyvalent Immune Fab (antidote)   (41-44)

Country of Origin :	UK	Class :	Polyclonal antibody
Originator :	Protherics	Type :	Antivenin
First Introduction :	US	Molecular Weight :	50 kDa
Introduced by :	Savage Laboratories	Expression system :	Immunized sheep
Trade Name :	CroFab	Manufacturer :	Protherics
CAS Registry No :	unknown		

Crotalidae Polyvalent Immune Fab (ovine) was launched as CroFab® in the US for the treatment of North American crotalid snake envenomation. It is the second antivenom to be commercialized over fifty years after Antivenin (Crotalidae) Polyvalent (equine) from Wyeth Labs. CroFab® is prepared by purification of Fab immunoglobulin fragments obtained from the blood of healthy sheep flocks immunized with one of the following North American snake venoms: *Crotalus atrox* (Western Diamondback rattlesnake), *Crotalus adamanteus* (Eastern Diamondback rattlesnake), *Crotalus scutulatus* (Mojave rattlesnake), and *Agkistrodon piscivorus* (Cottonmouth or Water Moccasin). The final preparation is achieved by mixing the four different monospecific antivenins after fractionating the immunoglobulin G from the ovine serum, cleaving the whole antibody into Fab and Fc fragments with papain and isolating each of the venom-specific Fab fragments on ion exchange and affinity chromatography columns. CroFab® acts by binding and neutralizing venom toxins, facilitating their redistribution away from target tissues and their elimination from the body. In infected mice, CroFab® was found to be 3-9 times more potent than Antivenin (Crotalidae) Polyvalent (equine) for the prevention of lethality of nine venoms. In a randomized clinical trial, initial control of envenomation was achieved in all patients treated with intravenous administrations of CroFab®. The risk of hypersensitivity is decreased with this mixture of specific antivenins compared to Antivenin (Crotalidae) Polyvalent (equine) due to the absence of the Fc components of the immunoglobulin. The elimination half-life of CroFab® ranged from approximately 12 to 23 hours, which is relatively short for an antibody.

Desloratadine (antihistamine)   (45-52)

Country of Origin : US
Originator :   Sepracor
First Introduction : UK
Introduced by :   Schering-Plough
Trade Name :   Clarinex
CAS Registry No. : 100643-71-8
Molecular Weight : 310.83

Desloratadine was launched last year as an improved version of Schering-Plough's Claritin® (Loratadine), for the treatment of nasal and non-nasal symptoms of seasonal allergic rhinitis (SAR). Desloratadine can be prepared by a 7-step synthesis starting with a Ritter reaction on 2-cyano-3-picoline and involving successively alkylation with 3-chlorobenzyl chloride, dehydration, Grignard reaction with the piperidine magnesium chloride, intramolecular cyclization in strong acidic medium, and finally demethylation of the piperidine. Desloratadine (the descarboethoxyloratadine) is a biologically active metabolite of the second-generation antihistamine loratadine. Desloratadine is a non-sedating competitive histamine H_1 receptor antagonist with increased potency and improved safety as compared to loratadine. When compared with other H_1 antagonists as an inhibitor of histamine-induced calcium flux in CHO cells, desloratadine was found to be more potent than most H_1 antagonists (such as the widely used terfenadine, fexofenadine, cetirizine, loratadine and astemizole) in this assay. Desloratadine is also a potent antagonist of muscarinic M_1 and M_3 receptors (not M_2) indicating anticholinergic activity. These effects may be the explanation for the unexpected decongestant effects of desloratadine reported in clinical trials. Desloratadine appears to be devoid of significant effects on potassium channels and does not appear to suffer from adverse interaction with cytochrome P450 inhibitors. Clinical studies have shown that it does not induce sedation or cardiac arrhythmias and does not potentiate the effects of alcohol. Apparent total body clearance is in the range of 114-201 l/h and the mean elimination half life is 19-34.6 h in human. Desloratadine is available as 5-mg tablets, which is the once-daily recommended dose for adults and children above 12 years. It may be taken without regard to food. The treatment not only improved symptoms of SAR but also improved patient ratings of nasal congestion. The FDA also has issued an "approvable" letter for this agent for the treatment of chronic idiopathic urticaria (CIU).

Drotrecogin alfa (antisepsis)   (53-57)

Country of Origin : US
Originator :   Lilly
First Introduction : US
Introduced by :   Lilly
Trade Name :   Xigris
CAS Registry No : 42617-41-4

Class :   Recombinant protein
Type :   Activated protein C
Expression system : Human kidney 293
                                    cells
Manufacturer :   Lilly

Drotrecogin alfa was introduced in the US as a new intravenous treatment for the reduction of mortality in adult patients with severe sepsis associated with acute organ dysfunction. Drotrecogin alfa is the first recombinant activated protein C expressed in human kidney 293 cells. Activated protein C, an endogenous serine protease, acts as an antithrombotic via inhibition of factor Va and VIIIa, promotes fibrinolysis via inactivation of plasminogen-activator inhibitor-1 and exerts an anti-inflammatory effect by inhibiting the production of inflammatory cytokines (tumor necrosis factor-alpha, interleukin-1 and 6) by monocytes. It was shown that the majority of patients with sepsis have reduced levels of activated protein C. In a baboon model of lethal *E. coli* sepsis, administration of activated protein C reduced mortality. In a phase III clinical trial, intravenous infusion of drotrecogin alfa at a dose of 24 µg/kg/h over 96 h significantly reduced the risk of death compared with placebo in patients with severe sepsis. However, an increased risk of serious bleeding has been associated with the administration of the drug.

Edaravone (neuroprotective)   (58-63)

Country of Origin : Japan
Originator :   Mitsubishi Pharma
First Introduction : Japan
Introduced by :   Mitsubishi Pharma
Trade Name :   Radicut
CAS Registry No. : 89-25-8
Molecular Weight : 174.20

Edaravone was marketed last year in Japan for improving neurologic recovery following acute brain infarction. Currently, several agents classified as neuroprotectants and acting by diverse mechanisms (inhibition of glutamate release, blockade of calcium channels, lazaroids) have been marketed for treating the outcomes of brain damage due to trauma, ischemia or cardiac arrest. Edavarone is the first antioxidant with free radical scavenging activity to be introduced for this pathology. This previously described molecule (in particular as norantipyrine, one of three metabolites of antipyrine in mammals) can be simply prepared by direct cyclization of phenylhydrazine with alkylacetoacetate. Edaravone is a lipophilic agent, readily accessible to brain tissue, that is capable of reducing edema in the brain following ischemia by blocking the arachidonic acid cascade triggering peroxidative neurodegeneration. Interestingly, this agent has been shown to quench active oxygen species in endothelial cell homogenate, as well as inhibiting *in vitro* lipid peroxidative disintegration of membranes, so making this compound effective during reperfusion following ischemic injury. As an additional indication, phase III trials started last year with edaravone for increasing the chance of recovery after subarachnoid hemorrhage.

Eletriptan (antimigraine)   (64-71)

Country of Origin : US
Originator :   Pfizer
First Introduction : Switzerland
Introduced by :   Pfizer
Trade Name :   Relpax
CAS Registry No. : 143322-58-1
Molecular Weight : 382.52

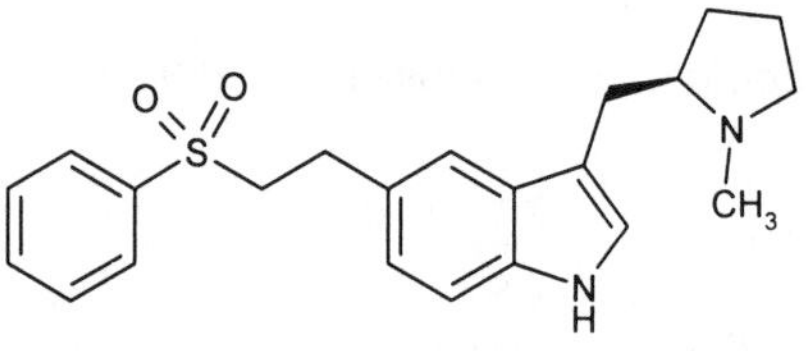

Eletriptan, the seventh member of the triptan class of antimigraine drugs was launched last year in Switzerland for the acute treatment of migraine. The 5-step synthesis of this conformationally restricted analog of sumatriptan involves the deprotonation of 5-bromoindole with Grignard reagent at the position C3 followed by a condensation with N-Cbz-D-proline acid chloride to afford the key Cbz-D-prolyl intermediate. After reduction with LiAlH$_4$, the sulfone moiety was introduced in postion 5 by a palladium cross-coupling Heck reaction using the phenylvinyl sulfone. Eletriptan is a 5-HT receptor agonist that binds to 5-HT$_{1B}$, 5HT$_{1D}$ and 5HT$_{1F}$ receptors with high potency. Activation of these receptors causes constriction of extracerebral intracranial vessels, abolition of the dural extravasation and neurogenic inflammation, and inhibition of trigeminal neuronal discharge. Eletriptan was found to be the most potent agonist at the 5-HT$_{1D}$ receptor compared to the other triptans with pEC$_{50}$ value of 9.2. In thousands of participants in clinical trials, eletriptan has acted more effectively and more rapidly than sumatriptan to relieve pain from mild to severe attacks of migraine. Eletriptan also reduced time lost for ordinary activity to patients with migraine attacks when compared to placebo and when compared to sumatriptan. Eletriptan was also superior to sumatriptan in terms of relieving functional disability, nausea, photophobia and phonophobia. Unlike other compounds of its class, eletriptan has a positive logD value, that could underlie its rapid and complete oral absorption and may be suggestive of a good brain penetration. The oral biovailability of Eletriptan is approximately 50% (14% for Sumatriptan) with a half life of 5.7h (2h for Sumatriptan). Eletriptan was well tolerated with mild to moderate adverse events including asthenia, somnolence, dizziness and nausea. Eletriptan is a new potent and fast-acting triptan for the treatment of migraine attacks.

Falecalcitriol (vitamin D)   (72-75)

Country of Origin : US
Originator :   University of Wisconsin
First Introduction : Japan
Introduced by :   Kissei/Sumitomo/Taisho
Trade Name :   Hornel, Fulstan
CAS Registry No. : 83805-11-2
Molecular Weight : 524.59

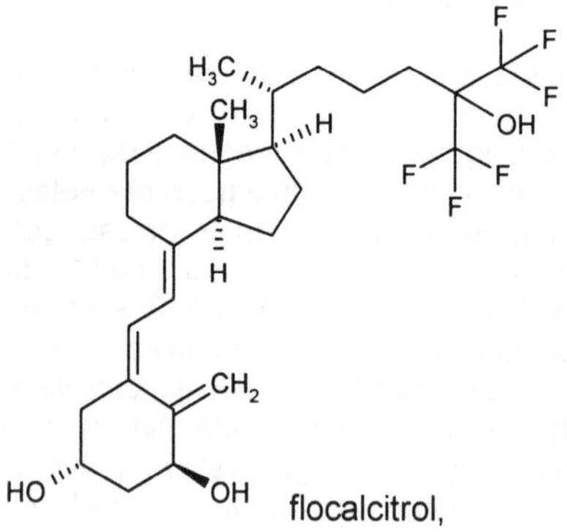

Falecalcitriol,   previously   known   as   flocalcitrol,   was launched last year by several codevelopers in Japan for the treatment of secondary hyperthyroidism (SHPT). This hexafluorinated analog of $1\alpha,25$-dihydroxyvitamin D_3 (calcitriol), the hormonally active form of vitamin D_3, can be obtained by several different synthetic routes from a conveniently protected cholestenol, a key step being an aldol reaction with hexafluoroacetone. Falecalcitriol is several times more active than $1,25(OH)_2D_3$ in regulating the proliferation of parathyroid cells and parathyroid hormone (PTH) synthesis that are believed to be mediated through binding to VDR, a nuclear receptor for vitamin D; furthermore, it was proposed that a bioactive 23S-hydroxylated metabolite, resistant to further metabolism, contributes to the retention of an active compound for longer in cells and so, to significantly lengthen the duration of action. In a comparative clinical study conducted in hemodialysis patients with moderate to severe SHPT, falecalcitriol was found to be more active than alfacalcidol in suppressing parathyroid hormone without triggering hypercalcemia.

Fudosteine (expectorant)   (76-79)

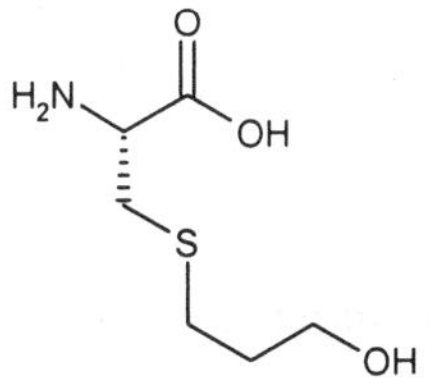

Country of Origin : Japan
Originator :   SS Pharmaceutical/Mitsubishi Pharma
First Introduction : Japan
Introduced by :   SS Pharmaceutical/Mitsubishi Pharma
Trade Name :   Cleanal
CAS Registry No. : 13189-98-5
Molecular Weight : 179.23

Fudostein was launched last year in Japan as a new mucoactive agent for the treatment of bronchitis and respiratory congestion. This cysteine derivative was obtained from L-cysteine by condensation with either allylic alcohol in the presence of potassium persulfate or the corresponding bromoalcohol in the presence of a base. Fudostein was shown to significantly reduce mucus glycoprotein hypersecretion and inhibit infiltration of airway mucosa by lymphocytes and inflammatory cells in bronchitic rats. When given to bronchitic rabbits, an oral dose of 500 mg/kg daily potently decreased the fucose/ N-acetylneuraminic acid in sputa, so exhibiting mucoregulatory properties. In another study with SO_2-exposed rabbits, fudostein suppressed blood flow of tracheal microvasculature increased by SO_2, partly due to scavenging of superoxide anion.

Imatinib mesilate (antineoplastic)   (80-87)

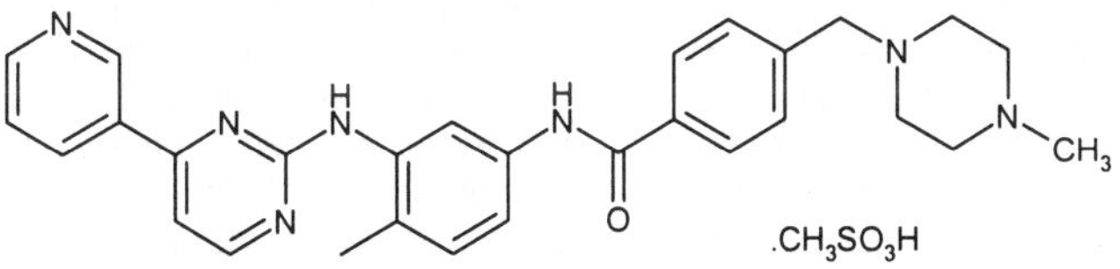

Country of Origin :   Switzerland   Trade Name :   Gleevec, Glivec
Originator :   Novartis   CAS Registry No. : 220127-57-1
First Introduction : US   Molecular Weight : 589.72
Introduced by :   Novartis

In May 2001, the FDA approved imatinib as a new cancer drug after a record review time of just 2.5 months. Imatinib was launched as Gleevec in the US for chronic myelogenous leukemia (CML) in blast crisis, accelerated phase or chronic phase after interferon-alpha failure. This compound can be prepared by a four step sequence from a condensation of the 1-(3-pyridyl)ethanone with dimethyl formamide dimethylacetal, followed by successive cyclization with the methyl-nitrophenyl guanidine, hydrogenolysis and condensation with the benzoyl chloride of the methylpiperazine. Imatinib is the first of a new class of anticancer drugs that are specifically designed to target the molecular pathways involved (oncogenic event) in the development of disease. The Brc-Abl oncoprotein is a constitutively active tyrosine kinase that causes CML. Imatinib is a competitive inhibitor of this tyrosine kinase as well as Abl, Kit and the PDGFR kinases. It binds to the ATP-binding site of the target kinase and prevents the transfer of phosphate from ATP to the tyrosine residues of various substrates and consequently blocks the proliferation of the leukemic cells. Phase II studies demonstrated that in chronic phase CML, over 90% of the patients had their leukocyte counts return to normal and 56% had a major cytogenic response. No phase III data is currently available. It is clear from the evidence available that imatinib has advantages over IFN-alpha, such as reduced toxicity, more rapid hematological response, higher rate of cytogenic response and oral administration. The drug is well tolerated, producing few side effects, classified as grade 1 nausea, muscle cramps, diarrhea, edema and vomiting. Imatinib is metabolized primarily by the CYP3A4 enzyme system and drugs capable of modulating this system would be expected to modify the patient's exposure. Novartis expects to launch imatinib for the treatment of gastrointestinal stromal tumors in 2002.

Levocetirizine (antihistamine)   (88-94)

Country of Origin : US
Originator :   Sepracor
First Introduction : Germany
Introduced by :   UCB
Trade Name :   Xusal
CAS Registry No : 130018-77-8
Molecular Weight : 388.89

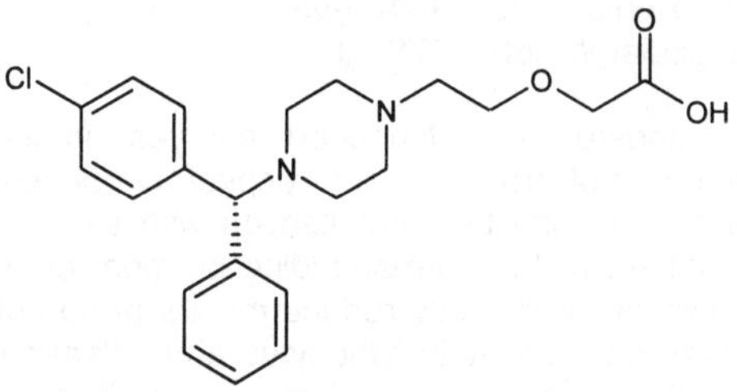

The (R)-enantiomer of the second-generation antihistamine cetirizine, levocetirizine, was first introduced in Germany for seasonal allergic rhinitis (including ocular symptoms), perennial allergic rhinitis and chronic idiopathic urticaria. The dihydrochloride salt can be prepared in four steps from optically active 4-chlorobenzhydrylamine obtained by resolution of its racemate with (+)-tartaric acid. Levocetirizine (eutomer) is a 2-fold more potent H_1 antagonist than cetirizine whereas the other enantiomer (distomer) is 10-fold less potent compared to levocetirizine. Pharmacodynamic studies on healthy volunteers showed that compared to cetirizine, half the dose of levocetirizine (5 mg) was necessary to obtain similar inhibitory effects in the skin test of histamine-induced wheal and flare as well as on histamine-induced nasal congestion and nasal resistance. There was no evidence of chiral inversion of levocetirizine in vivo in several species including human. The daily dose of drug is rapidly and extensively absorbed in human. Interestingly, its volume of distribution (0.41 kg/L) is smaller than that of the distomer (0.60 kg/L). The low volume of distribution is considered as favorable for an antihistamine both in terms of efficacy and safety. Due to its high metabolic stability and lack of effect on the activities of the major CYP isoenzymes, levocetirizine is unlikely to cause interactions with other administered drugs. No clinically relevant effect on electrocardiograms of healthy volunteers was detected.

Nesiritide (congestive heart failure)   (95-99)

Country of Origin : US	Class : Recombinant protein
Originator : Scios	Type : B-type natriuretic peptide
First Introduction : US	Molecular Weight : 3464
Introduced by : Scios/GlaxoSmithKline	Expression system : *E. coli*
Trade Name : Natrecor	Manufacturer : Scios
CAS Registry No : 124584-08-3	

Nesiritide was introduced in the US as a new intravenous treatment for patients with acutely decompensated congestive heart failure who have dyspnea at rest or with minimal activity. Nesiritide is the first human recombinant form of the potent vasodilatory B-type natriuretic peptide (also known as brain natriuretic peptide), a naturally occurring 32 amino acid peptide with a disulfide-bonded 17 amino acid ring structure. Nesiritide binds to the A-type natriuretic peptide receptor on the surface of endothelial and smooth muscle cells stimulating production of the second messenger cGMP which mediates predominantly vascular smooth muscle cell relaxation. It also facilitates the elimination of sodium and water by the kidney and decreases the secretion of certain hormones, such as adrenalin, angiotensin II, aldosterone and endothelin, which provoke long-term detrimental effects including blood vessel constriction and blood pressure elevation. In clinical trials with heart failure patients, nesiritide dose-dependently reduced pulmonary capillary wedge pressure, right atrial pressure and systemic vascular resistance and increased cardiac index without affecting heart rate. The major adverse effect observed was dose-dependent hypotension. In a phase III comparative trial, nesiritide was found to be superior to iv. nitroglycerine in its haemodynamic effects as well as easier to administer and better tolerated. Nesiritide is cleared by proteolytic cleavage by the enzyme neutral endopeptidase NEP24.11 and by binding to the C-type natriuretic peptide receptor followed by endocytosis and intracellular lysosomal hydrolysis. Since it has a short half-life (18 min), nesiritide is administered as a 2µg/kg bolus infusion followed by a continuous maintenance infusion at 0.01µg/kg/min usually over 24-48 hours.

OP-1 (osteoinductor)   (100-104)

Country of Origin : US	Class : Recombinant protein
Originator : Curis	Type : Bone morphogenic protein
First Introduction : Australia	Molecular Weight : 36 kDa
Introduced by : Stryker Biotech	Expression system : CHO cells
Trade Name : Novos	Manufacturer : Stryker Biotech

OP-1 implant was first marketed in Australia as an alternative to autograft in recalcitrant long bone nonunion fractures where use of autograft is unfeasible and alternative treatments have failed. The product is surgically implanted as a paste in the fracture gap. It is a combination of human recombinant osteogenic protein OP-1 (also called bone morphogenic protein BMP-7), the active processed mature disulfide-linked homodimer produced in CHO cells, and a bovine bone-derived collagen carrier. Osteogenic protein-1, is a cytokine of the tissue growth factor TGF-beta superfamily that initiates bone formation through the induction of cellular differentiation in mesenchymal cells which are recruited to the implant site from bone marrow, periosteum and muscle. Once bound at the cell surface, the active substance induces a cascade of cellular events leading to the formation of chondroblasts and osteoblasts, key role players in the bone formation process. In a clinical study for the treatment of tibial nonunion fractures, OP-1 implant gave similar healing results to those obtained with autologous hip bone graft with less blood loss, shorter operative times and without donor site complications. Since osteogenic protein-1 is also involved in the development of the kidney and the brain, several other applications of recombinant OP-1 are being studied.

Perospirone hydrochloride (neuroleptic)   (105-109)

Country of Origin : Japan
Originator :   Sumitomo Pharm.
First Introduction : Japan
Introduced by :   Sumitomo Pharm.
Trade Name :   Lullan
CAS Registry No. : 150915-41-6
Molecular Weight : 463.04

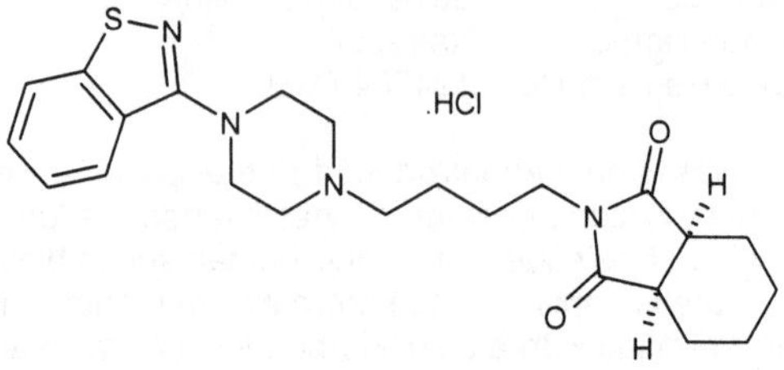

Perospirone hydrochloride was launched in Japan as a new treatment of schizophrenia and other psychoses. This structurally-related analog of ziprasidone can be classically prepared by alkylation of 1-(3-benzisothiazolyl)-piperazine with the appropriate N-(chlorobutyl)-succinimide. As a result of an *in vitro* determination of its binding profile, perospirone demonstrated a high affinity for 5-HT_2, 5-HT_{1A} and D_2 receptors as well as a significant but lower affinity for D_1 and α_1 receptors. Interestingly, these differential effects on neurotransmitters with a high *in vivo* occupancy of 5-HT_{2A} receptors with lower D_2 occupancy seems to provide this new agent with a significantly lower propensity for the development of extrapyramidal symptoms (EPS) and tardive schizophrenia, the main side-effects associated with classical antipsychotic therapy. Furthermore, perospirone exhibited anxiolytic-like effects in animal models. A long-term clinical trial (> 6 months) in a group of patients suffering from schizophrenia demonstrated the efficacy of perospirone over placebo for the treatment of the positive, negative and general symptoms of schizophrenia at doses of 12 mg to 96 mg t.i.d.

Tegaserod maleate (irritable bowel syndrome)   (110-115)

Country of Origin : Switzerland
Originator :   Novartis
First Introduction : Mexico
Introduced by :   Novartis
Trade Name :   Zelmac
CAS Registry No. : 189188-57-6
Molecular Weight : 417.46

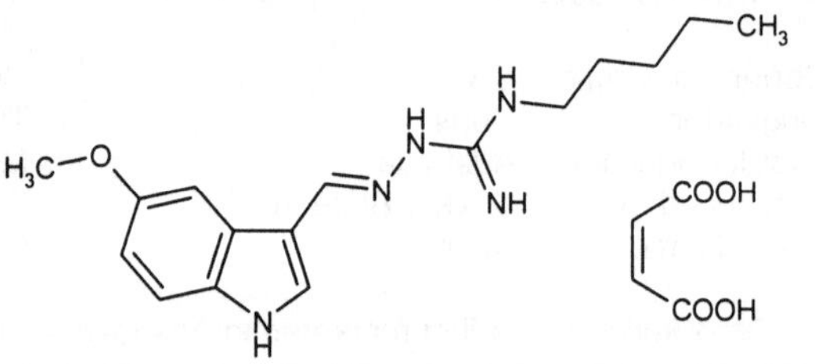

Tegaserod maleate was launched last year as a new oral treatment of constipation-predominant or diarrhea-predominant irritable bowel syndrome. It can be prepared by condensation of the 5-methoxy-3-formylindole with the appropriate aminoguanidine. Tegaserod is a selective partial agonist of the excitatory 5-HT_4 receptor (K_i = 18 nM in human caudate membranes) and was designed from the endogenous full agonist, serotonin. In the model of potentiation of the electrically stimulated contraction of longitudinal muscle of guinea pig ileum, Tegaserod exhibited a low-efficacy agonistic activity compared with serotonin. However, it appeared more potent in several studies on intestinal mucosal tissue from various species including man. A lack of effect on the QT_c interval was shown in rabbits then confirmed during clinical studies. Tegaserod is also clinically developed for the treatment of other functional GI disorders such as functional dyspepsia and gastroesophageal reflux disease (GERD).

Tenofovir disoproxil fumarate (antiviral)   (116-122)

Country of Origin : US
Originator :   Gilead Sciences
First Introduction : US
Introduced by :   Gilead
                 Sciences
Trade Name :   Viread
CAS Registry No. : 202138-50-9
Molecular Weight : 635.5

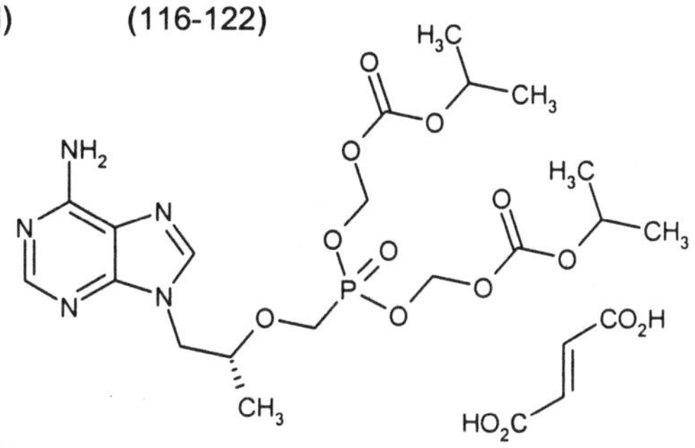

Tenofovir disoproxil fumarate (tenofovir DF) is the first nucleotide analog reverse transcriptase inhibitor (NRTI) to be launched in the US as a new oral treatment for HIV infection. This inhibitor can be prepared in six steps from (S)-glycidol by successive hydrogenation and intramolecular esterification to give the cyclic carbonate which reacts with adenine to afford the key hydroxypropyl adenine. The latter is transformed into the phosphonic acid [(R)-PMPA] condensed with the appropriate carbonate to give the phosphonic acid diester. Tenofovir DF is a water soluble diester that acts as a prodrug which is rapidly hydrolyzed to form the free tenofovir (analog of adenosine monophosphate) that acts only after two intracellular phosphorylation steps as a potent competitive inhibitor for reverse transcriptase. Tenofovir as the triphosphate form is then incorporated into DNA and causes DNA chain termination. In contrast to nucleoside analogs, tenofovir does not require initial intracellular phosphorylation, a limiting factor in resting cells. Tenofovir DF diffuses rapidly into cells due to its liphophilicity, unlike tenofovir, which requires an endocytosis transport process. The *in vitro* anti HIV activity of the oral prodrug is substantially greater than that of tenofovir alone (up to 100-fold), related to more rapid/extensive cellular uptake. Despite the fact that the prodrug is rapidly converted to free tenofovir in plasma after oral absorption, it is suggested that even small amounts of prodrug can provide enhanced antiviral activity. In clinical studies, tenofovir DF has been shown to reduce the level of HIV in the blood for up to 48 weeks when added to patient's existing antiretroviral regimens. The drug also reduced viral load even in patients whose HIV had developed resistance to currently available antiretroviral drugs. Tenofovir DF is unique in demonstrating efficacy against 3TC (lamivudine)-resistant HIV strains. Moreover, the long intracellular half-life (from 12 to 50h) of the tefonovir diphosphate allowing for once daily dosing is probably an important factor in the efficacy of this drug *in vivo*. The drug is eliminated by the kidney, is not metabolized by the liver and is not associated with P450 interactions. The bioavailability of the prodrug was increased significantly when taken with food from 27 to 40%. The oral bioavailabilty of tenofovir is < 10%.

Telithromycin (antibiotic)   (123-129)

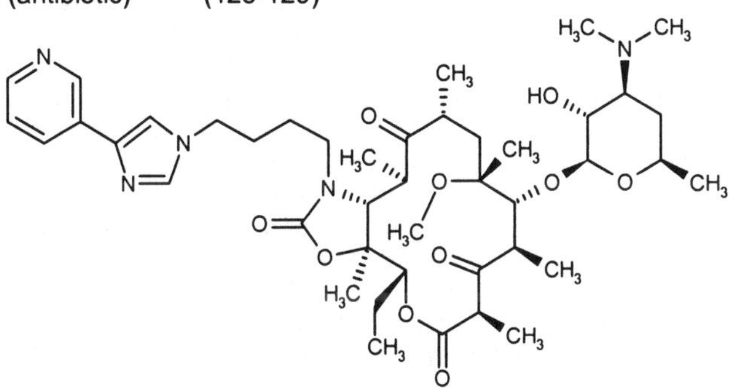

Country of Origin : France
Originator :   Aventis
First Introduction : Germany
Introduced by :   Aventis

Trade Name :   Ketek
CAS Registry No : 191114-48-4
Molecular Weight : 812.01

Telithromycin was first launched in Germany as a once-daily oral treatment for respiratory infections including community-acquired pneumonia, acute bacterial exacerbations of chronic bronchitis, acute sinusitis and tonsillitis/pharyngitis. This semisynthetic derivative of the natural macrolide erythromycin is the first marketed ketolide, a new class of antibiotics featuring a C3-ketone instead of the L-cladinose group. The 14-membered ring antibacterial agent prevents bacterial protein synthesis by binding to two domains of the 50S subunit of bacterial ribosomes. It shows potent *in vitro* activity against common respiratory pathogens including *Streptococcus pneumoniae*, *Haemophilus influenzae*, *Moraxella catarrhalis* and *Streptococcus pyogenes* as well as other atypical pathogens. The 3-keto group confers increased acidic stability and reduced induction of macrolide-lincosamide-streptogramin B resistance that is frequently observed with macrolides. The substituted C11-C12 carbamate residue appears not only to increase affinity for the ribosomal binding site but also to stabilize the compound against esterase hydrolysis and avoid resistance due to elimination of macrolides from the cell by an efflux pump encoded by the *mef* gene in certain pathogens. Telithromycin is both a competitive inhibitor and a substrate of CYP3A4. However, unlike several macrolides such as troleandomycin, it does not form a stable inhibitory CYP P-450 Fe^{2+}-nitrosoalkane metabolite complex which is potentially hepatotoxic. The drug is well tolerated and well distributed into pulmonary tissues, bronchial secretions, tonsils and saliva. It turns out to be highly concentrated in azurophil granules of polymorphonuclear neutrophils thereby facilitating its delivery to the phagocytosed bacteria.

Travoprost (antiglaucoma)   (130-134)

Country of Origin : US
Originator :   Alcon
First Introduction : US
Introduced by :   Alcon
Trade Name :   Travatan
CAS Registry No. : 157283-68-6
Molecular Weight : 500.55

Travoprost was launched in the US as Travatan®, an ophthalmic solution (0.004%) administered topically for the treatment of elevated intraocular hypertension (IOP) through open-angle glaucoma, a common optic neuropathy and a leading cause of blindness. Travoprost is the isopropyl ester of (+)-fluprostenol, a new prostaglandin derivative belonging to the $PGF_{2\alpha}$ analog class. This compound can be prepared in 8 steps from a bicyclic lactone aldehyde by a Wittig alkylidenation followed, after 2 ketonic reductions, by a lactol opening to prostenoic acid while protecting and deprotecting appropriately. Travoprost is a full agonist of FP receptors with a greater affinity than $PGF_{2\alpha}$ (K_i = 52 nM). Pharmacologic studies in rabbits treated daily for a week demonstrated a significant increase of the microvascular optic nerve head blood flow without significant alterations in the systemic flow. In several placebo-controlled clinical studies with hundreds of patients suffering from open-angle glaucoma or ocular hypertension, travoprost (1 to 4 pm daily) dose-dependently reduced IOP (by about 30% at 4 pm); it was shown equal or superior to latanoprost, another prostaglandin analog (5 pm daily) or the adrenergic ß-blocker timolol (500 pm bid). In addition, travoprost was safe and well tolerated with a low incidence of conjunctival hyperaemia.

Trimegestone (progestogen)   (135-142)

Country of Origin : France
Originator :   Aventis
First Introduction : Sweden
Introduced by :   Wyeth Pharmaceuticals
Trade Name :   Totelle Sekvens, Ondeva
CAS Registry No. : 074513-62-5
Molecular Weight : 342.47

Trimegestone was launched in Sweden in combination with 17 beta-estradiol as hormone replacement therapy (HRT) for the oral treatment of menopausal vasomotor symptoms and the prevention of osteoporosis. Trimegestone can be synthesized in a multistep process starting with 3,3-(ethylenedioxy)estra-5(10),9(11)-dien-17-one and involving a final key regio- and enantioselective reduction of the 17beta-2-oxopropionyl side chain with *Saccharomyces cerevisiae* in sodium acetate buffer. The norpregnane progestin, trimegestone, exhibited high specificity and affinity for the progesterone receptor, no affinity for the estrogen receptor, and weak affinity for androgen, glucocorticoid and mineralcorticoid receptors. The relative binding affinity of trimegestone for the progestin receptor was 7 times that of progesterone, 4.5 and 1.5 times greater than norethindrone and medroxyprogesterone acetate, respectively. The decrease in circulating estrogen associated with menopause is thought to contribute to a variety of diseases in women, including osteoporosis, cancers, cardiovascular disease, stroke and cognitive decline. Estrogen conserves bone mass by reducing bone turnover. Estrogen replacement therapy (ERT) is recommended for all women at high risk for osteoporosis. However, estrogen therapy alone has been linked to an increase risk of endometrial cancer; thus progestin (such as trimegestone) is often prescribed in combination with estrogen for women who have not had a hysterectomy. The progestin blocks the estrogenic activity in the endometrium, thereby reducing the potential unwanted cell proliferation in response to estrogen administration. This action of progestin occurs without compromising the beneficial effects of estrogen on hot flashes and bone loss. A study in rats with osteopenia showed that treatment with trimegestone in combination with 17beta-estradiol for 2 months was superior to norethisterone in preventing bone loss. Treatment with trimegestone also more effectively prevented estradiol-induced uterine atrophy as compared to norethisterone. In clinical trials, trimegestone was found to be a highly effective oral progestone for endometrial protection and beneficial effects have been observed on anxiety, depression, somatic, and vasomotor menopausal symptoms. The combination provides improved and predictable cycle control and a better lipid profile in comparison with existing products. Minimal progestogenic adverse events (i.e. mastalgia, acne, nausea, leg cramps, seborrhea and bloatedness) were reported. Totelle Sekvens[®] employs a cyclic regimen of 14 days of 2mg of 17beta-estradiol alone and 14 days in combination with 500 µg of trimegestone.

Valganciclovir hydrochloride (antiviral)   (143-147)

Country of Origin : Switzerland
Originator :   Roche
First Introduction : US
Introduced by :   Roche
Trade Name :   Valcyte
CAS Registry No : 175865-59-5
Molecular Weight : 390.83

Valganciclovir hydrochloride, a prodrug of the antiviral ganciclovir, was launched in the US for the oral treatment of cytemegalovirus (CMV) retinitis, a sight-threatening complication in patients with AIDS. This L-valyl ester prodrug can be prepared in three steps from the nucleoside analog ganciclovir by trimethylsilyl-protection of the amino group, coupling with N-benzyloxycarbonyl-L-valine-N-carboxyanhydride, hydrolysis with hydrochloric acid and hydrogenolysis of the Cbz-protecting group. Valganciclovir is well absorbed and rapidly hydrolyzed to ganciclovir by intracellular esterases in the intestinal mucosal cells and by hepatic esterases. Unlike ganciclovir, valganciclovir was demonstrated to be actively transported by the intestinal peptide transporter PEPT1 in Caco-2 cells. As a consequence, its absolute bioavailability in human was 10-fold higher compared to ganciclovir (6%). In clinical trials, it was shown that a twice-daily 900 mg dose of valganciclovir resulted in similar systemic ganciclovir exposure to 5 mg/kg twice-daily intravenous injection of ganciclovir. Valganciclovir concentrations could not be quantified in most patients within three to four hours. In a randomized non-blind phase III clinical trial, oral valganciclovir (900 mg twice daily for three weeks then 900 mg once daily for one week) was as effective as intravenous ganciclovir (5 mg/kg twice daily for three weeks then 5 mg/kg once daily). Oral treatment with valganciclovir avoided catheter-related infection that sometimes occurred with intravenous ganciclovir.

References

1. The collection of new therapeutic entities first launched in 2001 originated from the following sources : (a) CIPSLINE, Prous database; (b) Pharmaprojects; (c) IDdb, Current Drugs database; (d) Drug Topics, W.M. Davis and M.C. Vinson, "New Drug Approvals of 2000", Part 1, Feb. 5, 2001 and Part 2, Mar. 5, 2001.
2. B. Gaudillière, P. Bernardelli and P. Berna, Ann. Rep. Med. Chem., 36, 293 (2001).
3. B. Gaudillière and P. Berna, Ann. Rep. Med. Chem., 35, 331 (2000).
4. B. Gaudillière, Ann. Rep. Med. Chem., 34, 317 (1999).
5. P. Galatsis, Ann. Rep. Med. Chem., 33, 327 (1998).
6. J.K. Lee, M. K. Lee, Y-P. Yun, Y. Kim, J.S. Kim, Y.S. Kim, K. Kim, S.S. Han and C-K. Lee, Int. Immunopharmamacol., 1, 1275 (2001).
7. D. Lai, Curr. Opin. Anti-Infect. Invest. Drugs, 2, 332 (2000).
8. A. Djeraba and P. Quere, Int. J. Immunopharmacol., 22, 365 (2000).
9. G K. Darryl, Rec. Res. Dev. Immunol., 1(Pt. 1), 209 (1999).
10. R.W. Stuart, D.L. Lefkowitz, J.A. Lincoln, K. Howard, M.P. Gelderman and S.S. Lefkowitz, Int. J. Immunopharmacol., 19, 75 (1997).
11. W.A. Gahl, N. Eng. J. Med., 345, 55 (2001).
12. G.M. Pastores and R . Thadhani, The Lancet, 358, 601 (2001).
13. C.M. Eng, N. Guffon, W.R. Wilcox, D.P. Germain, P. Lee, S. Waldek, L. Caplan, G.E. Linthorst and R.J. Desnick, N. Eng. J. Med., 345, 9 (2001).
14. R. Schiffmann, G.J. Murray, D. Treco, P. Daniel, M. Sellos-Moura, M. Myers, J.M. Quirk, G.C. Zirzow, M. Borowski, K. Loveday, T. Anderson, F. Gillespie, K.L. Oliver, N.O. Jeffries, E. Doo, T.J. Liang, C. Kreps, K. Gunter, K. Frei, K. Crutchfield, R.F. Selden and R.O. Brady, Proc. Nat. Ac. Sci., 97, 365 (2000).
15. A. Frustaci, C. Chimenti, R. Ricci, L. Natale, M.A. Russo, M. Pieroni, C.M. Eng and R.J. Desnick, N. Eng. J. Med., 345, 55 (2001).
16. A. Ferrajoli, S. O'Brien and M.J. Keating, Exp. Opin. Biol. Ther., 1, 1059 (2001).
17. J.A. Smith, Cancer Practice, 9, 211 (2001).
18. J.M. Flynn and J.C. Byrd, Curr. Opin. Oncol., 12, 574 (2000).
19. M.J. Keating, K. Rai and I. Flinn, Blood, 94, 705a (1999).
20. R.E. Small, M.A. Wixted and W.N. Roberts, Formulary, 36, 191 (2001).
21. P. Peichl, Curr. Opin. Anti-Inflammatory Immunomodulatory Invest. Drugs, 2, 56 (2000).
22. B.D. Freeman and T.G. Buchman, Exp. Opin. Biol. Ther., 1, 301 (2001).
23. A. Bendele, M. Colloton, M. Vrkljan, J. Morris and K. Sabados, J. Lab. Clin. Med., 125, 493 (1995).
24. L.A. Sorbera, P.A. Leeson, X. Rabasseda and J. Castañer, Drugs Fut., 26, 433 (2001).
25. S. Gandolfi, S.T. Simmons, R. Sturm, K. Chen and A.M. Vandenburgh, Adv. Therapy, 18, 110 (2001).

26. L.B. Cantor, Expert Opin. Invest. Drugs, 10, 721 (2001).
27. R.F. Brubacker, E.O. Schoff, C.B. Nau, S.P. Carpenter, K. Chen and A.M. Vandenburgh, Am. J. Ophthalmol., 131, 19 (2001).
28. M. Dirks, H. Dubiner, D. Cooke et al., Invest. Ophthalmol. Vis. Sci., 41 : S514, Abstr 2737 (2000).
29. R.G. Tilton, T.A. Brock and R.A.F. Dixon, Expert Opin. Invest. Drugs, 10, 1291 (2001).
30. N.E. Mealy, M. Bagès and M. del Fresno, Drugs Fut., 26, 1149 (2001).
31. S. Roux, V. Breu, S.I. Ertel and M. Clozel, J. Mol. Med., 77, 364 (1999).
32. V. Breu, S.I. Ertel, S. Roux and M. Clozel, Expert Opin. Invest. Drugs, 7, 1173 (1998).
33. K. Münter and M. Kirchengast, Emerging Drugs, 6, 3 (2001).
34. K. Fattinger, C. Funk, M. Pantze, C. Weber, J. Reichen, B. Stieger and P.J. Meier, Clin. Pharmacol. Ther., 69, 223 (2001).
35. P.A. Hunter, Drug News Perspect., 14, 440 (2001).
36. R. A. Fromtling, Drug News Perspect., 14, 181 (2001).
37. P.A. Hunter, Drug News Perspect., 14, 309 (2001)
38. G.M. Keating and B. Jarvis, Drugs, 61, 1121 (2001).
39. T.R. Rogers, Curr. Opin. Critical Care, 7, 238 (2001).
40. J.F. Meis and P.E. Verweij, Drugs, 61, 13 (2001).
41. R.C. Dart, S.A. Seifert, L.V. Boyer, R.F. Clark, E. Hall, P. McKinney, J. McNally, C.S. Kitchens, S.C. Curry, G.M. Bodgan, S.B. Ward and R.S. Porter, Arch. Intern. Med., 161, 2030 (2001).
42. R.C. Dart and J. McNally, Ann. Emerg. Med., 37, 181 (2001).
43. S.A. Seifert and L.V. Boyer, Ann. Emerg. Med., 37, 189 (2001).
44. P. Consroe, N.B. Egen, F.E. Russell, K. Gerrish, D.C. Smith, A. Sidki and J.T. Landon, Am. J. Trop. Med. Hyg., 53, 507 (1995).
45. P. Norman, A. Dihlmann and X. Rabasseda, Drugs Today, 37, 215 (2001).
46. W. Marvin Davis and M.C. Vinson, Drug Topics, 61 (2002).
47. A. Graul, P.A. Leeson and J. Castañer, Drugs Fut., 25, 339 (2000).
48. A. Graul, P.A. Leeson and J. Castañer, Drugs Fut., 26, 404 (2001).
49. D.K. Agrawal, Expert Opin. Invest. Drugs, 10, 547 (2001).
50. R.S. Geha and E.O. Meltzer, J. Allergy Clin. Immunol., 107, 751 (2001).
51. A.S. Nayak and E. Schenkel, Allergy, 56, 1077 (2001).
52. K. McClellan and B. Jarvis, Drugs, 61, 789 (2001).
53. L.A. Sorbera, L. Martín and R.M. Castañer, Drugs Fut., 26, 440 (2001).
54. S. Krisnagopalan and R.P. Dellinger, BioDrugs, 15, 645 (2001).
55. G.R. Bernard, J.-L. Vincent, P.-F. Laterre, S.P.LaRosa, J.-F. Dhainaut, A. Lopez-Rodriguez, J.S. Steingrub, G.E. Garber, J.D. Helterbrand, E.W. Ely and C.J. Jr. Fisher, N. Engl. J. Med., 344, 699 (2001).
56. B.D. Freeman and T.G. Buchman, Expert Opin. Invest. Drugs, 11, 69 (2002).
57. S. Kanji, J.W. Devlin, K.A. Piekos and E. Racine, Pharmacotherapy, 21, 1389 (2001).
58. M. Nakashima, Jap. Nippon Byoin Yakuzaishikai Zasshi, 37, 1493 (2001).
59. Y. Suzuki and K. Umemura, Nippon Yakurigaku Zasshi, 116, 379 (2000).
60. T.-W. Wu, L.-H. Zeng, J. Wu and K.-P. Fung, Life Sci., 67, 2387 (2000).
61. H. Takamatsu, K. Kondo, Y. Ikeda and K. Umemura, Eur. J. Pharmacol., 362, 137 (1998).
62. H. Kawai, H. Nakai, M. Suga, S. Yuki, T. Watanabe and K.-I. Saito, J. Pharmacol. Exp. Ther., 281, 921 (1997).
63. A. Graul and J. Castañer, Drugs Fut., 21, 1014 (1996).
64. P. Cole, and X. Rabasseda, Drugs Today, 37, 159 (2001).
65. A.I. Graul, Drug News Perspect., 15, 30 (2002).
66. J. Ngo, X. Rabasseda and J. Castañer, Drugs Fut., 25, 306 (2000).
67. P. Tfelt-Hansen, P. De Vries and P.R. Saxena, Drugs, 60, 1259 (2000).
68. J. Schoenen, Curr.Opin. Neurology, 10, 237 (1997).
69. H.C. Diener and A. McHarg, Int. J. Clin. Practice, 54, 670 (2000).
70. M.J. Gawel, Can. J. Clin. Pharmacol., 6 suplA, 20A (1999).
71. H.C. Diener, H. Kaube and V. Limmroth, J. Neurology, 246, 515 (1999).
72. Y. Imanishi, M. Inaba, H. Seki, H. Koyama, Y. Nishizawa, H. Morii and S. Otani, J. Steroid Biochem. Mol. Biol., 70, 243 (1999).
73. T. Akiba, F. Marumo, A. Owada, S. Kurihara, A. Inoue, Y. Chida, R. Ando, T. Shinoda; Y. Ishida and Y. Ohashi, Am. J. Kidney Dis., 32, 238 (1998).

74.   S. Komuro, M. Sato, H. Kanamaru, H. Kanako, I. Nakatsuka and A. Yoshitake, Xenobiotica, 29, 603 (1999).
75.   A.Graul and J. Castañer, Drugs Fut., 22, 473 (1997).
76.   T. Wroblewski, P. Leeson and J. Castañer, Drugs Fut., 23, 374 (1998).
77.   K. Takahashi, H. Kai, M. Otsuka, H. Mizuno, T. Koda and T. Miyata, Env. Tox. Pharm., 10, 89 (2001).
78.   K. Takahashi, M. Ishikawa, H. Kohya, H. Mizuno, H. Ohno, Y. Isohama, K. Takahama and T. Miyata, Jap. J. Pharmacol., 61(Suppl. 1), Abst P-194 (1993).
79.   K. Takahashi, H. Mizuno, H. Ohno, H. Kai, K. Takahama and T. Miyata, Jap. J. Pharmacol., 58(Suppl. 1), Abst O-54 (1992).
80.   P.M. Fernandez, Drugs Today, 37, 485 (2001).
81.   B.J. Druker, M. Talpaz, D.J. Resta, B. Peng, E. Buchdunger, J.M. Ford, N.B. Lydon, H. Kantarjian, R. Capdeville, S. Ohno-Jones and C.L. Sawyers, N. Engl. J. Med., 344, 1031 (2001).
82.   M. Habeck, Lancet, 3, 6 (2002).
83.   F. de Bree, L.A. Sorbera, R. Fernandez and J. Castañer, Drugs Fut., 26, 545 (2001).
84.   H. Joensuu and S. Dimitrijevic, Annals of Medecine, 33, 451 (2001).
85.   M. Levin, Drugs Today 37, 823 (2001).
86.   K. Lyseng-Williamson and B. Jarvis, Drugs, 61, 1765 (2001).
87.   A. I. Graul, Drugs News Perspect., 15, 38 (2002).
88.   L.M. Salmun, Expert Opin. Invest. Drugs, 11, 259 (2002).
89.   E. Baltes, R. Coupez, H. Giezek, G. Voss, C. Meyerhoff and M.S. Benedetti, Fundamental & Clinical Pharmacology, 15, 269 (2001).
90.   C.J. Opalka, T.E. D'Ambra, J.J. Faccone, G. Bodson and E. Cossement, Synthesis, 766 (1995).
91.   D.Y. Wang, F. Hanotte, C. De Vos and P. Clement, Allergy, 56, 339 (2001).
92.   J.L. Devalia, C. De Vos, F. Hanotte and E. Baltes, Allergy, 56, 50 (2001).
93.   J.-P. Tillement, Allergy, 50, 12 (1995).
94.   M.S. Benedetti, M. Plisnier, J. Kaise, L. Maier, E. Baltes, C. Arendt and N. McCracken, Eur. J. Clin. Pharmacol., 57, 571 (2001).
95.   R.E. Hobbs, R.M. Mills and J.B. Young, Expert Opin. Invest. Drugs, 10, 935 (2001).
96.   R.E. Hobbs and R.M. Mills, Expert Opin. Invest. Drugs, 8, 1063 (1999).
97.   W.S. Colucci, U. Elkayam, D.P. Horton, W.T. Abraham, R.C. Bourge, A.D. Johnson, L.E. Wagonier, M.M. Givertz, C.-S. Liang, M. Neibaur, H. Haught and T.H. LeJemtel, N. Engl. J. Med., 343, 246 (2000).
98.   G.C. Fonarow, Rev. Cardiov. Med., 2, S32 (2001).
99.   M.R. Burger and A.J. Burger, Curr. Opin. Invest. Drugs, 2, 929 (2001).
100.   S.D. Cook, S.L. Salkeld, L.P. Patron and D.C. Rueger, Biomaterials Engineering and Devices, 1, 267 (2000).
101.   E. H. J. Groeneveld and E.H. Burger, Eur. J. Endocrinol., 142, 9 (2000).
102.   P. Kellokumpu-Lehtinen and T. Tulijoki, Bone Morphogenet. Proteins, 47 (1996).
103.   U. Ripamonti and A.H. Reddi, Critical Reviews in Oral Biology and Medicine, 8, 154 (1997).
104.   S.D. Cook and D.C. Rueger, Clinical Orthopaedics and Related Research, 324, 29 (1996).
105.   S.V. Onrust and K. McClellan, CNS Drugs, 15, 329 (2001).
106.   I. Kusumi, Y. Takahashi, K. Suzuki, K. Kameda and T. Koyama, Jap. J. Neural Transm., 107, 295 (2000).
107.   H. Sakamoto, K. Matsumoto, Y. Ohno and M. Nakamura, Pharmacol. Biochem. Behav., 60, 873 (1998).
108.   Y. Ohno, K. Ishida-Tokuda, T. Ishibashi, H. Sakamoto, R. Tagashira, T. Horisawa, K. Yabuuti, K. Matsumoto, A. Kawabe and M. Nakamura, Pol. J. Pharmacol., 49, 213 (1997).
109.   Anonymous, Drugs Fut., 16, 122 (1991).
110.   M. Camilleri, Aliment. Pharmacol. Ther., 15, 277 (2001).
111.   F. De Ponti and M. Tonini, Drugs, 61, 317 (2001).
112.   P. Norman, Curr. Opin. Cent. Peripher. Nerv. Syst. Invest. Drugs, 2, 344 (2000).
113.   L. Scott and C. M. Perry, Drugs, 58, 491 (1999).
114.   A. Graul, J. Silvestre and J. Castañer, Drugs Fut., 24, 38 (1999).
115.   K.-H. Buchheit, R. Gamse, R. Giger, D. Hoyer, F. Klein, E. Kloeppner, H.-J. Pfannkuche and H. Mattes, J. Med. Chem., 38, 2331 (1995).
116.   L.A. Sorbera and J. Castañer, Drugs Fut., 26, 1217 (2001).
117.   L.A. Sorbera and J. Castañer, Drugs Fut., 25, 1309 (2001).

118.   L.A. Sorbera and J. Castañer, Drugs Fut., <u>23</u>, 1279 (1998).
119.   D. Gilden, AIDS treatment News, <u>364</u>, 2 (2001).
120.   T. Cihlar, G. Birkus, D.E. Greenwalt and M.J.M. Hitchcock, Antivir. Res., <u>54</u>, 37 (2002).
121.   M.D. Miller, N.A. Margot, K. Hertogs, B. Larder and V. Miller, Nucleosides, Nucleotides & Nucleic acids, <u>20</u>, 1025 (2001)
122.   P. Barditch-Crovo, S.G. Deeks, A. Collier, S. Safrin, D.F. Coakley, M. Miller, B.P. Kearney, R.L. Coleman, P.D. Lamy, J.O. Kahn, I. McGowan and P.S. Lietman, Antimicrob. Agents Chemother., <u>45</u>, 2733 (2001).
123.   Y.-Q. Xiong and T.P. Le, Drugs Today, <u>37</u>, 617 (2001).
124.   H.M. Yassin and L.L. Dever, Expert Opin. Invest. Drugs, <u>10</u>, 353 (2001).
125.   J.A. Barman Balfour and D.P. Figgitt, Drugs, <u>61</u>, 815 (2001).
126.   A. Bryskier, Clin. Microbiol. Infect., <u>6</u>, 661 (2000).
127.   D. Felmingham, Clin. Microbiol. Infect., <u>7(S3)</u>, 2 (2001).
128.   S. Douthwaite, Clin. Microbiol. Infect., <u>7(S3)</u>, 11 (2001).
129.   G. Labbe, M. Flor and B. Lenfant, J. Antimicrob. Chemother. , <u>44(SA)</u>, 61 (1999).
130.   L.A. Sorbera and J. Castañer, Drugs Fut., <u>25</u>, 41 (2000).
131.   P.A. Netland, T. Landry, E.K. Sullivan, R. Andrew, L. Silver, A. Weiner, S. Mallick, J. Dickerson, M.V. Bergamini, S.M. Robertson and A.A. Davis, Am. J. Ophthalm., <u>132</u>, 472 (2001).
132.   T.R. Dean, G.E. Barnes, B. Li and M.L. Chandler, Invest. Ophthalmol. Visual Sci., <u>40,</u> Abstr 2688 (1999).
133.   S. Robertson and L.H. Silver, 12th Congr. Eur. Soc. Ophthalmol., Abstr FP 153 (1999).
134.   R. Garadi, L. Silver, T. Landry and F.D. Turner, Invest. Ophthalmol. Visual Sci., <u>40,</u> Abstr 4378 (1999).
135.   L.A. Sorbera, P.A. Leeson and J. Castañer, Drugs Fut., <u>25</u>, 465 (2000).
136.   L.A. Sorbera, P.A. Leeson and J. Castañer, Drugs Fut., <u>26</u>, 515 (2001).
137.   A.I. Graul, Drug News Perspect., <u>14</u>, 19 (2000).
138.   S.G. Lundeen, Z. Zhang, Y. Zhu, J.M. Carver and R.C. Winneker, J. Steroid Biochem. Mol. Biol., <u>78</u>, 137 (2001).
139.   Z. Zhang, S.G. Lundeen, Y. Zhu, J.M. Carver and R.C. Winneker, Steroids, <u>65</u>, 637 (2000).
140.   A.I. Graul, L.A. Sorbera and J.R. Prous, Drugs Today, <u>37</u>, 703 (2001).
141.   Y. Bouali, M. Gaillard-Kelly and P.J. Marie, Gynecological Endocrinology, <u>15</u>, 48 (2001).
142.   D. Philibert, F. Bouchoux, M. Degryse, D. Lecaque, F. Petit, M. Gaillard, Gynecological Endocrinology, <u>13</u>, 316 (1999).
143.   P. Reusser, Expert Opin. Invest. Drugs, <u>10</u>, 1745 (2001).
144.   M. Curran and S. Noble, Drugs, <u>61</u>, 1145 (2001).
145.   M. Sugawara, W. Huang, Y.-J. Fei, F.H. Leibach, V. Ganapathy and M.E. Ganapathy, J. Pharm. Sci., <u>89</u>, 781 (2000).
146.   L.A. Sorbera, R. Castañer and J. Castañer, Drugs Fut., <u>25</u>, 474 (2000).
147.   H. Piper, T.A. Ciulla, R.P. Danis and L.M. Pratt, Exp. Opin. Pharmacother., <u>1</u>, 1343 (2001).

Chapter 27. Biosimulation: Dynamic Modeling of Biological Systems

Kevin Hall, Rebecca Baillie, and Seth Michelson
Entelos Inc.
4040 Campbell Avenue, Menlo Park, CA 94025

Introduction - The completion of the human genome project and the rapid development of proteomics promises to provide the complete list of ingredients for human life. But genomic and proteomic information alone cannot provide an understanding of human physiology. As an analogy, the parts list for an automobile does not elucidate its integrated function, much less its performance characteristics. Rather, in order to understand biological function it is necessary to unravel the complex interactions between the parts and understand their qualitative and quantitative dynamics as an integrated whole. The field of biosimulation directly addresses this challenge.

Biosimulation uses the language of mathematics to model biological components and their interactions. The mathematical equations are often highly nonlinear and too complex to solve analytically, so numerical algorithms for approximating the solution are implemented on a computer. Recent improvements in computing power, along with the accumulation of detailed biological knowledge, have made it possible to realistically simulate and predict the behavior of complex biological systems in both health and disease.

In a recent review, Ho assessed the applicability of biosimulation to each phase of the pharmaceutical research and development process (1). In the discovery and pre-clinical phases, biosimulation can be applied to finding and validating novel targets, pathways, and biomarkers, as well as assessing whether or not these discoveries are pertinent to human physiology. In the clinical phase, biosimulation can be applied to clinical trial design and the identification of predictive biomarkers. In the approval phase, biosimulation can be used to explain data variability and suggest optimal therapeutic strategies. Noble *et al.* echo Ho's evaluation and add that biosimulation can also be used to predict drug safety (2, 3).

This review highlights the recently published literature on biosimulation of systems at multiple levels of organization – from sub-cellular to multi-organ physiological systems. Biosimulation success stories as well as the challenges facing this promising new field are presented.

CELLULAR AND SUB-CELLULAR BIOSIMULATION

Cell Metabolism - The ubiquitous display of biochemical pathway charts in countless laboratories testifies that cellular metabolism has been extensively studied. However, the sheer complexity makes it difficult to predict the dynamics of cellular metabolism under different experimental circumstances. In other words, whereas the pathway diagram provides a static representation of the biochemical reactions, a dynamic, quantitative representation is needed to understand how cellular metabolism is regulated under varying environmental circumstances.

Given the biochemical reactions, it is straightforward to use the law of mass action to generate a set of ordinary differential equations that model the reaction dynamics (4). Often, the equations can be simplified using conservation principles, thermodynamics, and knowledge of relative reaction rates. Such simplifications give rise to typical biochemical reaction kinetics such as Michaelis-Menton kinetics.

In order to simulate the metabolic dynamics, it is necessary to specify the values of the kinetic parameters. Unfortunately, *in vivo* parameter values are generally unknown and accurate parameter estimation is a fundamental challenge for the field of biosimulation. For biosimulation of cellular metabolism, the usual procedure is to use kinetic parameter values measured *in vitro,* where the enzymes under investigation may be extracted from a variety of systems. Whether or not the parameter values are applicable to a particular *in vivo* system is the subject of much debate, especially considering issues like cellular compartmentalization, high intracellular protein content, and the possibility of unknown *in vivo* regulators.

A recent biosimulation paper by Teusink *et al.* directly addressed the applicability of *in vitro* kinetic parameters to *in vivo* systems by measuring glycolytic enzyme kinetics from yeast extracts (5). The authors applied the *in vitro* parameter measurements to a model of *in vivo* yeast glycolysis and compared the simulation results with experiments in an *in vivo* yeast system. Since the yeast glycolytic pathway is well understood, the authors were able to formulate a comprehensive model of the biochemical reaction kinetics. Thus, discrepancies between observed and simulated behavior were assumed to be a measure of the applicability of the *in vitro* parameter values to the *in vivo* system.

The kinetic parameters measured *in vitro* were unable to reproduce the metabolite concentrations and fluxes measured *in vivo*, with more than half of the fluxes differing by more than a factor of two. The authors varied the kinetic parameter values in their model to better understand the discrepancy between the *in vitro* measurements and their effective *in vivo* values. The kinetics of all enzymes required some degree of adjustment to accurately simulate *in vivo* behavior. For example, alcohol dehydrogenase was observed to be nine times more active *in vitro* than it appeared to be *in vivo*. The authors used biosimulation to suggest various potential *in vivo* regulatory mechanisms to explain the discrepancies, but further experiments are required to test these suggestions.

While several groups are developing methods for finding optimal kinetic parameters to fit *in vivo* experimental data (6, 7), Edwards *et al.* simulated the *E-coli* metabolic network using a method that does not involve kinetic rate equations (8-11). Rather, the authors used Flux Balance Analysis (FBA) to simulate pathway flux distributions given knowledge of the metabolic network, the stoichiometric coefficients for each enzyme, and exchange fluxes under experimental control (e.g., nutrient uptake rates). This information specifies a set of ordinary differential equations where the rate of change of each metabolite is defined by the difference between fluxes creating and utilizing the metabolite. Traditional biochemical modeling further specifies the fluxes as nonlinear functions of the metabolites and enzyme kinetic parameters in order to compute the metabolite concentrations as a function of time. But FBA does not compute metabolite concentrations. Rather, FBA computes the balanced steady state metabolic fluxes such that the metabolite concentrations do not change.

Under the steady state assumption, the differential equations are reduced to a set of linear algebraic equations for the metabolic fluxes. Since the number of fluxes is greater than the number of metabolites, the linear system of FBA equations is underdetermined. Thus, the metabolic fluxes can only be specified as a function of several parameters. Fortunately, the parameter space is constrained by thermodynamic considerations (e.g., knowledge of reaction reversibility or irreversibility) and flux capacities.

In order to solve for the steady state metabolic fluxes within the specified constraints, Edwards *et al.* used linear programming techniques to find the parameter values that maximized the fluxes generating biomass (8-10). The underlying assumption was that *E. coli* has evolved its metabolic parameters to maximize growth. The authors showed that for various experimental manipulations of the acetate and succinate exchange fluxes, measured *E. coli* metabolic fluxes did indeed correspond to the computed metabolic fluxes optimized for maximal growth (9). Furthermore, FBA correctly predicted metabolic shifts under different experimental conditions (10).

FBA has also been used to predict the growth characteristics of mutant strains of *E. coli*. Enzyme gene deletions were simulated by setting the corresponding metabolic flux to zero (8, 11). Whereas the FBA assumption of evolution towards optimal growth is reasonable for wild type *E. coli*, it does not apply to mutants. Rather, FBA generates a best-case scenario for mutant growth that is probably not realized experimentally. The likely sub-optimal growth of mutants prohibits using FBA to quantitatively predict mutant metabolic flux distributions. However, FBA can be used to determine if the optimal biomass flux of a mutant is greater than a minimum threshold required for growth. In 68 of 79 cases, FBA correctly predicted whether or not the mutant strains could grow under various experimental conditions, with most mutants being capable of growth (8). These results suggested that *E. coli* has a robust metabolic network with a large number of metabolically nonessential genes. Nevertheless, the few essential genes identified by FBA may correspond to anti-microbial drug targets.

While the steady state assumption of FBA is reasonable for microbes in a constant environment, application of FBA to higher organisms may be complicated by frequent variations in both substrate delivery and external stimuli. For example, rapid and substantial changes in metabolic fluxes and metabolite concentrations have been observed during feeding and physical activity in animals and humans. In such cases, dynamic models of metabolic regulation are required.

Naturally Occurring Gene Networks - Making mechanistic sense of the vast amount of genetic microarray data requires understanding the functional interactions between genes. To begin to quantitatively understand how naturally occurring gene networks behave, several realistic models have been constructed to simulate cell cycle control, regulation of the tryptophan operon, the lysis-lysogeny switch of phage lambda, phage T7 gene expression, and the spatial gene expression pattern observed in early *Drosophila* development (12-19).

There are two main challenges for biosimulation of gene networks. The first challenge is to recognize conditions when stochastic or random effects are important. Most gene network biosimulations use the paradigm of biochemical reaction kinetics modeled by deterministic differential equations. While this representation may be appropriate for some systems, it has been recognized that stochastic kinetics can produce significant variations in gene expression, especially considering the typically low concentrations of reactants (14, 15). However, simulating stochastic kinetics comes at a great computational cost and it may be necessary to create more efficient numerical algorithms for simulating large stochastic genetic regulatory networks.

The second challenge arises because all organisms presently have a significant fraction of genes with unknown biological function. Therefore, it is difficult to determine whether erroneous biosimulation results are due to the use of incorrect *in vivo* kinetic parameters or because an important unknown gene function is missing from the simulation. This difficulty was illustrated by von Dassow *et al.* in their attempt to simulate the *Drosophila* segment polarity network (19). In spite of an extensive numerical search for appropriate kinetic parameter values, the authors were initially unable to accurately simulate the experimentally observed gene expression pattern. The authors realized that their gene network model was missing fundamental, but as yet unknown, interactions. Interestingly, once those interactions were identified by the authors and included in the gene network model, the simulation results were accurate and robust for a wide range of kinetic parameters. This means that the connectivity of the gene network was the primary determinant of the system's behavior and that accurate values of the kinetic parameters were of secondary importance. The authors speculated that genetic regulatory networks have evolved their connectivity to become robust to alterations in kinetic parameters. From a drug discovery perspective, this is discouraging news since it implies that gene network behavior may be unaffected by pharmacologic modulation of kinetic parameters. Nevertheless, biosimulation can be used to predict the sensitivity of the network to parametric modulation of potential drug targets.

Synthetic Gene Networks - In contrast to simulating naturally occurring genetic regulatory circuits, several researchers are using biosimulation to help design synthetic gene networks with prescribed dynamical properties (20, 21). Such synthetic gene networks might someday be used to interact with naturally occurring gene networks and thereby modify overall cellular behavior.

Gardner et al. sought to construct a genetic toggle switch that could be turned on and off with a transient chemical stimulus or temperature change (20). They proposed a synthetic genetic regulatory circuit consisting of two promoters arranged in a mutually inhibitory network. The authors simulated this simple system as a pair of ordinary differential equations describing the rates of change of the repressor proteins as a nonlinear function of each other's concentration. The parameters in the equations correspond to each repressor's net synthesis rate and the kinetics of their mutual inhibitory effect on each other's transcription. Using biosimulation techniques, the authors designed particular biological components with kinetics that generate a system with two stable steady states corresponding to the on and off positions of the switch. The stable steady states were separated by an unstable steady state that acts as a switch threshold. Gardner et al. showed that when the gene circuit was expressed in E. coli, the network indeed possessed the desired switching property predicted by the biosimulation.

In a similar study, Elowitz et al. constructed a synthetic oscillatory genetic network consisting of three promoters for cyclic inhibitory repressors (21). The authors formulated a set of six coupled ordinary differential equations describing the rates of change of the three mRNA levels as well as the repressor protein concentrations. Like Gardner et al., these authors used biosimulation techniques to design network components with kinetic parameter values causing the desired property of oscillation. Recognizing the potential importance of stochastic effects in gene regulation, the authors formulated a stochastic version of their differential equations and accurately predicted the somewhat noisy oscillation observed experimentally when the network was expressed in E. coli.

Signal Transduction - Cells make sense of, and respond to, their environment via complex signal transduction mechanisms. Intensive drug discovery efforts are focused on creating therapeutic compounds that interact with signal transduction pathways involved in disease. While a great deal is known about the qualitative mechanisms of signal transduction, it is remains unclear how cells quantitatively sense and integrate various cues from their environment. Biosimulation can help researchers better understand the quantitative properties of signal transduction and the potential effects of pharmacologic modulation of these pathways.

For example, one of the most well understood signal transduction networks allows bacteria to sense nutrient gradients in order to direct their average motion towards the source of the nutrient, a phenomenon called chemotaxis. But how does a bacterium accomplish this complex task, especially since it is too small to directly detect a typical spatial chemical gradient? The answer lies in how bacteria move: a sequence of smooth movements interrupted by tumbles that randomly determine the direction of the next smooth movement (22). The frequency of tumbles is responsive to temporal changes in chemoattractant concentration such that tumbling frequency decreases as chemoattractant concentration increases. This creates a biased random walk that, on average, moves the bacterium towards the chemoattractant source. Interestingly, this behavior is independent of the initial chemoattractant concentration over several orders of magnitude, a phenomenon called perfect adaptation (23). Although the signal transduction network that modulates tumbling frequency has been well described, researchers were unclear as to how this network exhibits the property of perfect adaptation.

A pair of recent biosimulation papers directly addressed this mystery (24, 25). Barkai and Leibler created a mathematical model of the chemotaxis receptor complex in *E. coli* (24). By modeling the transitions between receptor phosphorylation and methylation states, the authors showed that perfect adaptation is a signal transduction network property that did not require fine-tuning of kinetic parameters. In other words, following a transient response to stepwise changes in chemoattractant concentration, the baseline receptor activity was restored for a wide range of kinetic parameters. Yi *et al.* showed that this robust network property could be interpreted as a classical integral controller, a concept familiar to engineers (25). Thus, biosimulation methods elucidated the mechanism of perfect adaptation and suggest a generic strategy that cells can use to perform integral control. With this information it may be possible to recognize integral control modules in diverse cell signaling networks (26, 27). Since integral control is robust for a wide range of kinetic parameters, drug targets aimed at modulating kinetics of cellular integral control modules are unlikely to be effective.

The above simulations of chemotaxis receptor activity used deterministic ordinary differential equations to model kinetic transitions between receptor complex states. This approach accurately simulated the ensemble average of bacteria chemotactic behavior. To simulate chemotaxis signaling of an individual bacterium, Morton-Firth *et al.* argued that the small numbers of molecules and inherent thermal fluctuations require methods that incorporate stochastic effects (28, 29). These authors created a novel methodology for simulating stochastic systems that keeps track of the number of molecules and their respective conformational states, and simulates the transitions between states using transition probabilities derived from kinetic parameters. Furthermore, because their simulations were aimed at the molecular level, the authors could compute free energies of the various molecular states based on rate constants and binding coefficients. Their molecular-level representation of the Barkai and Leibler model also exhibited perfect adaptation.

Cell Locomotion - The above biosimulations of bacterial chemotactic signal transduction do not attempt to explain how a bacterium ultimately ends up moving. In other words, how is the chemical energy inside an organism converted to the mechanical energy necessary for locomotion? While qualitative descriptions of locomotion abound, the typically large number of proteins involved in locomotion and the intricacy of their regulatory processes complicate any quantitative analysis. Bottino *et al.* recently constructed a quantitative biosimulation of the less typical, but simpler, crawling mechanism of nematode sperm (30).

Crawling cells use their cytoskeleton to protrude a part of the cell that adheres to the crawling surface. The cell body is then pulled forward by contraction. Whereas most cells use a complicated actin-based machinery, nematode sperm use a simple system of polymerization of major sperm protein (MSP) to generate a cellular protrusion. Depolymerization of MSP in the cell body generates contraction (31). The biosimulation of Bottino *et al.* used a two-dimensional, finite element representation of the MSP cytoskeleton and incorporated an intracellular pH gradient to modulate MSP filament polymerization and depolymerization.

A unique insight of the biosimulation was that the cell contraction mechanism is mechanistically related to protrusion. Using their biosimulation, the authors recognized that the energy for crawling is generated by polymerization and bundling of MSP at the front of the sperm where higher pH and other factors cause the protein to form a fibrous gel thereby storing elastic energy. Formation of the gel pushes a portion of the cell membrane forward and adhesion proteins attach the protrusion to the crawling surface. As polymerization continues at the front of the sperm, the previously formed fibrous MSP gel is displaced towards the back of the sperm where the decreased pH depolymerizes the MSP. Depolymerization releases the stored elastic energy in the gel causing contraction thereby pulling the cell body forward.

The model of Bottino *et al.* accurately simulated both the motion and shape of the crawling sperm under a variety of experimental conditions. While the MSP-based crawling is much simpler than the more typical actin-based crawling, the nematode sperm biosimulation provides a quantitative basis for understanding the locomotion of crawling cells.

DISEASE BIOSIMULATION

Cardiac Arrhythmia - Biosimulation of cardiac electrophysiology has a long history and has advanced in close collaboration with experiments. Realistic differential equation models have been constructed for almost every type of cardiac cell, incorporating a host of ion channels, exchangers, pumps, and dynamically regulated intracellular pools of sequestered calcium (32-38).

Recently, biosimulation has been used to better understand the cause of a cardiac arrhythmia, called long-QT syndrome, known to be associated with a mutant sodium channel. By incorporating the electrophysiologic properties of the mutant channel in a biosimulation of a ventricular cell, Clancy and Rudy were able to demonstrate how the mutant channel caused prolonged action potentials and dangerous changes in the cardiac electrical activity called early after-depolarizations (39).

In a similar study, Winslow *et al.* simulated the electrophysiologic consequences of heart failure by modeling the observed differential gene expression of ion channels, pumps, and exchangers (40). The simulated diseased action potentials and calcium dynamics accurately matched experimental observations of failing heart cells. Furthermore, the authors were able to determine the relative contribution of each of the altered protein levels to the observed electrophysiologic behavior.

While the above cellular simulations provide insight into cardiac electrophysiology in both health and disease, most life-threatening cardiac arrhythmias occur because of disrupted spatial propagation of the cardiac action potential. To better understand how electrical activity propagates throughout the heart, models of cellular electrophysiology have been coupled together in realistic spatial geometries. For example, Harrild and Henriquez recently simulated the upper chambers of the heart, called the atria (41). The authors created a realistic, three-dimensional, finite element representation of the human atria that included the complex geometry of major muscle bundles, pectinate muscles, and vessel openings. An electrophysiology model of a human atrial cell (32) was implemented on this complex scaffold. The biosimulation accurately simulated the anisotropic electrical propagation of the normal atrial activation pattern as well as activation patterns during electrical pacing. Realistic three-dimensional biosimulations of electrical propagation in the lower chambers of the heart, the ventricles, have also been constructed (42, 43). The eventual goal is to use such realistic three-dimensional organ models to better understand how alterations in cellular electrophysiology, due to disease or pharmacologic intervention, impact the spatial propagation of electrical activity. Hall *et al.* have also used biosimulation of spatial cardiac electrical propagation to test novel arrhythmia diagnostics and algorithms for locating abnormal sources of cardiac electrical activity (44, 45).

<u>Type 2 Diabetes</u> - The regulation of blood glucose by insulin is a classical endocrine control process. Glucose levels must be maintained within a narrow range in order to avoid metabolic complications. Thus, a complex glucose-insulin control mechanism has evolved to include multiple negative feedback loops spanning several organs. An increased glucose level causes the pancreas to increase insulin secretion. Elevated insulin acts to both increase glucose utilization (in tissues like muscle and fat) as well as decrease glucose production by the liver, thereby returning glucose levels to normal. In type 2 diabetes, these control mechanisms become defective and the net result is a dangerously high level of blood glucose.

A recent study by Topp *et al.* addressed the question of how normal glucose-insulin regulation might progress over months and years towards the dysregulation observed in type 2 diabetes (46). The authors simulated alterations in pancreatic function that might arise as a nonlinear function of the blood glucose level. In particular they hypothesized that long-term pancreatic function first increases, reaches a maximum, and then decreases with increasing glucose. The decreasing phase of pancreatic function was meant to reflect a cytotoxic effect of elevated glucose.

Topp *et al.* showed that their model has three steady states: one stable steady state with normal glucose and insulin levels, another stable steady state with a high glucose level and no insulin, and a third unstable steady state with intermediate glucose and insulin levels. The authors simulated three different pathways to diabetes. The first pathway occurred as a result of a defect in the long-term pancreatic adaptation to glucose such that a slightly higher glucose level was regulated. This pathway corresponded to a shift of the normal steady state. The second pathway involved loss of glucose regulation resulting from killing the pancreatic cells that secrete insulin. This pathway corresponded to changing the model parameters such that the normal steady state and the unstable steady state disappeared (*via* a process called a saddle-node bifurcation), leaving only the stable steady state with elevated glucose and no insulin. The third pathway to diabetes involved driving the system from the normal state to the other side of the unstable steady state. In this case, the unstable steady state acted like a switch threshold and the system migrated to the stable steady state with high glucose and no insulin.

The biosimulation of Topp *et al.* was the first to address the issue of longitudinal progression of type 2 diabetes. While their analysis is intriguing, the biosimulation did not include the interplay between glucose and fat metabolism and the overall state of energy balance, factors believed to play a significant role in the development of the disease. Since their model was aimed at long time scales and incorporated aggregate representations of physiological processes, it was not applicable to the simulation of short-term experimental protocols that are used to assess various diabetic defects.

Hall *et al.* constructed a comprehensive and quantitative biosimulation of the major physiologic systems involved in human metabolism and the defects responsible for type 2 diabetes (47). The authors first simulated the normal processes of digestion, absorption, storage, and oxidation of carbohydrate, fat, and protein, as well as the hormonal regulation of these processes. To simulate type 2 diabetes, the authors adjusted several model parameters in order to simulate diabetic defects in pancreatic cell function as well as defective insulin sensitivity of various tissues, including pancreas, muscle, adipose, and liver. The parametric defects were implemented in various combinations and degrees to simulate a heterogeneous population of "virtual patients" ranging from insulin resistant to severe diabetic. Simulation results were validated with *in vivo* measurements of normal and diabetic responses to various mixed meals, glucose tolerance tests, glucose-insulin clamps, and hormone infusions. This comprehensive biosimulation elucidated the relative contribution of each defect in the pathophysiology of type 2 diabetes and could be used to predict the *in vivo* consequences of modulating putative therapeutic targets.

Asthma – Allergic asthma is a chronic inflammatory disease of the lower airways characterized by airway obstruction and hyperresponsiveness. These characteristics of asthma are the result of complex interactions between airway tissues, inflammatory cells, chemical mediators, and the adaptive immune response. To better understand these interactions and how they contribute to the pathophysiology of asthma, Stokes *et al.* created a biosimulation of asthma that encompasses airway physiology and the inflammatory effector system (48). Simulations with this model reproduced the acute and chronic characteristics of established mild allergic asthma including both early- and late-phase airway obstruction following antigen challenge (Figure 1), airway hyperresponsiveness, and chronic eosinophilic inflammation, as well as many other factors. The biosimulation also exhibited characteristic responses to known therapeutics, including beta 2-agonists, glucocorticosteroids, leukotriene antagonists, and IgE antagonists. The asthma biosimulation has been used to investigate current questions in the field of asthma.

For example, the mechanisms driving the hallmark airway hyperresponsiveness to nonspecific stimuli are not well defined. To examine how changes in various cell and tissue characteristics contribute to airway hyperresponsiveness, Klinke *et al.* varied parameters specifying airway structure and function in the asthma biosimulation (49). They found that changes in airway smooth muscle contractility play a dominant role in airway hyperresponsiveness. Additionally, the airway hyperresponsiveness observed in asthmatic patients could be explained by cytokines secreted by T cells that modulate smooth muscle contractility. This result highlights the indirect synergistic effects of cytokines on smooth muscle contractility.

Lewis *et al.* applied the asthma biosimulation to study how polymorphic differences in the beta 2-adrenoceptor affect clinical response to beta 2-adrenoceptor agonist therapy (50). A polymorphism of the beta 2-adrenoceptor is associated with different degrees of agonist-induced downregulation. Counterintuitively, patients with reduced asthma control with regularly scheduled beta 2-agonist therapy often express the polymorphism that does not downregulate in response to an agonist to the same degree as patients who remain responsive. It was hypothesized that patients who remained responsive to chronic beta 2-agonist therapy would have already maximally downregulated beta 2-adrenoceptors prior to therapy due to the presence of endogenous catecholamines (51). Lewis *et al.* used the asthma biosimulation to examine beta 2-adrenoceptor kinetics and demonstrated that the proposed hypothesis was inconsistent with known receptor dynamics and could not explain the observed clinical response.

Perhaps the most compelling result of the asthma biosimulation was the prediction that a proposed therapy in clinical trials, an interleukin-5 (IL-5) antagonist, would not be effective for human asthma (52). A hallmark characteristic of asthma is eosinophilic inflammation, and IL-5 is a critical cytokine driving eosinophil differentiation, activation, and maturation. Based on animal studies, anti-IL-5 therapy should eliminate eosinophil infiltration into the airway and thereby reduce airway obstruction. Surprisingly, the asthma biosimulation predicted that anti-IL-5 therapy would have little effect on airflow obstruction (Figure 1). Although anti-IL-5 effectively reduced eosinophil number, various other resident and infiltrating airway cells continued to cause significant airway obstruction. Thus, the biosimulation predicted that the anti-IL-5 therapy would not be an effective therapy because of significant redundancy in the system. This prediction was confirmed by the results from an anti-IL-5 clinical trial (53). The above simulation results demonstrate that the asthma biosimulation has been effectively used to investigate the mechanisms underlying asthma pathophysiology and predict clinical outcomes of potential therapeutics.

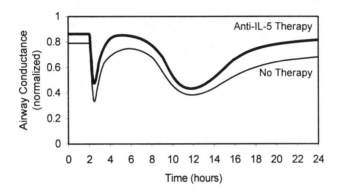

Figure 1. Biosimulation of the asthma response following an antigen challenge given at 2 hours. The airway conductance is normalized with respect to a non-asthmatic control. The asthma biosimulation reproduced the hallmark biphasic response to the antigen challenge, as well as the chronic impairment of airway conductance. Simulation of anti-IL-5 therapy predicted an insignificant improvement of the airway conductance compared to the untreated virtual patient. Adapted from (49).

Conclusion - Biosimulation offers a way to integrate knowledge of composite parts and their interactions in order to better understand physiology and predict the behavior of integrated systems. In a recent editorial on the future of biotechnology, Pollard stated that biosimulation, "...is the ultimate test of understanding molecular physiology" (54). As an integrative complement to last century's greatly successful reductionist biological program, biosimulation offers much hope for the future of biological science and medicine.

Acknowledgements – The authors would like to thank Annette Doherty, Michael French, Jill Fujisaki, Christy Kalb, Annette Lewis, David Polidori, Jeff Trimmer, and Barry Sudbeck for their thoughtful review of the manuscript.

References

1. R. L. Ho, Emerging Therapeutic Targets, 4, 699-714 (2000).
2. D. Noble and T. J. Colatsky, Emerging Therapeutic Targets, 4, 39-49 (2000).
3. D. Noble, J. Levin, W. Scott, Drug Discov. Today, 4, 10-16 (1999).
4. J. P. Keener and J. Sneyd, "Mathematical Physiology", Springer Verlag, 1998.
5. B. Teusink, J. Passarge, C. A. Reijenga, E. Esgalhado, C. C. van der Weijden, M. Schepper, M. C. Walsh, B. M. Bakker, K. van Dam, H. V. Westerhoff, J. L. Snoep, Eur. J. Biochem., 267, 5313-5329 (2000).
6. P. Mendes and D. Kell, Bioinformatics, 14, 869-883 (1998).
7. F. Hynne, S. Dano, P. G. Sorensen, Biophys. Chem., 94, 121-163 (2001).
8. J. S. Edwards and B. O. Palsson, BMC. Bioinformatics, 1, 1 (2000).
9. J. S. Edwards, R. U. Ibarra, B. O. Palsson, Nat. Biotechnol., 19, 125-130 (2001).
10. J. S. Edwards, R. Ramakrishna, B. O. Palsson, Biotechnol. Bioeng., 77, 27-36 (2002).
11. J. S. Edwards and B. O. Palsson, Proc. Natl. Acad. Sci. U.S.A., 97, 5528-5533 (2000).
12. Sveiczer, A. Csikasz-Nagy, B. Gyorffy, J. J. Tyson, B. Novak, Proc. Natl. Acad. Sci. U.S.A., 97, 7865-7870 (2000).
13. M. Santillan and M. C. Mackey, Proc. Natl. Acad. Sci. U.S.A., 98, 1364-1369 (2001).

14. A. Arkin, J. Ross, H. H. McAdams, Genetics 149, 1633-1648 (1998).
15. H. H. McAdams and A. Arkin, Proc.Natl.Acad.Sci.U.S.A 94, 814-819 (1997).
16. D. Endy, L. You, J. Yin, I. J. Molineux, Proc. Natl. Acad. Sci. U.S.A., 97, 5375-5380 (2000).
17. V. V. Gursky, J. Reinitz, A. M. Samsonov, Chaos, 11, 132-141 (2001).
18. L. Sanchez and D. Thieffry, J. Theor..Biol., 211, 115-141 (2001).
19. G. von Dassow, E. Meir, E. M. Munro, G. M. Odell, Nature, 406, 188-192 (2000).
20. T. S. Gardner, C. R. Cantor, J. J. Collins, Nature, 403, 339-342 (2000).
21. M. B. Elowitz and S. Leibler, Nature, 403, 335-338 (2000).
22. J. B. Stock and M. Surette in E. coli and S. typhimurium: Cellular and Molecular Biology, F. C. Neidhardt, Ed., American Society of Microbiology, Washington DC, 1996.
23. J. B. Stock in "Regulation of Cellular Signal Transduction Pathways by Desensitization and Amplification", D. R. Sibley and M. D. Houslay, Eds., Wiley, Chichester, 1994.
24. N. Barkai and S. Leibler, Nature 387, 913-917 (1997).
25. T. M. Yi, Y. Huang, M. I. Simon, J. Doyle, Proc. Natl. Acad. Sci. U.S.A., 97, 4649-4653 (2000).
26. D. A. Lauffenburger, Proc. Natl. Acad. Sci. U.S.A., 97, 5031-5033 (2000).
27. L. H. Hartwell, J. J. Hopfield, S. Leibler, A. W. Murray, Nature, 402, C47-C52 (1999).
28. C. A. J. M. Firth and D. Bray in "Computational Modeling of Genetic and Biochemical Networks", The MIT press, Cambridge, MA, 2001.
29. C. J. Morton-Firth, T. S. Shimizu, D. Bray, J. Mol. Biol., 286, 1059-1074 (1999).
30. D. Bottino, A. Mogilner, T. Roberts, M. Stewart, G. Oster, J. Cell Sci., 115, 367-384 (2002). Link to movies: http://jcs.biologists.org/cgi/content/full/115/2/367/DC1
31. J. E. Italiano, Jr., M. Stewart, T. M. Roberts, J. Cell Biol., 146, 1087-1096 (1999).
32. A. Nygren, C. Fiset, L. Firek, J. W. Clark, D. S. Lindblad, R. B. Clark, and W. R. Giles, Circ. Res., 82, 63-81 (1998).
33. M. Courtemanche, R. J. Ramirez, S. Nattel, Am. J. Physiol., 275, H301-H321 (1998).
34. D. S. Lindblad, C. R. Murphey, J. W. Clark, W. R. Giles, Am. J. Physiol., 271, H1666-H1696 (1996).
35. S. S. Demir, J. W. Clark, C. R. Murphey, W. R. Giles, Am. J. Physiol., 266, C832-C852 (1994).
36. C. H. Luo and Y. Rudy, Circ. Res., 74, 1097-1113 (1994).
37. M. S. Jafri, J. J. Rice, R. L. Winslow, Biophys. J., 74, 1149-1168 (1998).
38. R. L. Winslow, J. Rice, S. Jafri, E. Marban, B. O'Rourke, Circ. Res., 84, 571-586 (1999).
39. C. E. Clancy and Y. Rudy, Nature, 400, 566-569 (1999).
40. R. L. Winslow, J. L. Greenstein, G. F. Tomaselli, B. O'Rourke, Philosophical Transactions The Royal Society London A., 359, 1187-1200 (2001).
41. D. M. Harrild, R. C. Penland, C. S. Henriquez, J. Electrocardiol., 33, 241-251 (2000).
42. R. L. Winslow, D. F. Scollan, J. L. Greenstein, C. K. Yung, W. Baumgartner, Jr., G. Bhanot, D. L. Gresh, and B. E. Rogowitz, IBM Systems Journal, 40, 342-359 (2002).
43. D. Noble, Science, 295, 1678-1682 (2002).
44. K. Hall and L. Glass, Phys. Rev. Lett., 82, 5164-5167 (1999).
45. K. Hall and L. Glass, J. Cardiovasc. Electrophysiol., 10, 387-398 (1999).
46. B. Topp, K. Promislow, G. deVries, R. M. Miura, D. T. Finegood, J. Theor. Biol., 206, 605-619 (2000).
47. K. Hall, S. Q. Siler, J. Trimmer, and D. Polidori, Ann. Biomed. Eng., 29, S13. (2001).
48. C. L. Stokes, A. K. Lewis, K. Subramanian, D. J. Klinke, M. Okino, and J. M. Edelman, J. Allergy Clin. Immunol., 107, 933. (2001).   D. J. Klinke, A. K. Lewis, C. L. Stokes.
49. D. J. Klinke, A. K. Lewis, S.-P. Wong, and C. L. Stokes, Am. J. Resp. Crit. Care Med., 163, A832. (2001).
50. A. K. Lewis, D. J. Klinke, and C. L. Stokes, Am. J. Resp. Crit. Care Med., 163, A143. (2001).
51. E. Israel, J. M. Drazen, S. B. Liggett, H. A. Boushey, R. M. Cherniack, V. M. Chinchilli, D. M. Cooper, J. V. Fahy, J. E. Fish, J. G. Ford, M. Kraft, S. Kunselman, S. C. Lazarus, R. F. Lemanske, R. J. Martin, D. E. McLean, S. P. Peters, E. K. Silverman, C. A. Sorkness, S. J. Szefler, S. T. Weiss, and C. N. Yandava, Am. J. Respir. Crit. Care Med., 162, 75-80 (2000).
52. A. K. Lewis, T. Paterson, C. C. Leong, N. Defranoux, S. T. Holgate, and C. L. Stokes, Int. Arch. Allergy Immunol., 124, 282-286. (2001).
53. M. J. Leckie, A. ten Brinke, J. Khan, Z. Diamant, B. J. O'Connor, C. M. Walls, A. K. Mathur, H. C. Cowley, K. F. Chung, R. Djukanovic, T. T. Hansel, S. T. Holgate, P. J. Sterk, P. J. Barnes, Lancet, 356, 2144-2148 (2000).
54. T. D. Pollard, JAMA, 287, 1725-1727 (2002).

GENERIC NAME	INDICATION	YEAR INTRO.	ARMC VOL., PAGE
abacavir sulfate	antiviral	1999	35, 333
acarbose	antidiabetic	1990	26, 297
aceclofenac	antiinflammatory	1992	28, 325
acemannan	wound healing agent	2001	37, 259
acetohydroxamic acid	hypoammonuric	1983	19, 313
acetorphan	antidiarrheal	1993	29, 332
acipimox	hypolipidemic	1985	21, 323
acitretin	antipsoriatic	1989	25, 309
acrivastine	antihistamine	1988	24, 295
actarit	antirheumatic	1994	30, 296
adamantanium bromide	antiseptic	1984	20, 315
adrafinil	psychostimulant	1986	22, 315
AF-2259	antiinflammatory	1987	23, 325
afloqualone	muscle relaxant	1983	19, 313
agalsidase alfa	fabry's disease	2001	37, 259
alacepril	antihypertensive	1988	24, 296
alclometasone dipropionate	topical antiinflammatory	1985	21, 323
alemtuzumab	anticancer	2001	37, 260
alendronate sodium	osteoporosis	1993	29, 332
alfentanil HCl	analgesic	1983	19, 314
alfuzosin HCl	antihypertensive	1988	24, 296
alglucerase	enzyme	1991	27, 321
alitretinoin	anticancer	1999	35, 333
alminoprofen	analgesic	1983	19, 314
almotriptan	antimigraine	2000	36, 295
anakinra	antiarthritic	2001	37, 261
alosetron hydrochloride	irritable bowel syndrome	2000	36, 295
alpha-1 antitrypsin	protease inhibitor	1988	24, 297
alpidem	anxiolytic	1991	27, 322
alpiropride	antimigraine	1988	24, 296
alteplase	thrombolytic	1987	23, 326
amfenac sodium	antiinflammatory	1986	22, 315
amifostine	cytoprotective	1995	31, 338
aminoprofen	topical antiinflammatory	1990	26, 298
amisulpride	antipsychotic	1986	22, 316
amlexanox	antiasthmatic	1987	23, 327
amlodipine besylate	antihypertensive	1990	26, 298
amorolfine HCl	topical antifungal	1991	27, 322
amosulalol	antihypertensive	1988	24, 297
ampiroxicam	antiinflammatory	1994	30, 296
amprenavir	antiviral	1999	35, 334
amrinone	cardiotonic	1983	19, 314
amsacrine	antineoplastic	1987	23, 327
amtolmetin guacil	antiinflammatory	1993	29, 332
anagrelide HCl	hematological	1997	33, 328
anastrozole	antineoplastic	1995	31, 338
angiotensin II	anticancer adjuvant	1994	30, 296

GENERIC NAME	INDICATION	YEAR INTRO.	ARMC VOL., PAGE
aniracetam	cognition enhancer	1993	29, 333
APD	calcium regulator	1987	23, 326
apraclonidine HCl	antiglaucoma	1988	24, 297
APSAC	thrombolytic	1987	23, 326
aranidipine	antihypertensive	1996	32, 306
arbekacin	antibiotic	1990	26, 298
argatroban	antithromobotic	1990	26, 299
arglabin	anticancer	1999	35, 335
arotinolol HCl	antihypertensive	1986	22, 316
arteether	antimalarial	2000	36, 296
artemisinin	antimalarial	1987	23, 327
aspoxicillin	antibiotic	1987	23, 328
astemizole	antihistamine	1983	19, 314
astromycin sulfate	antibiotic	1985	21, 324
atorvastatin calcium	dyslipidemia	1997	33, 328
atosiban	preterm labor	2000	36, 297
atovaquone	antiparasitic	1992	28, 326
auranofin	chrysotherapeutic	1983	19, 314
azelaic acid	antiacne	1989	25, 310
azelastine HCl	antihistamine	1986	22, 316
azithromycin	antibiotic	1988	24, 298
azosemide	diuretic	1986	22, 316
aztreonam	antibiotic	1984	20, 315
balsalazide disodium	ulcerative colitis	1997	33, 329
bambuterol	bronchodilator	1990	26, 299
barnidipine HCl	antihypertensive	1992	28, 326
beclobrate	hypolipidemic	1986	22, 317
befunolol HCl	antiglaucoma	1983	19, 315
benazepril HCl	antihypertensive	1990	26, 299
benexate HCl	antiulcer	1987	23, 328
benidipine HCl	antihypertensive	1991	27, 322
beraprost sodium	platelet aggreg. inhibitor	1992	28, 326
betamethasone butyrate prospinate	topical antiinflammatory	1994	30, 297
betaxolol HCl	antihypertensive	1983	19, 315
betotastine besilate	antiallergic	2000	36, 297
bevantolol HCl	antihypertensive	1987	23, 328
bexarotene	anticancer	2000	36, 298
bicalutamide	antineoplastic	1995	31, 338
bifemelane HCl	nootropic	1987	23, 329
bimatoprost	antiglaucoma	2001	37, 261
binfonazole	hypnotic	1983	19, 315
binifibrate	hypolipidemic	1986	22, 317
bisantrene HCl	antineoplastic	1990	26, 300
bisoprolol fumarate	antihypertensive	1986	22, 317
bivalirudin	antithrombotic	2000	36, 298
bopindolol	antihypertensive	1985	21, 324
bosentan	antihypertensive	2001	37, 262

GENERIC NAME	INDICATION	YEAR INTRO.	ARMC VOL., PAGE
brimonidine	antiglaucoma	1996	32, 306
brinzolamide	antiglaucoma	1998	34, 318
brodimoprin	antibiotic	1993	29, 333
bromfenac sodium	NSAID	1997	33, 329
brotizolam	hypnotic	1983	19, 315
brovincamine fumarate	cerebral vasodilator	1986	22, 317
bucillamine	immunomodulator	1987	23, 329
bucladesine sodium	cardiostimulant	1984	20, 316
budipine	antiParkinsonian	1997	33, 330
budralazine	antihypertensive	1983	19, 315
bulaquine	antimalarial	2000	36, 299
bunazosin HCl	antihypertensive	1985	21, 324
bupropion HCl	antidepressant	1989	25, 310
buserelin acetate	hormone	1984	20, 316
buspirone HCl	anxiolytic	1985	21, 324
butenafine HCl	topical antifungal	1992	28, 327
butibufen	antiinflammatory	1992	28, 327
butoconazole	topical antifungal	1986	22, 318
butoctamide	hypnotic	1984	20, 316
butyl flufenamate	topical antiinflammatory	1983	19, 316
cabergoline	antiprolactin	1993	29, 334
cadexomer iodine	wound healing agent	1983	19, 316
cadralazine	hypertensive	1988	24, 298
calcipotriol	antipsoriatic	1991	27, 323
camostat mesylate	antineoplastic	1985	21, 325
candesartan cilexetil	antihypertension	1997	33, 330
capecitabine	antineoplastic	1998	34, 319
carboplatin	antibiotic	1986	22, 318
carperitide	congestive heart failure	1995	31, 339
carumonam	antibiotic	1988	24, 298
carvedilol	antihypertensive	1991	27, 323
caspofungin acetate	antifungal	2001	37, 263
cefbuperazone sodium	antibiotic	1985	21, 325
cefcapene pivoxil	antibiotic	1997	33, 330
cefdinir	antibiotic	1991	27, 323
cefditoren pivoxil	oral cephalosporin	1994	30, 297
cefepime	antibiotic	1993	29, 334
cefetamet pivoxil HCl	antibiotic	1992	28, 327
cefixime	antibiotic	1987	23, 329
cefmenoxime HCl	antibiotic	1983	19, 316
cefminox sodium	antibiotic	1987	23, 330
cefodizime sodium	antibiotic	1990	26, 300
cefonicid sodium	antibiotic	1984	20, 316
ceforanide	antibiotic	1984	20, 317
cefoselis	antibiotic	1998	34, 319
cefotetan disodium	antibiotic	1984	20, 317
cefotiam hexetil HCl	antibiotic	1991	27, 324
cefozopran HCl	injectable cephalosporin	1995	31, 339

GENERIC NAME	INDICATION	YEAR INTRO.	ARMC VOL., PAGE
cefpimizole	antibiotic	1987	23, 330
cefpiramide sodium	antibiotic	1985	21, 325
cefpirome sulfate	antibiotic	1992	28, 328
cefpodoxime proxetil	antibiotic	1989	25, 310
cefprozil	antibiotic	1992	28, 328
ceftazidime	antibiotic	1983	19, 316
cefteram pivoxil	antibiotic	1987	23, 330
ceftibuten	antibiotic	1992	28, 329
cefuroxime axetil	antibiotic	1987	23, 331
cefuzonam sodium	antibiotic	1987	23, 331
celecoxib	antiarthritic	1999	35, 335
celiprolol HCl	antihypertensive	1983	19, 317
centchroman	antiestrogen	1991	27, 324
centoxin	immunomodulator	1991	27, 325
cerivastatin	dyslipidemia	1997	33, 331
cetirizine HCl	antihistamine	1987	23, 331
cetrorelix	female infertility	1999	35, 336
cevimeline hydrochloride	anti-xerostomia	2000	36, 299
chenodiol	anticholelithogenic	1983	19, 317
CHF-1301	antiparkinsonian	1999	35, 336
choline alfoscerate	nootropic	1990	26, 300
cibenzoline	antiarrhythmic	1985	21, 325
cicletanine	antihypertensive	1988	24, 299
cidofovir	antiviral	1996	32, 306
cilazapril	antihypertensive	1990	26, 301
cilostazol	antithrombotic	1988	24, 299
cimetropium bromide	antispasmodic	1985	21, 326
cinildipine	antihypertensive	1995	31, 339
cinitapride	gastroprokinetic	1990	26, 301
cinolazepam	hypnotic	1993	29, 334
ciprofibrate	hypolipidemic	1985	21, 326
ciprofloxacin	antibacterial	1986	22, 318
cisapride	gastroprokinetic	1988	24, 299
cisatracurium besilate	muscle relaxant	1995	31, 340
citalopram	antidepressant	1989	25, 311
cladribine	antineoplastic	1993	29, 335
clarithromycin	antibiotic	1990	26, 302
clobenoside	vasoprotective	1988	24, 300
cloconazole HCl	topical antifungal	1986	22, 318
clodronate disodium	calcium regulator	1986	22, 319
clopidogrel hydrogensulfate	antithrombotic	1998	34, 320
cloricromen	antithrombotic	1991	27, 325
clospipramine HCl	neuroleptic	1991	27, 325
colesevelam hydrochloride	hypolipidemic	2000	36, 300
colestimide	hypolipidaemic	1999	35, 337
colforsin daropate HCl	cardiotonic	1999	35, 337
crotelidae polyvalent immune fab	antidote	2001	37, 263

GENERIC NAME	INDICATION	YEAR INTRO.	ARMC VOL., PAGE
cyclosporine	immunosuppressant	1983	19, 317
cytarabine ocfosfate	antineoplastic	1993	29, 335
dalfopristin	antibiotic	1999	35, 338
dapiprazole HCl	antiglaucoma	1987	23, 332
defeiprone	iron chelator	1995	31, 340
defibrotide	antithrombotic	1986	22, 319
deflazacort	antiinflammatory	1986	22, 319
delapril	antihypertensive	1989	25, 311
delavirdine mesylate	antiviral	1997	33, 331
denileukin diftitox	anticancer	1999	35, 338
denopamine	cardiostimulant	1988	24, 300
deprodone propionate	topical antiinflammatory	1992	28, 329
desflurane	anesthetic	1992	28, 329
desloratadine	antihistamine	2001	37, 264
dexfenfluramine	antiobesity	1997	33, 332
dexibuprofen	antiinflammatory	1994	30, 298
dexmedetomidine hydrochloride	sedative	2000	36, 301
dexrazoxane	cardioprotective	1992	28, 330
dezocine	analgesic	1991	27, 326
diacerein	antirheumatic	1985	21, 326
didanosine	antiviral	1991	27, 326
dilevalol	antihypertensive	1989	25, 311
dirithromycin	antibiotic	1993	29, 336
disodium pamidronate	calcium regulator	1989	25, 312
divistyramine	hypocholesterolemic	1984	20, 317
docarpamine	cardiostimulant	1994	30, 298
docetaxel	antineoplastic	1995	31, 341
dofetilide	antiarrhythmic	2000	36, 301
dolasetron mesylate	antiemetic	1998	34, 321
donepezil HCl	anti-Alzheimer	1997	33, 332
dopexamine	cardiostimulant	1989	25, 312
dornase alfa	cystic fibrosis	1994	30, 298
dorzolamide HCL	antiglaucoma	1995	31, 341
dosmalfate	antiulcer	2000	36, 302
doxacurium chloride	muscle relaxant	1991	27, 326
doxazosin mesylate	antihypertensive	1988	24, 300
doxefazepam	hypnotic	1985	21, 326
doxercalciferol	vitamin D prohormone	1999	35, 339
doxifluridine	antineoplastic	1987	23, 332
doxofylline	bronchodilator	1985	21, 327
dronabinol	antinauseant	1986	22, 319
drospirenone	contraceptive	2000	36, 302
drotrecogin alfa	antisepsis	2001	37, 265
droxicam	antiinflammatory	1990	26, 302
droxidopa	antiparkinsonian	1989	25, 312
duteplase	anticougulant	1995	31, 342
ebastine	antihistamine	1990	26 302
ebrotidine	antiulcer	1997	33, 333

GENERIC NAME	INDICATION	YEAR INTRO.	ARMC VOL., PAGE
ecabet sodium	antiulcerative	1993	29, 336
edaravone	neuroprotective	2001	37, 265
efavirenz	antiviral	1998	34, 321
efonidipine	antihypertensive	1994	30, 299
egualen sodium	antiulcer	2000	36, 303
eletriptan	antimigraine	2001	37, 266
emedastine difumarate	antiallergic/antiasthmatic	1993	29, 336
emorfazone	analgesic	1984	20, 317
enalapril maleate	antihypertensive	1984	20, 317
enalaprilat	antihypertensive	1987	23, 332
encainide HCl	antiarrhythmic	1987	23, 333
enocitabine	antineoplastic	1983	19, 318
enoxacin	antibacterial	1986	22, 320
enoxaparin	antithrombotic	1987	23, 333
enoximone	cardiostimulant	1988	24, 301
enprostil	antiulcer	1985	21, 327
entacapone	antiparkinsonian	1998	34, 322
epalrestat	antidiabetic	1992	28, 330
eperisone HCl	muscle relaxant	1983	19, 318
epidermal growth factor	wound healing agent	1987	23, 333
epinastine	antiallergic	1994	30, 299
epirubicin HCl	antineoplastic	1984	20, 318
epoprostenol sodium	platelet aggreg. inhib.	1983	19, 318
eprosartan	antihypertensive	1997	33, 333
eptazocine HBr	analgesic	1987	23, 334
eptilfibatide	antithrombotic	1999	35, 340
erdosteine	expectorant	1995	31, 342
erythromycin acistrate	antibiotic	1988	24, 301
erythropoietin	hematopoetic	1988	24, 301
esmolol HCl	antiarrhythmic	1987	23, 334
esomeprazole magnesium	gastric antisecretory	2000	36, 303
ethyl icosapentate	antithrombotic	1990	26, 303
etizolam	anxiolytic	1984	20, 318
etodolac	antiinflammatory	1985	21, 327
exemestane	anticancer	2000	36, 304
exifone	nootropic	1988	24, 302
factor VIIa	haemophilia	1996	32, 307
factor VIII	hemostatic	1992	28, 330
fadrozole HCl	antineoplastic	1995	31, 342
falecalcitriol	vitamin D	2001	37, 266
famciclovir	antiviral	1994	30, 300
famotidine	antiulcer	1985	21, 327
fasudil HCl	neuroprotective	1995	31, 343
felbamate	antiepileptic	1993	29, 337
felbinac	topical antiinflammatory	1986	22, 320
felodipine	antihypertensive	1988	24, 302
fenbuprol	choleretic	1983	19, 318
fenoldopam mesylate	antihypertensive	1998	34, 322

GENERIC NAME	INDICATION	YEAR INTRO.	ARMC VOL., PAGE
fenticonazole nitrate	antifungal	1987	23, 334
fexofenadine	antiallergic	1996	32, 307
filgrastim	immunostimulant	1991	27, 327
finasteride	5α-reductase inhibitor	1992	28, 331
fisalamine	intestinal antiinflammatory	1984	20, 318
fleroxacin	antibacterial	1992	28, 331
flomoxef sodium	antibiotic	1988	24, 302
flosequinan	cardiostimulant	1992	28, 331
fluconazole	antifungal	1988	24, 303
fludarabine phosphate	antineoplastic	1991	27, 327
flumazenil	benzodiazepine antag.	1987	23, 335
flunoxaprofen	antiinflammatory	1987	23, 335
fluoxetine HCl	antidepressant	1986	22, 320
flupirtine maleate	analgesic	1985	21, 328
flurithromycin ethylsuccinate	antibiotic	1997	33, 333
flutamide	antineoplastic	1983	19, 318
flutazolam	anxiolytic	1984	20, 318
fluticasone propionate	antiinflammatory	1990	26, 303
flutoprazepam	anxiolytic	1986	22, 320
flutrimazole	topical antifungal	1995	31, 343
flutropium bromide	antitussive	1988	24, 303
fluvastatin	hypolipaemic	1994	30, 300
fluvoxamine maleate	antidepressant	1983	19, 319
follitropin alfa	fertility enhancer	1996	32, 307
follitropin beta	fertility enhancer	1996	32, 308
fomepizole	antidote	1998	34, 323
fomivirsen sodium	antiviral	1998	34, 323
formestane	antineoplastic	1993	29, 337
formoterol fumarate	bronchodilator	1986	22, 321
foscarnet sodium	antiviral	1989	25, 313
fosfosal	analgesic	1984	20, 319
fosinopril sodium	antihypertensive	1991	27, 328
fosphenytoin sodium	antiepileptic	1996	32, 308
fotemustine	antineoplastic	1989	25, 313
fropenam	antibiotic	1997	33, 334
fudosteine	expectorant	2001	37, 267
gabapentin	antiepileptic	1993	29, 338
gadoversetamide	MRI contrast agent	2000	36, 304
gallium nitrate	calcium regulator	1991	27, 328
gallopamil HCl	antianginal	1983	19, 319
ganciclovir	antiviral	1988	24, 303
ganirelix acetate	female infertility	2000	36, 305
gatifloxacin	antibiotic	1999	35, 340
gemcitabine HCl	antineoplastic	1995	31, 344
gemeprost	abortifacient	1983	19, 319
gemtuzumab ozogamicin	anticancer	2000	36, 306
gestodene	progestogen	1987	23, 335
gestrinone	antiprogestogen	1986	22, 321

GENERIC NAME	INDICATION	YEAR INTRO.	ARMC VOL., PAGE
glatiramer acetate	Multiple Sclerosis	1997	33, 334
glimepiride	antidiabetic	1995	31, 344
glucagon, rDNA	hypoglycemia	1993	29, 338
GMDP	immunostimulant	1996	32, 308
goserelin	hormone	1987	23, 336
granisetron HCl	antiemetic	1991	27, 329
guanadrel sulfate	antihypertensive	1983	19, 319
gusperimus	immunosuppressant	1994	30, 300
halobetasol propionate	topical antiinflammatory	1991	27, 329
halofantrine	antimalarial	1988	24, 304
halometasone	topical antiinflammatory	1983	19, 320
histrelin	precocious puberty	1993	29, 338
hydrocortisone aceponate	topical antiinflammatory	1988	24, 304
hydrocortisone butyrate	topical antiinflammatory	1983	19, 320
ibandronic acid	osteoporosis	1996	32, 309
ibopamine HCl	cardiostimulant	1984	20, 319
ibudilast	antiasthmatic	1989	25, 313
ibutilide fumarate	antiarrhythmic	1996	32, 309
idarubicin HCl	antineoplastic	1990	26, 303
idebenone	nootropic	1986	22, 321
iloprost	platelet aggreg. inhibitor	1992	28, 332
imatinib mesylate	antineoplastic	2001	37, 267
imidapril HCl	antihypertensive	1993	29, 339
imiglucerase	Gaucher's disease	1994	30, 301
imipenem/cilastatin	antibiotic	1985	21, 328
imiquimod	antiviral	1997	33, 335
incadronic acid	osteoporosis	1997	33, 335
indalpine	antidepressant	1983	19, 320
indeloxazine HCl	nootropic	1988	24, 304
indinavir sulfate	antiviral	1996	32, 310
indobufen	antithrombotic	1984	20, 319
insulin lispro	antidiabetic	1996	32, 310
interferon alfacon-1	antiviral	1997	33, 336
interferon gamma-1b	immunostimulant	1991	27, 329
interferon, gamma	antiinflammatory	1989	25, 314
interferon, gamma-1α	antineoplastic	1992	28, 332
interferon, β-1a	multiple sclerosis	1996	32, 311
interferon, β-1b	multiple sclerosis	1993	29, 339
interleukin-2	antineoplastic	1989	25, 314
ioflupane	diagnosis CNS	2000	36, 306
ipriflavone	calcium regulator	1989	25, 314
irbesartan	antihypertensive	1997	33, 336
irinotecan	antineoplastic	1994	30, 301
irsogladine	antiulcer	1989	25, 315
isepamicin	antibiotic	1988	24, 305
isofezolac	antiinflammatory	1984	20, 319
isoxicam	antiinflammatory	1983	19, 320

GENERIC NAME	INDICATION	YEAR INTRO.	ARMC VOL., PAGE
isradipine	antihypertensive	1989	25, 315
itopride HCl	gastroprokinetic	1995	31, 344
itraconazole	antifungal	1988	24, 305
ivermectin	antiparasitic	1987	23, 336
ketanserin	antihypertensive	1985	21, 328
ketorolac tromethamine	analgesic	1990	26, 304
kinetin	skin photodamage/ dermatologic	1999	35, 341
lacidipine	antihypertensive	1991	27, 330
lafutidine	gastric antisecretory	2000	36, 307
lamivudine	antiviral	1995	31, 345
lamotrigine	anticonvulsant	1990	26, 304
lanoconazole	antifungal	1994	30, 302
lanreotide acetate	acromegaly	1995	31, 345
lansoprazole	antiulcer	1992	28, 332
latanoprost	antiglaucoma	1996	32, 311
lefunomide	antiarthritic	1998	34, 324
lenampicillin HCl	antibiotic	1987	23, 336
lentinan	immunostimulant	1986	22, 322
lepirudin	anticoagulant	1997	33, 336
lercanidipine	antihyperintensive	1997	33, 337
letrazole	anticancer	1996	32, 311
leuprolide acetate	hormone	1984	20, 319
levacecarnine HCl	nootropic	1986	22, 322
levalbuterol HCl	antiasthmatic	1999	35, 341
levetiracetam	antiepileptic	2000	36, 307
levobunolol HCl	antiglaucoma	1985	21, 328
levobupivacaine hydrochloride	local anesthetic	2000	36, 308
levocabastine HCl	antihistamine	1991	27, 330
levocetirizine	antihistamine	2001	37, 268
levodropropizine	antitussive	1988	24, 305
levofloxacin	antibiotic	1993	29, 340
levosimendan	heart failure	2000	36, 308
lidamidine HCl	antiperistaltic	1984	20, 320
limaprost	antithrombotic	1988	24, 306
linezolid	antibiotic	2000	36, 309
liranaftate	topical antifungal	2000	36, 309
lisinopril	antihypertensive	1987	23, 337
lobenzarit sodium	antiinflammatory	1986	22, 322
lodoxamide tromethamine	antiallergic ophthalmic	1992	28, 333
lomefloxacin	antibiotic	1989	25, 315
lomerizine HCl	antimigraine	1999	35, 342
lonidamine	antineoplastic	1987	23, 337
lopinavir	antiviral	2000	36, 310
loprazolam mesylate	hypnotic	1983	19, 321
loprinone HCl	cardiostimulant	1996	32, 312
loracarbef	antibiotic	1992	28, 333
loratadine	antihistamine	1988	24, 306

GENERIC NAME	INDICATION	YEAR INTRO.	ARMC VOL., PAGE
lornoxicam	NSAID	1997	33, 337
losartan	antihypertensive	1994	30, 302
loteprednol etabonate	antiallergic ophthalmic	1998	34; 324
lovastatin	hypocholesterolemic	1987	23, 337
loxoprofen sodium	antiinflammatory	1986	22, 322
Lyme disease	vaccine	1999	35, 342
mabuterol HCl	bronchodilator	1986	22, 323
malotilate	hepatoprotective	1985	21, 329
manidipine HCl	antihypertensive	1990	26, 304
masoprocol	topical antineoplastic	1992	28, 333
maxacalcitol	vitamin D	2000	36, 310
mebefradil hydrochoride	antihypertensive	1997	33, 338
medifoxamine fumarate	antidepressant	1986	22, 323
mefloquine HCl	antimalarial	1985	21, 329
meglutol	hypolipidemic	1983	19, 321
melinamide	hypocholesterolemic	1984	20, 320
meloxicam	antiarthritic	1996	32, 312
mepixanox	analeptic	1984	20, 320
meptazinol HCl	analgesic	1983	19, 321
meropenem	carbapenem antibiotic	1994	30, 303
metaclazepam	anxiolytic	1987	23, 338
metapramine	antidepressant	1984	20, 320
mexazolam	anxiolytic	1984	20, 321
mifepristone	abortifacient	1988	24, 306
miglitol	antidiabetic	1998	34, 325
milnacipran	antidepressant	1997	33, 338
milrinone	cardiostimulant	1989	25, 316
miltefosine	topical antineoplastic	1993	29, 340
miokamycin	antibiotic	1985	21, 329
mirtazapine	antidepressant	1994	30, 303
misoprostol	antiulcer	1985	21, 329
mitoxantrone HCl	antineoplastic	1984	20, 321
mivacurium chloride	muscle relaxant	1992	28, 334
mivotilate	hepatoprotectant	1999	35, 343
mizolastine	antihistamine	1998	34, 325
mizoribine	immunosuppressant	1984	20, 321
moclobemide	antidepressant	1990	26, 305
modafinil	idiopathic hypersomnia	1994	30, 303
moexipril HCl	antihypertensive	1995	31, 346
mofezolac	analgesic	1994	30, 304
mometasone furoate	topical antiinflammatory	1987	23, 338
montelukast sodium	antiasthma	1998	34, 326
moricizine HCl	antiarrhythmic	1990	26, 305
mosapride citrate	gastroprokinetic	1998	34, 326
moxifloxacin HCL	antibiotic	1999	35, 343
moxonidine	antihypertensive	1991	27, 330
mupirocin	topical antibiotic	1985	21, 330
muromonab-CD3	immunosuppressant	1986	22, 323

GENERIC NAME	INDICATION	YEAR INTRO.	ARMC VOL., PAGE
muzolimine	diuretic	1983	19, 321
mycophenolate mofetil	immunosuppressant	1995	31, 346
nabumetone	antiinflammatory	1985	21, 330
nadifloxacin	topical antibiotic	1993	29, 340
nafamostat mesylate	protease inhibitor	1986	22, 323
nafarelin acetate	hormone	1990	26, 306
naftifine HCl	antifungal	1984	20, 321
naftopidil	dysuria	1999	35, 344
nalmefene HCl	dependence treatment	1995	31, 347
naltrexone HCl	narcotic antagonist	1984	20, 322
naratriptan HCl	antimigraine	1997	33, 339
nartograstim	leukopenia	1994	30, 304
nateglinide	antidiabetic	1999	35, 344
nazasetron	antiemetic	1994	30, 305
nebivolol	antihypertensive	1997	33, 339
nedaplatin	antineoplastic	1995	31, 347
nedocromil sodium	antiallergic	1986	22, 324
nefazodone	antidepressant	1994	30, 305
neflinavir mesylate	antiviral	1997	33, 340
neltenexine	cystic fibrosis	1993	29, 341
nemonapride	neuroleptic	1991	27, 331
nesiritide	congestive heart failure	2001	37, 269
neticonazole HCl	topical antifungal	1993	29, 341
nevirapine	antiviral	1996	32, 313
nicorandil	coronary vasodilator	1984	20, 322
nifekalant HCl	antiarrythmic	1999	35, 344
nilutamide	antineoplastic	1987	23, 338
nilvadipine	antihypertensive	1989	25, 316
nimesulide	antiinflammatory	1985	21, 330
nimodipine	cerebral vasodilator	1985	21, 330
nipradilol	antihypertensive	1988	24, 307
nisoldipine	antihypertensive	1990	26, 306
nitrefazole	alcohol deterrent	1983	19, 322
nitrendipine	hypertensive	1985	21, 331
nizatidine	antiulcer	1987	23, 339
nizofenzone fumarate	nootropic	1988	24, 307
nomegestrol acetate	progestogen	1986	22, 324
norfloxacin	antibacterial	1983	19, 322
norgestimate	progestogen	1986	22, 324
OCT-43	anticancer	1999	35, 345
octreotide	antisecretory	1988	24, 307
ofloxacin	antibacterial	1985	21, 331
olanzapine	neuroleptic	1996	32, 313
olopatadine HCl	antiallergic	1997	33, 340
omeprazole	antiulcer	1988	24, 308
ondansetron HCl	antiemetic	1990	26, 306
OP-1	osteoinductor	2001	37, 269
orlistat	antiobesity	1998	34, 327

GENERIC NAME	INDICATION	YEAR INTRO.	ARMC VOL., PAGE
ornoprostil	antiulcer	1987	23, 339
osalazine sodium	intestinal antinflamm.	1986	22, 324
oseltamivir phosphate	antiviral	1999	35, 346
oxaliplatin	anticancer	1996	32, 313
oxaprozin	antiinflammatory	1983	19, 322
oxcarbazepine	anticonvulsant	1990	26, 307
oxiconazole nitrate	antifungal	1983	19, 322
oxiracetam	nootropic	1987	23, 339
oxitropium bromide	bronchodilator	1983	19, 323
ozagrel sodium	antithrombotic	1988	24, 308
paclitaxal	antineoplastic	1993	29, 342
panipenem/betamipron	carbapenem antibiotic	1994	30, 305
pantoprazole sodium	antiulcer	1995	30, 306
paricalcitol	vitamin D	1998	34 327
parnaparin sodium	anticoagulant	1993	29, 342
paroxetine	antidepressant	1991	27, 331
pefloxacin mesylate	antibacterial	1985	21, 331
pegademase bovine	immunostimulant	1990	26, 307
pegaspargase	antineoplastic	1994	30, 306
pemirolast potassium	antiasthmatic	1991	27, 331
penciclovir	antiviral	1996	32, 314
pentostatin	antineoplastic	1992	28, 334
pergolide mesylate	antiparkinsonian	1988	24, 308
perindopril	antihypertensive	1988	24, 309
perospirone HCL	neuroleptic	2001	37, 270
picotamide	antithrombotic	1987	23, 340
pidotimod	immunostimulant	1993	29, 343
piketoprofen	topical antiinflammatory	1984	20, 322
pilsicainide HCl	antiarrhythmic	1991	27, 332
pimaprofen	topical antiinflammatory	1984	20, 322
pimobendan	heart failure	1994	30, 307
pinacidil	antihypertensive	1987	23, 340
pioglitazone HCL	antidiabetic	1999	35, 346
pirarubicin	antineoplastic	1988	24, 309
pirmenol	antiarrhythmic	1994	30, 307
piroxicam cinnamate	antiinflammatory	1988	24, 309
pivagabine	antidepressant	1997	33, 341
plaunotol	antiulcer	1987	23, 340
polaprezinc	antiulcer	1994	30, 307
porfimer sodium	antineoplastic adjuvant	1993	29, 343
pramipexole HCl	antiParkinsonian	1997	33, 341
pramiracetam H_2SO_4	cognition enhancer	1993	29, 343
pranlukast	antiasthmatic	1995	31, 347
pravastatin	antilipidemic	1989	25, 316
prednicarbate	topical antiinflammatory	1986	22, 325
prezatide copper acetate	vulnery	1996	32, 314
progabide	anticonvulsant	1985	21, 331
promegestrone	progestogen	1983	19, 323

GENERIC NAME	INDICATION	YEAR INTRO.	ARMC VOL., PAGE
propacetamol HCl	analgesic	1986	22, 325
propagermanium	antiviral	1994	30, 308
propentofylline propionate	cerebral vasodilator	1988	24, 310
propiverine HCl	urologic	1992	28, 335
propofol	anesthetic	1986	22, 325
pumactant	lung surfactant	1994	30, 308
quazepam	hypnotic	1985	21, 332
quetiapine fumarate	neuroleptic	1997	33, 341
quinagolide	hyperprolactinemia	1994	30, 309
quinapril	antihypertensive	1989	25, 317
quinfamide	amebicide	1984	20, 322
quinupristin	antibiotic	1999	35, 338
rabeprazole sodium	gastric antisecretory	1998	34, 328
raloxifene HCl	osteoporosis	1998	34, 328
raltitrexed	anticancer	1996	32, 315
ramatroban	antiallergic	2000	36, 311
ramipril	antihypertensive	1989	25, 317
ramosetron	antiemetic	1996	32, 315
ranimustine	antineoplastic	1987	23, 341
ranitidine bismuth citrate	antiulcer	1995	31, 348
rapacuronium bromide	muscle relaxant	1999	35, 347
rebamipide	antiulcer	1990	26, 308
reboxetine	antidepressant	1997	33, 342
remifentanil HCl	analgesic	1996	32, 316
remoxipride HCl	antipsychotic	1990	26, 308
repaglinide	antidiabetic	1998	34, 329
repirinast	antiallergic	1987	23, 341
reteplase	fibrinolytic	1996	32, 316
reviparin sodium	anticoagulant	1993	29, 344
rifabutin	antibacterial	1992	28, 335
rifapentine	antibacterial	1988	24, 310
rifaximin	antibiotic	1985	21, 332
rifaximin	antibiotic	1987	23, 341
rilmazafone	hypnotic	1989	25, 317
rilmenidine	antihypertensive	1988	24, 310
riluzole	neuroprotective	1996	32, 316
rimantadine HCl	antiviral	1987	23, 342
rimexolone	antiinflammatory	1995	31, 348
risedronate sodium	osteoporosis	1998	34, 330
risperidone	neuroleptic	1993	29, 344
ritonavir	antiviral	1996	32, 317
rivastigmin	anti-Alzheimer	1997	33, 342
rizatriptan benzoate	antimigraine	1998	34, 330
rocuronium bromide	neuromuscular blocker	1994	30, 309
rofecoxib	antiarthritic	1999	35, 347
rokitamycin	antibiotic	1986	22, 325
romurtide	immunostimulant	1991	27, 332
ronafibrate	hypolipidemic	1986	22, 326

GENERIC NAME	INDICATION	YEAR INTRO.	ARMC VOL., PAGE
ropinirole HCl	antiParkinsonian	1996	32, 317
ropivacaine	anesthetic	1996	32, 318
rosaprostol	antiulcer	1985	21, 332
rosiglitazone maleate	antidiabetic	1999	35, 348
roxatidine acetate HCl	antiulcer	1986	22, 326
roxithromycin	antiulcer	1987	23, 342
rufloxacin HCl	antibacterial	1992	28, 335
RV-11	antibiotic	1989	25, 318
salmeterol hydroxynaphthoate	bronchodilator	1990	26, 308
sapropterin HCl	hyperphenylalaninemia	1992	28, 336
saquinavir mesvlate	antiviral	1995	31, 349
sargramostim	immunostimulant	1991	27, 332
sarpogrelate HCl	platelet antiaggregant	1993	29, 344
schizophyllan	immunostimulant	1985	22, 326
seratrodast	antiasthmatic	1995	31, 349
sertaconazole nitrate	topical antifungal	1992	28, 336
sertindole	neuroleptic	1996	32, 318
setastine HCl	antihistamine	1987	23, 342
setiptiline	antidepressant	1989	25, 318
setraline HCl	antidepressant	1990	26, 309
sevoflurane	anesthetic	1990	26, 309
sibutramine	antiobesity	1998	34, 331
sildenafil citrate	male sexual dysfunction	1998	34, 331
simvastatin	hypocholesterolemic	1988	24, 311
SKI-2053R	anticancer	1999	35, 348
sobuzoxane	antineoplastic	1994	30, 310
sodium cellulose PO4	hypocalciuric	1983	19, 323
sofalcone	antiulcer	1984	20, 323
somatomedin-1	growth hormone insensitivity	1994	30, 310
somatotropin	growth hormone	1994	30, 310
somatropin	hormone	1987	23, 343
sorivudine	antiviral	1993	29, 345
sparfloxacin	antibiotic	1993	29, 345
spirapril HCl	antihypertensive	1995	31, 349
spizofurone	antiulcer	1987	23, 343
stavudine	antiviral	1994	30, 311
succimer	chelator	1991	27, 333
sufentanil	analgesic	1983	19, 323
sulbactam sodium	β-lactamase inhibitor	1986	22, 326
sulconizole nitrate	topical antifungal	1985	21, 332
sultamycillin tosylate	antibiotic	1987	23, 343
sumatriptan succinate	antimigraine	1991	27, 333
suplatast tosilate	antiallergic	1995	31, 350
suprofen	analgesic	1983	19, 324
surfactant TA	respiratory surfactant	1987	23, 344
tacalcitol	topical antipsoriatic	1993	29, 346

GENERIC NAME	INDICATION	YEAR INTRO.	ARMC VOL., PAGE
tacrine HCl	Alzheimer's disease	1993	29, 346
tacrolimus	immunosuppressant	1993	29, 347
talipexole	antiParkinsonian	1996	32, 318
taltirelin	CNS stimulant	2000	36, 311
tamsulosin HCl	antiprostatic hypertrophy	1993	29, 347
tandospirone	anxiolytic	1996	32, 319
tasonermin	anticancer	1999	35, 349
tazanolast	antiallergic	1990	26, 309
tazarotene	antipsoriasis	1997	33, 343
tazobactam sodium	β-lactamase inhibitor	1992	28, 336
tegaserod maleate	irritable bowel syndrome	2001	37, 270
teicoplanin	antibacterial	1988	24, 311
telithromycin	antibiotic	2001	37, 271
telmesteine	mucolytic	1992	28, 337
telmisartan	antihypertensive	1999	35, 349
temafloxacin HCl	antibacterial	1991	27, 334
temocapril	antihypertensive	1994	30, 311
temocillin disodium	antibiotic	1984	20, 323
temozolomide	anticancer	1999	35, 349
tenofovir disoproxil fumarate	antiviral	2001	37, 271
tenoxicam	antiinflammatory	1987	23, 344
teprenone	antiulcer	1984	20, 323
terazosin HCl	antihypertensive	1984	20, 323
terbinafine HCl	antifungal	1991	27, 334
terconazole	antifungal	1983	19, 324
tertatolol HCl	antihypertensive	1987	23, 344
thymopentin	immunomodulator	1985	21, 333
tiagabine	antiepileptic	1996	32, 319
tiamenidine HCl	antihypertensive	1988	24, 311
tianeptine sodium	antidepressant	1983	19, 324
tibolone	anabolic	1988	24, 312
tilisolol HCl	antihypertensive	1992	28, 337
tiludronate disodium	Paget's disease	1995	31, 350
timiperone	neuroleptic	1984	20, 323
tinazoline	nasal decongestant	1988	24, 312
tioconazole	antifungal	1983	19, 324
tiopronin	urolithiasis	1989	25, 318
tiquizium bromide	antispasmodic	1984	20, 324
tiracizine HCl	antiarrhythmic	1990	26, 310
tirilazad mesylate	subarachnoid hemorrhage	1995	31, 351
tirofiban HCl	antithrombotic	1998	34, 332
tiropramide HCl	antispasmodic	1983	19, 324
tizanidine	muscle relaxant	1984	20, 324
tolcapone	antiParkinsonian	1997	33, 343
toloxatone	antidepressant	1984	20, 324
tolrestat	antidiabetic	1989	25, 319
topiramate	antiepileptic	1995	31, 351

GENERIC NAME	INDICATION	YEAR INTRO.	ARMC VOL., PAGE
topotecan HCl	anticancer	1996	32, 320
torasemide	diuretic	1993	29, 348
toremifene	antineoplastic	1989	25, 319
tosufloxacin tosylate	antibacterial	1990	26, 310
trandolapril	antihypertensive	1993	29, 348
travoprost	antiglaucoma	2001	37, 272
tretinoin tocoferil	antiulcer	1993	29, 348
trientine HCl	chelator	1986	22, 327
trimazosin HCl	antihypertensive	1985	21, 333
trimegestone	progestogen	2001	37, 273
trimetrexate glucuronate	*Pneumocystis carinii* pneumonia	1994	30, 312
troglitazone	antidiabetic	1997	33, 344
tropisetron	antiemetic	1992	28, 337
trovafloxacin mesylate	antibiotic	1998	34, 332
troxipide	antiulcer	1986	22, 327
ubenimex	immunostimulant	1987	23, 345
unoprostone isopropyl ester	antiglaucoma	1994	30, 312
valaciclovir HCl	antiviral	1995	31, 352
vaglancirclovir HCL	antiviral	2001	37, 273
valrubicin	anticancer	1999	35, 350
valsartan	antihypertensive	1996	32, 320
venlafaxine	antidepressant	1994	30, 312
verteporfin	photosensitizer	2000	36, 312
vesnarinone	cardiostimulant	1990	26, 310
vigabatrin	anticonvulsant	1989	25, 319
vinorelbine	antineoplastic	1989	25, 320
voglibose	antidiabetic	1994	30, 313
xamoterol fumarate	cardiotonic	1988	24, 312
zafirlukast	antiasthma	1996	32, 321
zalcitabine	antiviral	1992	28, 338
zaleplon	hypnotic	1999	35, 351
zaltoprofen	antiinflammatory	1993	29, 349
zanamivir	antiviral	1999	35, 352
zidovudine	antiviral	1987	23, 345
zileuton	antiasthma	1997	33, 344
zinostatin stimalamer	antineoplastic	1994	30, 313
ziprasidone hydrochloride	neuroleptic	2000	36, 312
zofenopril calcium	antihypertensive	2000	36, 313
zoledronate disodium	hypercalcemia	2000	36, 314
zolpidem hemitartrate	hypnotic	1988	24, 313
zomitriptan	antimigraine	1997	33, 345
zonisamide	anticonvulsant	1989	25, 320
zopiclone	hypnotic	1986	22, 327
zuclopenthixol acetate	antipsychotic	1987	23, 345

GENERIC NAME	INDICATION	YEAR INTRO.	ARMC VOL.,	PAGE
gemeprost	ABORTIFACIENT	1983	19,	319
mifepristone		1988	24,	306
lanreotide acetate	ACROMEGALY	1995	31,	345
nitrefazole	ALCOHOL DETERRENT	1983	19,	322
tacrine HCl	ALZHEIMER'S DISEASE	1993	29,	346
quinfamide	AMEBICIDE	1984	20,	322
tibolone	ANABOLIC	1988	24,	312
mepixanox	ANALEPTIC	1984	20,	320
alfentanil HCl	ANALGESIC	1983	19,	314
alminoprofen		1983	19,	314
dezocine		1991	27,	326
emorfazone		1984	20,	317
eptazocine HBr		1987	23,	334
flupirtine maleate		1985	21,	328
fosfosal		1984	20,	319
ketorolac tromethamine		1990	26,	304
meptazinol HCl		1983	19,	321
mofezolac		1994	30,	304
propacetamol HCl		1986	22,	325
remifentanil HCl		1996	32,	316
sufentanil		1983	19,	323
suprofen		1983	19,	324
desflurane	ANESTHETIC	1992	28,	329
propofol		1986	22,	325
ropivacaine		1996	32,	318
sevoflurane		1990	26,	309
levobupivacaine hydrochloride	ANESTHETIC, LOCAL	2000	36,	308
azelaic acid	ANTIACNE	1989	25,	310
betotastine besilate	ANTIALLERGIC	2000	36,	297
emedastine difumarate		1993	29,	336
epinastine		1994	30,	299
fexofenadine		1996	32,	307
nedocromil sodium		1986	22,	324
olopatadine hydrochloride		1997	33,	340
ramatroban		2000	36,	311
repirinast		1987	23,	341

GENERIC NAME	INDICATION	YEAR INTRO.	ARMC VOL., PAGE	
suplatast tosilate		1995	31,	350
tazanolast		1990	26,	309
lodoxamide tromethamine	ANTIALLERGIC	1992	28,	333
loteprednol etabonate	OPHTHALMIC	1998	34,	324
donepezil hydrochloride	ANTI-ALZHEIMERS	1997	33,	332
rivastigmin		1997	33,	342
gallopamil HCl	ANTIANGINAL	1983	19,	319
cibenzoline	ANTIARRHYTHMIC	1985	21,	325
dofetilide		2000	36,	301
encainide HCl		1987	23,	333
esmolol HCl		1987	23,	334
ibutilide fumarate		1996	32,	309
moricizine hydrochloride		1990	26,	305
nifekalant HCl		1999	35,	344
pilsicainide hydrochloride		1991	27,	332
pirmenol		1994	30,	307
tiracizine hydrochloride		1990	26,	310
anakinra	ANTIARTHRITIC	2001	37,	261
celecoxib		1999	35,	335
meloxicam		1996	32,	312
leflunomide		1998	34,	324
rofecoxib		1999	35,	347
amlexanox	ANTIASTHMATIC	1987	23,	327
emedastine difumarate		1993	29,	336
ibudilast		1989	25,	313
levalbuterol HCl		1999	35,	341
montelukast sodium		1998	34,	326
pemirolast potassium		1991	27,	331
seratrodast		1995	31,	349
zafirlukast		1996	32,	321
zileuton		1997	33,	344
ciprofloxacin	ANTIBACTERIAL	1986	22,	318
enoxacin		1986	22,	320
fleroxacin		1992	28,	331
norfloxacin		1983	19,	322
ofloxacin		1985	21,	331
pefloxacin mesylate		1985	21,	331
pranlukast		1995	31,	347
rifabutin		1992	28,	335
rifapentine		1988	24,	310
rufloxacin hydrochloride		1992	28,	335
teicoplanin		1988	24,	311

GENERIC NAME	INDICATION	YEAR INTRO.	ARMC VOL., PAGE	
temafloxacin hydrochloride		1991	27,	334
tosufloxacin tosylate		1990	26,	310
arbekacin	ANTIBIOTIC	1990	26,	298
aspoxicillin		1987	23,	328
astromycin sulfate		1985	21,	324
azithromycin		1988	24,	298
aztreonam		1984	20,	315
brodimoprin		1993	29,	333
carboplatin		1986	22,	318
carumonam		1988	24,	298
cefbuperazone sodium		1985	21,	325
cefcapene pivoxil		1997	33,	330
cefdinir		1991	27,	323
cefepime		1993	29,	334
cefetamet pivoxil hydrochloride		1992	28,	327
cefixime		1987	23,	329
cefmenoxime HCl		1983	19,	316
cefminox sodium		1987	23,	330
cefodizime sodium		1990	26,	300
cefonicid sodium		1984	20,	316
ceforanide		1984	20,	317
cefoselis		1998	34,	319
cefotetan disodium		1984	20,	317
cefotiam hexetil hydrochloride		1991	27,	324
cefpimizole		1987	23,	330
cefpiramide sodium		1985	21,	325
cefpirome sulfate		1992	28,	328
cefpodoxime proxetil		1989	25,	310
cefprozil		1992	28,	328
ceftazidime		1983	19,	316
cefteram pivoxil		1987	23,	330
ceftibuten		1992	28,	329
cefuroxime axetil		1987	23,	331
cefuzonam sodium		1987	23,	331
clarithromycin		1990	26,	302
dalfopristin		1999	35,	338
dirithromycin		1993	29,	336
erythromycin acistrate		1988	24,	301
flomoxef sodium		1988	24,	302
flurithromycin ethylsuccinate		1997	33,	333
fropenam		1997	33,	334
gatifloxacin		1999	35,	340
imipenem/cilastatin		1985	21,	328
isepamicin		1988	24,	305
lenampicillin HCl		1987	23,	336
levofloxacin		1993	29,	340

GENERIC NAME	INDICATION	YEAR INTRO.	ARMC VOL., PAGE	
linezolid		2000	36,	309
lomefloxacin		1989	25,	315
loracarbef		1992	28,	333
miokamycin		1985	21,	329
moxifloxacin HCl		1999	35,	343
quinupristin		1999	35,	338
rifaximin		1985	21,	332
rifaximin		1987	23,	341
rokitamycin		1986	22,	325
RV-11		1989	25,	318
sparfloxacin		1993	29,	345
sultamycillin tosylate		1987	23,	343
telithromycin		2001	37,	271
temocillin disodium		1984	20,	323
trovafloxacin mesylate		1998	34,	332
meropenem	ANTIBIOTIC,	1994	30,	303
panipenem/betamipron	CARBAPENEM	1994	30,	305
mupirocin	ANTIBIOTIC, TOPICAL	1985	21,	330
nadifloxacin		1993	29,	340
alemtuzumab	ANTICANCER	2001	37,	260
alitretinoin		1999	35,	333
arglabin		1999	35,	335
bexarotene		2000	36,	298
denileukin diftitox		1999	35,	338
exemestane		2000	36,	304
gemtuzumab ozogamicin		2000	36,	306
letrazole		1996	32,	311
OCT-43		1999	35,	345
oxaliplatin		1996	32,	313
raltitrexed		1996	32,	315
SKI-2053R		1999	35,	348
tasonermin		1999	35,	349
temozolomide		1999	35,	350
topotecan HCl		1996	32,	320
valrubicin		1999	35,	350
angiotensin II	ANTICANCER ADJUVANT	1994	30,	296
chenodiol	ANTICHOLELITHOGENIC	1983	19,	317
duteplase	ANTICOAGULANT	1995	31,	342
lepirudin		1997	33,	336
parnaparin sodium		1993	29,	342
reviparin sodium		1993	29,	344

GENERIC NAME	INDICATION	YEAR INTRO.	ARMC VOL., PAGE	
lamotrigine	ANTICONVULSANT	1990	26,	304
oxcarbazepine		1990	26,	307
progabide		1985	21,	331
vigabatrin		1989	25,	319
zonisamide		1989	25,	320
bupropion HCl	ANTIDEPRESSANT	1989	25,	310
citalopram		1989	25,	311
fluoxetine HCl		1986	22,	320
fluvoxamine maleate		1983	19,	319
indalpine		1983	19,	320
medifoxamine fumarate		1986	22,	323
metapramine		1984	20,	320
milnacipran		1997	33,	338
mirtazapine		1994	30,	303
moclobemide		1990	26,	305
nefazodone		1994	30,	305
paroxetine		1991	27,	331
pivagabine		1997	33,	341
reboxetine		1997	33,	342
setiptiline		1989	25,	318
sertraline hydrochloride		1990	26,	309
tianeptine sodium		1983	19,	324
toloxatone		1984	20,	324
venlafaxine		1994	30,	312
acarbose	ANTIDIABETIC	1990	26,	297
epalrestat		1992	28,	330
glimepiride		1995	31,	344
insulin lispro		1996	32,	310
miglitol		1998	34,	325
nateglinide		1999	35,	344
pioglitazone HCl		1999	35,	346
repaglinide		1998	34,	329
rosiglitazone maleate		1999	35,	347
tolrestat		1989	25,	319
troglitazone		1997	33,	344
voglibose		1994	30,	313
acetorphan	ANTIDIARRHEAL	1993	29,	332
crotelidae polyvalent immune fab	ANTIDOTE	2001	37,	263
fomepizole		1998	34,	323
dolasetron mesylate	ANTIEMETIC	1998	34,	321
granisetron hydrochloride		1991	27,	329

GENERIC NAME	INDICATION	YEAR INTRO.	ARMC VOL.,	PAGE
ondansetron hydrochloride		1990	26,	306
nazasetron		1994	30,	305
ramosetron		1996	32,	315
tropisetron		1992	28,	337
felbamate	ANTIEPILEPTIC	1993	29,	337
fosphenytoin sodium		1996	32,	308
gabapentin		1993	29,	338
levetiracetam		2000	36,	307
tiagabine		1996	32,	320
topiramate		1995	31,	351
centchroman	ANTIESTROGEN	1991	27,	324
caspofungin acetate	ANTIFUNGAL	2001	37,	263
fenticonazole nitrate		1987	23,	334
fluconazole		1988	24,	303
itraconazole		1988	24,	305
lanoconazole		1994	30,	302
naftifine HCl		1984	20,	321
oxiconazole nitrate		1983	19,	322
terbinafine hydrochloride		1991	27,	334
terconazole		1983	19,	324
tioconazole		1983	19,	324
amorolfine hydrochloride	ANTIFUNGAL, TOPICAL	1991	27,	322
butenafine hydrochloride		1992	28,	327
butoconazole		1986	22,	318
cloconazole HCl		1986	22,	318
liranaftate		2000	36,	309
flutrimazole		1995	31,	343
neticonazole HCl		1993	29,	341
sertaconazole nitrate		1992	28,	336
sulconizole nitrate		1985	21,	332
apraclonidine HCl	ANTIGLAUCOMA	1988	24,	297
befunolol HCl		1983	19,	315
bimatroprost		2001	37,	261
brimonidine		1996	32,	306
brinzolamide		1998	34,	318
dapiprazole HCl		1987	23,	332
dorzolamide HCl		1995	31,	341
latanoprost		1996	32,	311
levobunolol HCl		1985	21,	328
travoprost		2001	37,	272
unoprostone isopropyl ester		1994	30,	312

GENERIC NAME	INDICATION	YEAR INTRO.	ARMC VOL., PAGE	
acrivastine	ANTIHISTAMINE	1988	24,	295
astemizole		1983	19,	314
azelastine HCl		1986	22,	316
cetirizine HCl		1987	23,	331
desloratadine		2001	37,	264
ebastine		1990	26,	302
levocabastine hydrochloride		1991	27,	330
levocetirizine		2001	37,	268
loratadine		1988	24,	306
mizolastine		1998	34,	325
setastine HCl		1987	23,	342
alacepril	ANTIHYPERTENSIVE	1988	24,	296
alfuzosin HCl		1988	24,	296
amlodipine besylate		1990	26,	298
amosulalol		1988	24,	297
aranidipine		1996	32,	306
arotinolol HCl		1986	22,	316
barnidipine hydrochloride		1992	28,	326
benazepril hydrochloride		1990	26,	299
benidipine hydrochloride		1991	27,	322
betaxolol HCl		1983	19,	315
bevantolol HCl		1987	23,	328
bisoprolol fumarate		1986	22,	317
bopindolol		1985	21,	324
bosentan		2001	37,	262
budralazine		1983	19,	315
bunazosin HCl		1985	21,	324
candesartan cilexetil		1997	33,	330
carvedilol		1991	27,	323
celiprolol HCl		1983	19,	317
cicletanine		1988	24,	299
cilazapril		1990	26,	301
cinildipine		1995	31,	339
delapril		1989	25,	311
dilevalol		1989	25,	311
doxazosin mesylate		1988	24,	300
efonidipine		1994	30,	299
enalapril maleate		1984	20,	317
enalaprilat		1987	23,	332
eprosartan		1997	33,	333
felodipine		1988	24,	302
fenoldopam mesylate		1998	34,	322
fosinopril sodium		1991	27,	328
guanadrel sulfate		1983	19,	319
imidapril HCl		1993	29,	339
irbesartan		1997	33,	336
isradipine		1989	25,	315
ketanserin		1985	21,	328

GENERIC NAME	INDICATION	YEAR INTRO.	ARMC VOL., PAGE	
lacidipine		1991	27,	330
lercanidipine		1997	33,	337
lisinopril		1987	23,	337
losartan		1994	30,	302
manidipine hydrochloride		1990	26,	304
mebefradil hydrochloride		1997	33,	338
moexipril HCl		1995	31,	346
moxonidine		1991	27,	330
nebivolol		1997	33,	339
nilvadipine		1989	25,	316
nipradilol		1988	24,	307
nisoldipine		1990	26,	306
perindopril		1988	24,	309
pinacidil		1987	23,	340
quinapril		1989	25,	317
ramipril		1989	25,	317
rilmenidine		1988	24,	310
spirapril HCl		1995	31,	349
telmisartan		1999	35,	349
temocapril		1994	30,	311
terazosin HCl		1984	20,	323
tertatolol HCl		1987	23,	344
tiamenidine HCl		1988	24,	311
tilisolol hydrochloride		1992	28,	337
trandolapril		1993	29,	348
trimazosin HCl		1985	21,	333
valsartan		1996	32,	320
zofenopril calcium		2000	36,	313
aceclofenac	ANTIINFLAMMATORY	1992	28,	325
AF-2259		1987	23,	325
amfenac sodium		1986	22,	315
ampiroxicam		1994	30,	296
amtolmetin guacil		1993	29,	332
butibufen		1992	28,	327
deflazacort		1986	22,	319
dexibuprofen		1994	30,	298
droxicam		1990	26,	302
etodolac		1985	21,	327
flunoxaprofen		1987	23,	335
fluticasone propionate		1990	26,	303
interferon, gamma		1989	25,	314
isofezolac		1984	20,	319
isoxicam		1983	19,	320
lobenzarit sodium		1986	22,	322
loxoprofen sodium		1986	22,	322
nabumetone		1985	21,	330
nimesulide		1985	21,	330

GENERIC NAME	INDICATION	YEAR INTRO.	ARMC VOL., PAGE	
oxaprozin		1983	19,	322
piroxicam cinnamate		1988	24,	309
rimexolone		1995	31,	348
tenoxicam		1987	23,	344
zaltoprofen		1993	29,	349
fisalamine	ANTIINFLAMMATORY,	1984	20,	318
osalazine sodium	INTESTINAL	1986	22,	324
alclometasone dipropionate	ANTIINFLAMMATORY,	1985	21,	323
aminoprofen	TOPICAL	1990	26,	298
betamethasone butyrate propionate		1994	30,	297
butyl flufenamate		1983	19,	316
deprodone propionate		1992	28,	329
felbinac		1986	22,	320
halobetasol propionate		1991	27,	329
halometasone		1983	19,	320
hydrocortisone aceponate		1988	24,	304
hydrocortisone butyrate propionate		1983	19,	320
mometasone furoate		1987	23,	338
piketoprofen		1984	20,	322
pimaprofen		1984	20,	322
prednicarbate		1986	22,	325
pravastatin	ANTILIPIDEMIC	1989	25,	316
arteether	ANTIMALARIAL	2000	36,	296
artemisinin		1987	23,	327
bulaquine		2000	36,	299
halofantrine		1988	24,	304
mefloquine HCl		1985	21,	329
almotriptan	ANTIMIGRAINE	2000	36,	295
alpiropride		1988	24,	296
eletriptan		2001	37,	266
lomerizine HCl		1999	35,	342
naratriptan hydrochloride		1997	33,	339
rizatriptan benzoate		1998	34,	330
sumatriptan succinate		1991	27,	333
zolmitriptan		1997	33,	345
dronabinol	ANTINAUSEANT	1986	22,	319
amsacrine	ANTINEOPLASTIC	1987	23,	327
anastrozole		1995	31,	338

GENERIC NAME	INDICATION	YEAR INTRO.	ARMC VOL., PAGE	
bicalutamide		1995	31,	338
bisantrene hydrochloride		1990	26,	300
camostat mesylate		1985	21,	325
capecitabine		1998	34,	319
cladribine		1993	29,	335
cytarabine ocfosfate		1993	29,	335
docetaxel		1995	31,	341
doxifluridine		1987	23,	332
enocitabine		1983	19,	318
epirubicin HCl		1984	20,	318
fadrozole HCl		1995	31,	342
fludarabine phosphate		1991	27,	327
flutamide		1983	19,	318
formestane		1993	29,	337
fotemustine		1989	25,	313
gemcitabine HCl		1995	31,	344
idarubicin hydrochloride		1990	26,	303
imatinib mesylate		2001	37,	267
interferon gamma-1α		1992	28,	332
interleukin-2		1989	25,	314
irinotecan		1994	30,	301
lonidamine		1987	23,	337
mitoxantrone HCl		1984	20,	321
nedaplatin		1995	31,	347
nilutamide		1987	23,	338
paclitaxal		1993	29,	342
pegaspargase		1994	30,	306
pentostatin		1992	28,	334
pirarubicin		1988	24,	309
ranimustine		1987	23,	341
sobuzoxane		1994	30,	310
toremifene		1989	25,	319
vinorelbine		1989	25,	320
zinostatin stimalamer		1994	30,	313
porfimer sodium	ANTINEOPLASTIC ADJUVANT	1993	29,	343
masoprocol	ANTINEOPLASTIC,	1992	28,	333
miltefosine	TOPICAL	1993	29,	340
dexfenfluramine	ANTIOBESITY	1997	33,	332
orlistat		1998	34,	327
sibutramine		1998	34,	331
atovaquone	ANTIPARASITIC	1992	28,	326
ivermectin		1987	23,	336

GENERIC NAME	INDICATION	YEAR INTRO.	ARMC VOL., PAGE	
budipine	ANTIPARKINSONIAN	1997	33,	330
CHF-1301		1999	35,	336
droxidopa		1989	25,	312
entacapone		1998	34,	322
pergolide mesylate		1988	24,	308
pramipexole hydrochloride		1997	33,	341
ropinirole HCl		1996	32,	317
talipexole		1996	32,	318
tolcapone		1997	33,	343
lidamidine HCl	ANTIPERISTALTIC	1984	20,	320
gestrinone	ANTIPROGESTOGEN	1986	22,	321
cabergoline	ANTIPROLACTIN	1993	29,	334
tamsulosin HCl	ANTIPROSTATIC HYPERTROPHY	1993	29,	347
acitretin	ANTIPSORIATIC	1989	25,	309
calcipotriol		1991	27,	323
tazarotene		1997	33,	343
tacalcitol	ANTIPSORIATIC, TOPICAL	1993	29,	346
amisulpride	ANTIPSYCHOTIC	1986	22,	316
remoxipride hydrochloride		1990	26,	308
zuclopenthixol acetate		1987	23,	345
actarit	ANTIRHEUMATIC	1994	30,	296
diacerein		1985	21,	326
octreotide	ANTISECRETORY	1988	24,	307
adamantanium bromide	ANTISEPTIC	1984	20,	315
drotecogin alfa	ANTISEPSIS	2001	37,	265
cimetropium bromide	ANTISPASMODIC	1985	21,	326
tiquizium bromide		1984	20,	324
tiropramide HCl		1983	19,	324
argatroban	ANTITHROMBOTIC	1990	26,	299
bivalirudin		2000	36,	298
defibrotide		1986	22,	319
cilostazol		1988	24,	299
clopidogrel hydrogensulfate		1998	34,	320
cloricromen		1991	27,	325

GENERIC NAME	INDICATION	YEAR INTRO.	ARMC VOL., PAGE	
enoxaparin		1987	23,	333
eptifibatide		1999	35,	340
ethyl icosapentate		1990	26,	303
ozagrel sodium		1988	24,	308
indobufen		1984	20,	319
picotamide		1987	23,	340
limaprost		1988	24,	306
tirofiban hydrochloride		1998	34,	332
flutropium bromide	ANTITUSSIVE	1988	24,	303
levodropropizine		1988	24,	305
benexate HCl	ANTIULCER	1987	23,	328
dosmalfate		2000	36,	302
ebrotidine		1997	33,	333
ecabet sodium		1993	29,	336
egualen sodium		2000	36,	303
enprostil		1985	21,	327
famotidine		1985	21,	327
irsogladine		1989	25,	315
lansoprazole		1992	28,	332
misoprostol		1985	21,	329
nizatidine		1987	23,	339
omeprazole		1988	24,	308
ornoprostil		1987	23,	339
pantoprazole sodium		1994	30,	306
plaunotol		1987	23,	340
polaprezinc		1994	30,	307
ranitidine bismuth citrate		1995	31,	348
rebamipide		1990	26,	308
rosaprostol		1985	21,	332
roxatidine acetate HCl		1986	22,	326
roxithromycin		1987	23,	342
sofalcone		1984	20,	323
spizofurone		1987	23,	343
teprenone		1984	20,	323
tretinoin tocoferil		1993	29,	348
troxipide		1986	22,	327
abacavir sulfate	ANTIVIRAL	1999	35,	333
amprenavir		1999	35,	334
cidofovir		1996	32,	306
delavirdine mesylate		1997	33,	331
didanosine		1991	27,	326
efavirenz		1998	34,	321
famciclovir		1994	30,	300
fomivirsen sodium		1998	34,	323
foscarnet sodium		1989	25,	313

GENERIC NAME	INDICATION	YEAR INTRO.	ARMC VOL., PAGE	
ganciclovir		1988	24,	303
imiquimod		1997	33,	335
indinavir sulfate		1996	32,	310
interferon alfacon-1		1997	33,	336
lamivudine		1995	31,	345
lopinavir		2000	36,	310
neflinavir mesylate		1997	33,	340
nevirapine		1996	32,	313
oseltamivir phosphate		1999	35,	346
penciclovir		1996	32,	314
propagermanium		1994	30,	308
rimantadine HCl		1987	23,	342
ritonavir		1996	32,	317
saquinavir mesylate		1995	31,	349
sorivudine		1993	29,	345
stavudine		1994	30,	311
tenofovir disoproxil fumarate		2001	37,	271
valaciclovir HCl		1995	31,	352
zalcitabine		1992	28,	338
zanamivir		1999	35,	352
zidovudine		1987	23,	345
cevimeline hydrochloride	ANTI-XEROSTOMIA	2000	36,	299
alpidem	ANXIOLYTIC	1991	27,	322
buspirone HCl		1985	21,	324
etizolam		1984	20,	318
flutazolam		1984	20,	318
flutoprazepam		1986	22,	320
metaclazepam		1987	23,	338
mexazolam		1984	20,	321
tandospirone		1996	32,	319
flumazenil	BENZODIAZEPINE ANTAG.	1987	23,	335
bambuterol	BRONCHODILATOR	1990	26,	299
doxofylline		1985	21,	327
formoterol fumarate		1986	22,	321
mabuterol HCl		1986	22,	323
oxitropium bromide		1983	19,	323
salmeterol hydroxynaphthoate		1990	26,	308
APD	CALCIUM REGULATOR	1987	23,	326
clodronate disodium		1986	22,	319
disodium pamidronate		1989	25,	312
gallium nitrate		1991	27,	328
ipriflavone		1989	25,	314

GENERIC NAME	INDICATION	YEAR INTRO.	ARMC VOL., PAGE	
dexrazoxane	CARDIOPROTECTIVE	1992	28,	330
bucladesine sodium	CARDIOSTIMULANT	1984	20,	316
denopamine		1988	24,	300
docarpamine		1994	30,	298
dopexamine		1989	25,	312
enoximone		1988	24,	301
flosequinan		1992	28,	331
ibopamine HCl		1984	20,	319
loprinone hydrochloride		1996	32,	312
milrinone		1989	25,	316
vesnarinone		1990	26,	310
amrinone	CARDIOTONIC	1983	19,	314
colforsin daropate HCL		1999	35,	337
xamoterol fumarate		1988	24,	312
cefozopran HCL	CEPHALOSPORIN, INJECTABLE	1995	31,	339
cefditoren pivoxil	CEPHALOSPORIN, ORAL	1994	30,	297
brovincamine fumarate	CEREBRAL VASODILATOR	1986	22,	317
nimodipine		1985	21,	330
propentofylline		1988	24,	310
succimer	CHELATOR	1991	27,	333
trientine HCl		1986	22,	327
fenbuprol	CHOLERETIC	1983	19,	318
auranofin	CHRYSOTHERAPEUTIC	1983	19,	314
taltirelin	CNS STIMULANT	2000	36,	311
aniracetam	COGNITION ENHANCER	1993	29,	333
pramiracetam H_2SO_4		1993	29,	343
carperitide	CONGESTIVE HEART FAILURE	1995	31,	339
nesiritide		2001	37,	269
drospirenone	CONTRACEPTIVE	2000	36,	302
nicorandil	CORONARY VASODILATOR	1984	20,	322
dornase alfa	CYSTIC FIBROSIS	1994	30,	298
neltenexine		1993	29,	341

GENERIC NAME	INDICATION	YEAR INTRO.	ARMC VOL., PAGE	
amifostine	CYTOPROTECTIVE	1995	31,	338
nalmefene HCL	DEPENDENCE TREATMENT	1995	31,	347
ioflupane	DIAGNOSIS CNS	2000	36,	306
azosemide	DIURETIC	1986	22,	316
muzolimine		1983	19,	321
torasemide		1993	29,	348
atorvastatin calcium	DYSLIPIDEMIA	1997	33,	328
cerivastatin		1997	33,	331
naftopidil	DYSURIA	1999	35,	343
alglucerase	ENZYME	1991	27,	321
erdosteine	EXPECTORANT	1995	31,	342
fudosteine		2001	37,	267
agalsidase alfa	FABRY'S DISEASE	2001	37,	259
cetrorelix	FEMALE INFERTILITY	1999	35,	336
ganirelix acetate		2000	36,	305
follitropin alfa	FERTILITY ENHANCER	1996	32,	307
follitropin beta		1996	32,	308
reteplase	FIBRINOLYTIC	1996	32,	316
esomeprazole magnesium	GASTRIC ANTISECRETORY	2000	36,	303
lafutidine		2000	36,	307
rabeprazole sodium		1998	34,	328
cinitapride	GASTROPROKINETIC	1990	26,	301
cisapride		1988	24,	299
itopride HCL		1995	31,	344
mosapride citrate		1998	34,	326
imiglucerase	GAUCHER'S DISEASE	1994	30,	301
somatotropin	GROWTH HORMONE	1994	30,	310
somatomedin-1	GROWTH HORMONE INSENSITIVITY	1994	30,	310

GENERIC NAME	INDICATION	YEAR INTRO.	ARMC VOL., PAGE	
factor VIIa	HAEMOPHILIA	1996	32,	307
levosimendan	HEART FAILURE	2000	36,	308
pimobendan		1994	30,	307
anagrelide hydrochloride	HEMATOLOGIC	1997	33,	328
erythropoietin	HEMATOPOETIC	1988	24,	301
factor VIII	HEMOSTATIC	1992	28,	330
malotilate	HEPATOPROTECTIVE	1985	21,	329
mivotilate		1999	35,	343
buserelin acetate	HORMONE	1984	20,	316
goserelin		1987	23,	336
leuprolide acetate		1984	20,	319
nafarelin acetate		1990	26,	306
somatropin		1987	23,	343
zoledronate disodium	HYPERCALCEMIA	2000	36,	314
sapropterin hydrochloride	HYPERPHENYL-ALANINEMIA	1992	28,	336
quinagolide	HYPERPROLACTINEMIA	1994	30,	309
cadralazine	HYPERTENSIVE	1988	24,	298
nitrendipine		1985	21,	331
binfonazole	HYPNOTIC	1983	19,	315
brotizolam		1983	19,	315
butoctamide		1984	20,	316
cinolazepam		1993	29,	334
doxefazepam		1985	21,	326
loprazolam mesylate		1983	19,	321
quazepam		1985	21,	332
rilmazafone		1989	25,	317
zaleplon		1999	35,	351
zolpidem hemitartrate		1988	24,	313
zopiclone		1986	22,	327
acetohydroxamic acid	HYPOAMMONURIC	1983	19,	313
sodium cellulose PO4	HYPOCALCIURIC	1983	19,	323
divistyramine	HYPOCHOLESTEROLEMIC	1984	20,	317
lovastatin		1987	23,	337

GENERIC NAME	INDICATION	YEAR INTRO.	ARMC VOL., PAGE	
melinamide		1984	20,	320
simvastatin		1988	24,	311
glucagon, rDNA	HYPOGLYCEMIA	1993	29,	338
acipimox	HYPOLIPIDEMIC	1985	21,	323
beclobrate		1986	22,	317
binifibrate		1986	22,	317
ciprofibrate		1985	21,	326
colesevelam hydrochloride		2000	36,	300
colestimide		1999	35,	337
fluvastatin		1994	30,	300
meglutol		1983	19,	321
ronafibrate		1986	22,	326
modafinil	IDIOPATHIC HYPERSOMNIA	1994	30,	303
bucillamine	IMMUNOMODULATOR	1987	23,	329
centoxin		1991	27,	325
thymopentin		1985	21,	333
filgrastim	IMMUNOSTIMULANT	1991	27,	327
GMDP		1996	32,	308
interferon gamma-1b		1991	27,	329
lentinan		1986	22,	322
pegademase bovine		1990	26,	307
pidotimod		1993	29,	343
romurtide		1991	27,	332
sargramostim		1991	27,	332
schizophyllan		1985	22,	326
ubenimex		1987	23,	345
cyclosporine	IMMUNOSUPPRESSANT	1983	19,	317
gusperimus		1994	30,	300
mizoribine		1984	20,	321
muromonab-CD3		1986	22,	323
mycophenolate mofetil		1995	31,	346
tacrolimus		1993	29,	347
defeiprone	IRON CHELATOR	1995	31,	340
alosetron hydrochloride	IRRITABLE BOWEL	2000	36,	295
tegasedor maleate	SYNDROME	2001	37,	270

GENERIC NAME	INDICATION	YEAR INTRO.	ARMC VOL., PAGE	
sulbactam sodium	β-LACTAMASE INHIBITOR	1986	22,	326
tazobactam sodium		1992	28,	336
nartograstim	LEUKOPENIA	1994	30,	304
pumactant	LUNG SURFACTANT	1994	30,	308
sildenafil citrate	MALE SEXUAL DYSFUNCTION	1998	34,	331
gadoversetamide	MRI CONTRAST AGENT	2000	36,	304
telmesteine	MUCOLYTIC	1992	28,	337
interferon ß-1a	MULTIPLE SCLEROSIS	1996	32,	311
interferon ß-1b		1993	29,	339
glatiramer acetate		1997	33,	334
afloqualone	MUSCLE RELAXANT	1983	19,	313
cisatracurium besilate		1995	31,	340
doxacurium chloride		1991	27,	326
eperisone HCl		1983	19,	318
mivacurium chloride		1992	28,	334
rapacuronium bromide		1999	35,	347
tizanidine		1984	20,	324
naltrexone HCl	NARCOTIC ANTAGONIST	1984	20,	322
tinazoline	NASAL DECONGESTANT	1988	24,	312
clospipramine hydrochloride	NEUROLEPTIC	1991	27,	325
nemonapride		1991	27,	331
olanzapine		1996	32,	313
perospirone hydrochloride		2001	37,	270
quetiapine fumarate		1997	33,	341
risperidone		1993	29,	344
sertindole		1996	32,	318
timiperone		1984	20,	323
ziprasidone hydrochloride		2000	36,	312
rocuronium bromide	NEUROMUSCULAR BLOCKER	1994	30,	309
edaravone	NEUROPROTECTIVE	1995	37,	265
fasudil HCL		1995	31,	343
riluzole		1996	32,	317

GENERIC NAME	INDICATION	YEAR INTRO.	ARMC VOL., PAGE	
bifemelane HCl	NOOTROPIC	1987	23,	329
choline alfoscerate		1990	26,	300
exifone		1988	24,	302
idebenone		1986	22,	321
indeloxazine HCl		1988	24,	304
levacecarnine HCl		1986	22,	322
nizofenzone fumarate		1988	24,	307
oxiracetam		1987	23,	339
bromfenac sodium	NSAID	1997	33,	329
lornoxicam		1997	33,	337
OP-1	OSTEOINDUCTOR	2001	37,	269
alendronate sodium	OSTEOPOROSIS	1993	29,	332
ibandronic acid		1996	32,	309
incadronic acid		1997	33,	335
raloxifene hydrochloride		1998	34,	328
risedronate sodium		1998	34,	330
tiludronate disodium	PAGET'S DISEASE	1995	31,	350
verteporfin	PHOTOSENSITIZER	2000	36,	312
beraprost sodium	PLATELET AGGREG. INHIBITOR	1992	28,	326
epoprostenol sodium		1983	19,	318
iloprost		1992	28,	332
sarpogrelate HCl	PLATELET ANTIAGGREGANT	1993	29,	344
trimetrexate glucuronate	*PNEUMOCYSTIS CARINII* PNEUMONIA	1994	30,	312
histrelin	PRECOCIOUS PUBERTY	1993	29,	338
atosiban	PRETERM LABOR	2000	36,	297
gestodene	PROGESTOGEN	1987	23,	335
nomegestrol acetate		1986	22,	324
norgestimate		1986	22,	324
promegestrone		1983	19,	323
trimegestone		2001	37,	273
alpha-1 antitrypsin	PROTEASE INHIBITOR	1988	24,	297
nafamostat mesylate		1986	22,	323

GENERIC NAME	INDICATION	YEAR INTRO.	ARMC VOL., PAGE	
adrafinil	PSYCHOSTIMULANT	1986	22,	315
finasteride	5α-REDUCTASE INHIBITOR	1992	28,	331
surfactant TA	RESPIRATORY SURFACTANT	1987	23,	344
dexmedetomidine hydrochloride	SEDATIVE	2000	36,	301
kinetin	SKIN PHOTODAMAGE/ DERMATOLOGIC	1999	35,	341
tirilazad mesylate	SUBARACHNOID HEMORRHAGE	1995	31,	351
APSAC	THROMBOLYTIC	1987	23,	326
alteplase		1987	23,	326
balsalazide disodium	ULCERATIVE COLITIS	1997	33,	329
tiopronin	UROLITHIASIS	1989	25,	318
propiverine hydrochloride	UROLOGIC	1992	28,	335
Lyme disease	VACCINE	1999	35,	342
clobenoside	VASOPROTECTIVE	1988	24,	300
falecalcitriol	VITAMIN D	2001	37,	266
maxacalcitol		2000	36,	310
paricalcitol		1998	34,	327
doxercalciferol	VITAMIN D PROHORMONE	1999	35,	339
prezatide copper acetate	VULNERARY	1996	32,	314
acemannan	WOUND HEALING AGENT	2001	37,	257
cadexomer iodine		1983	19,	316
epidermal growth factor		1987	23,	333